Rechargeable Ion Batteries

Rechargeable Ion Batteries

Materials, Design, and Applications of Li-Ion Cells and Beyond

Edited by Katerina E. Aifantis, R. Vasant Kumar, and Pu Hu

WILEY-VCH

Editors

Prof. Katerina E. Aifantis
Department of Mechanical & Aerospace Engineering
University of Florida, Gainesville
1064 Center Drive
Florida
United States

Prof. R. Vasant Kumar
Materials Science + Metallurgy
University of Cambridge
27 Charles Babbage Road
CB3 0FS Cambridge
United Kingdom

Prof. Pu Hu
Wuhan Institute of Technology
School of Material Sciences & Engineering
No.206, Guanggu 1st road
430205 Wuhan
China

Cover: © Black_Kira/Shutterstock

Library of Congress Card No.: applied for

British Library Cataloguing-in-Publication Data
A catalogue record for this book is available from the British Library.

Bibliographic information published by the Deutsche Nationalbibliothek
The Deutsche Nationalbibliothek lists this publication in the Deutsche Nationalbibliografie; detailed bibliographic data are available on the Internet at <http://dnb.d-nb.de>.

Print ISBN: 978-3-527-35018-6
ePDF ISBN: 978-3-527-83669-7
ePub ISBN: 978-3-527-83671-0
oBook ISBN: 978-3-527-83670-3

Typesetting Straive, Chennai, India
Printing and Binding CPI Group (UK) Ltd, Croydon, CR0 4YY

C118246_041122

Katerina Aifantis would like to dedicate this book to her parents, Maria & Elias for their neverending support, and Fr. Symeon Krayiopoulos, who was asking her when she would begin the second book ever since the first one got published.

R. Vasant Kumar would like to dedicate this book to Gill, Vijay, Kailo, and Anastasia.

Pu Hu would like to dedicate this book to his wife Ping and his daughter Tong.

Katarina Afrontis would like to dedicate this book to her parents, Maria & Filip for their neverending support, and Fr. Symeon Kravlopoulos, who was asking her when she would begin the second book ever since the first one got published.

R. Vasant Kumar would like to dedicate this book to Gill, Vijay, Kevin, and Anastasia.

Pu Hu would like to dedicate this book to his wife Fang and his daughter Ting.

Contents

Preface

The development of electrochemical energy storage systems is a fascinating field that bridges many disciplines, such as electrochemistry, nanotechnology, synthesis, engineering, materials science, mechanics and recycling. Our motivation in writing this book was to inspire students and young researchers from all these fields to lend their expertise to the battery community in reaching one of the main goals that seem to unite our society: build a future that will limit carbon emissions and allow us to retain the sustainability of our world. Battery developers will also find our book as a useful guide since we summarize the current state of the art for the electrodes and electrolytes of different electrochemical systems.

We began working on this book in 2013, as a second edition to *High Energy Density Lithium Batteries, Materials, Engineering, Applications* (Editors: Aifantis, Hackney, Kumar), which was published in 2010. However, as additional rechargeable battery systems have emerged, we decided to write a new book that covers not only lithium-ion, but also other next-generation systems such as lithium–sulfur, sodium-ion, magnesium-ion, and zinc-ion. Apart from lithium–sulfur batteries, which are targeted for future vehicle applications, the other ion batteries (aside from Li-ion) are considered for grid storage applications.

To make our book accessible to a wide audience, we start with introductory chapters on the working principles and development of batteries, and as we progress, we go into greater depths on the open issues that must be addressed for the most promising rechargeable ion systems. To ensure continuity throughout the book we did not only act as editors, but each chapter except for Chapter 7 is co-written by one of us, and we decided together on the layout of each chapter. The lead editor in particular supervised closely all chapters, ensuring the uniformity and consistency of an authored, rather than edited scientific book.

We are very close collaborators with friendly scientific and technological engagements among ourselves and with the authors, and hope that the enthusiasm and joy that we had in preparing our book is transmitted to the readers. In ending, we would like to specially thank our publisher Martin Preuss for his continuous advice and encouragement.

4 July 2022 *Katerina E. Aifantis, R. Vasant Kumar, Pu Hu*

1

Introduction to Electrochemical Cells

R. Vasant Kumar[1] *and Thapanee Sarakonsri*[2]

[1] *University of Cambridge, Department of Materials Science and Metallurgy, 27 Charles Babbage Road, Cambridge CB3 0FS, UK*
[2] *Chiang Mai University, Department of Chemistry, Faculty of Science, Chiang Mai 50200, Thailand*

1.1 What are Batteries?

The purpose of this chapter is to provide basic knowledge on batteries, which will allow for their general understanding. Therefore, after defining their components and structure, an overview of the quantities that characterize these storage devices will be given. Scientifically, batteries are referred to as electrochemical or galvanic cells, due to the fact that they store electrical energy in the form of chemical energy and because the electrochemical reactions that take place are also termed galvanic. Galvanic reactions are thermodynamically favorable (the free-energy difference, ΔG, is negative) and occur spontaneously when two materials of different positive standard reduction potentials are connected by an electronic load (meaning that a voltage is derived). The material with the lower positive standard reduction potential undergoes oxidation providing electrons by the external circuit to the material with the higher positive standard reduction potential, which in turn undergoes a reduction reaction. These half-reactions occur concurrently and allow for the conversion of chemical energy to electrical energy by means of electron transfer through an external circuit. It follows that the material with the lower positive standard reduction potential is called the negative electrode or anode on discharge (since it provides electrons), while the material with the higher positive standard reduction is called the positive electrode or cathode on discharge (since it accepts electrons). It follows that the discharge process occurs in the electrochemical cells upon operation of the devices they power.

In addition to the electrodes, the two other constituents that are required for such reactions to take place are the electrolyte phase/solution and the separator. The electrolyte is an ion-conducting material, which can be in the form of an aqueous, molten salt, or solid solution, while the separator is a membrane that physically prevents direct contact between the two electrodes and allows ions but not electrons to pass through; it, therefore, ensures electronic insulation for charge neutralization in both the anode and cathode once the reaction is completed and

Rechargeable Ion Batteries: Materials, Design, and Applications of Li-Ion Cells and Beyond, First Edition. Edited by Katerina E. Aifantis, R. Vasant Kumar, and Pu Hu.

prevents internal short-circuiting of electrons. Internal short-circuiting implies the movement of electrons from the anode to cathode through the electrolyte, which dissipates the chemical energy without providing electrical potential in the external circuit. When the electrolyte is solid, it simultaneously functions as a membrane (separator) and an ionic conductor. Two final parts required to complete a commercial galvanic cell are the terminals. They are necessary when applying the batteries to electrical appliances with specific holder designs to prevent a short circuit from the reverse installation of the battery, and they are shaped to match the receptacle facilities provided in the appliances. For example, in cylindrical batteries, the negative terminal is either designed to be flat, or to protrude out of the battery end, while the positive terminal extends as a pip at the opposite end. A simple galvanic cell is illustrated in Figure 1.1a, while Figure 1.1b shows terminal designs for cylindrical batteries.

To meet the voltage or current used in specific appliances, cylindrical galvanic cells are connected in series or parallel. Figure 1.2a,brepresents parallel and series connections; parallel connections allow for the current to be doubled, while series connections allow for the voltage to be doubled.

In addition to cylindrical battery cells, as those shown in Figures 1.1 and 1.2, flat battery configurations are also quite common. The biggest impetus for these configurations came from the rapid growth of portable radios since the flat cells use the space of the battery box more efficiently than cylindrical ones. The electrodes are made in the form of flat plates, which are suspended in the electrolyte and are held immobilized in a microporous separator (Figure 1.3a). The separator

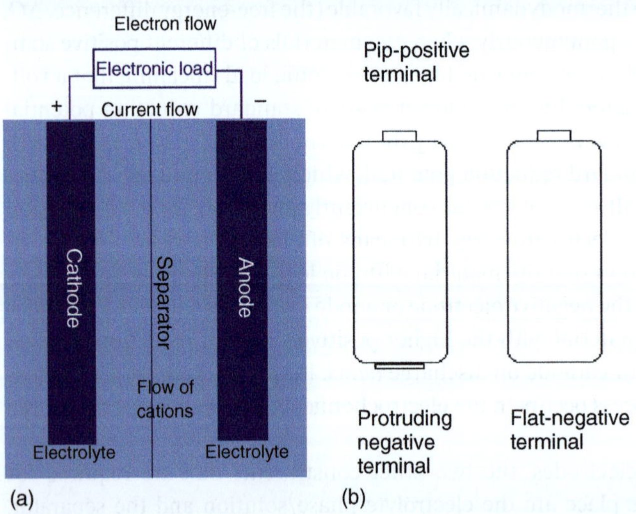

Figure 1.1 (a) The schematic diagram of a simple galvanic cell. (b) Terminal designs for cylindrical batteries.

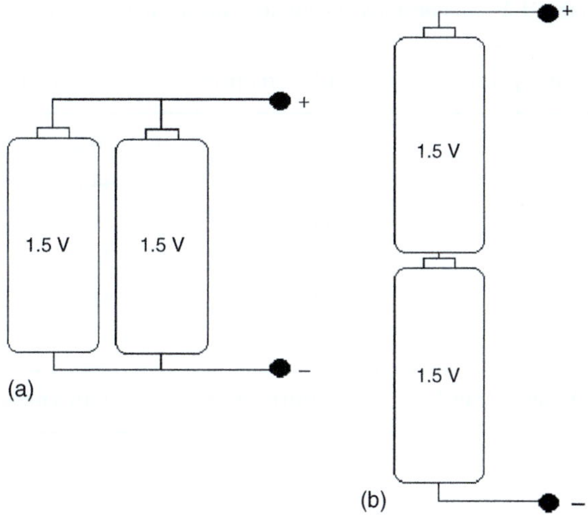

Figure 1.2 (a) An illustration of batteries connected in parallel to obtain double current. (b) An illustration of batteries connected in series to obtain 3 V.

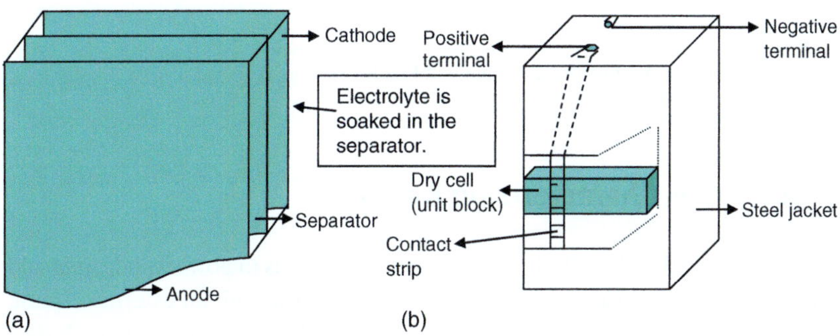

Figure 1.3 (a) Single-flat-cell configuration; (b) composite-flat-cell configuration.

also helps in isolating the electrodes, preventing any short-circuiting whereby ions can directly move internally between the anode and cathode. Short-circuiting will result in capacity loss, parasitic reactions, and heat generation. This can also lead to catastrophic situations causing fires, explosions, leakage of materials, and accidents. The configuration of Figure 1.3a can be scaled up to very large sizes, for high currents and large storage capacities, by placing each cell inside a plastic envelope and stacking them inside a steel jacket. Connector strips are used to collect and connect the positive and the negative electrodes to a common positive and negative terminal; a sketch of such cell compaction is shown in Figure 1.3b.

Both cylindrical and flat cells come in various sizes so that they can fit a wide range of portable appliances and devices. Table 1.1 summarizes the various battery sizes that are available commercially.

Table 1.1 Dimensions of commercially available battery sizes [1].

Battery size	Diameter (mm)	Height (mm)
N	12	30.2
AAA	10.5	44.5
AA	14.5	50.5
C	26.2	50
D	34.2	61.5
F	32.0	91.0
Length (mm)	**Width (mm)**	**Thickness (mm)**
Flat cells		
24	13.5	6.0
43	43	6.4
Rectangular cells		
48.5	26.5	17.5

Source: Republished with permission of Linden and Reddy [1], Copyright Clearance Center, Inc.

1.2 Quantities Characterizing Batteries

Upon operation of galvanic cells, implying that the device is in power mode, it is said that the galvanic cell is discharged and electrons flow, through an external circuit, from the anode to the cathode. As a result, the cathode attains a negative charge, while the anode becomes positively charged. Consequently, cations are attracted from the anode to the cathode (and vice versa for the anions) and diffuse through the electrolyte. Typical electrochemical redox reactions that may take place upon operation of batteries are shown in Table 1.2, whereas the quantities that characterize batteries are defined in Table 1.3.

To better understand the differences between various battery chemistries, some of the quantities in Table 1.3 are further elaborated on below.

1.2.1 Voltage

The theoretical standard cell voltage, E^0(cell), can be determined using the electrochemical series and is given by the difference between the standard electrode potential at the cathode, E^0(cathode), and the standard electrode potential at the anode, E^0(anode) [2] as

$$E^0(\text{cathode}) - E^0(\text{anode}) = E^0(\text{cell}) \tag{1.1}$$

The standard electrode potential, E^0, for an electrode reaction, written (by convention) as a reduction reaction (i.e. involving consumption of electrons), is the potential

Table 1.2 Standard electrode potentials in aqueous electrolyte at 298 K (written as reduction reactions by convention).

Reaction	E^0 (V)
$Li^+ + e^- \rightarrow Li$	−3.10
$Na^+ + e^- \rightarrow Na$	−2.71
$Mg^{2+} + 2e^- \rightarrow Mg$	−2.36
$½H_2 + e^- \rightarrow H^-$	−2.25
$Mn^{2+} + 2e^- \rightarrow Mn$	−1.18
$MnO_2 + 2H_2O + 4e^- \rightarrow Mn + 4OH^-$	−0.98
$2H_2O + 2e^- \rightarrow H_2 + 2OH^-$	−0.83
$Cd(OH)_2 + 2e^- \rightarrow Cd + 2OH^-$	−0.82
$Zn^{2+} + 2e^- \rightarrow Zn$	−0.76
$Ni(OH)_2 + 2e^- \rightarrow Ni + 2OH^-$	−0.72
$Fe^{2+} + 2e^- \rightarrow Fe$	−0.44
$Cd^{2+} + 2e^- \rightarrow Cd$	−0.40
$PbSO_4 + 2e^- \rightarrow Pb + SO_4^{2-}$	−0.35
$Ni^{2+} + 2e^- \rightarrow Ni$	−0.26
$MnO_2 + 2H_2O + 4e^- \rightarrow Mn(OH)_2 + 2OH^-$	−0.05
$2H^+ + 2e^- \rightarrow H_2$	0.00
$Cu^{2+} + e^- \rightarrow Cu^+$	+0.16
$Ag_2O + H_2O + 2e^- \rightarrow 2Ag + 2OH^-$	+0.34
$Cu^{2+} + 2e^- \rightarrow Cu$	+0.34
$O_2 + 2H_2O + 4e^- \rightarrow 4OH^-$	+0.40
$2NiOOH + 2H_2O + 2e^- \rightarrow 2Ni(OH)_2 + 2OH^-$	+0.48
$NiO_2 + 2H_2O + 2e^- \rightarrow Ni(OH)_2 + 2OH^-$	+0.49
$MnO_4^{2-} + 2H_2O + 2e^- \rightarrow MnO_2 + 4OH^-$	+0.62
$2AgO + H_2O + 2e^- \rightarrow Ag_2O + 2OH^-$	+0.64
$Fe^{3+} + e^- \rightarrow Fe^{2+}$	+0.77
$Hg^{2+} + e^- \rightarrow Hg^+$	+0.80
$Ag^+ + e^- \rightarrow Ag$	+0.80
$2Hg^{2+} + 2e^- \rightarrow Hg^+$	+0.91
$O_2 + 4H^+ + 4e^- \rightarrow 2H_2O$	+1.23
$ZnO + H_2O + 2e^- \rightarrow Zn + 2OH^-$	+1.26
$Cl_2 + 2e^- \rightarrow 2Cl^-$	+1.36
$PbO_2 + 4H^+ + 2e^- \rightarrow Pb^{2+} + 2H_2O$	+1.47
$PbO_2 + SO_4^{2-} + 4H^+ + 2e^- \rightarrow PbSO_4 + 2H_2O$	+1.70
$F_2 + 2e^- \rightarrow 2F^-$	+2.87

Table 1.3 Battery characteristics.

Battery characteristics	Definition	Unit
Open-circuit voltage	Maximum voltage in the charged state at zero current	Volt (V)
Current	Low currents are characterized by activation losses, while the maximum current is normally determined by mass-transfer limitations	Ampere (A)
Energy density	The energy that can be derived per unit volume of the weight of the cell	Watt-hours per liter ($Wh\,m^{-3}$)
Specific energy density	The energy that can be derived per unit weight of the cell (or sometimes per unit weight of the active electrode material)	Watt-hours per kilogram ($Wh\,kg^{-1}$)
Power density	The power that can be derived per unit weight of the cell	Watt per kilogram ($W\,kg^{-1}$)
Capacity	The theoretical capacity of a battery is the quantity of electricity involved in the electrochemical reaction	Ampere-hours per gram ($Ah\,g^{-1}$).
Shelf life	The time a battery can be stored inactive before its capacity falls to 80%	Years
Service life	The time a battery can be used at various loads and temperatures	Hours (usually normalized for ampere per kilogram ($A\,kg^{-1}$) and ampere per liter ($A\,l^{-1}$))
Cycle life	The number of discharge/charge cycles it can undergo before its capacity falls to 80%	Cycles

Source: Reproduced with permission from Weal et al. [2]/University of Cambridge.

generated by that reaction under the condition that the reactants and the products are in their standard state in relation to a reference electrode. (A reactant or product is defined to be in its standard state when the component in a condensed phase is at unit activity and any component in the gas phase is at a partial pressure of 1 atm.) In aqueous systems, the standard hydrogen potential is taken as the universal reference electrode, whose potential is defined as zero. In practical terms, the standard hydrogen electrode can be constructed as follows: (i) a high surface area of platinum is deposited on a platinum foil or plate, which is then dipped into an acid solution of unit activity of H^+ ions, corresponding to 1 M acid solution, then (ii) pure hydrogen at one atmosphere is passed over this electrode. A list containing selected standard electrode potentials at 298 K in an aqueous solution is given in Table 1.2 and these refer to equilibrium potentials at zero net current values at each of the electrodes. The batteries that make use of these materials as electrodes will be described in Chapter 2. It should be noted that the standard electrode potential for a reduction reaction in an aqueous solution is relative to the hydrogen electrode, which is taken as zero. Thus, potentials that are defined for half-cells are represented as reduction reactions. In a battery, two half-cells are present such that reduction takes place on one electrode and oxidation on the other.

To obtain a true estimate of the actual open-circuit cell voltage, E_{eq} in the fully charged state for operation of the battery; the theoretical cell voltage is modified by the Nernst equation, which takes into account the nonstandard state of the reacting component as

$$E_{eq} = E^0 - RT \ln Q \tag{1.2}$$

where T is the operating temperature in kelvin (k), $Q = a_{\text{products}}/a_{\text{reactants}}$ is the chemical quotient for the overall cell reaction, and R is the gas constant (8.314 J k^{-1} mol^{-1}). Q is represented in the same way as the equilibrium constant K, except that the activities (a) and partial pressures (p) in Eq. (1.2) reflect the actual nonstandard values prevailing in the system. For example, for the electrode reaction

$$M^{2+} + 2e^- = M \tag{1.3}$$

the actual Nernstian electrode potential (E), also referred to as open-circuit voltage (OCV) under equilibrium (for net zero current) is

$$E = E^o - \frac{RT}{2F} \ln \left(\frac{a_M}{a_{M^{2+}}} \right) \tag{1.4}$$

Notation E_e (rather than E) is also often used to denote that the actual equilibrium potential of the electrode is determined by the Nernst equation and the standard electrode potential E^0 refers to a very specific situation of reaction species being held in their "standard states," The Nernstian potential in Eq. (1.4) will change with time due to any self-discharge by which the activity (or concentration) of the electroactive component in the cell is modified. Thus, the nominal voltage is determined by the cell chemistry at any given point of time. M and M^{2+} refer to the effective concentrations of the two components in the phase within which they are present. (For the hydrogen reaction, as an example where a gas phase is involved, the activity of metal is replaced by the partial pressure term, pH_2.)

The operating voltage produced is further modified as a result of discharge reactions actually taking place and will always be lower than the thermodynamically calculated theoretical voltage (also referred to as equilibrium potential) by the Nernst equation (OCV) due to polarization losses (these arise from overpotentials for overcoming activation barrier and/or diffusional limitations) and the resistance losses (IR drop) of the battery as the voltage is dependent on the current, I, drawn by an external load and the cell resistance, R, in the path of the current. Specifically, polarization losses arise to overcome any activation energy for the electrode reaction and/or concentration gradients near the electrode(s). The factors determining the overpotentials are dependent upon electrode kinetics from rates of electrodic reactions and diffusional rates of one or more active components, and, thus, vary with temperature, state of charge, and with the age of the cell. It is important to note that the actual voltage appearing at the terminal needs to be sufficient for the intended application.

1.2.2 Electrode Kinetics (Polarization and Cell Impedance)

Before continuing to the other quantities indicated in Table 1.3, the electrode kinetics, which was previously shown to affect the voltage, must be described. Thermodynamics expressed in terms of the electrode potentials can tell us the theoretical and open-circuit cell voltage, as well as how feasible it is for a cell reaction to occur. However, it is necessary to consider kinetics to obtain a better understanding of what the actual cell voltage maybe, since the charge transfer, the rates of the reactions at the electrodes and diffusional barriers are usually the limiting factors. In continuing, therefore, the main kinetic issues that affect battery performance are summarized.

1.2.2.1 Electrical Double Layer

When a metal electrode is in an electrolyte, the charge on the metal will attract ions of opposite charge in the electrolyte, and the dipoles in the solvent will align. This forms a layer of charge in both the metal and the electrolyte, called the *electrical double layer*, as shown in Figure 1.4. The electrochemical reactions take place in this layer, and all atoms or ions that are reduced or oxidized must pass through this layer. Thus, the ability of ions to pass through this layer controls the kinetics, and is, therefore, the limiting factor in controlling the electrode reaction. The energy barrier toward the electrode reaction, described as the activation energy of the electrochemical reaction, lies across this double layer.

1.2.2.2 Rate of Reaction

The rates of the chemical reactions are governed by the Arrhenius relationship, such that the rate of reaction, k, is

$$k \propto \exp(-Q^*/RT) \tag{1.5}$$

where Q^* is the activation energy for the reaction, T is the temperature in Kelvin, and R is the universal gas constant.

In this case, the rate of the reaction can be measured by the current produced, since current is the amount of charge produced per unit amount of time, and therefore proportional to the number of electrons produced per unit amount of time that is proportional to the rate of the reaction.

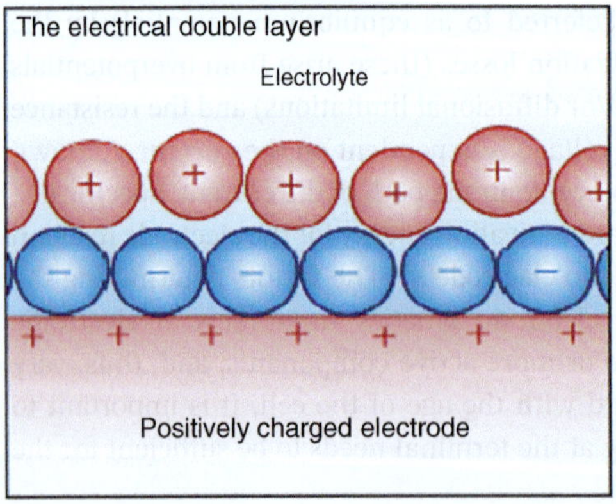

Figure 1.4 Illustration of double layer. Source: Reproduced with permission from Weal et al. [2]/ University of Cambridge.

1.2.2.3 Electrodes Away from Equilibrium

When an electrode is not at the equilibrium potential, an *overpotential* exists, given by

$$\eta = E - E_e \tag{1.6}$$

where η is the overpotential, E is the actual potential, and E_e is the equilibrium potential, calculated using the Nernst equation. Overpotential is used synonymously with polarization potential and described as arising from "polarization process" at a given electrode.

1.2.2.4 The Tafel Equation

The Tafel equation provides a relationship between the current and the overpotential during the oxidation or reduction reaction of an electrode. Consider a general reaction for the oxidation of a metal anode:

$$M \rightarrow M^{z+} + ze^- \tag{1.7}$$

where z is the number of cations/electrons. The rate of this reaction, k_a, is governed by the Arrhenius relationship:

$$k_a = A\ \exp(-Q/RT) \tag{1.8}$$

where A is a frequency factor, which takes into account the rate of collision between the electroactive species and the electrode surface. From Faraday's law, one can express the rate in terms of the exchange current density at the anode, $i_{0,a}$:

$$i_{0,a} = zFk_a = zFK\ \exp(-Q/RT) \tag{1.9}$$

where $F = 96\,540\ \mathrm{C\ mol^{-1}}$ is Faraday's constant. If an overpotential η_a ($\eta_a = E_a - E_e$; where E_e is the Nernst potential for the oxidation half-cell) is now applied in the anodic direction, the activation energy of the reaction becomes

$$Q = \alpha zF\eta_a \tag{1.10}$$

where α is the "symmetry factor" of the electrical double layer, nominally taken as 0.5, assuming symmetrical behavior in both directions.

Therefore the anodic current density, i_a, is

$$i_a = zFK\ \exp(-[Q - \alpha zF\eta_a]/RT) = zFK\ \exp(-Q/RT)\ \exp(\alpha zF\eta_a/RT) \tag{1.11}$$

which by Eq. (1.9) reduces to

$$i_a = i_{0,a}\ \exp(\alpha zF\eta_a/RT) \tag{1.12}$$

The subscript *a* here refers to process at the anode. Equation (1.12) is known as the *Tafel equation*. By taking natural logs and rearranging them, Eq. (1.12) can be written as

$$\eta_a = (RT/\alpha zF)\ \ln(i_a/i_{0,a}) \tag{1.13}$$

By setting $RT/(\alpha zF) = b_a$ and $\ln i_0 = -a_a/b_a$, Eq. (1.13) can be rewritten as

$$\eta_a = a_a + b_a\ \ln i_a \tag{1.14}$$

Or in terms of the anode potential, E_a,

$$\ln(i_a) = \ln(i_{0,a}) + (E_a - E_e)\,\alpha zF/RT \quad (1.15)$$

Solving Eq. (1.15) for E_a, gives

$$\eta_a = b_a \log(i_a/i_{0,a}) \quad (1.16)$$

where b_a is the anodic Tafel slope. Similarly, we can consider the reduction of metal ions at a cathode:

$$M^{z+} + ze^- \rightarrow M \quad (1.17)$$

The activation energy will be decreased by (1–α)$zF\eta_c$ (subscripts c indicate cathode), giving the cathodic current density as

$$i_c = i_{0,c} \exp([1-\alpha]zF\eta_c/RT) \quad (1.18)$$

and

$$\eta_c = \left(RT/([1-\alpha]zF) \ln(i_c/i_{0,c})\right) \quad (1.19)$$

Therefore, the cathode overpotential is $\eta_c = E_c - E_e$, where E_e is the Nernst potential, and E_c is expressed as

$$E_c = b_c \log(i_c/i_{0,c}) \quad (1.20)$$

where b_c is the cathodic Tafel slope. A typical representation of a Tafel plot – a plot of log i vs. E – is shown in Figure 1.5. Thus, for an applied potential, the current density, i, can be found from the Tafel plot in an electrolytic cell when the battery is being charged or discharged.

1.2.2.5 Example: Plotting a Tafel Curve for a Copper Electrode

Let us consider an electrode made of copper immersed in a half-cell containing copper ions at a 1 M concentration, referring to an aqueous solution. The half-cell reaction for copper is

$$Cu^{2+} + 2e^- \rightarrow Cu \qquad E^0 = +0.34\,V \quad (1.21)$$

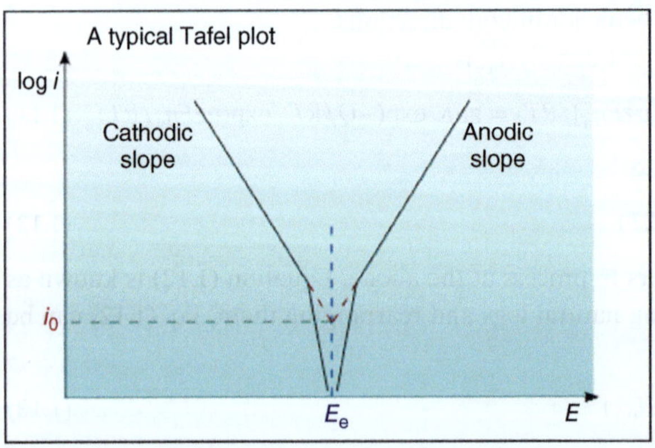

Figure 1.5 A typical Tafel plot. Source: Reproduced with permission from Weal et al. [2]/ University of Cambridge.

The exchange current density at 1 M concentration of copper ions, for the above reaction, is $i_0 = 1\,\text{A m}^{-2}$, reflecting the current density at zero overpotential (thus at $E_{e,}$ in this example $E_e = E^0$, as both Cu and Cu^{2+}(aq) are in their standard states), that is, at zero net reaction (and thus at zero net current). Therefore, the magnitude of the exchange current density is a reflection of the reversibility of a given electrode reaction and signifies the rate at which equilibrium is established on being disturbed away from a given equilibrium condition.

For the Tafel equation

$$\eta = a + b\ \log(i) = a + b\ \log(i/i_0) \tag{1.22}$$

the general expression for the Tafel slope is

$$b = (\pm)\ 2.303RT/\alpha zF \tag{1.23}$$

Taking $T = 300\,\text{K}$, and allowing for copper $\alpha = 0.5$ and $z = 2$, the Tafel slopes are calculated as $b_a = 0.059\,\text{V decade}^{-1}$ of current and $b_c = -0.059\ \text{V decade}^{-1}$ (of log current). Furthermore, for the anodic curve

$$\eta_a = E_a - E_e;\ \ E_a = E_e + \eta_a \tag{1.24}$$

and for the cathodic curve

$$\eta_c = E_e - E_c;\ \ E_c = E_e - \eta_c \tag{1.25}$$

The corresponding Tafel plot for copper is shown in the diagram in Figure 1.6. For example, during discharge, if the redox reaction is in the direction opposite of Eq. (1.21), where Cu is oxidized to copper ions in the solution, the electrode potential will be less than 0.34 V along the polarization line. The greater the operating current density, the lower the electrode potential. This in-effect contributes to the reduction in the cell potential during discharge as a result of overpotential losses, signifying the energy barrier for the electron-transfer reaction. On the other hand, during charging, the electrode potential increases with the applied current, thus increasing the potential required for charging the cell back to its original state (by electrochemical reduction in this example). For a faster charging rate, a higher

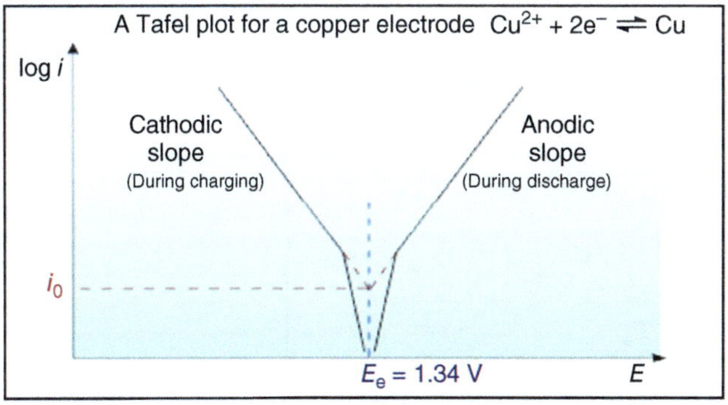

Figure 1.6 Tafel plot for a copper electrode. Source: Reproduced with permission from Weal et al. [2]/University of Cambridge.

current density is desirable, but this can arise only at the expense of a higher applied voltage (higher energy) to overcome the increasing overpotentials. (Note that E^0 is modified by the Nernst equation to obtain a value for E_e, for species in nonstandard states.)

1.2.2.6 Other Limiting Factors

At very high currents, a limiting current may be reached as a result of the concentration overpotential, η_c, restricting mass-transfer rates to the diffusion rate of the electroactive species. A limiting current arises, which can be derived from Fick's first law of diffusion, under the condition that the electrode surface is depleted of the ion, and the recovery of the ion concentration is limited by ion transport through the electrolyte diffusion boundary layer.

The limiting current is diffusion limited, and can be determined by Fick's law of diffusion as

$$i_L = zFDC/\delta \tag{1.26}$$

where i_L is the limiting current density over a boundary layer, D is the diffusion coefficient of metal cations in the electrolyte, C is the concentration of metal cations in the bulk electrolyte, and δ is the thickness of the boundary layer. Typical values for Cu^{2+}, for example, would be $D = 2\times10^{-9}\ m^2\ s^{-1}$, $C = 0.05\times10^4\ kg\ m^{-3}$, and $\delta = 6\times10^{-4}$ m; these values give a limiting current density of $i_L = 3.2\times10^2\ A\ m^{-2}$.

The concentration overpotential, thus, represents the difference between the cell potential at the electrolyte concentration and the cell potential at the surface concentration because of depletion (or accumulation) at high-current densities, given by

$$\eta_c(\text{conc}) = 2.303RT/zF\ \ln(i/i_L) \tag{1.27}$$

A Tafel curve showing this diffusion limiting of the current is depicted in Figure 1.7.

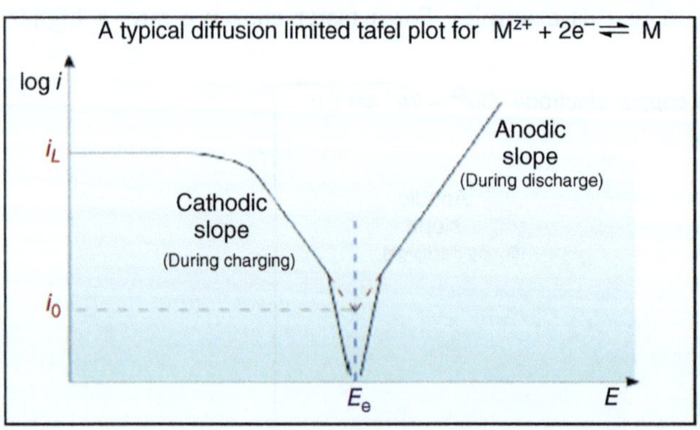

Figure 1.7 Diffusion limited current for the cathodic reaction. Source: Reproduced with permission from Weal et al. [2]/University of Cambridge.

1.2.2.7 Tafel Curves for a Battery

In a battery, there are two sets of Tafel curves present, one for each electrode material. During discharge, one material will act as the anode (termed as the negative [−] electrode being at the lower potential) and the other as the cathode (termed as the positive [+] electrode being at the higher potential). During charging, the roles will be reversed such that at the negative [−] electrode, cathodic reactions take place and at the positive [+] electrode, anodic reactions occur by an externally applied potential difference to recover species back to the state before discharge. The actual potential difference between the two electrodes for a given current density can be found in the Tafel curve. The total cell potential is the difference between the anodic potential, E_a, and the cathodic potential, E_c.

In a galvanic cell, the actual potential, $V'_{\text{cell, discharge}}$, is less than the Nernst potential

$$V'_{\text{cell,discharge}} = E_c - E_c - \mid \eta_c \mid + E_\alpha - \mid \eta_a \mid \tag{1.28}$$

E_a, $\eta_{c,}$ E_c, and η_a are defined in Section 1.2.2.3. Upon discharge, the cell potential may be further decreased by the ohmic drop due to the internal resistance of the cell, r. Thus, the actual cell potential is given by

$$V_{\text{cell,discharge}} = V'_{\text{cell,discharge}} - iAr \tag{1.29}$$

where A is the geometric area relevant to the internal resistance and i is the cell current density.

Similarly, on charging, the applied potential is greater than the Nernstian potential and can be calculated by the equation

$$V'_{\text{cell,charge}} = E_c + \mid \eta_c \mid + E_\alpha + \mid \eta_a \mid \tag{1.30}$$

The cell-charging potential may now be increased by the ohmic drop, and the final actual cell-charging potential is given by

$$V_{\text{cell,charge}} = V'_{\text{charge,harge}} + iAr \tag{1.31}$$

In summary, it can be stated that to maximize the power density (P with units W m^{-2}), which is the product of the cell potential and the current density, it is important to achieve the most optimum value of the cell potential at the lowest overpotential and internal resistance. Usually at low-current densities, overpotential losses arise from an activation energy barrier related to electron-transfer reactions, while at a high-current density, the transport of ions becomes rate-limiting, giving rise to a current limit. Ohmic losses (iAR) arising from the current (current density $i \times$ surface area A), flowing through a resistor R, arise with increasing current and can be further increased due to additional ohmic resistance by the formation of insulating phases during the progress of charging or discharging and also from any non-redox parasitical chemical reactions. Power is the product of voltage and current; therefore, decreasing the current density by increasing the true surface area can also, in principle, result in a high-power density. However, unwanted side reactions may also be enhanced.

1.2.3 Capacity

The bar graph of Figure 1.8 shows the difference between the theoretical and actual capacities in $mAh\,g^{-1}$ for various battery systems.

The theoretical molar capacity of a battery is the quantity of electricity involved in the electrochemical reaction. It is denoted as Q_{charge} and is given by

$$Q_{charge} = xnF \tag{1.32}$$

where x is the number of moles of a chosen electroactive component that take place in the reaction, and n is the number of electrons transferred per mole of reaction. The mass of the electroactive component can be calculated as

$$M = x\,M_r \tag{1.33}$$

where M denotes the mass of the electroactive component in the cell and M_r the molecular mass of the same component. The capacity is conventionally expressed as $Ah\,kg^{-1}$ (numerically equal to $mAh\,g^{-1}$) thus given in terms of mass, often called specific capacity, $C_{specific}$, and it is expressed as

$$C_{specific} = nF/M_r \tag{1.34}$$

If the specific capacity is multiplied by the mass of the electroactive component in the cell, one will obtain the rated capacity of a given cell. It is important to note that the mass may refer to the final battery mass including packaging or it may be reported with respect to the mass of the electroactive components alone. It is quite straightforward to recalculate the capacity in terms of the mass of the cell by dividing the rated capacity by the total mass of the cell.

In practice, the full battery capacity can never be realized, as there is a significant mass contribution from nonreactive components, such as binders, conducting

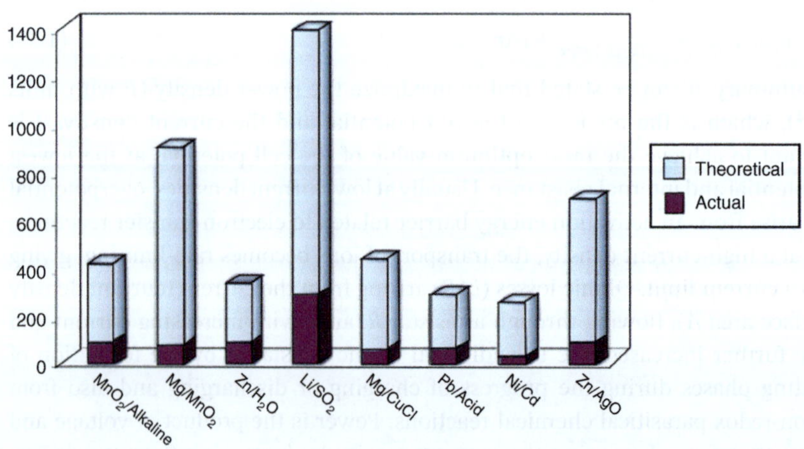

Figure 1.8 Theoretical and actual voltages of various battery systems. Source: Reproduced with permission from Weal et al. [2]/University of Cambridge.

particles, separators, electrolytes, current collectors, substrates, and packaging. Additionally, the chemical reactions cannot be carried out to completion; either due to unavailability of the reactive components and inaccessibility of the active materials or due to poor reactivity at the electrode/electrolyte interface. The capacity is strongly dependent upon the load and can decrease rapidly at high-drain rates as defined by the magnitude of current drawn, due to increased overpotential losses and ohmic losses that can exacerbate the problems with completion of the reaction. At higher drain rates denoting high-operating currents, a battery will be discharged faster.

1.2.4 Shelf Life

A cell may be subject to self-discharge in addition to discharge during operation. Self-discharge is caused by parasitic reactions, such as corrosion, which occur even when the cell is not in use. Thus, the chemical energy may slowly decrease with time. Further energy loss may occur as a result of discharge where insulating products may be formed or the electrolyte may be consumed. Therefore, shelf life (or storage from manufacturing to use in electronic devices) is limited by factors relating to both non-use and normal usage.

1.2.5 Discharge Curve/Cycle Life

The discharge curve is a plot of the voltage against the percentage of the capacity discharged. A flat discharge curve is desirable as this means that the voltage remains constant as the battery is used up. Some discharge curves are illustrated in Figure 1.9, where the potential is plotted against time as the battery is discharged through a fixed load. In the ideal mode, the cell potential remains steady with time until the capacity is fully exhausted at the same steady rate and then it falls off to a low level. Some of the primary lithium cells display this type of nearly ideal flat discharge characteristics. In most other real batteries, the voltage may slope down gently with time as in primary alkaline cells or do so in two or more stages during discharge as in Leclanché cells (see Table 1.4).

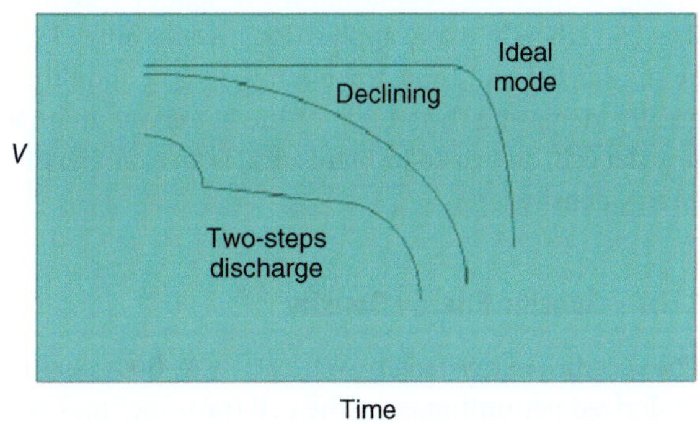

Figure 1.9 Change of voltage with time behavior in different cells. Source: Reproduced with permission from Weal et al. [2]/ University of Cambridge.

Table 1.4 History of electrochemical cell development tabulated with years and inventors.

Type	Year	Inventor	Battery
Primary batteries	1800	Alessandro Volta	Voltaic pile
	1836	John Frederic Daniel	Daniel cell
	1844	William Robert Grove	Grove cell
	1860	Callaud	Gravity cell
	1866	Georges-Lionel Leclanché	Leclanché wet cell
	1888	Carl Gassner	Zinc–carbon dry cell
	1955	Lewis Urry	Alkaline battery
	1970	No information	Zinc-air battery
	1975	Sanyo Electric Co	Lithium–manganese cell
	2004	Panasonic Corporation	Oxyride battery
Secondary batteries	1859	Raymond Gaston Planté	Planté lead–acid cell
	1881	Camille Alphonse Faure	Improved lead–acid cell
	1899	Waldemar Jungner	Nickel–cadmium cell
	1899	Waldemar Jungner	Nickel–iron cell
	1946	Union Carbide Company	Alkaline manganese secondary cell
	1970	Exxon laboratory	Lithium–titanium disulfide
	1980	Moli Energy	Lithium–molybdenum disulfide
	1990	Samsung	Nickel–metal hydride
	1991	Sony	Lithium-ion
	1999	Sony	Lithium polymer

1.2.6 Energy Density

The energy density is the energy that can be derived per unit volume of the cell and is often quoted as $\mathrm{Wh\,l^{-1}}$, (watt-hour per liter). This value is dependent upon the density of the components and the design by which the various materials are interfaced together. In many applications, availability of space for placing a battery must be minimized and thus, the energy density should be as high as possible without greatly increasing the weight of the battery to attain a given energy level. The battery flat cell, described in Figure 1.2, is an example of efficient design that increases energy density.

1.2.7 Specific Energy Density

The specific energy density, $\mathrm{Wh\,kg^{-1}}$(watt-hour per kilogram), is the energy that can be derived per unit mass of the cell (or sometimes per unit mass of the active electrode material). It is the product of the specific capacity and the operating voltage in one full discharge cycle. Both the current and the voltage may vary within a discharge cycle and, therefore, the specific energy derived is calculated by integrating the product of the current and the voltage over time

$$\text{Specific Energy} = \int V \cdot I \, dt \tag{1.35}$$

The discharge time is related to the maximum and minimum voltage thresholds and is dependent upon the state of availability of the active materials and/or the avoidance of an irreversible state for a rechargeable battery. The maximum voltage threshold may be related to an irreversible drop of voltage in the first cycle, after which that part of the cycle is not available. The minimum threshold voltage may be determined by a lower limit below which the voltage is deemed to be too low for practical use or set the limit for some irreversible losses, such that the system can only inadequately provide energy and power.

An active component may be less available due to side reactions, such as (i) zinc reacting with the electrolyte in alkaline or silver-oxide–zinc batteries, (ii) dendrite formation in rechargeable batteries, and (iii) formation of passivation layers on the active components. Since batteries are used mainly as energy-storage devices, the amount of energy (Wh) per unit mass (kg) is the most important property quoted for a battery. It must be noted that the quoted values only apply to the typical rates at which a particular type of battery is discharged. The specific- energy-density values vary typically between 40 and 300 Wh kg^{-1} for primary batteries (single-use as these are not designed for recharging) and nominally 30 and 260 Wh kg^{-1} for secondary (rechargeable) batteries, used commercially.

1.2.8 Power Density (Wh g^{-1})

The power density is the power that can be derived per unit mass of the cell. At higher drains, signifying higher currents relating to high-power densities, the specific energy tends to fall off rapidly, hence, decreasing the capacity. This trade-off between power and energy density is best expressed in a Ragone plot, an idealized version of which is given in Figure 1.10. It is obvious that a certain battery has a range of values for specific energy and power, rather than a battery having a specific

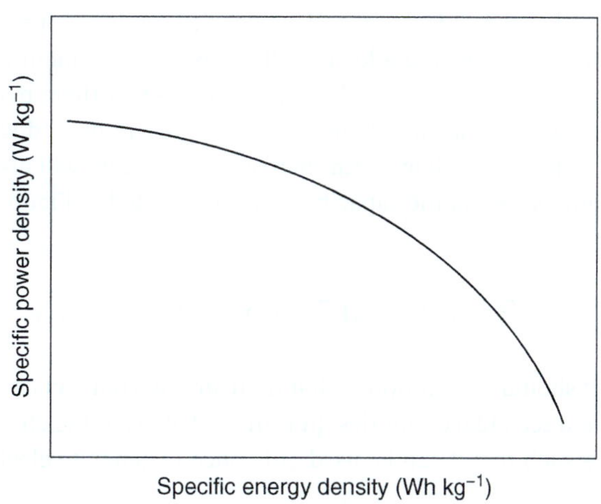

Figure 1.10 Ideal Ragone plot.

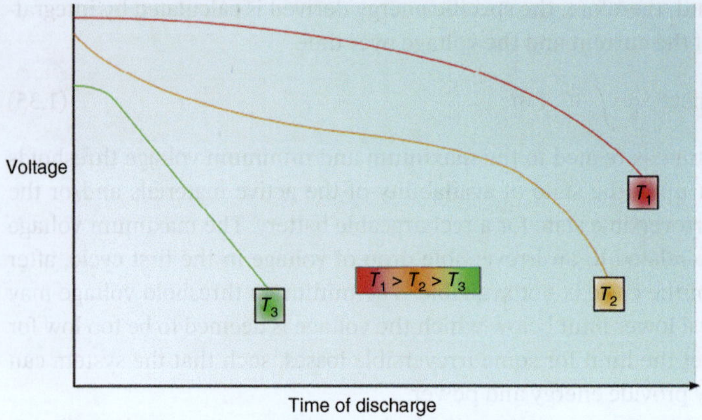

Figure 1.11 Effects of temperature on battery capacity. Source: Reproduced with permission from Weal et al. [2]/University of Cambridge.

value of energy and power. To derive the maximum amount of energy, the current or the power drain must be at the lowest practical level. For given cell chemistry, increasing the surface area of the electrodes can increase the cell's current at a given current density and, thus, deliver more power. The most efficient way to deliver a high-power density is to increase the effective surface area of an electrode while keeping the nominal geometric area constant. It is important to consider any increase in parasitic reactions that may be enhanced due to the increase in the effective surface area. For example, in systems where corrosion is a concern, simply increasing the surface area may enhance the corrosion reactions while depleting the active material. Under these circumstances, the cell capacity will decrease along with the shelf life.

1.2.9 Service Life/Temperature Dependence

The rate of the reaction in the cell is temperature dependent according to kinetics theories. The internal resistance also varies with temperature; low temperatures give a higher internal resistance. At very low temperatures, the liquid electrolyte may freeze giving a lower voltage as ion movement is impeded. At very high temperatures, the chemicals may decompose, or there may be enough energy available to activate unwanted and reversible reactions, reducing the capacity. The rate of decrease of voltage with increasing discharge will also be higher at lower temperatures, as will the capacity; this is illustrated in Figure 1.11.

1.3 Primary and Secondary Batteries

It should be mentioned that there are two main classifications of batteries: primary and secondary batteries. In primary batteries, the chemical energy stored in the cell is such that it can be used only once to generate electricity; that is, once the cell is fully discharged it cannot be of further use. In secondary batteries, the reverse redox reaction (also referred to as electrolysis or simply as charging) can occur when the current is applied at a potential higher than the cell potential (E_{cell}) and the battery can be used reversibly numerous times. During charging, electrons flow to the negative [−] electrode through the external circuit for facilitating a cathodic reaction (electrochemical reduction) typically with cations (e.g. Cu^{2+} ions) from the positive [+] electrode diffusing through the electrolyte. The external current is maintained by an anodic reaction (electrochemical oxidation) at the [+] electrode generating electrons flowing through the external circuit. The power supply maintains the voltage difference between the [+] and the [−] electrodes to sustain the charge movement.

In Table 1.4, a timeline depicts the historic development of batteries. Presently, the most advanced technology of primary batteries is the Oxyride battery, developed by Panasonic Corporation but not yet widely used. For the secondary batteries, the lithium-polymer battery is the most commercially advanced, mostly used as a power backup in laptop computers and slim and lightweight mobile phones. The most common secondary battery used commonly in communication devices, such as cellular phones, is the lithium-ion battery.

Batteries are also sometimes classified in terms of the mode in which they are used:

- **Portable batteries:** These cover a wide range of batteries from those used in toys to those used in mobile phones and laptops.
- **Transport batteries:** The largest application of these is in the starting, lighting, and ignition (SLI) for cars or in electrical vehicles (e.g. in e-scooters and hybrid electrical vehicles).
- **Stationary batteries:** Include applications for standby power, backup in computers, telecommunications, emergency lighting, and load-leveling with renewable energy, such as solar cells during darkness and windmills during very calm weather.
- **Electrical Vehicle (EV) Batteries:** Anticipated to grow rapidly starting from this decade in competition with IC-engine-based vehicles and ultimately fully replacing the IC-engine vehicles.
- **Energy-Storage Systems (ESS):** Growth in this area will parallel the growth in renewable energy sources applications as in solar farms and wind turbines. The EV and the ESS batteries will dominate battery applications for decades assuming development proceeds as planned.

Thus, batteries are correctly perceived as a critical enabling technology and future improvements are continuously sought. Various battery chemistries will be examined in detail in the following chapters.

1.4 Conclusions

Now that a general overview of batteries has been given, starting from their main components and configuration, to how they function and are disposed of, specific battery systems will be examined. The first two chapters that follow, present a historic timeline describing the development of batteries from the first cells of the 1800s to the commercial cells of the twentieth-first century. It will be seen that Li batteries are the most promising high-energy-storage devices and therefore the majority of the following chapters focus on the next-generation Li cathodes, anodes, and electrolytes that will significantly increase their lifetime and applications.

References

1 Linden, D. and Reddy, T.B. (2002). *Handbook of Batteries*, 3e. New York: McGraw-Hill.

2 Weal, E., Tan, J.C., and Kumar, R.V. (2005). *University of Cambridge DoITPoMS Teaching and Learning Packages*. Cambridge: University of Cambridge http://www.doitpoms.ac.uk/tlplib/batteries/index.php.

2

Primary Batteries

Thapanee Sarakonsri[1] *and R. Vasant Kumar*[2]

[1] *Chiang Mai University, Department of Chemistry, Faculty of Science, Chiang Mai 50200, Thailand*
[2] *University of Cambridge, Department of Materials Science and Metallurgy, 27 Charles Babbage Road, Cambridge CB3 0FS, UK*

2.1 Introduction

As mentioned in Chapter 1, primary batteries are not easily or safely recharged. This is due to the fact that the electrochemical reactions that occur are not easily reversible, and the cell is operated until the active component in one or both of the electrodes is either exhausted or more frequently unable to continue the reactions any further. Any attempt for reversing the reaction via recharging in a primary cell is dangerous and can cause the battery to heat up, catch fire and even explode. Therefore, once they are discharged, they are disposed of. Most primary cells are "dry cells" – the electrolyte is not a liquid but a paste or a gel.

The electrical resistance in primary cells is usually high; thus, even if charging was possible, it would be a slow process. At normal practical charging rates, a large proportion of the current would have most likely dissipated as heat, causing further safety hazards. Primary batteries are therefore designed to operate at low currents and have a relatively long lifetime. Generally, primary batteries typically (but not always) have a higher capacity ($Ah\,kg^{-1}$), a higher specific energy ($Wh\,kg^{-1}$) and may also have a higher initial voltage than secondary (rechargeable) batteries of comparable chemistries. They are used in portable devices, toys, watches, hearing aids, and medical implants. Some examples of commercially used primary batteries are shown in Table 2.1.

Rechargeable Ion Batteries: Materials, Design, and Applications of Li-Ion Cells and Beyond, First Edition.
Edited by Katerina E. Aifantis, R. Vasant Kumar, and Pu Hu.

Table 2.1 Primary battery chemistries used today [1].

System cathode/anode	Nominal cell voltage(V)	Specific energy (Wh kg^{-1})	Advantages	Disadvantages	Applications
Carbon/Zinc	1.50	65	Lowest cost; variety of shapes and sizes	Low-energy density; poor low-temperature performance	Torches; radios; electronic toys and games
Mg/MnO_2	1.60	105	Higher capacity than C/Zn; good shelf-life	High gassing on discharge; delayed voltage	Military and aircraft receiver-transmitters
Zn/Alk/MnO_2	1.50	95	Higher capacity than C/Zn; good low-temperature performance	Moderate cost	Personal stereos; calculators; radio; television
Zn/HgO	1.35	105	High-energy density; flat discharge; stable voltage	Expensive; energy density only moderate	Hearing aids; pacemakers; photography; military sensors/detectors
Cd/HgO	0.90	45	Good high and low-temperature performance; good shelf-life	Expensive; low-energy density	
Zn/Ag_2O	1.50	130	High-energy density, good high-rate performance	Expensive (but cost effective for niche applications)	Watches; photography; larger space applications
Zn/air	1.50	290	High-energy density; long shelf-life	Dependent on environment; limited power output	Watches; hearing aids; railway signals; electric fences
Li/$SOCl_2$	3.60	300	High-energy density; long shelf-life	Only low-to-moderate rate applications	Memory devices; standby electrical power devices; medical devices
Li/SO_2	3.00	280	High-energy density; best low-temperature performance; long shelf-life	High-cost pressurized system	Military and special industrial needs
Li/MnO_2	3.00	200	High-energy density; good low-temperature performance; cost effective	Small in size, only low-drain applications	Electrical medical devices; memory circuits; fusing

2.2 The Early Batteries

Before describing in detail a selection of the main battery systems indicated in Table 2.1, the predecessors of today's batteries are introduced. Historically, the first battery was invented by Alessandro Volta, an Italian physicist, in 1800, who observed Luigi Galvani's (also an Italian physicist) experiment on frog legs, in 1780, by connecting two different metals in series with the frog's leg and to one another. The leg twitched when it was touched with the iron scalpel (Figure 2.1) and, therefore, Galvani believed that frog legs can generate electricity and called it "animal electricity." However, Volta believed that the electricity came from the connection of the two metals and proved his idea by constructing a battery consisting of alternate copper and zinc disks; a cloth or cardboard soaked in brine (NaCl) was placed in between the copper and zinc metal layers. Due to this pile-up design, Volta's cell was called a "Voltaic Pile" (Figure 2.2), and even though it was stable

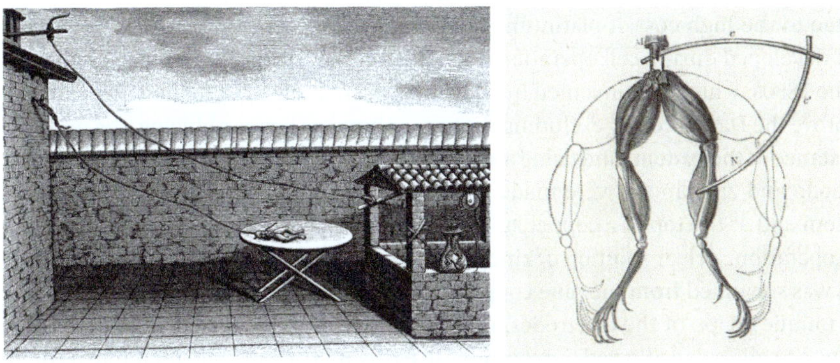

Figure 2.1 Galvani's experiment on frog legs. Source: Reproduced with permission from Roberge [2].

Figure 2.2 An illustration of Volta's cell, called a "Voltaic Pile". Source: Reproduced with permission from Miller et al. [3].

during continuous draining, electrolyte leaking was encountered. Furthermore, a thin layer of hydrogen gas (H_2) formed on the copper surface increasing the internal resistance of the battery. As a result, the "Voltaic Pile" had a short life (~1 hours) and the voltage per element was 0.75 V.

In 1836, John Frederic Daniel, a British chemist, improved the Volta cell by constructing a battery that used the same electrode-base metals, but a different configuration. The zinc was immersed in a sulfuric acid (H_2SO_4) solution that was held in a porous earthenware container. This compartment, which functioned as the anode, was then immersed in a copper sulfate ($CuSO_4$) solution that was in a copper pot that acted as the cathode. This cell, referred to as the Daniel cell, provided 1.1 V and was used for powering the telegraph system.

Eight years later in 1844, the Grove cell was invented by William Robert Grove, a British lawyer, judge, and physical scientist. The cell consisted again of a zinc anode, but the copper cathode was replaced by platinum. The anode and cathode compartments were separated by porous earthenware, and each electrode was immersed, respectively, in sulfuric and nitric acid (HNO_3) solutions.

Due to the high cost of platinum metal and the toxicity of nitric oxide (NO) gas that developed during cell operation, the Grove cell was replaced by the gravity cell in the 1860s, which was invented by Callaud, a French scientist. Callaud was able to improve the Daniel cell by excluding the porous barrier (hence reducing, the internal resistance of the system) and using a simple glass jar as the cell container. The copper cathode and zinc anode were made in a crow's foot shape and were placed at the bottom and at the top of a copper sulfate solution, respectively (Figure 2.3). During cell operation, a clear solution of zinc sulfate ($ZnSO_4$) was formed around the anode and was separated from the blue copper sulfate solution by its lower density. Due to the unique shape of the electrodes, this cell was also called a "crowfoot cell." The open-top cell design allowed the solutions to easily mix or spill, and therefore it could be only used for stationary applications. The voltage of this battery was 1.08 V, and its specific lifetime was longer than that of the voltaic pile.

Figure 2.3 Illustration of a crowfoot cell. Source: Reproduced with permission from Miller et al. [3].

2.3 The Zinc/Carbon Cell

Although the initial electrochemical cells consisted of zinc and copper electrodes, in the development of the first advanced battery, copper was replaced by carbon and manganese dioxide. The zinc-based cell is not only the first industrial prototype battery invented but also one of today's most widely used batteries, due to its low production cost. Its main components have not changed over the centuries, and only minor alterations were introduced before reaching today's form. In the first zinc-based cell the anode was zinc–amalgam, instead of pure zinc and the cathode contained manganese dioxide and carbon; this was the famous Leclanché cell.

2.3.1 The Leclanché Cell

The first primary battery was invented in 1866 by French telegraphic engineer Georges-Lionel Leclanché. It was the first cell to contain only one low-corrosive fluid electrolyte with a solid cathode. This gave it a low self-discharge – a capacity loss without the electrodes being connected – in comparison to previously attempted batteries. The cell was called the Leclanché *wet* cell because an ammonium chloride (NH_4Cl) aqueous solution was used as the electrolyte. The other cell components were a zinc–amalgam (Zn/Hg) bar anode and a cathode comprising a mixture of manganese dioxide (MnO_2) and a small amount of carbon (C) powders. The cell was designed by placing the cathode material in a porous pot and subsequently placing a carbon rod in the middle of the cathode; then the whole electrode was put in a square glass container. After inserting the zinc–amalgam bar into the same glass container the container was filled with ammonium chloride electrolyte to complete the formation of a single functional cell. An illustration of an original Leclanché cell is shown in Figure 2.4a. Leclanché improved his cell in 1876 by heating compressed manganese dioxide and carbon with resin at 100 °C to form a hard solid cathode; the porous pot was therefore not needed. The cathode was also referred to as an "agglomerate block," since a pair of agglomerate blocks were attached to the carbon plate with rubber bands and then placed in the glass container along with the zinc bar. Leclanché used his invention widely in his telegraphic office for telegraphy signaling. Commercially, the most popular use of the Leclanché cell was for house doorbells but later in the 1920s, for portable radios, flashlights, and telephones. The voltage of this cell was 1.5 V with a nominal capacity was 224 Ah kg^{-1}.

2.3.2 The Gassner Cell

In 1888, Carl Gassner, a German scientist, further improved the Leclanché cell by using ferric hydroxide [FeO(OH)] mixed with manganese dioxide and a small amount of carbon as the cathode material, and instead of using a zinc–amalgam bar, a hollowed zinc cylindrical container (bobbin) was used as the anode. The liquid ammonium chloride electrolyte was replaced by an electrolyte paste that consisted

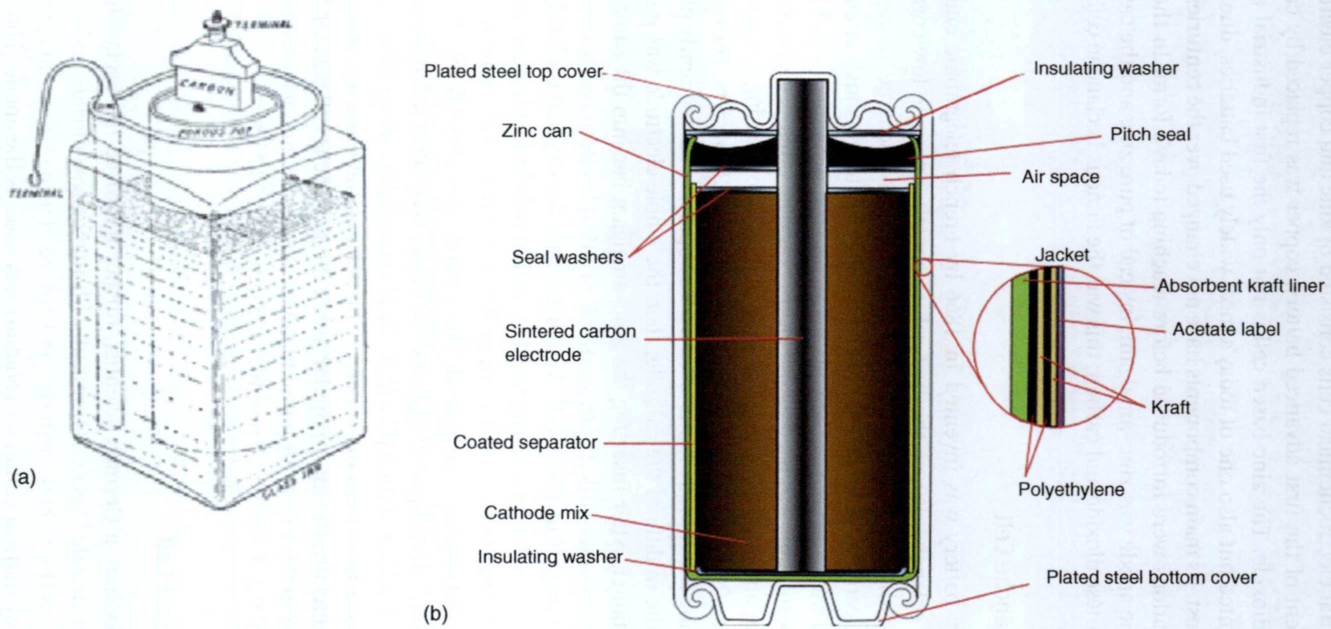

Figure 2.4 (a) Original Leclanché cell. Source: Republished with permission of Linden and Reddy [4]/McGraw-Hill. (b) The commercial design structure of a zinc–carbon cell. Source: Reproduced with permission from Weal et al. [1]/University of Cambridge.

of zinc chloride ($ZnCl_2$), ammonium chloride, water, and wheat flour as a binder. The battery was therefore no longer a "wet" cell but a "dry" cell. The electrolyte paste was then mixed with the manganese dioxide/carbon mixture (cathode) and poured into the zinc anode container with the carbon rod placed in the center of the paste. The advantage of this dry cell over the Leclanché wet cell was the ease of handling within a more durable design. The successful improvement by Gassner resulted in the commercialization of this cell as the zinc–carbon dry cell by the National Carbon Company (USA) in 1896. This cell provided a potential of 1.5 V, a gravimetric energy density of 55–77 Wh kg^{-1}, a volumetric energy density of 120–152 Wh dm^{-3}, and a capacity of 146–202 Ah kg^{-1}.

2.3.3 Current Zinc/Carbon Cell

Figure 2.4b depicts the most common commercial design of the zinc–carbon battery. The Zn serves as both the battery casing and the anode. The cathode consists of manganese dioxide (MnO_2), powdered black carbon, and some electrolyte (the electrolyte is ammonium chloride, while for extra heavy-duty applications, the electrolyte is zinc chloride mixed with a minor amount of ammonium chloride). The MnO_2 and carbon mixture is wetted with electrolyte and shaped into a cylinder with a small hole in the center. A carbon rod is inserted into the center, which serves as a current collector. The current collector plays a key role in batteries as it is used to conduct electricity between the reacting parts of the electrode and the terminal and, hence, needs to be in good contact with the electrode. The carbon current collector was porous to allow gases to escape while providing structural support. The separator was made from natural cellulose fibers, or cereal paste coated with heavy paper or treated absorbent kraft paper (the kind of brown paper used to make large envelopes or grocery bags). The MnO_2 to C ratios vary between 10 : 1 and 3 : 1, with a 1 : 1 mixture being used for photoflash batteries, as this gives a better performance for intermittent use at high bursts of current. It should be noted that even though these cells are commonly termed Zn/C, the actual cathode active material is MnO_2, derived from natural ore, while C is added for conductivity and to provide mechanical stability to the electrode–electrolyte system.

The voltage of these batteries is between 1.5 and 1.7 V, their shelf-life is 1–2 years at room temperature and their service life is 110 minutes of continuous use. Figure 2.5 depicts a typical discharge curve where the voltage decreases with time at a selected constant current of 500 mA. The capacity is around 40 Ah kg^{-1} at low drains and the energy density is 55–77 Wh kg^{-1}. These cells are generally sensitive to external factors, such as temperature, and are more effective when discharged intermittently, since this allows for diffusion to take place to avoid polarization (Chapter 1, Section 1.3) that can occur from discharge – the separation and grouping of reactants and products that may result in increasing the internal resistance in the cell. Generally, the voltage slopes with time during discharge due to cell chemistry changes with discharge but also from internal leakage. To minimize leakage, in modern cells, the electrolyte is produced in a gel form or held as a viscous paste in a porous separator.

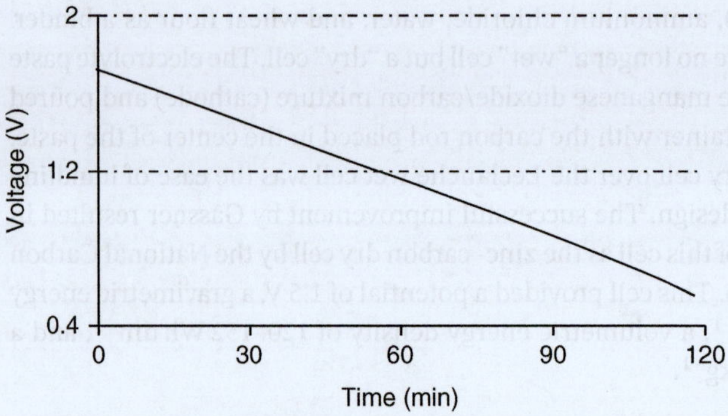

Figure 2.5 Discharge curve for a Zn/C cell at 500 mA. Source: Reproduced with permission from Weal et al. [1]/University of Cambridge.

In addition to cylindrical Zn/C cells, flat Zn/C cells are also quite common. These Zn/C flat cells were introduced in the 1920s for use in portable radios. The zinc anode was painted with a carbon-conductive paint, which acted as the conductor at the anode–electrolyte interface for minimizing the electrical resistance. The cathode was made from a mixture of MnO_2, graphite or acetylene black carbon, and electrolyte paste. Each cell was placed in a plastic envelope and stacked inside a steel jacket (as shown in Figure 1.3 in Chapter 1). Common connector strips of copper were used to collect and connect the positive and the negative electrodes to a common positive and a negative terminal. Zn/C flat cells provide a potential of 1.5 V, a capacity of 238 Ah kg^{-1}, and are still used today for toys and some flashlights.

2.3.3.1 Electrochemical Reactions

As mentioned previously the zinc/carbon cell uses zinc and manganese dioxide as the anode and cathode active materials respectively, meaning they only take place in the reaction. The carbon is added to the cathode to increase conductivity and retain moisture.

The overall reaction in the cell is

$$Zn + 2MnO_2 \rightarrow ZnO + Mn_2O_3 \quad +1.54\,V$$

The exact mechanism for this is complicated, and not fully understood; however, the approximate half-cell reactions are

Anode:

$$Zn \rightarrow Zn^{2+} + 2e^- \quad +0.76\,V$$

Cathode:

$$2NH_4^+ + 2MnO_2 + 2e^- \rightarrow Mn_2O_3 + H_2O + 2NH_3 \quad +0.78\,V$$

The electrolyte in which the above reactions take place is ammonium chloride. However, this is complicated by the fact that the ammonium ion produces two gaseous products:

$$2NH_4^+ + 2e^- \rightarrow 2NH_3 + H_2$$

These products must be absorbed to prevent buildup of pressure in the vessel, therefore zinc chloride is added to the ammonium chloride for the following two reactions to occur:

$$ZnCl_2 + 2NH_3 \rightarrow Zn(NH_3)_2Cl_2$$

$$2MnO_2 + H_2 \rightarrow Mn_2O_3 + H_2O$$

2.3.3.2 Components

Historically the black carbon used was in the form of graphite; however, acetylene black carbon is often used in modern batteries as it can hold more electrolytes. Various forms of MnO_2 are available: (i) natural manganese dioxide; ores occur naturally in Gabon, Greece, and Mexico with 70–85% MnO_2, (ii) activated manganese dioxide, (iii) chemically synthetic manganese dioxide (90–95% MnO_2), and (iv) electrolytic manganese dioxide (EMD) that is widely used in industrial cells but mainly in alkaline batteries as it allows for a higher cell capacity, higher rate capabilities, and less polarization.

Pure Zn sometimes alloyed with minor alloying elements such as Pb is used as the can material thus serving both as the anode and the container. A zinc-corrosion inhibitor is also added, which forms an oxide layer. This inhibitor is usually mercuric oxide or mercurous chloride. A typical electrolyte composition is 26.0% NH_4Cl, 8.8% $ZnCl_2$, 65.2% H_2O, 0.25–1.0% corrosion inhibitor. While the oxide layer prevents corrosion, thus prolonging the shelf-life, it increases the electrical resistance permitting use only under low currents.

The carbon rod is inserted into the cathode and acts as a current collector. It also provides structural support and vents hydrogen gas that evolves as the reactions proceed. The as-produced rods are very porous, so they must be treated with waxes or oils to prevent loss of water, but at the same time remain porous enough to allow hydrogen gas to pass through. Ideally, they should also prevent oxygen from entering the cell, as this would aid corrosion of the zinc.

The separator physically separates the anode and the cathode but allows ionic conduction to occur in the electrolyte. The following two types of separators are used: (i) a gelled paste that is placed into the zinc can, and once the cathode is inserted, the paste is forced up the sides of the can between the zinc and the MnO_2 cathode; (ii) kraft paper coated with cereal paste, or another gelling agent, rolled into a cylinder and along with a circular bottom sheet is added to the can. The cathode is added, and the rod is inserted, pushing the paper against the walls of the cans. This compression releases some electrolytes from the cathode mix, soaking the paper. As the paste is relatively thick, more electrolytes can be held by the paper than the paste, giving an increased capacity; thus, paper is usually the preferred separator.

The seal of the battery can be asphalt pitch, wax/resin mix, or plastic (usually polyethylene or polypropylene). Air space is usually left between the seal and the cathode to allow for expansion. The function of the seal is to prevent evaporation of the electrolyte and to prevent oxygen from entering the cell and corroding the zinc.

The jacket provides strength and protection and will hold the manufacturer's label. It contains various components, which can be metal, paper, plastic, mylar (biaxially oriented polyethylene terephthalate [boPET] polyester film), cardboard (sometimes asphalt-lined), and foil. It should be noted here that the jacket is not an active material in the cell but serves as the cathode current collector. The terminals of the battery aid conductivity and prevent exposure of the zinc and are tin-plated steel or brass.

2.3.4 Disadvantages

There are various shapes and sizes of zinc–carbon dry cells, but the most common ones fabricated are of size D (see Table 1.1 in Chapter 1). They have been used in flashlights (familiar name "flashlight cell"), portable radios, photoflashes, and toys since 1896 and are still manufactured today (e.g. Energizer Holdings, Inc.). The reason for producing large sizes is the physical limitation by which cell efficiency increases by decreasing the current density; this corresponds to large cell size [5]. Disadvantages of zinc–carbon batteries arise from the fact that electrolyte appears in the overall cell reaction, resulting in a decreasing electrolyte concentration during cell operation, and causing corrosion of the zinc anode by acidic NH_4^+ ions [6]. As a result, such batteries have an unstable voltage and short battery shelf-life.

2.4 Alkaline Batteries

Research on zinc–carbon dry cells has helped to improve their shelf-life and power density for heavy-duty devices, such as portable lighting and motors, while meeting the environmental benign requirements. The improvements that were performed on zinc–carbon batteries to make them more efficient resulted in the development of alkaline-manganese dioxide batteries, which are the most common commercial dry cells today.

Lewis Urry developed the small alkaline battery at the Eveready company laboratory in 1949 and this cell design was further improved by Samuel Ruben in 1953. The first modern alkaline cell was commercialized in 1959 by Ever Ready, and by 1970 it was produced all over the world. Billions of alkaline cells are used worldwide each year and the alkaline battery market is expected to grow at a growth rate of US$ 570.00 million. They use the same anode and cathode as the zinc/carbon cells, but their electrolyte contains an alkaline solution (aqueous KOH), and, hence, their name is alkaline cells. Their capacity at over 65 Ah kg^{-1} is 25–30% higher than the zinc/carbon cell. Furthermore, they offer good performance at high discharge rates and a continuous discharge profile at low temperatures.

Their active materials used are also essentially the same as in the Gassner cell: zinc and manganese dioxide along with the conducting material, carbon. However, the electrolyte is potassium hydroxide, which is more conductive, resulting in a low internal electrical resistance in the cell due to the higher ionic conductivity from the fast-moving hydroxyl ions. In particular, synthetic manganese dioxide (EMD or chemically synthetic manganese dioxide [CMD]) has replaced natural manganese dioxide ore due to the increased capacity and higher activity over the natural material, while black carbon has replaced graphite to take advantage of the superior conducting and absorption properties [7]. Furthermore, to increase the surface area of the anode, the cell design utilizing a zinc can (which serves as a conductor and a mechanically strong container) as the anode was replaced with an external steel jacket and a gel of amalgamated zinc (Zn/Hg) powder mixed with electrolyte as the anode (Figure 2.6). This resulted in better ionic conduction within the anode mix and reduced the mass-transport polarization losses. The current collector for this anode is a brass (CuZn) pin that is inserted through the middle of the gel and connected to the anode terminal. The steel can be nickel-plated or has a conductive carbon coating placed on the inner surface (Figure 2.6a).

It can be seen by comparing Figures 2.4b and 2.6a that alkaline cells are "inside out" compared to the Zn/C cells since the manganese dioxide cathode is external to the zinc anode, giving better diffusion properties, and lower internal resistance. The container is made of steel, which is readily passivated in an alkaline solution and serves as both a conductor and a mechanically strong container. The entire cell is hermetically sealed with nylon.

Additional advantages of alkaline batteries are: they have a longer service delivery over zinc–carbon cells; their capacity is constant during varying discharge loads; they can be heavily drained for relatively long periods of time. The operating temperature is in the range of −20 to +70 °C, the gravimetric energy density is 66–99 Wh kg^{-1}, the volumetric energy density is 122–268 Wh dm^{-3}, and the total capacity can range from 700 mAh to 10 Ah for multi-cell batteries with different voltages; multi-cell batteries are made by connecting miniature or cylindrical cells in series and/or parallel circuits. A typical discharge curve is shown in Figure 2.7. The commercial service capacity is higher than the zinc–carbon battery (whose gravimetric and volumetric energy densities are 55–77 Wh kg^{-1} and 120–152 Wh dm^{-3}). The commercial sizes of alkaline batteries vary from N, AA, AAA, C, and D. Small portable electronic devices such as MP3 players, digital cameras, and scientific calculators are examples of devices employing general AA- or AAA-size alkaline batteries. Alkaline cells are not manufactured in a flat cell configuration, but they are also available as button cells.

2.4.1 Electrochemical Reactions

The following half-cell reactions take place in the cell:

At the anode:

$$Zn(s) + 2OH^-(aq) \rightarrow Zn(OH)_2(s) + 2e^- \qquad +1.25\ V$$

$$Zn(OH)_2 + 2OH^- \rightarrow [Zn(OH)_4]^{2-}(aq)$$

At the cathode:

$$2MnO_2(s) + H_2O + 2e^- \rightarrow Mn_2O_3(s) + 2OH^- \qquad +0.4\ V$$

For full discharge:

$$MnO_2 + 2H_2O + 2e^- \rightarrow Mn(OH)_2(s) + 2OH^-$$

Overall:

$$Zn + 2MnO_2 \rightarrow Zn(OH)_2 + Mn_2O_3$$

or

$$Zn + MnO_2 + 2H_2O \rightarrow Mn(OH)_2 + Zn(OH)_2 \qquad +1.65\ V$$

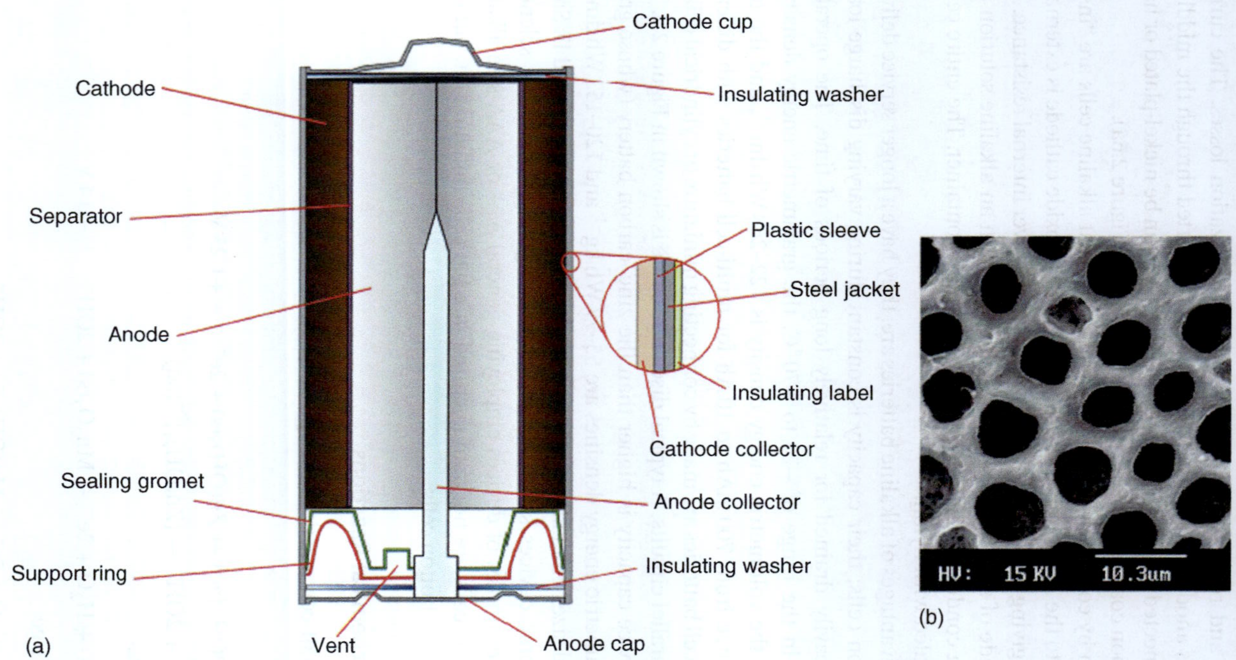

Figure 2.6 (a) The commercial design structure of an alkaline cell (b) Microporous separator. Source: (a) Reproduced with permission from Weal et al. [1]/University of Cambridge, (b) Reproduced with permission from Weal et al. [1]/University of Cambridge.

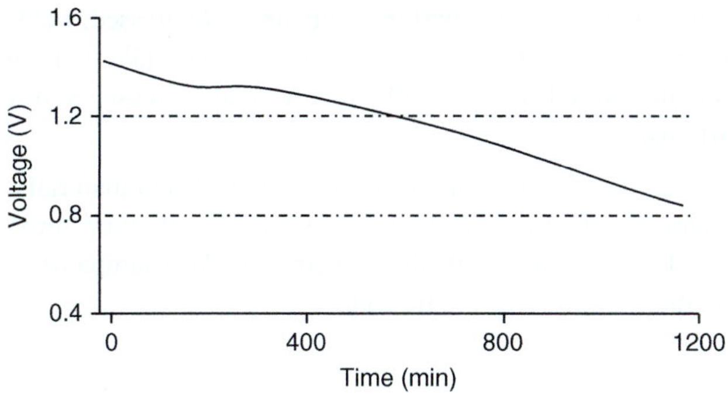

Figure 2.7 Discharge curve at 500 mA for an alkaline cell. Source: Reproduced with permission from Weal et al. [1]/University of Cambridge.

It is still not possible to describe the cathodic reaction on discharge in a simple unambiguous way, despite a lot of research. The reality is more complicated than described in the two reactions shown above. In fact, the discharge curve has two fairly distinct sections corresponding to the change in the oxidation state of Mn from +4 to +3 and then to +2 during the reduction of MnO_2.

2.4.2 Components

For an alkaline cell, electrochemically produced MnO_2 must be used. The ore rhodochrosite ($MnCO_3$) is dissolved in sulfuric acid, and electrolysis is carried out under carefully controlled conditions using titanium, lead alloys, or carbon for the electrode onto which the oxide is deposited. This gives the highest possible purity, typically $92 \pm 0.3\%$. The cathode itself also contains approximately 10% graphite – a higher %C results in more powerful batteries. A typical composition of a cathode would be 70% MnO_2 (of which 10% is water), ~10% graphite, and 1–2% acetylene black carbon, with some binding agents and electrolytes for aiding wetting with the electrolyte by decreasing the surface tension.

The zinc anode must be very pure (99.85–99.90%) and is produced by electroplating or distilling. A small amount of lead is sometimes added to help prevent corrosion (usually ~0.05%). The zinc is powdered by discharging a small stream of molten zinc into a jet of air "atomizing" it. The powder contains particles between 0.0075 and 0.8 mm. In particular, there are two methods for the formation of the anodes from the powder:

- **Gelled anodes:** These contain approximately 76% zinc, 7% mercury, 6% sodium carboxymethyl cellulose, and 11% KOH solution, which is extruded into the cell as the viscosity is high. In very small cells NaOH is added to reduce creepage around the seal area. However, this mixture is not ideal as it does not fully utilize the zinc at high current densities. Two-phase anodes have, therefore, been developed, consisting of a clear gel phase and a more compact zinc-powder gel phase, which enables 90% zinc usage.

- **Porous anodes:** The zinc powder is wetted with mercury and cold pressed, welding the particles together. The porosity can be controlled by materials such as NH_4Cl or plastic binders if required, which can be removed later. These anodes can carry very high currents.

The separators are "macroporous" and are made from woven or felted materials. When non-woven separators made from synthetic fibers are used, they must be selected to resist high pH values of the alkaline electrolyte. An example of a polymeric membrane separator is shown in Figure 2.6b.

2.4.3 Disadvantages

Zinc can progressively corrode in the alkaline electrolyte, releasing hydrogen gas, which results in a pressure increase and the consequent expansion of the cell. Such a reaction can be suppressed by alloying Zn with elements such as Hg or Pb; however, both are notorious for their toxicity. While the alkaline cells are superior to the Zn–C dry cells, for their steadier voltage output, longer life, and higher current capability, they still have a fairly high internal resistance. Their components also have a high thermal coefficient of resistivity. As a result of these factors, the faster an alkaline cell is drained, the higher the percentage of the load that is dissipated as heat. The capacity is, therefore, greatly reduced. For example, with an AA-sized alkaline battery, the nominal capacity is >3000 mAh kg^{-1} at low drains (<500 mA), but at a current drain of >1000 mA (as may be typically required in a digital camera), the capacity is decreased to <750 mAh kg^{-1}.

2.5 Button Batteries

In addition to cylindrical batteries, there is another category of battery shape that is known as a button battery. The common feature between material chemistries used in button batteries is that they allow for low discharge rates, and therefore provide a longer lifetime, making them suitable for scientific and biomedical implantable devices. Their most common design is shown in Figure 2.8, while description of a button cells is given below.

2.5.1 Mercury Oxide Battery

The first button cell was the mercuric oxide (HgO) battery, which was developed during World War II for military applications and was then used in scientific instruments. This cell comprises a zinc anode, a mercuric oxide cathode, and zinc oxide and potassium hydroxide/sodium hydroxide (NaOH) electrolyte (the zinc oxide makes up 7% of the weight of the electrolyte). Because of its high performance, such as high capacity per volume (300–500 W dm^{-3}), stable voltage upon discharge, and long shelf-life, this cell was used in hearing aids, watches, cameras, and scientific calculators. There are two battery designs: a 1.35 V cell,

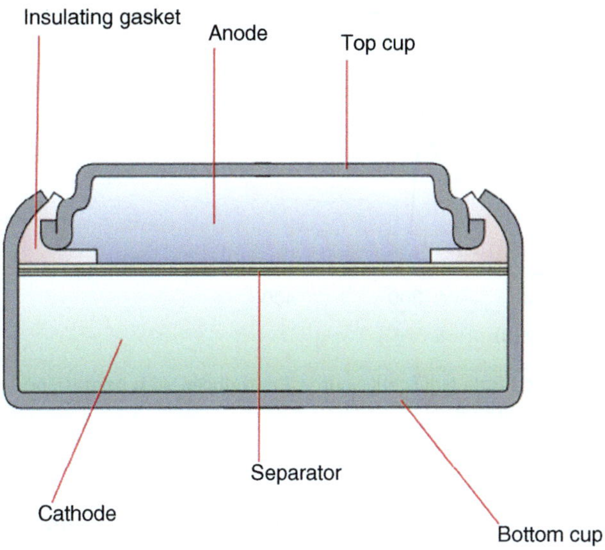

Figure 2.8 Configuration of a button battery. Source: Reproduced with permission from Weal et al. [1]/University of Cambridge.

whose cathode is 100% mercuric oxide, and a 1.4 V cell, whose cathode is a mixture of mercuric oxide and manganese dioxide. The first cell is suitable for stable voltage and long storage applications such as sonobuoys (underwater sonic wave receivers) and early satellites, while the second one is most suitable for long-term continuous low-drain applications, such as hearing aids and watches. Due to the environmental concern of using mercury metal, this battery was replaced by zinc/silver, zinc–air, and lithium batteries. In addition to button cells, there also exist flat-pellet and cylindrical configurations, but they are expensive and used only for special applications.

2.5.1.1 Electrochemical Reactions

The reactions that take place upon discharge of these cells are as follows.

At the anode:

$$Zn + 2OH^- \rightarrow ZnO + H_2O + 2e^- \qquad +1.25\,V$$

At the cathode:

$$HgO + H_2O + 2e^- \rightarrow Hg + 2(OH)^- \qquad +0.0977\,V$$

Overall:

$$Zn + HgO \rightarrow ZnO + Hg \qquad +1.35\,V$$

A typical discharge curve is shown in Figure 2.9.

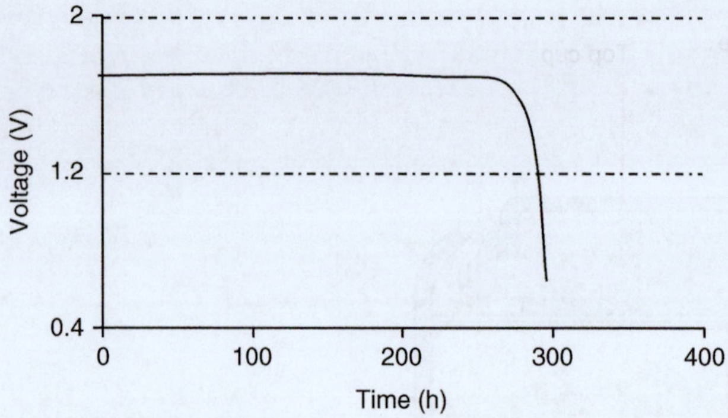

Figure 2.9 Discharge curve at 6.5 kΩ for a HgO cell. Source: Reproduced with permission from Weal et al. [1]/University of Cambridge.

2.5.2 Zn/Ag_2O Battery

The first zinc/silver oxide battery was developed in the late 1930s by G. André by using a suitable separator that could clearly separate the two electrodes. These cells have a relatively high specific energy density of 150 Wh kg^{-1}, and can deliver current at a very high rate, at a constant voltage of 1.6 V. However, the construction materials are of high cost, so such battery systems are limited to the production of button cells, for use in calculators, watches, hearing aids, and other such devices that require small batteries and long service life. A typical discharge curve is shown in Figure 2.10.

The cathode is generally composed of monovalent silver oxide with added graphite to improve conductivity. The anode is zinc powder mixed with a gelling agent, which is then immersed in an alkaline electrolyte; these batteries however are not referred to as alkaline cells, although scientifically they are also a type of alkaline cell. The two electrodes are separated by a combination of layers of grafted plastic membrane, treated cellophane, and nonwoven absorbent fibers. The top cup (negative terminal) is made up of laminated layers of copper, tin, steel, and nickel, and the bottom

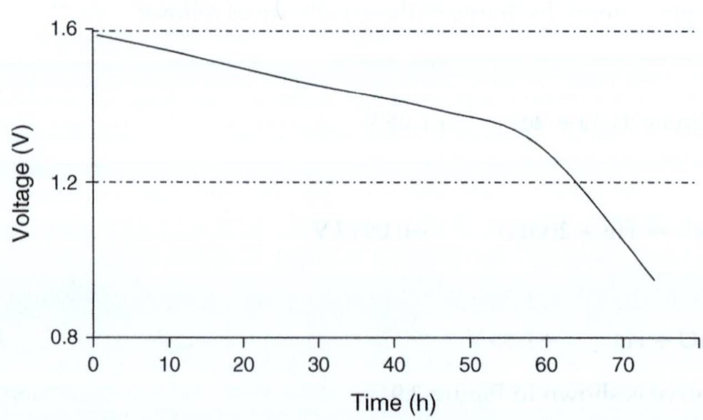

Figure 2.10 Discharge curve at 1 kΩ for a Zn/Ag cell. Source: Reproduced with permission from Weal et al. [1]/University of Cambridge.

cup (positive terminal) is nickel-plated steel. An insulating gasket prevents contact between the two. The cell looks similar to the HgO button cell shown in Figure 2.9.

2.5.2.1 Electrochemical Reactions

The reactions that occur in this cell are indicated below.

At the anode:

$$Zn + 2OH^- \rightarrow Zn(OH)_2 + 2e^- \qquad +1.25\ V$$

At the cathode:

$$Ag_2O + H_2O + 2e^- \rightarrow 2Ag + 2OH^- \qquad +0.35\ V$$

Overall:

$$Ag_2O + H_2O + Zn \rightarrow 2Ag + Zn(OH)_2 \qquad +1.6\ V$$

The voltage of these batteries is around 1.6 V, and it is linearly dependent on temperature. These cells can be used continuously for several thousands of hours and their shelf-life can be several years at room temperature. Their discharge characteristics are relatively flat with a specific capacity of 80–90 Ah kg^{-1}.

2.5.3 Metal–Air Batteries

When a reactive metal anode is electrochemically coupled with an air (oxygen) cathode, the cathode provides an exhaustible amount of nearly weightless reactive component. A basic metal–air battery consists of a metal anode, an oxygen cathode, and an electrolyte through which ions can diffuse and electrochemically oxidize the anode. While oxygen is consumed at the cathode, it is quickly replenished from the surrounding air; thus, there is no net material consumption at the cathode. Oxygen diffuses through the porous cathode and is electrochemically reduced in the cell:

$$O_2 + 2H_2O + 4e^- \rightarrow 4OH^- \quad +0.401\ V$$

The hydroxide ions then diffuse through the aqueous electrolyte (usually neutral or alkaline solutions) and oxidize the metallic anode. Such cells have a high value of theoretical energy density and provide a flat discharge voltage curve. The capacity limit is determined by the anodic capacity and the presence of any parasitic reactions. Strictly speaking, these cells are primary cells as they cannot be easily electrochemically recharged once the anode is exhausted. However, the spent anode can be replaced mechanically with a new fresh anode; thus, the term mechanically rechargeable battery is often used in the literature. Since air is used as an oxidant, and a metal anode is considered as a solid fuel, the term metal fuel cell is also sometimes used to describe these cells.

Considerable progress has been made in the development of air electrodes. A metallic mesh is used as the current collector and as a support for the cathode. An air access hole allows access to oxygen from the surrounding air. The rate of oxygen transfer into the cathode is controlled by the area of the hole and/or the porosity of a diffusion membrane deposited on the cathode surface. The operating current achieved in the cell is directly proportional to the oxygen consumption at

Table 2.2 Metal–air cells.

Metal anode	Theoretical capacity (Ah g^{-1})	Theoretical cell voltage (V)	Theoretical specific energy (Wh g^{-1})	Operating voltage (V)
Li	3.86	3.4	13	2.4
Al	2.98	2.7	8.1	1.6
Mg	2.20	3.1	6.8	1.4
Ca	1.34	3.4	4.6	2.0
Zn	0.82	1.6	1.3	1.2
Fe	0.96	1.3	1.2	1.0

the cathode. Thus, the current can be increased by increasing the porosity until the reaction rate of the above reaction becomes rate limiting. The catalyst layer itself consists of manganese oxide in a carbon matrix (a small platinum loading may also be provided) and is made hydrophobic by admixing with fine Teflon particles. Thus, a dry boundary is maintained between the gas phase and the aqueous electrolyte.

A number of metallic electrodes have been developed for metal–air cells. Table 2.2 provides values for theoretical cell voltages, theoretical specific capacities, and theoretical specific energies for selected metal anodes. The practical operating voltages are often much lower due to polarization effects at the electrodes.

In the subsequent subsections, two different metal/air batteries are described.

2.5.3.1 Zn/Air Battery

The zinc/air system has attracted much attention as a candidate for electrical vehicle (EV) applications. It is also seen as an environmental-friendly cell. The development of this battery began in the 1930s and was used for military radios in World War II. It was finalized in today's commercial design in the early 1970s in N and D sizes, which are mostly used for paging equipment. Later on, the button cell was introduced for hearing aids and watches. The chemistry design of this cell was adopted from the research on zinc–air fuel cells, which comprised a zinc granular powder and electrolyte mixture anode, a hydrophobic cathode composite (low surface black carbon and Teflon (polytetrafluoroethylene, C_nF_{2n} [PTFE]) particles), a catalyst-coated nickel screen layer, an air diffusion Teflon layer, a hydrophobic Teflon layer, and an air distribution membrane. This cell has air access holes allowing oxygen from the ambient atmosphere to enter the cell and perform the electrochemical reaction at the cathode. These batteries are commercially available in 1.4 V single-cell, and 5.6 and 8.4 V multicell batteries. The energy density of the 1.4 V batteries ranges from 442 to 970 Wh dm^{-3}. These batteries, therefore, are not only preferable from an environmental point of view over zinc/silver but also more efficient. In addition to the button configuration, Zn/air batteries are also produced in the flat cell configuration.

Electrochemical Reactions The electrochemical reactions that take place in this cell are as follows:

At the anode:

$$Zn \rightarrow Zn^{2+} + 2e^- \qquad +1.25\,V$$

$$Zn^{2+} + 2OH^- \rightarrow Zn(OH)_2$$

$$Zn(OH)_2 \rightarrow ZnO + H_2O$$

At the cathode:

$$\tfrac{1}{2}O_2 + H_2O + 2e^- \rightarrow 2OH^- \qquad +0.40\,V$$

Overall:

$$Zn + \tfrac{1}{2}O_2 \rightarrow ZnO \qquad +1.65\,V$$

In addition to providing high-energy densities, the zinc/air cells have a flat discharge voltage, after the initial fall from 1.4 to 1.2 V (Figure 2.11) and a long shelf-life. The button cell configurations have been successfully applied in hearing aids, watches, and other low-power applications such as memory preservation. For electric vehicle applications, modest progress has been made hampered by some operational difficulties.

Disadvantages The service life of a Zn/air battery is affected by the "water vapor transfer." It arises from the inevitable vapor pressure difference between the atmosphere and over the electrolyte. At low humidity, the electrolyte can get concentrated causing cell failure due to electrolyte loss, while at high humidity the electrolyte is diluted resulting in lower conductivity. Under sustained water gain, the cathode catalyst is easily flooded accompanied by a loss of electrochemical activity. Direct oxidation of the zinc anode by oxygen entering the cell will result in a loss of capacity, accentuated at higher temperatures. The alkaline solution is also prone to carbonation due to direct reaction between the alkaline electrolyte and the atmospheric carbon dioxide (400+ ppm), which can aggravate the water loss reaction and impede air access at the cathode.

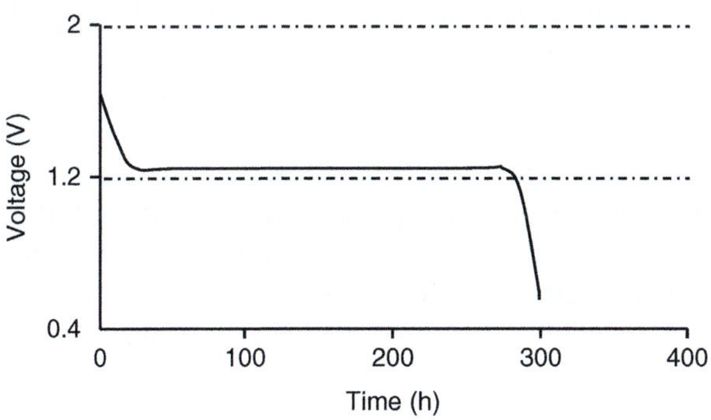

Figure 2.11 Discharge curve at 620 Ω for a Zn/air cell. Source: Reproduced with permission from Weal et al. [1]/University of Cambridge.

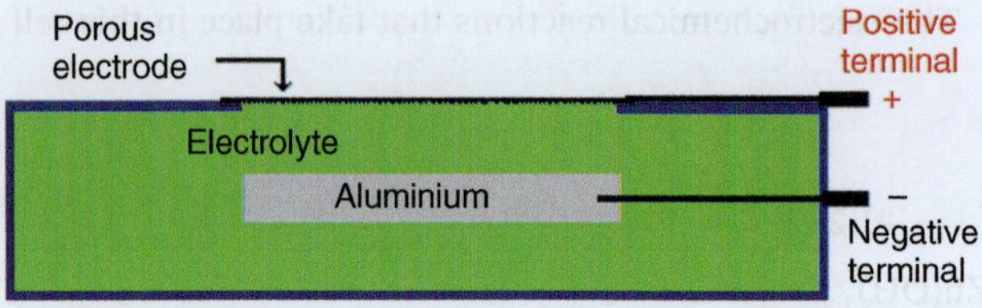

Figure 2.12 Schematic depiction of a standard aluminum–air battery. Source: Reproduced with permission from Weal et al. [1]/University of Cambridge.

2.5.3.2 Aluminum/Air Batteries

The structure of a standard aluminum/air battery is illustrated in Figure 2.12. A piece of aluminum is immersed in an electrolyte near a porous electrode. This porous electrode has air on one side and electrolyte on the other. The electrolyte can be an aqueous NaCl solution (or seawater) or an alkali solution such as potassium hydroxide.

The aluminum reacts with OH^- ions to form aluminum hydroxide and release three electrons. The OH^- ions are present either because the electrolyte is an alkali solution, or because they are produced at the other electrode (see below). In particular, the reaction is

$$Al + 3OH^- \rightarrow Al(OH)_3 + 3e^- \quad +2.30\ V$$

At a very high pH, the $Al(OH)_3$ may dissolve in the electrolyte and the corresponding electrochemical reaction may be expressed as

$$Al + 4OH^- \rightarrow Al(OH)_4{}^- + 3e^- \quad +2.31\ V$$

At the porous electrode (cathode), which is often made of carbon, the water in the electrolyte reacts with oxygen from the air and absorbs the electrons produced at the aluminum electrode as

$$O_2 + 2H_2O + 4e^- \rightarrow 4OH^- \quad +0.401\ V$$

With a saltwater electrolyte, the open-circuit voltage of the cell is only about 1.2 V, but the normal operating voltage is much lower at about 0.7 or 0.8 V. With the KOH solution, the voltages are about 0.5 V higher and further optimized air electrodes for use with alkali solution are commercially available.

The energy density of the aluminum/air battery is excellent, but it is not yet in extensive use. The main reason for this is that corrosion reactions can take place between the electrolyte and the aluminum, leading to hydrogen evolution. If this happens at a high rate during operation, then clearly the main power-generating function of the battery will be impaired, but even a low rate of reaction will cause damage to the anode over time, and this can happen during storage. However, if the electrolyte is stored separately, and added when power is needed, then this problem can be avoided. This type of battery is usually called a reserve battery, and this is one market in which aluminum/air batteries have had some success. It has been reported that corrosion of aluminum can be minimized by using special alloying elements and/or by using special additives in the electrolyte.

2.6 Li Primary Batteries

The demand for high-energy density power sources to be used in small size and lightweight equipment, led researchers to explore Li-based cells since Li is the lightest metal. The first Li-battery system was introduced in 1958 by W.S. Harris, who studied the solubility and conductivity of inorganic salts in cyclic esters, ethylene carbonate, propylene carbonate, gamma-butyrolactone, and gamma-Valero lactone. Li is the anode, while some form of C (as the current conductor) is mixed with the electroactive cathode component, such as MnO_2 or thionyl chloride. [8].

Apart from the fact that Li is the lightest of all metals and it can float on water, it is also the most electropositive metal with a very low standard reduction potential of −3.05 V at 298 K; theoretically, a very high capacity of 3860 Ah kg^{-1} can be obtained from Li. Other battery-base metals offer much lower capacities; sodium provides 1160 Ah kg^{-1}, while zinc 820 Ah kg^{-1}. The drawback with Li, however, is that it can react very violently with water and air, prohibiting, thus, the use of aqueous electrolytes under normal conditions. Due to the reactivity of Li, all Li battery systems pay particular attention to abuse-free sealing (the consumer cannot easily tamper with the sealing by accident or on purpose) and rely mainly on organic electrolytes, which do not have as high conductivity as aqueous electrolytes. Lower conductivity is advantageous since it allows for a low self-discharge and, therefore, long-life batteries can be obtained, which are particularly useful in biomedical implantable devices. With many electrolytes such as propylene carbonate, liquid sulfur dioxide, thionyl chloride, and some molten salts, Li forms a passivating layer called the solid electrolyte interface, which prevents corrosion but not cell discharge. Finding suitable mixtures of organic liquids and metal salts that are safe to use took many years. This research resulted in the production of many different kinds of cells with a wide range of applications, from millimeter size batteries used in watches to batteries large enough to power small cars. As mentioned previously, Li is the anode, while some form of C (used as the current conductor) is mixed with the electroactive cathode component such as MnO_2 or thionyl chloride. Various cathode materials can be used but special *graphite* with fluorine treatment and manganese dioxide is among the most commonly used commercially for nonchargeable cells.

The first Li primary battery was commercially available in the form of a cylindrical lithium (anode)/manganese dioxide (cathode) cell by Sanyo Electric Co. in 1975. It was used in cameras and similar devices. Another variation uses iron sulfide as the cathode. This powerful combination provides a 1.5 V battery that can be directly substituted for alkaline or Leclanché (carbon–zinc) cells.

Several lithium systems have been developed over the past three decades. The anode always comprises lithium metal while alternate materials have been employed as cathodes. Examples of lithium primary battery systems are lithium–sulfur dioxide (SO_2), lithium–manganese dioxide (MnO_2), lithium–copper fluoride (CuF_2), lithium–silver chromate (Ag_2CrO_4), lithium–lead bismuthate ($Bi_2O_3 \cdot 2PbO$), lithium–polycarbon monofluoride ($(CF)_x$), lithium–iodine (I). However, lithium primary batteries cannot compete with alkaline batteries in the market, due to the fact that they provide the same cycle life as alkaline batteries, but at much higher manufacturing costs, and at the expense of safety issues. In particular, the manufacturing of lithium cells and batteries is a high technology business. The cells must be assembled in rooms with relative humidity below 3% and preferably equal to or less than 1%. Battery manufacturers have reported factory fires in the past when the materials of the batteries were unintentionally mishandled. Two primary Li battery systems are described below.

2.6.1 Lithium/Thionyl Chloride Batteries

In early 1968, a Li-primary cell was developed that allowed for a higher energy density than the previous Li cells and had a service life of 15–20 years [5]. It consists of a lithium foil anode and a thionyl chloride ($SOCl_2$) cathode; both electrodes are placed in an organic electrolyte that contains lithium-ion electrolyte (lithium tetrachloroaluminate, $Li(AlCl_4)$). The thionyl chloride is blended with a specially chosen carbon, such that it can be used as the cathode in a lithium cell without a separator, a more flexible feature in comparison with an aqueous battery. This allows a very low internal cell resistance in a very high-energy package. The discharge reaction for the Li/thionyl battery can be described as

$$2Li \rightarrow 2Li^+ + 2e^- \qquad 0\ V$$

$$SOCl_2 + 2Li^+ + 2e^- \rightarrow 2LiCl + ½S + ½SO_2 \qquad +3.6\ V$$

The overall cell reaction is

$$2Li + SOCl_2 \rightarrow 2LiCl + ½S + ½SO_2 \qquad +3.6\ V$$

where the thionyl chloride acts not only as the cathode material but also as the electrolyte when combined with a Li salt such as $LiClO_4$ or $LiAlCl_4$. A carbon black electrode is used as the conductor at the cathode.

A typical discharge curve is shown in Figure 2.13, illustrating that a flat discharge is obtained over a long period of time, allowing such lithium batteries to produce

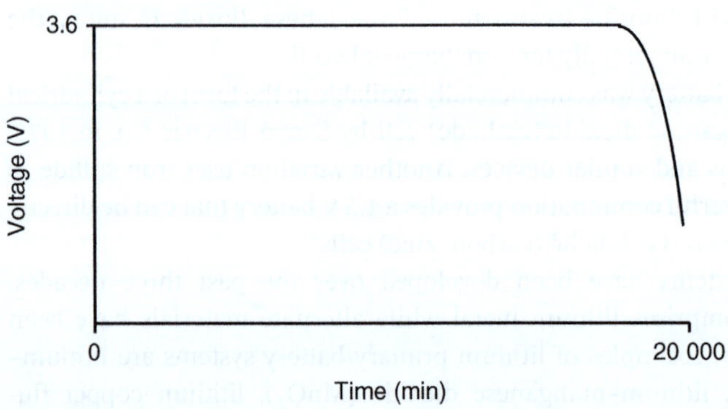

Figure 2.13 Discharge curve at 36 kΩ for a lithium/thionyl cell. Source: Reproduced with permission from Weal et al. [1]/University of Cambridge.

a very high current for fairly long periods of time. They also have a very high capacity of 450 Ah kg^{-1}, an energy density of 330 Wh kg^{-1}, and a long shelf-life with low-to-moderate drains making them well suited for medical devices, such as pacemakers for low current drains, drug pumps, and neuro-stimulators with medium power requirements. Their flat discharge curve for a long period of time also makes them useful for computer backup applications. Drawbacks for these cells are that they are expensive and must be assembled very carefully to avoid accidents, as they can get very hot or even catch fire if they are not assembled and handled properly.

2.6.2 Lithium/Sulfur Dioxide Cells

The lithium/sulfur dioxide cell was developed in the 1960s by Honeywell Inc. and the Mallory Battery Company [5]. As the name suggests the anode and cathode are Li and SO_2 respectively, while the electrolyte is highly conducting and comprises acetonitrile with dissolved lithium bromide (LiBr) and sulfur dioxide (SO_2). The SO_2 participates in the cathodic reaction on a porous carbon current collector as

$$2SO_2\,(l) + 2e^- \rightarrow S_2O_4^{\;2-}\,(l) \quad 3.0\,V$$

High-power cells are designed to deliver pulse loads of over 30 A at a voltage of 2 V. Practical energy densities of greater than 300 Wh kg^{-1} with capacities greater than 400 Ah kg^{-1} are easily obtained from this chemistry. Overheating and thermal runaway can be a problem in such a high-current/high-capacity cell; thus, use of shutdown separators and other external safety devices, such as fuses or current limiters, is important.

2.7 Oxyride Batteries

The most recent primary battery, launched in 2004, is the "Oxyride battery" by the Panasonic Corporation in Japan. This cell is suitable for high-drain devices such as digital cameras and cars [9]. In the same year, a 2.9 m special "oxyride car" with a 50 kg passenger was powered by only two oxyride batteries. The difference of this cell from alkaline cells is that a "Tablet Mix Control System" was employed to increase the amount of cathode material, which is an oxy nickel hydroxide (NiOOH) solution mixed with a finer-grained manganese dioxide and graphite for better cell performance. A vacuum-pouring technique is applied for packing a high amount of electrolyte into the cell to increase the surface area of the electrode. Higher power, longer battery life, and higher battery durability are therefore obtained. The operating voltage of this cell is in the range of 1.5 to 1.7 V, while the capacity of the cell is 28 Ah g^{-1} for AA sizes and 25 Ah g^{-1} for AAA sizes (up to two times longer than traditional alkaline batteries). The schematic representation of the cathode and anode characteristics of alkaline compared to oxyride batteries is shown in Figure 2.14. The active element indicated in the figure is finer-grained

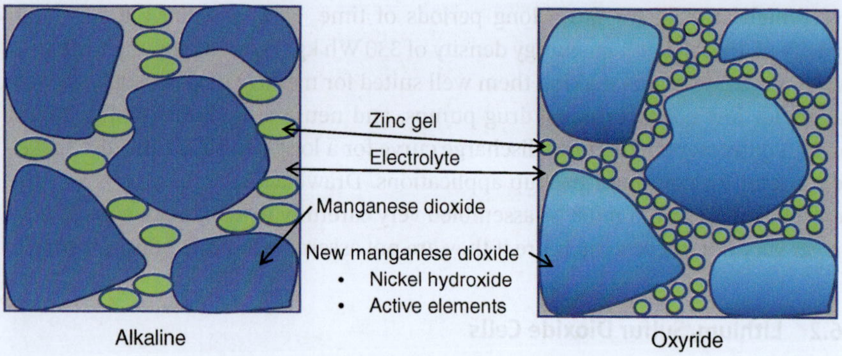

Figure 2.14 The schematic presentation of cathode and anode characteristics of alkaline and oxyride batteries. Source: Reproduced with permission from Wise GEEK [10].

graphite powder. As the oxyride batteries are under patent the discharge curves and half-reactions that occur are disclosed.

2.8 Damage in Primary Batteries

Ideally, the lifetime of batteries is dictated by their discharge curves, and they can be used efficiently until the galvanic cell has been drained completely. Over the lifetime of the cell, however, the morphology of the components will change; these morphological changes affect detrimentally the electrochemical properties of the cell. In particular, the following phenomena can be distinguished:

- During operation of the battery the temperature increases. When crystalline metals are exposed to elevated temperatures, their grains grow larger, which results in an increase in the impedance of the cell, due to loss of contact, thus decreasing the capacity and energy density.
- The diffusion of leaking electrons through the electrolyte to the anode surface causes dendrite formation on the anode, which as a result increases the anode volume and forces the two electrodes closer together. The self-discharge of the battery, therefore, increases resulting in a shorter lifetime.
- A metal (anode) may dissolve in the electrolyte and then reform at the cathode as dendrites, as shown for zinc in an alkaline cell in Figure 2.15. Similar images of dendrites are shown in Figure 2.16 [11]. The scanning electron microscopy and transmission electron microscopy images of the InSb anode material for lithium-ion batteries were prepared by the solution route method. The dendrites were nucleated and grown on zinc particles, which served as a reducing agent. The electron transfer process in growing the InSb, therefore, is analogous to the electron transfer process giving rise to the dendrites demonstrated in Figure 2.15.
- When dendrite formation is severe, dendrites may penetrate the separator giving a very high self-discharge, or even allowing a short circuit of direct electron flow between the anode and the cathode bypassing the electrolyte (electron flow should

Figure 2.15 Dendrite formation on Zn anode. Source: Reproduced with permission from Weal et al. [1]/University of Cambridge.

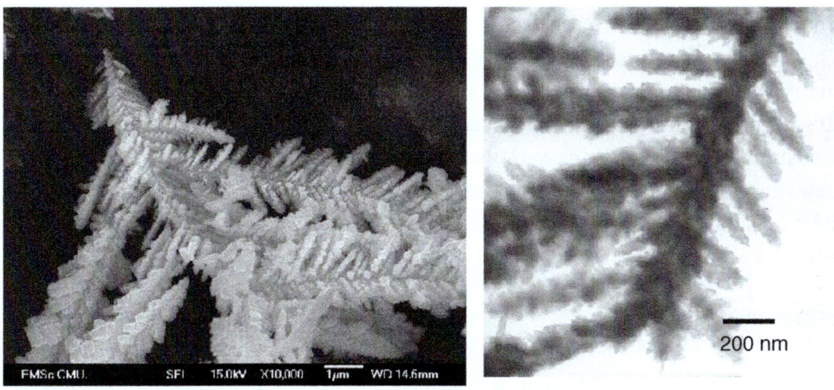

Figure 2.16 SEM and TEM images of InSb dendrite; the InSb anode material was prepared by the solution route method for a lithium-ion battery. Source: Reproduced with permission from Sarakonsri et al. [11]/IOP Publishing.

occur only through the external circuit, not through the electrolyte). Short circuits in a battery is highly undesirable, as this can produce large local transient currents leading to overheating and unleashing destructive power in the battery. Specifically, accidental short-circuiting in cells can cause extremely high currents and heating that can lead to excessive temperatures at which runaway reactions can occur between the anode and cathode and the components of the cell. The breakdown of chemical compounds can result in a rapid build-up of pressure within the cell resulting in explosions, fires, and serious secondary damage.

Internal short circuits do not only develop due to dendrite growth on the electrodes but can also occur by damaged separators, as follows: In the event of overheating, the separators that are often made of polymers may be distorted or may melt (or both phenomena can take place) resulting in a short-circuiting and increasing the rate of the runaway reaction; as more heat is produced, the reactions occur faster and faster.

To avoid fires that may take place in such cases two precautions are taken: (i) external battery circuits are available that can cut off the device on detecting a sudden drop in voltage, (ii) chemical inhibitors are added to electrolytes to make them flame retarding or self-extinguishing by acting as sinks for heat evolved in the case that runaway reactions occur. To prevent such deficiencies from occurring, suitable separators have to be designed during manufacturing. For example, separators are made rigid so as not to readily distort or melt on heating or whose impedance increases rapidly on heating resulting in the shutdown of the battery. It is also common to provide for vents to avoid a dangerous buildup of pressure inside the cell due to gassing. In general, batteries contain reactive components and have the potential to be dangerous if they are not carefully designed or handled.

2.9 Conclusions

Primary batteries are normally used in low-drain applications. They are not suitable for high-drain operations due to their limited lifetime and the need to be replaced. Primary batteries reproduce typically only 2% of the energy used during the manufacturing process and therefore are very inefficient from the point of energy usage and sustainability, but their easy-to-use nature makes them attractive in many applications. In the next chapter more evolved electrochemical cells that can be recharged will be presented.

References

1 Weal, E., Tan, J.C., and Kumar, R.V. (2005). *DoITPoMS Teaching and Learning Packages*. University of Cambridge http://www.doitpoms.ac.uk/tlplib/batteries/index.php.

2 Pierre R. Roberge (2013) Corrosion doctors, http://www.corrosion-doctors.org/Biographies/GalvaniBio.htm (accessed 3 November 2014).

3 Miller, K., Patterson, G., Thom, C. et al. (1919). *Cyclopedia of Telephony and Telegraphy*, vol. 1, the Project Gutenberg Ebook.

4 Linden, D. and Reddy, T.B. (2002). *Handbook of Batteries*, 3e. New York: McGraw-Hill.

5 Besenhard, J.O. (1999). *Handbook of Battery Materials*. Wienhiem: Wiley-VCH Verlag GmbH.

6 van Schalkwijk, W.A. and Scrosati, B. (2002). *Advances in Lithium-Ion Batteries*. Dordrecht: Klumwer Academic Publishers.

7 Silberberg, M.S. (2000). *Chemistry: The Molecular Nature of Matter and Change*, 2e. New York: McGraw-Hill.

8 Crompton, T.R. (2000). *Battery Reference Book*, 3e. Oxford: Newnes.

9 Pistoia, G. (2005). *Batteries for Portable Devices*. Amsterdam: Elsevier Science B.V.

10 Wise GEEK homepage (2003-2014). http://www.wisegeek.com/what-is-an-oxyride-battery.htm (accessed August 2014).

11 Sarakonsri, T., Choksawatpinyo, K., Seraphin, S., and Tunkasiri, T. (2005). Processing parameter studies on solution route synthesis of dendrite InSb powders, anode material for lithium-ion batteries. *CMU J. Special Issue on Nanotechnology* 4: 73.

3

A Review of Materials and Chemistry for Secondary Batteries

R. Vasant Kumar[1] *and Thapanee Sarakonsri*[2]

[1] *University of Cambridge, Department of Materials Science and Metallurgy, 27 Charles Babbage Road, Cambridge CB3 0FS, UK*
[2] *Chiang Mai University, Department of Chemistry, Chiang Mai 50200, Thailand*

The previous chapter was concerned with the development of electrochemical cells, whose material chemistries allow only discharge to take place. As described, however, in Chapter 1, the most advanced battery chemistries are those which allow for recharge to occur and, hence, the cell can be used multiple times. These reusable cells are, therefore, termed rechargeable or secondary batteries and will be described in the sequel. To the nonexpert, the term "secondary" is a misnomer in the sense that secondary batteries are not inferior to primary; the coinage of terms relates to history, as it can be seen in the timeline presented in Chapter 1.

A secondary battery can effectively be reused many times after it is discharged by applying electrical power to the cell electrodes to bring about the reverse reaction. In an electrochemical cell, the forward reaction is spontaneous while the reverse reaction used in charging is thermodynamically unfavorable (ΔG = positive). To reverse the reaction, the minimum applied potential required can be calculated from $\Delta G = -nFE$. For the Zn/Cu Daniel cell (also known as the gravity or crowfoot cell), described in Chapter 2, the voltage given off upon cell operation is $E^{\circ}_{\text{cell}} = -1.10\,\text{V}$. Thus, if an electrical potential more than 1.10 V is applied to the cell, the following charging reaction can take place:

$$Zn^{2+}(aq) + Cu(s) \rightarrow Zn(s) + Cu^{2+}(aq)$$

After being fully charged, the electrodes attain their original state, allowing the discharge reactions to occur spontaneously and, thus, the cell can be used again to provide energy to an electric appliance/device. This charge–discharge process is referred to as an electrochemical "cycle" and due to irreversibilities in the cell system, it cannot be repeated indefinitely, and eventually the cell must be discarded. Reasons leading to the irreversibility of battery reactions are: (i) the formation of an electrochemically inaccessible permanent phase during and/or at the end of discharge, (ii) the loss of mechanical integrity and fracture that result from the buildup stresses that occur during cycling, and (iii) the formation of

Rechargeable Ion Batteries: Materials, Design, and Applications of Li-Ion Cells and Beyond, First Edition. Edited by Katerina E. Aifantis, R. Vasant Kumar, and Pu Hu.

Table 3.1 Some typical properties of secondary batteries.

System	Nominal cell voltage (V)	Specific energy (Wh kg^{-1})	Cycle life (up to 80% initial capacity)	Charge time (h)	Self-discharge per month (%)
Lead–acid	2	30–50	200–350	8–16	5
Ni–Cd	1.25	45–80	1500	1	20
Ni–MH	1.25	60–120	300–500	2–4	30
Li-ion	3.6	110–180	500–1000	2–4	10
Li–polymer	3.6	100–130	300–500	2–4	10

Table 3.2 Global battery market for 2019.

Type of battery	Global demand, total US$ 126 billion
Li-ion	**$44 billion**
Lead–acid	$47 billion
Ni–MH	$4 billion
Ni–Cd	$0.5 billion
Other secondary	**$1.5 billion**
Alkaline primary	$15 billion
Li primary	$6 billion
Zn/C and another primary	$6 billion

dendrites primarily during charging resulting in fracture, loss of contact, or worse short-circuiting of the two electrodes. Due to these irreversible transformations and structural/morphological changes, there is a capacity loss in each cycle and eventually the capacity tends toward zero.

The key properties of the main commercial secondary batteries are shown in Table 3.1, while the commercial global market demand for these chemistries is shown in Table 3.2.

The secondary battery market is predicted to grow by 20–25% by the end of this decade, driven by the growth in the emerging hybrid electrical vehicles (HEVs), electrical vehicle (EV), and e-bike sectors, the green-energy-storage markets and reinforcement of emergency standby power in electrical grids and local/regional microgrids distributing power to customers. The European Union emission target of 130 kg CO_2 per km in 2015 for passenger car vehicles was met mainly propelled by the increased use of the so-called AGM–VRLA batteries (absorptive glass mat/valve regulated lead–acid [AGM–VRLA]) for micro-hybrid electrical vehicles (micro-HEVs) leveraging the existing 12–14 V architecture by using the auto start–stop function. The next target is 95 kg km^{-1}, set to be achieved in 2021–2022 for >95% of passenger cars; this target will drive the growth in electrical vehicles.

3.1 The Lead–Acid Battery (LAB)

The first secondary (rechargeable) battery system was invented in 1859 by the French physicist Raymond Gaston Planté, for powering the lights in train carriages. Both electrodes consisted of lead, while the electrolyte was acidic, and therefore this prototype rechargeable battery is known as the lead–acid battery [1]. In vehicles, it is also called the SLI battery since it controls the starting, lighting, and ignition of automobiles. There are other versions, such as traction batteries, used in electrified vehicles, such as in golf carts, cranes, and two/three wheelers. Increasing the usage of LABs in energy storage results in an uninterrupted power supply which has led to the development of new specifications, designs, and innovations.

The initial versions of these cells were composed of a single unit that contained two lead electrode sheets separated by a spiral rolled coarse cloth spacer, which was later replaced by more stable rubber tapes and strips; the cell was then filled with a sulfuric acid electrolyte solution [2]. With continuous electrochemical cycling, the electrolyte corroded the lead producing lead dioxide at one electrode and spongy lead at the other, therefore increasing the electrode surface area that could participate in the reaction. As a result, the capacity of the battery increased after the first cycle. At that time (in 1859), this cell was also trialed for load leveling [1].

After continuous research, an improved lead–acid cell with increased capacity was developed later by Planté's pupil, Camille Alphonse Faure in 1880–1882 using a lead foil as the anode and a plate of lead dioxide–sulfuric acid (PbO_2–H_2SO_4) paste pressed on a lead grid lattice as the cathode. This original idea of paste-formed cathodes has been more or less retained to the present day. Following the Faure design, several improved lead–acid battery configurations were developed, such as the lead–antimony alloy grid that was introduced by Sellon in 1881, which increased the mechanical strength of the electrodes considerably. The lead–calcium alloy grid was designed by H.E. Harring and U.B. Thomas in 1935, resulting in greater safety from reduced corrosion and gassing. The basic architecture of modern lead–acid automotive batteries resembles the Faure design.

Despite the arrival of several other secondary batteries in the last century, the lead–acid battery has retained the top end of the market spot in secondary batteries with a market value approaching US$ 47 billion per year (Table 3.2). Currently, there are no available economic alternatives to automotive starting, lighting, and ignition applications. Lead–acid batteries are also the power source in several hundred million e-bikes (Figure 3.1) on the road [3], and have the potential to make a major impact in hybrid and electrical vehicles in the near future estimated annual market value of US$ 40 billion in 2020 expanding to US$ 70 billion by 2027. Lead–acid battery-based bikes dominate the market due to the high demand in China, India, and other developing economies. Li-ion battery e-bikes are more popular in North America and Europe but are 3–10 times more expensive.

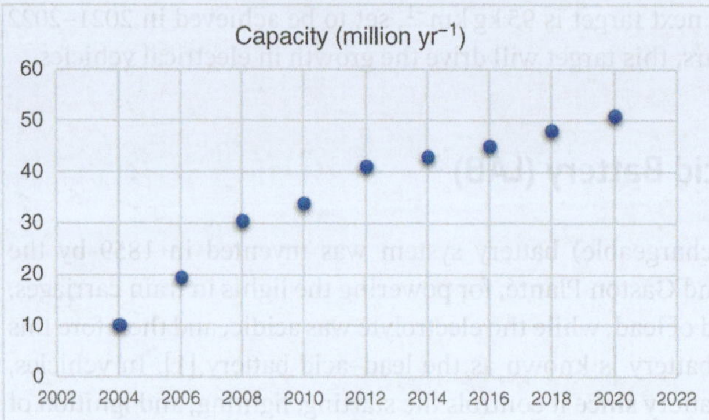

Figure 3.1 Growth of e-bikes (note the number of e-bikes produced was less than 0.1 million in 2000). Source: Reproduced with permission from Weal et al. [3]/University of Cambridge.

A lead–acid battery-powered electric car, the Reva manufactured by the Reva Electric Car Company in India, had attracted some attention nearly 2 decades ago. Mahindra took control of the company manufacturing small 3-door hatchback cars, which were powered by 8 lead–acid batteries (each providing 6 V) that are connected in series to obtain 48 V (9.6 kWh) with six hours of charge time. The maximum range for driving was 50 mi. The Reva cars were discontinued in 2019, as Mahindra introduced newer Li-ion battery-based EVs.

The European Union emission 2015 target (<130 kg CO_2 per km) for vehicles is already driving the increased use of the so-called AGM–VRLA batteries for micro-HEVs. The vehicle's small electric motor is powered onboard by a 12–14 V AGM–VRLA battery for immediate restart after stopping in the idle, stop, and go sequence. The engine restart that occurs after each automatic stop results in a large number of high-rate load phases during the battery cycle life. The start–stop cycle in the micro-HEV may be expected to decrease battery cycle life marginally; hence, the replacement market will rapidly increase in subsequent years. If the battery life is shorter, more batteries will be required for the same number of vehicles. A number of major car manufacturers are planning to convert up to 80% of their new hybrid vehicle fleets into micro-HEVs in the near future, and by the end of this decade, it is predicted that all HEVs will be micro-HEVs. In addition to the automotive market, lead–acid cells have an established presence in standby emergency power for medical, telecommunication, and memory backup applications. The next more stringent target in the EU of 95 kg CO_2 per km set for the 2021–2023 period will help accelerate the growth in the manufacturing and sale of Li-ion battery-based EVs in numbers; however, the lead–acid batteries use in cars will not diminish globally for many years into the future, although their % share is likely to decrease.

3.1.1 Electrochemical Reactions

The positive electrode mass consists of porous PbO_2, while the negative electrode is made of porous Pb, both derived from a leady oxide (PbO-containing metallic Pb)

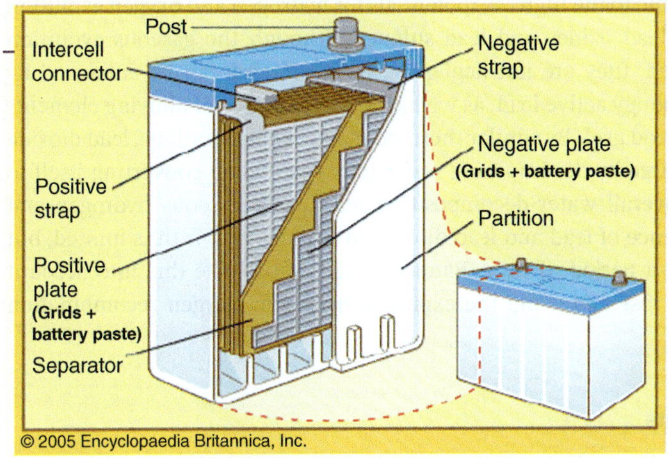

Figure 3.2 A cut-away version of a lead–acid battery; over 80% lead is consumed in making batteries. Source: Reproduced with permission from Weal et al. [3]/University of Cambridge.

precursor. The leady oxide is pasted on both electrodes with other supporting additives and then converted into the active charged state. For both electrodes, the active materials are present as pasted components on lead–alloy grids, as shown in Figure 3.2 [3], and the following nominal reactions take place during discharge:

Reaction at the anode:

$$\mathrm{Pb(s) + SO_4^{2-}(aq) \rightarrow PbSO_4(s) + 2e^-} \qquad E = 0.35\ \mathrm{V}$$

Reaction at the cathode:

$$\mathrm{PbO_2(s) + SO_4^{2-}(aq) + 4H_3O^+(aq) + 2e^-}$$
$$\mathrm{\rightarrow PbSO_4(s) + 6H_2O(l)} \qquad E = 1.70\ \mathrm{V}$$

Overall reaction:

$$\mathrm{Pb(s) + PbO_2(s) + 2SO_4^{2-}(aq) + 4H_3O^+(aq)}$$
$$\mathrm{\rightarrow 2PbSO_4(s) + 6H_2O} \qquad E = 2.05\ \mathrm{V}$$

The above reactions are simplified versions of the reality since sulfuric acid is dissociated with bisulfate ions rather than sulfate ions in the acid solution:

$$\mathrm{H_2SO_4 \Leftrightarrow H^+ + HSO_4{}^-}$$

Thus, the anodic, cathodic, and the overall reactions involve the bisulfate ions rather than the sulfate ions, giving rise to the following overall reaction:

$$\mathrm{Pb(s) + PbO_2(s) + 2H^+ + 2HSO_4{}^- \Leftrightarrow 2PbSO_4 + 2H_2O}$$

The above equation is known as the "theory of double sulfation" and was laid down in 1882 by Gladstone and Tribe. Thermodynamically, Pb should corrode in the acid solution to form lead sulfate while liberating hydrogen gas. Thankfully, this reaction is limited due to the high overpotentials for hydrogen and oxygen evolution on both the lead, lead oxides and lead sulfate. Although the gaseous evolution reactions are limited, they are not negligible, however, and can be enhanced by impurities in the spongy active lead, as well as by the presence of alloying elements (such as Sb) in the lead grid. By similar thermodynamic considerations, lead dioxide can also self-discharge by decomposing water to oxygen while converting itself to lead sulfate. The overall water decomposition reaction to gaseous hydrogen and oxygen in the presence of lead and lead dioxide in sulfuric acid is thus limited, but not negligible over a period of time. Battery designs must take this into account to minimize water loss and avoid the explosive hydrogen–oxygen recombination reaction.

3.1.2 Components

Both the anode and cathode active materials are made porous so as to increase the high-specific active surface over which the electrode reactions can occur, thus allowing the electrode reactions to take place more efficiently. In particular, the volumetric porosity of the active masses is over 50% at both electrodes. During the reactions, sulfuric acid (H_2SO_4) is consumed and lead sulfate ($PbSO_4$) is produced at both electrodes. The discharge capacity is directly increased as a result of increased availability of active components and a decrease in the current density. Initially, the electrodes comprise PbO, which is converted into porous Pb and porous PbO_2 on initial charging. The lead grids provide mechanical support for the electrode pastes and also act as electronic conductors between the electrodes and the load. The combination of a grid and the paste on it is referred to the plate. A separator, usually made of microporous polyethylene, is used between the positive and negative plates to prevent electrical shorting. The container and the lid are made of polypropylene, which is light, strong, does not embrittle at freezing temperatures, and most importantly is resistant to attack by sulfuric acid. A schematic cut-away diagram is shown in Figure 3.2 [3]. The specific electrical conductivity of the H_2SO_4 aqueous solution electrolyte in a lead–acid battery is of the order $0.6–0.8\,\Omega^{-1}\,cm^{-1}$, while that of Pb is $0.5\times10^5\,\Omega^{-1}\,cm^{-1}$, and that of PbO_2 is $0.7\times10^3\,\Omega^{-1}\,cm^{-1}$. As a result of discharge, the resistance of the active masses at both electrodes increases due to sulfation. In normal discharge, for the negative active mass, the decrease in conductivity is about 30%, while for the positive active mass, it is around 90%. Despite the large decrease in conductivity of the active masses, the resistance of the electrolyte is still the dominant factor in ohmic loss during operation.

Until two decades ago, to achieve rigidity in the grid, it comprised Pb–10 wt% Sb. But Sb is prone to dissolution in the acid, resulting in loss of water through dissociation into hydrogen. Thus, to maintain the battery, water addition was frequently required to compensate for the loss of water, to hydrogen and oxygen, during electrolysis; these batteries were, therefore, also known as "flooded batteries." The acid was, hence, contaminated and hydrogen was also generated, requiring controlled venting to avoid explosions. As a result, the battery life was around two years.

Modern lead–acid batteries are made to be maintenance free and do not need water addition throughout their life, which is extended to four to five years. The amount of Sb is decreased to 1.5 wt%, while hardening of the Pb is achieved using 0.1–0.5 wt% Ca in the Pb alloy. The previously mentioned valve-regulated lead–acid (VRLA) cells, which dominate the micro-HEV industry, minimize the need to refill with water by suitably changing the composition of the alloy grid. The polyethylene separator is replaced by an absorptive glass mat (AGM) separator, which absorbs the electrolyte, avoiding flooding and slippage, while the hydrogen and oxygen that evolve during electrolysis are allowed to recombine back into H_2O on catalyst plugs, thus avoiding water loss. At the same time, a sealed valve vents gas out only if the internal pressure exceeds a certain critical value. The porous separator can also allow for oxygen to diffuse from the positive PbO_2 electrode to the negative Pb electrode, where it can react with the Pb to form PbO; this prevents the anode from reaching a potential where hydrogen is easily produced. During charging, PbO is reduced back to Pb.

However, there are unavoidable limitations of lead–acid batteries, as service life is shortened by: (i) Corrosion on the positive plate caused by operating the battery at high temperatures above 21 °C or by leaving the battery uncharged. A schematic of a corroded grid is shown in Figure 3.3 [4]. (ii) Irreversible sulfation on the negative plate occurs when the battery is left unused for a period of time. Lead sulfate, the product of the discharge reactions, will irreversibly recrystallize to larger particles. Low surface area and nonconducting lead sulfate particles consequently form during discharge and subsequently block the conductive path needed for recharging; their formation is, therefore, associated with a capacity loss. Moreover, lead sulfate has a 37% higher volume than lead or lead alloys which will result in the expansion and deformation of the grid. Appearance of short circuits at the edges of the plates and through the separators also causes sulfation of the plates.

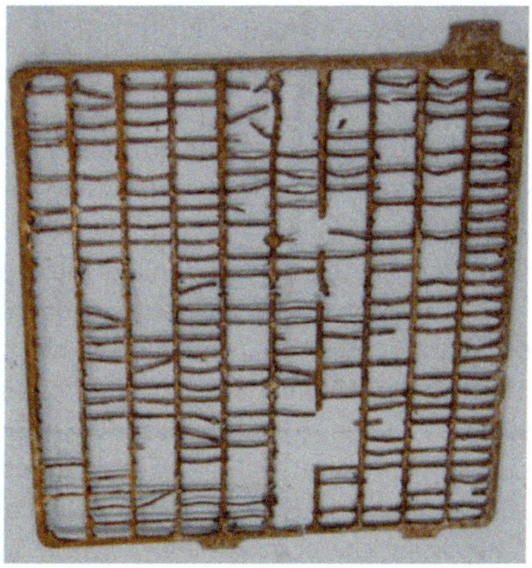

Figure 3.3 A corroded conventional lead grid on the positive plate. Source: Reproduced with permission from Firefly International Energy [4].

SLI batteries are not designed for deep discharge and they should be kept at open circuit to minimize corrosion, but they should be charged regularly to prevent sulfation. They are designed to contain a large number of relatively thin plates to maximize the surface area and current output. Before the automobile is started, an SLI battery is normally in the fully charged state and any lost charge is replaced by the alternator and the battery is returned to its fully charged state. Discharge below 50% will damage the plates permanently. Once the battery is covered in sulfates, the battery is dead and will not respond to recharging.

The latest commercially available lead battery is the bipolar battery design, which comprises a sponged powder lead metal anode pasted on lead grids and a cathode consisting of lead dioxide powder pasted on lead grids; the cell is filled with 4.5 M H_2SO_4 electrolyte gel instead of liquid. The battery consists of six cells that are placed in series inside the casing such that the positive and negative active materials are applied to each side of the electrode plates and interspersed with insulating separators. Every single cell provides a voltage of 2 V, which makes it a 12 V battery.

By reducing the weight by 50%, a lead–acid battery is capable of achieving an energy density approaching that of Ni–Cd or Ni–MH batteries. The greatest current challenge, therefore, is to replace the lead alloy grids with other electronically light materials. To achieve this weight reduction, the following approaches are being trialed: (i) a porous carbon-foam/carbon-felt matrix filled with lead deposits; (ii) carbon particles bonded within a polymer film; (iii) conductive titanium oxide particles forming a composite with an epoxy resin; (iv) Ti–SnO_2 composite electrodes.

3.1.3 New Components

Recently there was a report on a new invention of the Microcell™ foam by Firefly energy (research and development laboratories of Caterpillar, Inc). The Microcell foam is a corrosion-resistant material with superior physical and chemical properties over conventional lead-based materials. The Microcell foam consists of impregnated lead oxide slurry on grids, which is then further converted to a lead and lead dioxide sponge (shown in Figure 3.4) [4] that contains hundreds of spherical microcells. These microcells result in a diffusion path on the micron scale, which enhances the reaction rate; in conventional batteries, millimeter-scale diffusion takes place. The ISL battery employing the Microcell is reported to survive in very low temperatures, such as −18 °C. The comparison of the Microcell foam electrode with a conventional lead electrode is shown in Figure 3.5 [4].

Advances of this new technology introduced by Firefly are as follows:

The conventional lead electrode encounters a volume change of the active material during cycling that is approximately 37%. This new electrode, however, was designed to accommodate these volume changes, through its microfoam structure, as shown in Figure 3.6 [4], indicating a constant volume that does not change during electrochemical cycling. Dispensing with the volume changes inhibits electrode pulverization, and therefore, material failure is prevented, prolonging battery life.

Replacing the conventional negative lead metal-based plate with a Microcell composite foam electrode in a 3D architectural technology offers a cycle life that

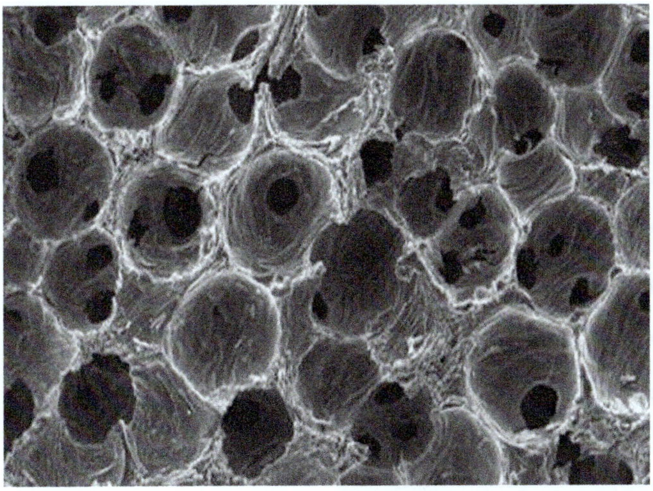

Figure 3.4 SEM image of Microcell™ foam electrode by Firefly energy. Source: Reproduced with permission from Firefly International Energy [4].

is four times longer than that of conventional lead–acid batteries, due to the fact that sulfation is reversible in 3D cells. The Microcell foam electrode material is not reactive in the acid and, hence, does not corrode. Reversible sulfation, un-corroded material, and low self-discharge rates result in longer storage times and longer periods of unused battery of about three to six months. The commercial battery of Firefly is called the Oasis battery and a schematic diagram is shown in Figure 3.7 [4]. The maximum theoretical capacity of the lead–acid cell is approximately 170 Wh kg^{-1}.

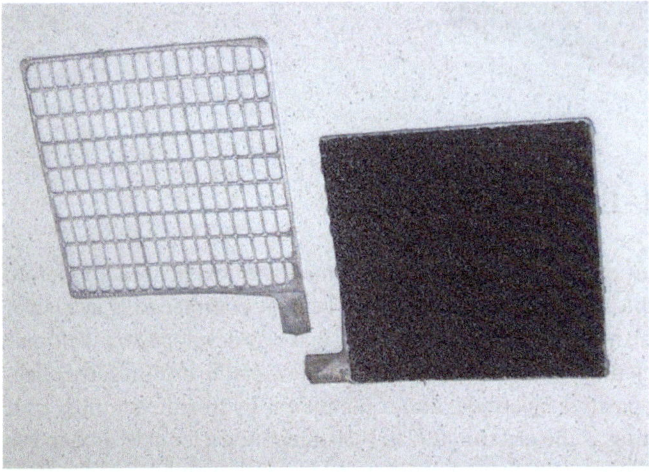

Figure 3.5 The comparison of Microcell™ foam electrode (right) and conventional lead electrode (left). Source: Reproduced with permission from Firefly International Energy [4].

Figure 3.6 Microcell™ foams do not experience mechanical damage during electrochemical cycling. (a) An un-cycled Microcell™ foam, (b) Microcell™ foam after usage. Source: Reproduced with permission from Firefly International Energy [4].

Figure 3.7 Oasis battery from Firefly energy. Source: Reproduced with permission from Firefly International Energy [4].

3.2 The Nickel–Cadmium Battery

After the development of the lead–acid battery, it took a whole generation before the second rechargeable battery, the Ni–Cd battery, became commercially available. Its predecessor was the nickel–iron alkaline battery that was developed by Waldmar Jungner, a Swedish scientist, in 1899 [5]. It consisted of a metallic iron-negative electrode, a nickel-oxide-positive electrode, and a potassium hydroxide solution with lithium hydroxide acting as the electrolyte. This cell generated a lot of hydrogen gas during the charging process, which, as a consequence, required an unsealed battery design. A direct consequence was that leaking could occur easily and therefore, short battery life was encountered. The gravimetric energy of these batteries was 30 Wh kg^{-1}, while the cell voltage was 1.4 V and the specific capacity was 21 Ah kg^{-1}.

In the same year (1899), Junger invented the nickel–cadmium battery, which was successfully commercialized in 1910 as a pocket-plate battery. Nickel hydroxide mixed with graphite acted as the cathode precursor, while cadmium hydroxide or cadmium oxide mixed with iron or iron compounds acted as the anode precursor; potassium hydroxide solution was used as the electrolyte. During formation, the positive plate of NiO(OH) and a negative plate with metallic Cd are formed. On discharge, the NiO(OH) in the positive electrode material changes to $Ni(OH)_2$, while the negative electrode material Cd oxidizes to $Cd(OH)_2$, and water is lost. At the end of every charge, some water can be decomposed into hydrogen and oxygen. In some batteries, gas evolution during charge and overcharge is prevented by special design features. For example, the battery is constructed with excess capacity in the Cd electrode; thus, hydrogen is not generated. To limit the amount of immobilized electrolytes, flat pockets of perforated nickel-plated steel strips were used to hold the active materials; the steel, however, made the battery too heavy for applications in portable devices. The gravimetric energy of these batteries was 27 Wh kg^{-1}, while the voltage and the specific capacities were 1.25 V and 20 Ah kg^{-1}, respectively. This cell also generated hydrogen gas during charge but in a smaller amount than the nickel–iron battery, and therefore was manufactured in both sealed and open or vented designs. In the sealed design, hydrogen and oxygen gases that were generated at the positive and negative electrodes during charging were recombined back into water. The open design allowed the release of gases and required the periodic addition of water to replace the vented gases.

In 1932, the sintered-plate nickel–cadmium battery was invented by Shlecht and Ackermann, two German scientists. The gravimetric energy of these batteries was 30–37 Wh kg^{-1}, the voltage 1.25 V, and the specific capacity 25–31 Ah kg^{-1}. The sintered plate (Figure 3.8) was a high surface area porous substrate impregnated with nickel and cadmium particles, and it was fabricated by a wet-slurry process [5]:a viscous slurry carbonyl nickel was applied on both sides of the nickel-plate grid, which was then sintered at 800–1000 °C [5]. These cells also exist in both vented and sealed designs. The vented design gives a high rate performance which is suitable for engine ignition (aircraft and diesel), but expensive due to the large amount of nickel used. The sealed design is suitable for portable devices, such as cameras, phones, computers, and camcorders, due to the advantage of being maintenance-free (it is not necessary to regularly add water).

The further-improved sealed nickel–cadmium battery is a jelly-roll design in which the cathode and anode layers are rolled into a cylindrical shape. This latest commercial design of the nickel–cadmium battery is mostly used in portable devices, such as mobile phones, digital cameras, and power tools. The components of this cell are a cadmium metal anode, a nickel oxide hydroxide or nickel (III) oxide (Ni_2O_3) cathode, and a potassium hydroxide (with a small amount of lithium hydroxide) electrolyte solution. The gravimetric energy of these batteries is 40–60 Wh kg^{-1}, the voltage is 1.2 V, and the specific capacity is 50 Ah kg^{-1}.

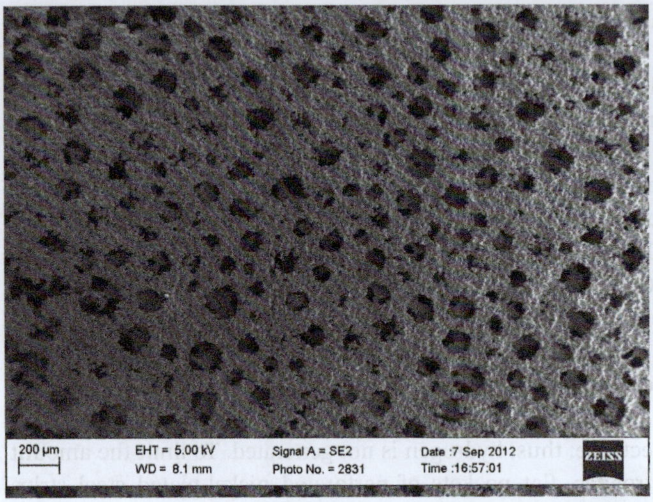

Figure 3.8 Micrograph of a sintered metallic plate.

The drawback of all Ni–Cd rechargeable battery systems is that they need to be fully discharged before the charging process can be initiated to obtain maximum cell voltage. Otherwise, upon charging, the voltage at which discharge ceased will be attained, instead of the maximum voltage possible. This deficiency is referred to as the "memory effect." The memory effect usually is more pronounced as the number of charge–discharge cycles is increased. The effect can be erased in most normal circumstances by a full discharge followed by a full charge. It is reported that the memory effect is connected with the changes in the crystal size involving the reaction between Ni and Cd in the negative electrode material and the large surface area of the sintered nickel support.

3.2.1 Electrochemical Reactions

The reaction at the anode is

$$\mathrm{Cd\,(s) + 2OH^-(aq) \rightarrow Cd(OH)_2(s) + 2e^-} \qquad E = 0.82\ \mathrm{V}$$

The reaction at the cathode is

$$\mathrm{2NiO(OH)(s) + 2H_2O\,(l) + 2e^- \rightarrow 2Ni(OH)_2(s) + 2OH^-(aq)} \qquad E = 0.48\ \mathrm{V}$$

The overall reaction is

$$\mathrm{Cd(s) + 2NiO(OH)(s) + 2H_2O\,(l) \rightarrow Cd(OH)_2(l) + 2Ni(OH)_2(s)} \qquad E = 1.3\ \mathrm{V}$$

The reversible potential for the overall cell reaction for a Ni–Cd battery is 1.3 V and has a higher energy density and a longer life than a Pb–acid battery. For a typical 25% depth of discharge (DOD), the cell can accept recharging several thousand times. Even at a very high DOD approaching 80%, the cell can be recharged a few hundred times. This is because degradation reactions in the cell are not massive. Over a long period of usage, however, the NiOOH is converted into Ni dioxide hydroxide which is not reversed back into usable active material. A major problem with Ni–Cd batteries has been the use of toxic Cd, and using this battery in toys is not appropriate.

3.3 Nickel–Metal Hydride (Ni–MH) Batteries

The Ni–MH batteries began their life as nickel–hydrogen (Ni–H) cells in the 1970s, mainly driven by research on hydrogen storage in metallic alloys. The first Ni–MH batteries were commercially available in 1989 as a variation of the Ni–H cells and their development is associated with the name of the inventor Masahiko Oshitani from the Yuasa Battery Company. As the name suggests, the cathode in a Ni–MH battery is identical to the one in a Ni–Cd battery, while the anode is a compound of alloys expressed as AB_5 where A comprises an alloy of rare earth: La, Ce, Nd, and Pr, while B is an alloy of Ni, Co, Mn, and Al. Thus, the MH electrode represents the alloy hydride which can reversibly absorb hydrogen during charging and desorb hydrogen into the electrolyte during discharging. The electrochemical reactions during discharge are as follows:

Anode:

$$\mathrm{MH(s) + OH^-(aq) \rightarrow H_2O + e^- + M} \qquad E = 0.72\,\mathrm{V}$$

Cathode:

$$\mathrm{NiO(OH)(s) + H_2O(l) + e^- \rightarrow Ni(OH)_2(s) + OH^-(aq)} \qquad E = 0.48\,\mathrm{V}$$

Overall reaction:

$$\mathrm{NiO(OH)(s) + MH \rightarrow Ni(OH)_2(s) + M} \qquad E_{\mathrm{cell}} = 1.2\,\mathrm{V}$$

The electrolyte is alkaline as in a Ni–Cd cell, and most frequently it is KOH (aqueous). The anodic compound is also sometimes made of AB_2, where A is Ti and/or V, and B is Zr (or Ni) with Co, Cr, Fe, and Mn.

A Ni–MH battery has nearly double the energy density of Ni–Cd at 70–80 Wh kg^{-1} with a peak-power density ranging between 200 and 1000 W kg^{-1}. The nominal cell voltage is 1.2 V, giving a specific capacity of 60–70 Ah kg^{-1}. The Ni–MH battery is promising for aerospace applications and was also used in the first HEV applications. As a result of the higher energy density and the use of nontoxic elements, the Ni–MH battery has replaced the Ni–Cd battery in many applications in low-end electronics, mobile computing, and wireless communications and processing. In general, the Ni–MH battery is less durable than the Ni–Cd battery and can suffer from high self-discharge, poor cycling ability, and lower shelf-life at higher ambient temperatures. The memory effect of Ni–MH is similar to that described for the Ni–Cd battery. It should be emphasized that damage in these cells can occur during charging, and therefore, specially designed chargers or special charging methods need to be used.

In conclusion, it is of interest to note that just as foams are the most promising electrodes in the newly designed lead–acid batteries described in Section 3.1.2, extensive research is being performed to use open-cell Ni foams as cathodes.

3.4 Secondary Alkaline Batteries

The competitor of the Ni–Cd battery is the alkaline manganese secondary battery, which was although first launched in the market by the Union Carbide Company (USA) in 1946, has not retained dominance in the secondary battery market. This Union Carbide everyday cell is available in D and G sizes. The cell consists of a zinc gel (mixed with electrolyte) anode and a cathode comprising manganese dioxide and 10% graphite mixture mixed with aqueous potassium hydroxide (KOH) electrolyte to form a paste; hence, the electrolyte and the cathode are present together in intimate contact. All components are packed in special hermetically sealed cells in D and G sizes.

The Rayovac Company (USA) started producing zinc-manganese dioxide RAM (Rechargeable Alkaline Manganese dioxide-zinc batteries) cells in 1993 with an energy density of 80 Wh kg^{-1} or 220 Wh dm^{-3}, which is twice that of Ni–Cd batteries. Moreover, the self-discharge loss is only 0.01% per year compared to Ni–Cd, which is 1% per year, and most importantly they are environmentally benign and exhibit excellent charge-retention properties over Ni–Cd. RAM cells are currently available in AAA, AA, C, and D sizes, and they are mostly used in toys, games, personal compact disks, cassette players, and tape recorders.

3.4.1 Components

Zn powder mixed with gelled KOH is used as the negative electrode. MnO_2 mixed with graphite and doped with 10% Bi_2O_3, to allow the reversible process to occur, is used as the positive electrode [3]. $BaSO_4$ and Ag catalyst are also added to the cathode to increase the power and the cycle life. The Ag helps to recombine the H_2 and O_2 gases produced by Zn corrosion. Otherwise, the chemistry and geometry of this battery are similar to those of a primary alkaline battery (see Chapter 2). These cells operate as secondary batteries with a 1.5 V window.

Despite some effort, these cells have not penetrated the market beyond a small segment. In particular, the high internal resistance is restrictive in achieving load currents less than 400 mA, making the cells only adequate for low-power-demand applications. The cycle life is rather poor at 10–50 cycles, depending upon the depth of discharge. Despite the low purchase cost, the cost per cycle is much higher than the well-established secondary cells (lead–acid, Li-ion, Li–polymer, and Ni–MH). However, the rechargeable alkaline cell is more economical than the primary alkaline cell. The self-discharge is very low, thus making the alkaline rechargeable battery useful for storing as a standby for up to a decade. Examples of companies that currently produce these cells are the Pierce Energy Battery Company, Alcava range by the Young Poong Corporation, and Zhongyin (Ningbo) Battery Co., Ltd [6].

3.5 Secondary Lithium Batteries

It did not take long for researchers to realize that lithium-based cells were promising candidates for secondary batteries. Due to its low-molecular weight, lithium has a specific capacity of 3860 Ah kg^{-1} in comparison with 260 Ah kg^{-1} for Pb. Li is also one of the most electropositive metals, giving rise to very high electrode potentials; thus, the theoretical energy density achieved by Li is much more favorable in comparison with base metals and transition metals. Since Li is more electropositive than H and is very reactive with water, the electrolyte must be nonaqueous and aprotic, and typically consists of a stable organic solvent (such as ethylene carbonate) mixed with a soluble Li–salt (such as $LiPF_6$).

The first commercial secondary Li cells were manufactured by Exxon Company (USA) in the 1970s with a $LiTiS_2$ cathode put forth by M. Stanley Whittingham. In 1980, Moli Energy (Canada, now E-One Moly Energy Ltd.) commercialized Li batteries with a $LiMoS_2$ cathode, while a few years, later Taridan (Israel) used a $Li_{0.3}MnO_2$ cathode. These three batteries used a Li foil as the anode and various liquid organic electrolytes. The gravimetric energy of the Li–MoS_2 battery is 61 Wh kg^{-1}, while that of the Li–MnO_2 battery is 230 Wh kg^{-1}. The first two, aforementioned, battery systems are not available presently due to their short life (a few cycles), while the Li–MnO_2 system is not available due to safety issues that resulted from the high reactivity of the metallic lithium anode with the electrolyte (a solution of $LiAsF_6$ in dioxolane as the solvent). Furthermore, the use of pure Li as an anode caused safety issues, as Li dendrites are easily formed under a normal rate of charging.

The Nippon Telegraph and Telephone Corporation (NTT) commercialized an AA-size Li battery with a LiV_2O_5 cathode and a Li–Al alloy anode that is still used today. The energy density of these secondary batteries varies from 64 to 135 Wh kg^{-1} and their cycle life is between 150 and 400 cycles.

3.5.1 Lithium-Ion Batteries

The most efficient secondary battery, though, is the lithium-ion battery which was commercially introduced by Sony in 1991 after pioneering research by Goodenough *et al.* [6]. This cell chemistry included a $LiCoO_2$ cathode, a carbon anode, and a 1 M lithium hexafluorophosphate ($LiPF_6$) in 1 : 1 weight mixture of EC:DEC (ethylene carbonate:diethyl carbonate) electrolyte, whose open-circuit potential after being fully charged is 4.2 V. During electrochemical cycling, the Li-ions are intercalated in the anode or cathode. The face-centered cubic structure of $LiCoO_2$ met the requirement of small-volume changes (about 2%) during the Li-insertion and deinsertion that took place upon cycling. The delithiated state is believed to maintain its layered structure, which is similar to a hexagonal close-packed structure. The cell potential arising from the voltage difference between delithiated $LiCoO_2$ and lithiated carbon is largely the reason for making it a high-energy-density battery. Graphite is a suitable anode material since its layered structure with hexagonal vacant sites allows lithium ions to be inserted with a relatively small volume change (approximately 11%) taking place and can return back to its initial volume after the delithiation process. Graphite and other carbon anode materials can readily achieve this at a potential that is still relatively close to that of pure Li (0.1–0.5 V). While the rate of charging, cyclability, and safety is significantly improved with a carbon anode, in comparison with a pure Li, the self-discharge rate is much higher. The gravimetric capacity of this cell is 372 Ah kg^{-1} and a volumetric capacity of 800 mAh dm^{-3}.

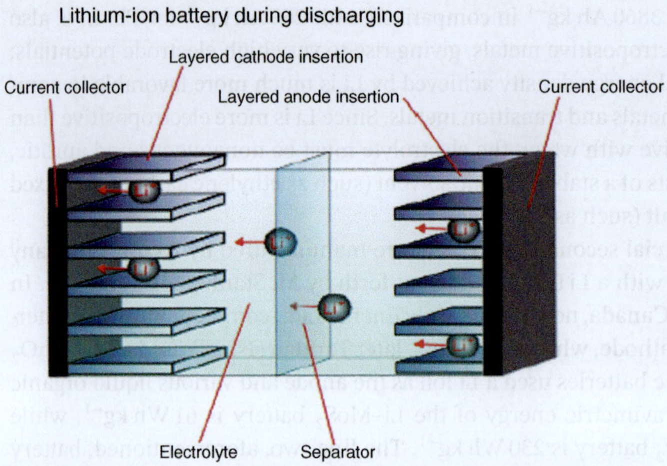

Figure 3.9 Schematic representation of discharge in a Li-ion battery. Source: Reproduced with permission from Weal et al. [3]/University of Cambridge.

A typical Li-ion battery is schematically represented in Figure 3.9 [3], while the working principle is depicted in Figure 3.10 [3]. In addition to the Co-based compounds, the Li cathode can comprise other chemistries based on Mn, Ni, P, and Fe and various combinations. Table 3.3 provides a comparison between different cathode materials.

Upon full lithium utilization, the cell capacity, simply calculated by Faraday's laws, is 3860 Ah kg^{-1} of Li in the cell. Thus, the actual rated capacity of the cell in Ah is determined by the weight of lithium in the cell. The specific capacity, on the other hand, takes into account the total mass of the cell and is the ratio of the rated capacity and the mass of the cell. In a general case, the cell mass can be calculated as

$$W_{\text{cell}} = w_{\text{Li}} f_{\text{A}} + w_{\text{Li}} f_{\text{C}} + w_{\text{aux}},$$

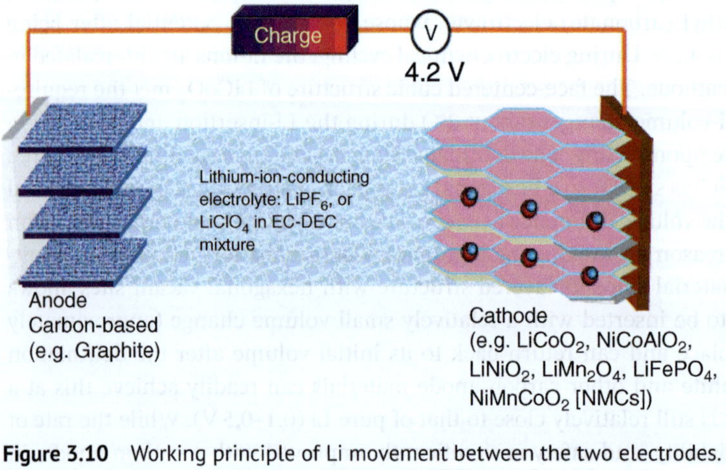

Figure 3.10 Working principle of Li movement between the two electrodes. Source: Reproduced with permission from Weal et al. [3]/University of Cambridge.

Table 3.3 Comparison of properties of different cathode chemistries.

Chemistry	Nominal V (V)	Charge V limit (V)	Charge and discharge C-rates	Energy density (Wh kg^{-1})	Applications	Note
Cobalt	3.60	4.2	1 C limit	110–190	Cell phones, cameras, laptops	Since the 1990s, it has been most commonly used for portable devices; has a high-energy density
Manganese (spinel)	3.7–3.8	4.2	10 C cont. 40 C pulse	110–120	Power tools, medical equipment	Low internal resistance; offers a high-current rate and fast charging but a lower energy density
NCM (nickel–cobalt manganese)	3.7	4.1[a)]	~5 C cont. 30 C pulse	95–130	Power tools, medical equipment	Provides a compromise between high current and high capacity
Phosphate (A 123 system)	3.2–3.3	3.6[a)]	35 C cont.	95–140	Power tools, medical equipment	High-current rate, long-cycle life, high-charge V, increased capacity but shorter cycle life

a) Higher voltage provides more capacity but reduce cycle life.
Source: Adapted from Ogasawara et al. [7].

where w_{Li} is the mass of lithium in the cell, f_A is the multiplier for the anode mass, f_C is the multiplier for the cathode mass, and w_{aux} is the auxiliary mass of the cell, which accounts for approximately 20% of the mass of the active parts of the cell. These parts include the electrolyte, the separator, the connectors, and other remaining components [3]. Thus, as an approximation,

$$W_{cell} \sim 1.2\, w_{Li}\, (f_A + f_C)$$

and the specific capacity is given by

$$w_{Li}\, 3860/1.2\, w_{Li}\, (f_A + f_C) = 3860/1.2\, (f_A + f_C).$$

The specific energy density, which is the energy density divided by the weight of the cell, is the product of the specific capacity and the operating voltage in one full discharge cycle. The values may be quoted for an average voltage or the maximum voltage in the discharge sequence. For a constant current drain, I, the specific energy density, E, can be written as

$$E = \left[I\, w_{Li} \int V\, dt \right] / w_{cell},$$

with the integration being carried out over the full specified discharge time. The specified discharge time is related to a maximum and minimum voltage threshold value to allow for recharging the battery. Outside the threshold values, the battery may attain an irreversible state. Furthermore, the specific energy density is related to the specific capacity, Q, as follows (see above for description of anode and cathode multipliers (f_A and f_C)):

$$E = QV_{\text{average}} = w_{\text{Li}}\ 3860/1.2\, w_{\text{Li}}\ (f_A + f_C)V_{\text{average}} = 3860/1.2\,(f_A + f_C)\, V_{\text{average}}.$$

To maximize the specific energy density, it is desirable to minimize the weight of the cell while maximizing the ratio of the weight of Li to the weight of the cell. For the Li-ion cell, for example, the theoretical stoichiometric value of the anodic multiplier (f_A) is equal to 10.3 while for the Co-based cathode (f_C), it is 25; thus, the maximum theoretical specific energy density for a maximum 4.2 V cell (with discharge the voltage decreases) is calculated to be between 380 and 460 Wh kg^{-1} depending upon whether the weight of the auxiliary components is taken into account or not. In practice, Li availability in both the anode and the cathode is more than halved; thus, the multipliers are typically $f_A > 21$ and $f_C > 50$, the excess value is arising not only from the weight of the binder and other additives, but also from the structural limitations imposed by the layered electrodes for fast intercalation and deintercalation during cycling, especially at high rates.

The operating voltage during discharge is decreased from a maximum value of 4.2 V to a cutoff value of 2.8 V, giving an average value of 3.35 V over the discharge cycle. The practical specific energy density is, therefore, in the region of 160–220 Wh kg^{-1} for a Li-ion cell. The one practical method for marginally increasing the specific energy density of a Li-ion cell is to decrease the weight of the auxiliary components of the cell further down; if, additionally, the Li-availability can be decreased to the maximum stoichiometric region for both the electrodes, then the specific energy density could be doubled. Changing the anode by incorporating Si with C and changing the cathode to different NMCs (Ni, Mn, and Co oxides in varying ratios) has helped to increase the capacity and energy density further.

The average specific energy density achieved by a Li-ion battery, due to innovations, is currently in the region of 160–280 Wh kg^{-1}. It is widely believed that with a considerable amount of research and development, the maximum specific energy density that can be achieved for a Li-ion cell within the next few years could be 250–320 Wh kg^{-1} per cell. The volumetric energy density is around 500 Wh l^{-1}. For example, the well-known 18650 Li-ion cells are cylindrical 45 g weight cells within a steel can, with a 2.2 Ah rated capacity and a volume of 0.0165 l, providing nominal gravimetric and volumetric energy densities of 195 Wh kg^{-1} and 514 Wh l^{-1}, respectively. The cell weight is distributed as: 22% anode, 55% cathode, and the rest taken up by the electrolyte/separator/current collectors and packaging. An 18,650 cell will run for one hour draining at 2.2 A and for 100 hours at 22 mA drain rate.

Considerable research and development programs have greatly helped the incremental development in the performance of Li-ion batteries (Figure 3.11) which is accompanied by a massive increase in volume and reduction in cost (Figure 3.12). The ratio of volumetric to gravimetric energy densities reflects the density of packing, that is, the weight of materials packed per unit volume. For Li-ion batteries, the density of packing varies between 2.4 and 3.2 kg l^{-1}.

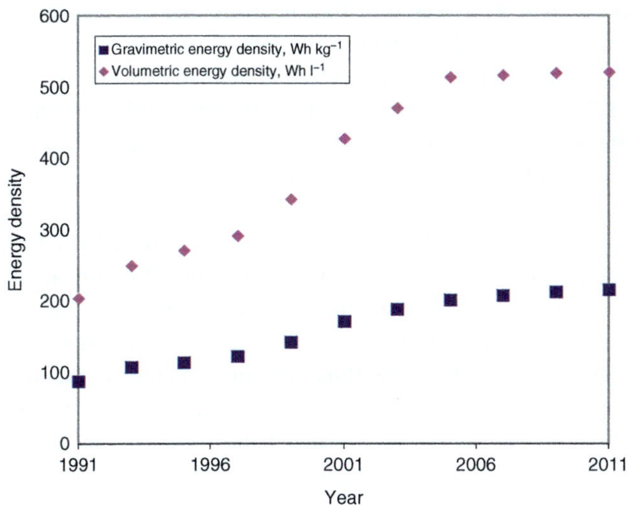

Figure 3.11 Energy density evolution of Li-ion batteries.

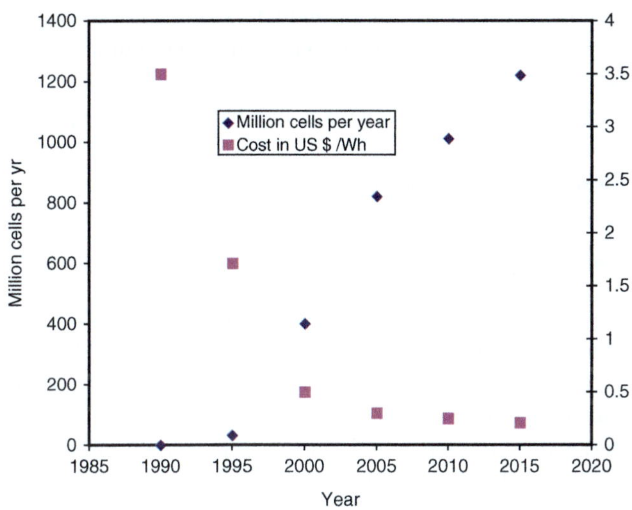

Figure 3.12 Evolution of volume and cost for Li-ion batteries.

It should be noted that the Nobel prize in chemistry for 2019 was given to John B. Goodenough, Michael Stanley Whittingham, and Akira Yoshino for their contributions to developing Li-ion batteries.

3.5.2 Li–Polymer Batteries

A safety concern arises from the high reactivity between the lithium and the liquid organic electrolyte in the deep charging state and may result in leaks and fire initiation. The use of a gel polymer electrolyte can mitigate if not prevent such risks. This gel electrolyte is used as a very thin film to overcome the resistance arising from the low conductivity of the polymer. In 1999, therefore, secondary Li batteries were developed that had the same electrode chemistry as Li-ion, but with the liquid/gel Li-ion electrolyte being replaced with a polymer, and are termed Li–polymer. Although the same electrodes as those used in Li-ion cells can be employed in Li–polymer cells, other materials may be more efficient.

As the name suggests, a polymer electrolyte can be used to replace a separator soaked in a liquid electrolyte. The use of a dry polymer electrolyte can lead to higher safety and reduced flammability and offers simplicity with respect to fabrication and geometry. Very-thin-film-type batteries are made possible with a polymer electrolyte. A true polymer electrolyte (such as polyethylene oxide) complexed with Li–salt (such as $LiPF_6$) has lower conductivity, thus, offering a higher resistance to charge movement that inhibits the rate of discharge and charging. It should be noted that as a way of compromise, the Li–polymer batteries do not as of yet commercially use true polymers but instead use soft polymer solids of the gel-type electrolytes.

Future developments in using composite electrolytes of inorganic oxides and polymers are making massive strides in research and commercial development activities. Inorganic oxides are also available which are Li-ion conductors and thus a good synergy of properties is available from oxide–polymer composites. These innovations and prospects are discussed in Chapter 7.

Another opportunity may arise from developments in Li-S batteries, with a potential to provide very high-specific-energy densities in the near future 5–10 years. Chapter 8 is devoted to these promising systems.

3.5.3 Lithium/Air Batteries

The lithium–air battery or Li–air in short was introduced in 1996 by Abraham et al. [8] but did not attract much attention until 2006 when it was revisited [7] and more focused research on these systems began. Lithium–air batteries have an extremely high-specific energy (11,140 Wh kg^{-1}), which meets the demand for high-energy-density batteries for automotive and aerospace power supplies. This specific energy is up to 5–15 times higher than that of current lithium-ion batteries and is comparable to that of gasoline. There are four types of lithium–air batteries based on the electrolyte types, which are: nonaqueous (aprotic) solvents, aqueous solvents, hybrid (nonaqueous/aqueous) solvents, and all solid-state electrolytes. For all these types, the anode is always lithium metal, while the cathode is always oxygen gas (from air). The anode undergoes oxidation during discharge

$$Li \rightarrow Li^+ + e^-,$$

and the cathode undergoes a reduction reaction as the lithium-ions recombine with oxygen from the air. Metal catalysts, such as manganese oxide (MnO_2), cobalt oxide (Co_3O_4), and iron oxide (Fe_2O_3 and Fe_3O_4), supported on mesoporous carbon play an important role for the reduction reaction to occur on the carbon layer at cathode.

In an organic-electrolyte-based cell, the products at the cathode are lithium oxides

$$Li^+ + e^- + O_2 \rightarrow LiO_2 \qquad 3.00\ V\ vs.Li/Li^+$$

$$2LiO_2 \rightarrow Li_2O_2 + O_2$$

$$LiO_2 + Li^+ + e^- \rightarrow Li_2O_2 \qquad 2.96\ V\ vs.Li/Li^+$$

The products of the oxygen reduction reaction are solid Li_2O_2 particles which are precipitated on the carbon support pores and can block the diffusion of oxygen gas into the electrode. The pore size of the carbon materials should be optimized to enhance the performance of lithium–air batteries.

For aqueous solvent-type cells, lithium hydroxide is produced when using an alkaline aqueous electrolyte

$$2Li^+ + 1/2O_2 + H_2O \rightarrow 2LiOH \qquad 3.44\ V$$

In an acidic aqueous electrolyte, lithium–ions and water are the products

$$2Li + 1/2O_2 + 2H^+ \rightarrow 2Li^+ + H_2O \qquad 4.27\ V$$

Aqueous cells require noble metal catalysts and an anode-protecting layer, such as a glass–ceramic layer (LiSICON, $LiM_2(PO_4)_3$) [9] to keep the stability of lithium ions in water. Therefore, the organic (nonaqueous) cell is more practical in terms of applications and can be developed more than the aqueous cell.

3.5.4 Evaluation of Li Battery Materials and Chemistry

The following chapters are dedicated to next-generation electrodes and electrolytes for Li batteries. The standard way to test the ability of materials to act as rechargeable Li electrodes is to electrochemically cycle them between the charged and discharged states to determine the battery power capability and "lifetime." The lifetime of a rechargeable battery depends on the stability of the voltage–current–capacity relationship as a function of the number of cycles. Such testing can be performed in several ways, such as by using flooded and button cells. Button cell testing is more standard and is used in most laboratories. Since one of the most active research areas in the battery sector is concerned with new material chemistries that can be used as Li electrodes, electrochemical testing will be briefly described here. First of all, the active electrode material needs to be mixed with a binder (e.g. 10% PVDF) and a small amount (approximately 10%) of carbon to make a laminate; carbon is always added to ensure that the electrode has a high conductivity. Then, the laminate is coated onto a metal foil (e.g. Cu) by a doctor blade; the metal acts as the current collector. After drying a solvent out in the oven, it is punched out into a circular shape and weighed to obtain the active material used in the cell. The cell assembly is then performed in a dry room (glove box) using a lithium foil as the counter electrode. An example of liquid electrolyte chemistry is 1 M $LiPF_6$ in 1 : 1 EC:DEC. If it is electrolyte candidates that are to be tested, standard electrode materials are used and the electrolyte is substituted. The maximum possible energy in the electrochemical cell is determined by measuring the open-circuit voltage (OCV) as a

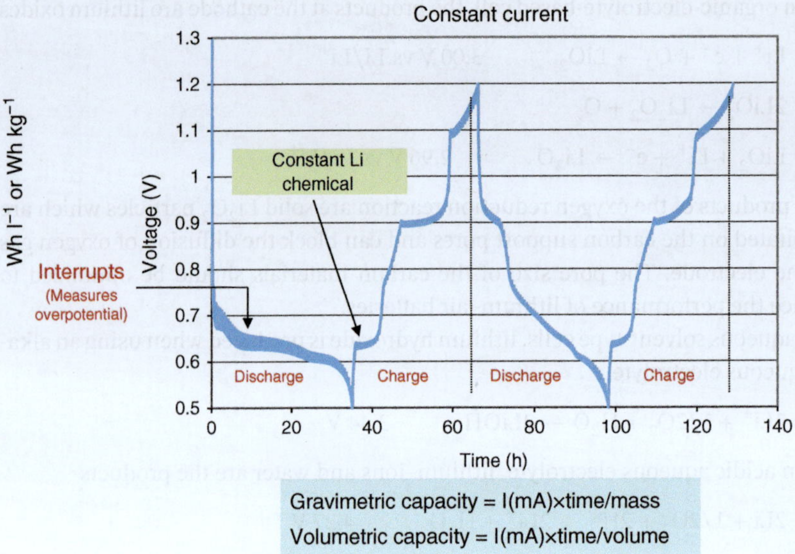

Figure 3.13 Curve obtained from electrochemical cycling.

function of the state of discharge. This will act as a benchmark in understanding the amount of internal resistance associated with mass and charge transport. After OCV measurement, the button cell is put in the cycler (battery test station). The testing procedure starts by discharging (or charging) the cell at a certain current between the maximum and minimum voltages, and once it reaches a minimum voltage, it will automatically charge and so on. The rate of discharge–charge can be varied by varying the current. Then the software gives a plot between voltage and time, as shown in Figure 3.13. Multiple plots of voltage vs. time for the charge–discharge cycle can also be obtained. The capacity of the tested cell can then be calculated for each discharge and charge cycle by measuring the area under the curve and dividing by the active material weight to obtain the capacity in $mAh\ g^{-1}$.

The important factors to consider in a practical battery electrode are: (i) increase in capacity for Li uptake; (ii) greater Li-ion diffusivity so that higher discharge currents can be achieved; (iii) greater chemical and electrochemical stability; (iv) structural and mechanical stability over a wide range of lithiation and delithiation; and (v) maintenance of planar surface thus avoiding lithium dendrites. For the electrolyte in a Li-ion battery, the factors that are critical are: (i) high electrical conductivity to help achieve higher current density with low joule heating; (ii) excellent chemical and electrochemical stability in contact with the cathode and the anode; (iii) high transport number for Li^+ ions, preferably equal to unity; and (iv) chemically inert with the separator. Ideally, the electrolyte/separator together has the mechanical properties of a solid but the electrical properties of a liquid. Typically, highly polar but small molecules, such as propylene carbonate (PC), are used to plasticize the host polymer chain to induce flexibility and segmented motion, as well as to solvate the Li^+ ions. This decreases ion–ion interaction. In the cell design, it is important to maximize the electrode/electrolyte interfacial area without blocking the pores. Li salts impart ionic conductivity to the electrolyte. $LiClO_4$ can provide good conductivity, but is avoided in commercial batteries for safety reasons. $LiAsF_6$ can generate toxic fumes as a result of internal decomposition reactions, especially during thermal runaways. Thus, the main focus is on $LiPF_6$ or $LiCF_3SO_3$ salts despite their relatively lower conductivity and stability.

Details regarding the cathodes, anodes, and electrolytes of Li-ion batteries are given in Chapters 5–7, while Chapter 4 focuses on their applications.

3.6 Battery Market

Table 3.4 shows the global battery market data for 2010, which is when secondary batteries consistently began to dominate over primary batteries reversing the historical situation until then. By 2020, as can be seen in the same table, the primary battery share shrunk to 26% market share while the secondary battery market share was 74% and is predicted to continue to increase. During the same period of 10 years the global market for all batteries increased by 50% and in 2019 it was worth over US$ 120 billion, with an annual growth forecast of over 14%, with the rapid growth in demand coming from EVs and energy-storage systems (ESS). Currently, more than 50% of the primary market is from alkaline cells, while half of the secondary market is still dominated by lead–acid batteries. Within the secondary battery market trend, lithium-based batteries will begin to take a dominant position with the anticipated growth in the EV–ESS sectors as early as 2023.

In the secondary battery market, several new cells have been introduced just within the last two decades, for example, the Ni–MH cell (1990), Li-ion cell (1991), rechargeable alkaline cells (1992), Li–polymer (1999), and also the concept of mechanically rechargeable Zn–air batteries (2001). The new batteries, especially those based on the Li chemistries, have not only met some of the existing demand, but can also be said to have revolutionized the battery market by accelerating demand for laptops, mobile phones, and electric vehicles.

Table 3.4 Global battery market data for 2010 - 2020.

Type of battery	2010 - Total value US$ 80 billion (%)	2020 - Total value US$ 120 (%)
Primary	Primary total: 42	Primary total: 26
Li-primary	12	10
Alkaline	22	14
C-Zinc and others	8	2
Secondary	Secondary total: 54	Secondary Total: 74
Lead–acid	28	37
Ni–MH and others	12	3
Li-ion	14	34

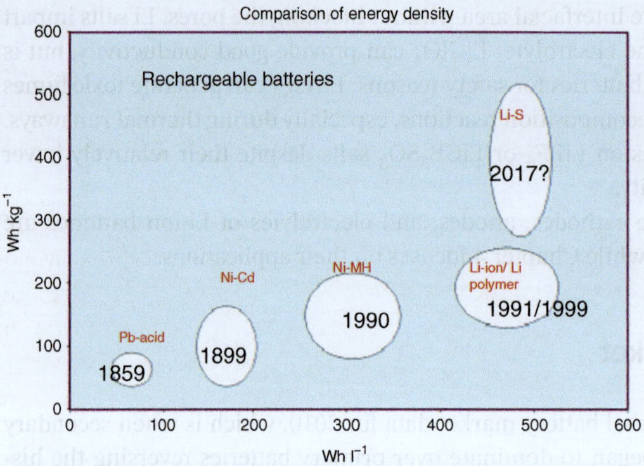

Figure 3.14 Comparison of energy density for various battery chemistries.

An inspection of Figure 3.14 clearly reveals that, with time, the specific energy (Wh kg^{-1}) and the energy density (Wh l^{-1}) have continued to grow as batteries advance. It is this combination of high values of the gravimetric and volumetric energies that have been the key factors heralding this rapid growth.

It should be noted that Li batteries did not replace the existing lead–acid or Ni–Cd batteries in the market. Instead, these new batteries have accelerated the development of portable computers, cellular telephones, and cordless hand-held tools to a degree that was impossible to imagine in the so-called mature market of rechargeable batteries before the 1990s. Both the Ni–MH and the Li-ion cells are produced in the region of 1 billion cells per year each with a market value of approximately US$ 4 billion and US$ 44 billion per year, respectively. This is a staggering growth considering these cells were not available at all 28 years ago. Ni–MH and Li-ion batteries are not only lighter due to their high-specific-energy density (Wh kg^{-1}) than the Ni–Cd and the Pb–acid rechargeable batteries but are also smaller due to their high-volumetric-energy density expressed in Wh l^{-1}. It is the high values of Wh kg^{-1} and Wh l^{-1} that have been the key factors in this rapid growth. It should also be noted that in the early 1990s, the Ni–MH system made good use of the Ni–Cd type of facilities, and in the late 1990s, the Li–polymer system utilized the Li-ion methods successfully to cut down on manufacturing lead time.

3.7 Recycling and Safety Issues

The batteries consumed in households amount to approximately 4 billion cells per year, and even though they are important for everyday life, they can be a source of health issues and contamination. Besides the caution of using the right battery for the right function, their proper disposal is also a critical issue that needs to be urgently considered.

It is important to note that choosing to purchase rechargeable batteries will decrease global battery wastes. It is a fact that 600 primary batteries can be replaced by only two AA nickel–cadmium secondary batteries. Rechargeable batteries also come out as USB batteries, which are convenient and do not need a charger.

In some cases, the hazardous heavy metals, contained in batteries, were replaced with time. Alkaline batteries are manufactured as a mercury-free battery since 1992. The mercuric oxide button cell was finally replaced by mercury-free batteries, such as silver oxide or zinc–air batteries. However, the nickel–cadmium battery has not been totally replaced by other nonhazardous batteries yet, but regulatory steps are in progress. Moreover, some appliances have the nickel–cadmium battery built in, which is difficult for battery recycling.

The annual number of batteries discarded runs in billions worldwide; safe disposal and recycling are crucial in this important energy-storage chain. Many countries, such as the USA, the UK, Germany, France, Turkey, Italy, and Poland, enforce by legislations the construction of battery recycling plants. Batteries are classified by the US federal government into nonhazardous and hazardous wastes. Used Ni–Cd, lead–acid, silver oxide, and button cells containing mercuric oxide, silver oxide, lithium, and zinc–air are classified as hazardous wastes and need to be recycled. Used alkaline, carbon–zinc, lithium-ion, nickel–metal hydride, and rechargeable alkali manganese batteries are considered as nonhazardous waste and can still be thrown away in the trash. These nonhazardous batteries, however, must also be recycled, since throwing these batteries together with other wastes is dangerous as some of them may not be completely dead and may generate charge and sparks which can lead to a fire and an explosion. They also should not be burned or incinerated as this can result in an explosion and contamination of the environment. In the near future, not recycling batteries will not be an available option.

3.7.1 Recycling of Lead–Acid Batteries

Lead–acid batteries will continue to remain the mainstay of secondary batteries and have found widespread industrial use as a result of low cost, simplicity of design, relative safety, and good specific power density. It is estimated that over 600 million rechargeable lead–acid batteries are produced each year worldwide. Of the nearly 15 million tons per year (mtpy) of lead produced worldwide, more than 80% is used in batteries. Over half of the total lead is derived from recovering lead from used batteries with lead sulfide ores contributing the rest. It can be expected that in the near future most of the global lead (20 mtpy by 2025) will be used in batteries for vehicles and emergency power supply and most of this will be met by recovering lead from used batteries.

Lead is the leading material in recycling rate, defined in terms of % of lead recycled with respect to the amount of lead available for recycling. Recycling rates of over 90% are achieved in many parts of the world, such as in North America, Western Europe, and Japan. Emerging markets are also witnessing improved trends in recycling lead–acid batteries, since lead is relatively easy to recycle and recover as metallic lead. Currently, lead–acid batteries are recycled mainly using a high-temperature, pyrometallurgical process. A 10,000 tons per annum plant would be deemed small, with some plants processing 100,000 tons or even 200,000 tons of secondary lead–acid cells per year.

Lead–acid batteries are, in fact, one of the most efficiently recycled products in the world – albeit with varying environmental standards. In developed nations, a variety of organized schemes are in operation. For example, Sweden, Germany, and Italy operate levy systems related to the lead market: when the lead price is so low that battery recovery is not economical, a levy is imposed on new batteries to finance recycling of used batteries. Italy and Ireland also have national collection and recycling schemes. In the United States, many states require retailers to accept used car batteries when customers purchase new batteries. Several US states even require a cash deposit on new battery purchases, which is refunded to the consumer once the used battery is returned to the retailer. Elsewhere, battery collection is normally market driven. In the UK, for example, where organized systems do not exist, car repair shops and scrap metal dealers collect used batteries.

The estimated RIR figures in Table 3.5 are based on official data collected from governments around the world. As such they are likely to underestimate true lead recovery rates because in many countries, especially in the developing world, the informal nature of lead recycling makes obtaining reliable data on secondary lead usage impossible. This is particularly true for Asian countries: lead is so highly priced there that it is likely that almost 100% of the metal from batteries is reused. Since formal data may not be fully available, the details are thus anecdotal.

The recycling record in other types of batteries has still a long way to go to catch up with what is achieved for the lead–acid system. All spent batteries represent a valuable resource of metals and materials, such as Ni, Cd, Ag, Co, Cu, polymers, carbon, and oxides. New legislations are being steadily introduced to encourage recycling of batteries. For example, in 2006 Directives on batteries and accumulators were adopted in the European Union. The Directive has set new national targets for nations within the EU for collection of consumer batteries. For example, according to this Directive, the UK is required to collect 46% by weight of batteries sold by 2016. New infrastructures have to be made available that allow consumer batteries to be safely collected, sorted, and recycled.

Some countries are leading the way in recycling consumer batteries, such as Belgium which has already achieved a collection rate that is over 65%.

Table 3.5 Apparent recycling input ratios (RIR) for lead.

Region	RIR (%)
Europe	62.7
Americas	74.2
Asia	15.7
World	41.6

Disposal of some batteries, such as Ni–Cd, can cause major environmental problems. Cd is toxic and can seep into the water supply. Ni is also semitoxic. Primary Li batteries are fire hazards, especially if they are not fully discharged. Most secondary Li batteries contain the toxic element Co. Both primary and secondary Li batteries contain toxic and flammable electrolytes. It must be stated, however, that, in general, recovery of metals from batteries is energy intensive. There is pressing need for further research for developing environment- friendly recycling technologies that do not give rise to an unsustainable energy penalty.

3.7.2 Details on the Recycling Process of Lead–Acid Batteries

In particular, the lead–acid battery recycling process stages are as follows:

- **Collection**: At repair workshops, garages, decontamination centers, clean points. The batteries and other lead-containing scrap are consolidated at intermediate agents or scrap dealers.
- **Transportation to appropriate plants.**
- **Materials preparation and sorting.**
- **Transportation to appropriate plants.**
- **Breakage**: Lead–acid batteries are composed of a variety of materials (Table 3.6), so they must be taken apart. Scrap batteries are broken apart in a hammer mill. The plastic casing (typically made of polypropylene, ebonite, and/or PVC) is shredded and recycled at extrusion plants. The sulfuric acid is also recovered and used in processes, such as the manufacture of gypsum. Meanwhile, paste comprising lead sulfate ($PbSO_4$), as well as the lead metallic grids, is extracted. In conventional lead recycling, the grids are normally added to the smelter; however, this is not always necessary, and instead, they can be melted and refined at lower temperatures. Table 3.7 shows the typical composition of the lead–acid paste.
- **Neutralization**: Sodium hydroxide is added to the discharged battery paste to neutralize residual sulfuric acid and desulfurize, which avoids sulfur dioxide (SO_2) emissions during smelting. Aqueous sodium sulfate, possibly containing some residual dissolved lead(II) together with colloidal particles, is then discharged.

Table 3.6 Typical composition of lead–acid battery.

Component	wt%
Lead–antimony alloy components (i.e. grids and poles)	25–30
Electrode paste	35–45
Sulfuric acid	10–15
Polypropylene	4–8
Other plastics (e.g. PVC and PE)	2–7
Ebonite	1–3
Other (e.g. glass)	<0.5

Source: Ref. [3]/with permission of University of Cambridge.

Table 3.7 Typical composition of lead–acid battery paste.

Material	wt%
Lead sulfate	55–65
Lead dioxide	15–40
Lead monoxide	5–25
Metallic lead	1–5
Carbon black, plastics, fibers, and other sulfates	1–4

Depending on the country, the consent concentrations for this discharge to sewers range from a few ppm down to zero; the upper limits are likely to be decreased in the future.

- **Smelting or electrowinning**: The dry neutralized paste is usually smelted, along with the metallic grids, with a reducing agent in a high-temperature Isasmelt or reverberatory or rotary furnace to produce lead. Coal or coke is the source of energy. The smelting takes place either in stand-alone reactors or more often with lead concentrates derived from lead sulfide ores. Any lead or antimony slag is reheated with refining dross in a rotary furnace producing lead–antimony (Pb/Sb) alloy and waste slag. Instead of smelting, some large-scale operations dissolve the lead sludge/filter cake in powerful acids, such as HCl, H_2SiF_6, or HBF_4, and recover lead by electrowinning.
- **Refining and blending**: A finished alloy is produced by adding the Pb/Sb alloy to the almost-pure lead metal in a refining and blending process.
- **Transport**: Assuming recycling and manufacturing plants are separate entities, the lead must be transported from the former to the latter.
- **New battery production**: Recovered lead is reoxidized to PbO granules. Sulfuric acid (H_2SO_4) is added to the granules to form a paste applied to lead grids to manufacture new battery electrodes. After battery assembly, electrical energy is used to convert PbO to Pb in the anode and PbO_2 in the cathode. An overview of lead–acid battery recycling process is given in Figure 3.15.

The majority of lead in used batteries is in the form of lead sulfate ($PbSO_4$). Decomposition of $PbSO_4$ requires the use of a relatively high temperature using coal or coke as the source of energy. Thus, the smelting route is associated with energy usage of 2–10 kWh per kg of lead along with CO_2 greenhouse gas emission, pollutants emission of lead dust, NO_x, and SO_2. Gas treatment for removing harmful emissions is expensive and legally binding and continues to get stricter. More advanced smelting technologies, such as Isasmelt, use Fe or soda to fix the S in the furnace by forming FeS-PbS matte or a slag containing Na_2SO_4. Disposal of leachable slags is also expensive and harmful to the environment. Shipping spent batteries to other distant locations is not permitted by most global treaties.

Several research methods have been advanced enough to overcome some of the problems, especially with sulfur encountered in the smelting route. A common

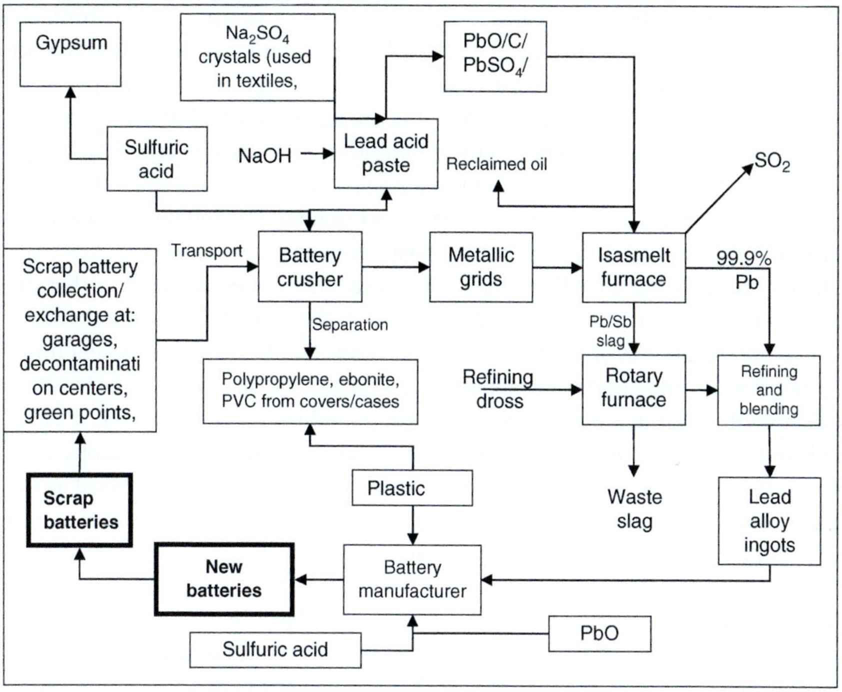

Figure 3.15 Overview of the current lead–acid battery recycling process flow. Source: Reproduced with permission from Weal et al. [3]/University of Cambridge.

factor is the use of NaOH (aq) or Na_2CO_3 (aq) solutions to fix S as soluble Na_2SO_4(aq) which can be crystallized out as a saleable by-product. The insoluble $PbCO_3$ or $Pb(OH)_2$ collected as sludge or filter cake is then smelted. It is not easy to fully avoid the retention of some S in the sludge/filter cake and the associated SO_2 emission.

To evade smelting altogether, some of the new processes have dissolved the lead sludge/filter cake in powerful acids, such as HCl, H_2SiF_6, or HBF_4, to recover lead by electrowinning. Electrowinning is capital intensive and is often suitable only for large-scale operations entailing large movement of hazardous raw materials. It is also energy intensive at 5–12 KWh per kg of lead. Finally, it can be polluting, if the electricity for such a process is generated from fossil-fuel power plants.

Once lead is recovered, it is then re-oxidized to PbO granules for making pastes with H_2SO_4(aq) to apply to the lead grids in the manufacturing of the battery electrodes. The energy density of a lead battery is 50 Wh calculated with one kg of Pb as the base. This amount is 2–3 orders of magnitude lower than the energy required to reproduce that lead by the current recovery methods. Thus, less than 0.5% of the energy is made available. It is true that batteries are convenient sources of energy; but these calculations show that with respect to energy, it is unsustainable. But not recovering lead from used batteries is not an option!

A new process of recovering lead directly from spent battery paste as PbO precursor for making new paste has been developed in the University of Cambridge [10] and commercialized by Ever Resource Ltd and iQ Power. In this new environmentally sound, paste-to-paste process, the net-carbon, SO_2, and lead dust emissions are very low toward making new batteries without the use of high-temperature pyrometallurgy or the energy-intensive electrowinning processes. By reacting spent lead battery paste with organic reagents and then combusting the organic crystallites, high surface area PbO is directly available for making pastes for new batteries.

In brief, the method comprises dissolving in a leaching–precipitation process, the lead-bearing active components of battery paste in an aqueous solution of carboxylic acid, and precipitating a lead-bearing organic precursor. Many forms of carboxylic acids and related compounds are routinely used in the food industry and domestic use, making the process of leaching very innocuous. Unlike other hydrometallurgical processes currently under development, lead is not fully dissolved in the leachate for electrowinning, but is only transient toward crystallization. For example, if acetic acid is used as the leaching agent, since lead acetate is soluble in the aqueous solution, lead can be fully leached. However, if citric acid ($C_6H_8O_7 \cdot H_2O$) is used as the reagent, since lead citrate has limited solubility in an aqueous solution, lead can be dissolved but is readily crystallized. A scanning electron microscopy (SEM) image of the organic crystallite is shown in Figure 3.16 [10].

In the subsequent step, the lead organic citrate is subject to heating for combusting and removing the organics embodied in the crystals. The combustion of the organics liberates a large quantity of heat that can be used for melting and refining the lead–alloy grid from a spent battery. The combustion reaction is accompanied by the liberation of lead in the form of PbO/Pb; the degree of oxidation is dependent upon the operating conditions. Another novel feature of the new process is the ability to produce the PbO product in a nanocrystalline form, as shown in the SEM and transmission electron microscopy (TEM) images in Figures 3.17 and 3.18, respectively [10].

Figure 3.16 SEM image of the lead citrate product from PbO. Source: Reproduced with permission from Kumar et al. [10]/Open Access Publisher, Licensed under CC BY 4.0.

Figure 3.17 SEM image of nano-PbO from combustion–calcination. Source: Reproduced with permission from Kumar et al. [10]/Open Access Publisher, Licensed under CC BY 4.0.

Figure 3.18 TEM image of nanocrystalline PbO. Source: Reproduced with permission from Kumar et al. [10]/Open Access Publisher, Licensed under CC BY 4.0.

As much of the energy required is met directly by the combustion of the organics used in the process, the overall energy requirement based on thermal analysis data is estimated to be in the order of 250 Wh per kg of Pb equivalent. This new process is not only significantly more sustainable from an energy and materials point of view, but also has a more important environmental impact due to significantly lowering emissions and discharge.

Under conventional approaches, consideration has only been given to the sustainability of materials and not to the energy in the recovery of electroactive components of a battery. The spent aqueous solution containing Na_2SO_4 can be subjected to the documented process for crystallizing Na_2SO_4 by evaporation/crystallization/centrifugation as a useful by-product for the chemical industry. A simplified flow chart is shown in Figure 3.19. The energy demand for calcination

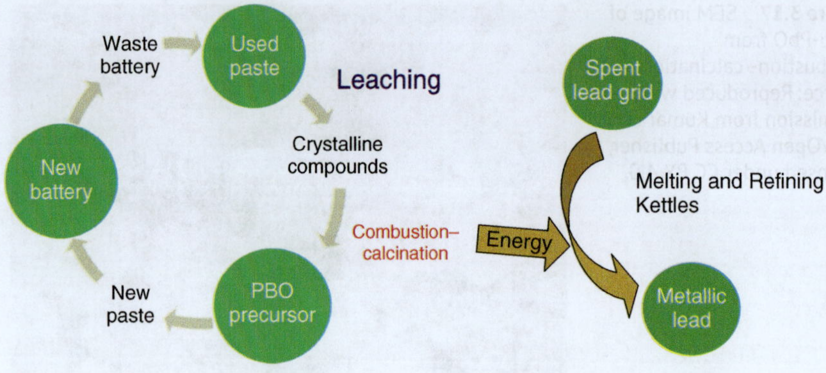

Figure 3.19 Schematic flowsheet for recycling lead battery waste.

and for melting and refining the grid alloys can be met by the organics in a net low-carbon process.

3.8 Conclusions

The purpose of this chapter was to provide an overview of the evolution of secondary batteries. As was seen, the most promising electrochemical cells that are currently used commercially are Li-ion and Li–polymer. Therefore, the subsequent chapters of this book will describe the material candidates to be used as cathodes, anodes, and electrolytes, in next-generation Li batteries. While Li–S batteries are not yet fully commercialized, they are promising batteries for the near future. A chapter is devoted to developments in Li–S batteries in the later part of this book. In concluding this book, more novel electrochemical storage devices, such as Na-ion, Zn, and Mg, will also be summarized, even though the latter two, in particular, are in their early developmental stages.

References

1 Gaston, P. (1859). The Storage of Electrical Energy, contributed by Osmania University, Paul Bedford, India. http://www.archive.org/details/storageofelectri031249mbp (accessed 6 November 2007).

2 DOE (1995). Primer on lead-acid storage battery, DOE handbook, DOE-HDBK-1084-95. http://www.archive.org/details/storageofelectri031249mbp (accessed 6 November 2007).

3 Weal, E., Tan, J.C., and Kumar, R.V. University of Cambridge (copyright 2004–2013). DoITPoMS Teaching and Learning Packages. http://www.doitpoms.ac.uk/tlplib/batteries/index.php (accessed 15 November 2021).

4 Firefly's composite foam-base battery technology. www.fireflyenergy.com (accessed 2 December 2014).

5 Linden, D. and Reddy, T.B. (2002). *Handbook of Batteries*, 3e. New York: McGraw-Hill.

6 Mizushima, K., Jones, P.C., Wiseman, P.J., and Goodenough, J.B. (1980). *Mater. Res. Bull.* 15: 783.

7 Ogasawara, T., Debart, A., Holzapfel, M. et al. (2006). *J. Am. Chem. Soc.* 128: 1390.

8 Abraham, K.M. (1996). *J. Electrochem. Soc.* 143: 1.

9 Visco, S.J., Katz, B.D., Nimon, Y.S., and De Jonghe, L.C. (2007). Protected active metal electrode and battery cell structure with non-aqueous interlayer architecture. US Patent 7282295.

10 Lead Recycling, PCT filed on 6 Nov 2007. RV Kumar, S Sonmez and VP Kotzeva; PCT/GB2007/004222; EU 07824458.9; RU 2009117620; US 12/513707; CN 200780041628.4; IN 2216/KOLNP/2009.

4

Applications of Lithium Batteries

Haokun Deng[1] *and Katerina E. Aifantis*[2]

[1] *Eve Energy Co., Ltd., Huifeng 7th Rd, Huizhou 516006, China*
[2] *University of Florida, Mechanical and Aerospace Engineering, 1064 Center, Gainesville, FL 32611, USA*

As described in Chapter 4, secondary Li batteries are the most rapidly growing high energy storage devices in industry. The advantages of Li-ion cells over other rechargeable electrochemical cells are: they do not exhibit a memory effect, they provide a higher energy density per unit mass, and their self-discharge rate is less than half compared to the next best available solutions, such as NiCd and NiMH cells. Furthermore, the lifetime of current commercial Li batteries is over 1000 cycles, while their shelf-life is more than 10 years [1]. On the negative side, the lithium-ion battery materials are more fragile and require a protection circuit to ensure safe operation; however, the recently developed Li-polymer cells are easier to handle with safety [2]. The purpose of the present chapter is to give an overview of the main current and future applications that Li batteries have in our lives, as well as those that Na batteries will have, so as to motivate the reader to focus on the following chapters about improving their technology. In Chapter 3, examples of hybrid and electric vehicles were given and the superiority of Li-powered vehicles was shown. Here, more details will be given concerning this endeavor.

4.1 Portable Electronic Devices

Presently, the main application of rechargeable Li-ion batteries is in portable electronic devices, such as cellular phones, digital cameras, global positioning system (GPS) devices, tablets, and laptop computers. As the key feature of these devices is portability, in addition to energy efficiency requirements, the size and weight of the Li-ion cells are a main concern, as in some cases the Li-ion batteries consume most of the weight of the device.

Cell phones have evolved from "ordinary" phones to digital cameras, MP3 players, wireless internet access devices, notepads, and notebooks. To accommodate all these features, more advanced Li-ion cells than those used in conventional cell phones must be developed, as the energy demand is much higher in these new

Rechargeable Ion Batteries: Materials, Design, and Applications of Li-Ion Cells and Beyond, First Edition. Edited by Katerina E. Aifantis, R. Vasant Kumar, and Pu Hu.

multifunction mobile phones. Continuous research for improving Li-ion batteries is, therefore, mandatory and as essential as microelectronic advancement. To illustrate the significance of battery capacity in device performance, the following example is given: the 2000 model of the Motorola StarTAC used a 680 mAh Li-ion cell, which allowed for a talk time of approximately four hours and a wait time of about 75 hours. A more recent Motorola model, the Razor V3 employs a higher capacity Li-ion cell of 900 mAh, which allows for about seven hours of talk time and a standby time of 290 hours [3]. The Li-ion polymer cell developed by Samsung SDI Co., Ltd possesses even a higher capacity of 2320 mAh [4], which is necessary for smartphones running an Android operating system. The latest Li-ion cell for the Huawei Mate30pro, smartphone released in 2019, has a capacity of 4200–4500 mAh. In addition to the huge capacity increase and much longer shelf life, a lot more features have been included due to the progress of battery technologies, e.g. faster-charging rate, less heat generation, higher safety, and a combined power module that allows smart power management.

Among these features, the quick-charge (QC) technique is the most critical one, since it is the bottleneck that affects the performance of a smartphone. Strictly speaking, QC is not a specialized technique, but a combination of a series of technical improvements on Li-ion batteries aiming to enhance charge speed; for example, coating graphite (the anode active material) with amorphous carbon can enhance the lithiation process at the graphite particle surface, and therefore, the charge speed can be increased. Another example is that the adjustment of the electrolyte recipe can also increase the charge speed. Adding carbon nanotubes into the anode material, typically 0.5–1.5% weight percentage of the anode material, can also improve the conductivity of the anode material, hence increasing the charge speed. Beginning in 2010, the evolution of the smartphone market triggered the demand for more chips and higher performance, as well as larger display screens. All of these demand a higher capacity of the Li-ion cells and use of the QC technique. In 2012, the 5V1A configuration, which corresponds to a 5W QC, became a popular charge configuration for cell phones, since the anode materials for Li-ion cells at the time were unable to sustain a high-charge power. Now, 120–200W QC dominates the market, to make sure that a phone can be fully charged in 8–15 minutes. Although the volume reduction of this product is not significant, its design is appropriate for use in smartphones with a large touch screen and small thickness. Figure 4.1 illustrates some widely used commercial Li-ion batteries through the years.

The other most popular application of Li batteries is in tablets and laptop computers. Although Li-ion cells are the most widely used, Li-polymer cells (employing polymer base electrolytes, although some do not consider them fully polymer) are starting to be preferred for some laptops and other electronic applications. The drawback of commercially available Li-polymer cells is that they are more expensive and require more time to manufacture. The advantages are that they can be shaped with greater ease so as to occupy more efficiently the available space in the device, they are resistant to damage and there is less danger of flammability [2]. These Li-polymer cells have found their initial applications in the computer industry. Apple was able to develop the very thin Mac Book Air (1.7 cm thick), by using

Figure 4.1 (a) 400 mAh Li-ion battery by BYD Company Ltd. In 2015. (b) 2500 mAh ($75 \times 52 \times 5$ mm, 128.2 Ah l^{-1}) Li-ion battery made by Netgear. (c) 1800 mAh Li-ion polymer battery developed by Samsung SDI Co., Ltd.

an integrated 49.9 Wh Li–polymer battery that allows for a maximum battery performance setting of approximately 18 hours [5]. However, the space saved in thickness is gained in length and width as this laptop is 32.5 cm long and 22.7 cm wide. Current research in next-generation electrolytes is described in detail in Chapter 7.

In 2019, the 5G network started to expand. This evolution in the IT industry means every single electronic terminal unit will have access to the internet for realizing the transformation of necessary data and help create a historical improvement in the user experience. In the next decade, we will see the commercialization of many innovational portable and wearable electronic devices. Some recent examples include smartwatches, electronic glasses (for virtual reality and augmented reality applications), true wireless stereo (TWS) earphones, smart bracelets, electronic cigarettes, and electronic bikes. Furthermore, power tools are believed to have a huge-potential market in the future. Thanks to the progress in very large-scale integrated circuits (VLSIC) and manufacturing, miniaturization of these power tools makes them practically efficient and affordable. These power tools represent a new category from other conventional portable electronics, as their purposes are completely different and their design language focuses more on portability instead of performance. Examples include the I-watch, smart band/bracelet, and exoskeleton. These devices are extremely volume- and weight-sensitive, as they cannot penetrate the market if they burden the consumers with heavy loads. Therefore, Li-ion batteries with the highest energy density are almost the only candidate

Table 4.1 History sales and forecast of some electronics market.

Product	Smartphone	Power tool	E-bike	TWS earphone
LIB average capacity	0.01 kWh	0.1 kWh	0.5–1 kWh	0.74–2.59 Wh 0.15–0.25 Wh
Capacity ratio (number of LIBs used in a single unit)	1	10	50	1
2015 sales	1 320 000 000 cells sold 13.2 GWh total energy	63 600 000 cells sold 6.4 GWh total energy	2 500 000 cells sold 2.3 GWh total energy	/
2019 sales	1 200 000 000 cells sold 12.0 GWh	106 900 000 cells sold 10.7 GWh	6 500 000 cells sold 6.6 GWh	129 000 000 cells sold 0.258 GWh
2025 forecast sales	1 190 000 000 cells sold 11.9 GWh	200 000 000 cells sold 22.0 GWh	16 000 000 cells sold 18.0 GWh	500 000 000 cells sold 1 GWh

Source: Adapted from Ref. [6]

for these power tools. The performance of Li-ion batteries directly determines the service life of the power tools, and any breakthrough in Li-ion batteries might give rise to their revolutionary popularity. Table 4.1 lists the market information for some of these applications in the past decade [6]. We can see that the market size of phones actually reaches a peak point, as the use of smartphones versus conventional cell phones stops growing gradually, but the demand for Li-ion cells increases as new electronics are commercialized. Therefore, we will be able to see more growth in the market of small Li-ion cells due to the trend of everything being connected to the internet.

4.2 Hybrid and Electric Vehicles

Despite the fact that electric vehicles have been gaining significant attention over the past decade, the first electric car was built in 1800 shortly after the invention of batteries, and about 50 years before the development of combustion engine vehicles. Specifically, Robert Anderson used early galvanic cells to power his carriage. The first rechargeable electric vehicle was produced in 1890 by Morrison Electric (named after William Morrison), whose structure was similar to a six-passenger wagon. It was powered by 24 lead–acid cells, which provided a total potential of 58 V, allowed for a driving range of 100 mi at a speed of ~9–20 km h^{-1}, and required 10 hours to recharge. Several electric vehicles were developed afterward even by Fernid and Porsche of the well-known Porche car company. In 1914, Ford partnered with Edison to use Ni–MH cells in electric vehicles; however, by 1912, gasoline-powered vehicles were almost three times cheaper than electric ones (US\$650 vs. US\$1,750), which therefore disappeared from the US market by 1935.

However, with environmental concerns rising and the forecasted shortage of fossil fuels, the interest in electric cars was revived in the 1970s with the US Congress passing the Electric and Hybrid Vehicle Research, Development, and Demonstration Act in 1976. Even NASA chose an electric vehicle, the Lunar rover, to be the first manned vehicle on the moon in 1971. Prototype lead–acid powered vehicles were produced by GM at the time, with a range of 40 mi before requiring recharge, and a top speed of 72 km h^{-1}. Their limited mileage and non-efficient charging did not allow them to compete with conventional vehicles.

In further attempts to promote sustainable vehicles, the US passed, in 1990, the Clean Air Act Amendment and in 1992 the Energy Policy Act. Despite this, it was not until new Ni–MH, and Li-ion systems were developed that they started to become viable commercially.

Although commercial secondary Li-ion batteries cover the needs of the portable electronic industry satisfactorily, the future of electric vehicles depends on the further development of Li-ion battery technology. Electric motors have already reached a level where they convert ~77% of the electrical energy needed to charge the battery into kinetic energy. This is significantly higher than the 12–30% of the chemical energy in gasoline that internal combustion vehicles convert into kinetic.

The recent environmental public awareness, in combination with the major scientific warnings about the excess production of CO_2, have paved the way for hybrid and electric vehicles. As a leading party in the auto industry, the European Commission declared in November 2017 that all new cars must reduce CO_2 emissions by 15% and 30% by 2025, and 2030, respectively. Following up, in December 2018, the European Commission declared that the 2030 target of 30% reduction in CO_2 emissions was increased to 37.5% and that by 2021, CO_2 emissions must be below 95 and 147 g km^{-1} for passenger and business cars, respectively. A 95 euro penalty would be applied for every 1 g of CO_2 over-emissions. Other countries, such as China, Korea, and Japan, also put in place similar policies, which strongly support the growth of electric vehicles and suppress conventional ones. In December 2019, the Chinese Ministry of Industry and Information Technology published a development plan for a sustainable vehicle industry between 2021 and 2035, which states that by 2025, EV sales will increase to 25% of the overall market. Furthermore, in September 2020, China announced during the 75th general assembly of the United Nations that by 2030, peak CO_2 emissions will be realized, meaning that CO_2 emissions will stop growing, and by 2060, CO_2 neutrality will be realized (CO_2 emissions will be 0).

Hybrid-electric vehicle s (HEVs) are defined as vehicles that use two distinct power sources to operate the vehicle; namely a combustion engine system together with an electric propulsion system, which initially was powered by Ni–MH batteries. Some major Ni–MH battery manufacturers include Panasonic, Cobasys, and Johnson Controls Inc. Li-ion batteries slowly started entering the HEV market after

2008 when $LiFePO_4$ cathodes started to become commercially available. A123 was the first $LiFePO_4$-based Li-ion cell manufacturer. After 2010, $Li(NiCoMn)O_2$ cathode materials became another appropriate candidate for Li-ion cells.

Depending on the primary power source during operation, HEVs can be categorized as parallel HEVs, electric-assisted or mild HEVs, and plug-in-HEVs (PHEVs).

Parallel HEVs use the electric motor during low-demand traveling (low speed, low acceleration), and the internal combustion engine (ICE) is activated when there is a higher energy demand to move the car.

Electric-assisted or mild HEVs use the electric motor only during heavy operation demands; for example, when a high acceleration is needed.

When referring to HEVs, the main category that comes to mind is the plug-in electric vehicle (PEV), due to the fact that they can be recharged from a regular electric plug or charged on the go from the engine. These vehicles depend mainly on the electric motor that is powered by the Li-ion battery energy storage and use the gasoline engine in case of emergency to maintain operation during discharge of the battery.

HEVs, therefore, provide the same performance as conventional vehicles, but through a significantly reduced dependence on fuel. Their environmental friendliness is not only attributed to the consumption of less fuel but also to the lower carbon emissions that are achieved through the need for smaller ICEs and the use of energy-saving technologies, such as regenerative braking, by which energy is captured during the braking of a HEV to recharge the battery. Moreover, there is a significant noise reduction in HEVs in comparison to internal combustion vehicles.

The first HEV was the Prius developed by Toyota in 1997, but it was not released worldwide until 2000, making Honda's Insight the first hybrid to be released in the US (1999) in the modern era. A little over a decade later, race car companies unveiled their HEVs (Ferrari Laferrari, McLaren P1, and Porsche 918 Spyder), which allowed for top speeds over 321 $km\,h^{-1}$.

Electric vehicles are the more evolved version of HEVs as they operate solely on an electric propulsion system and do not require any form of fuel. The first commercial electric EV of the modern era became available in 1996 by GM (EV1) and was powered by lead–acid batteries, leading to a driving range of 100 mi, while later Ni–MH batteries allowed for a driving range of 140 mi. Its high cost and low sales resulted in it going out of production in 1999. To avoid repairs of future defects, GM recalled all EV1s in 2003.

Although commercially available Li-ion cells are not the most appropriate for EVs, the first commercial EV powered solely by Li-ion cells, the Tesla Roadster was manufactured in 2008, by enabling elaborate power-control and monitoring electronics for the battery cells. The Tesla Roadster battery is a custom microprocessor-controlled Li-ion battery comprised by 6831 individual cells and weighs 992 lb. A full charge takes about 3.5 hours using the Tesla Motors High Power Connector, which is plugged in a normal household plug [7]. The Tesla Roadster can accelerate from 0 to 60 mph in less than 4 seconds and has an electronically limited top speed of 125 mph. The range of the vehicle under a full charge is about 220 mi (based on EPA combined city/highway cycle) and the battery life is 5 years or 100 000 mi. Initially, Tesla produced luxury electric sports cars, but in 2012, its sedan production began with the Model S, which used over 7000 individual standard 18 650 lithium cells manufactured by Panasonic, allowing for a driving range of 412 mi. The improved battery performance and safety features are attributed to application of a more advanced cathode material, which is nickel–cobalt–aluminum (NCA) with enhanced compaction density, and more efficient power-management technologies. The 85D model released in 2014, with an 85 kWh battery module, has a range of 270 mi; it takes 4.4 seconds to accelerate from 0 to 60 mph and the maximum speed is 155 mph (249 km h^{-1}) [7]. The progress of the Li-ion batteries employed for Tesla EVs is illustrated in Table 4.2 [8].

Following, Tesla's example, almost all major car companies have been producing EVs, including: Chevy's Volt by GM released in 2008, whose 2019 version has a mileage range of 259 mi and can reach 0–60 in 6.5 seconds [9]; Mustang's Mach E, whose premium model has a range of 300 mi and reaches 0–60 mph in 4.8 seconds [10]; the Volvo XC40 with a 208 mileage range and 4.7 seconds to reach 0–60 mph, and Volkswagen's ID.4 [11], which has a good mileage range of 250 mi but requires 7.6 seconds to get from 0–60 mph. The Nissan Leaf requires even more time, 8.4 seconds to go from 0 to 60, and its mileage range is 226 [12].

European brands, such as BMW, Mercedes, and Audi, also have EVs. BMW has released many EV models, including the i3, i4, and i8, and provides PHEV versions

Table 4.2 Different generations of Panasonic 18650 cells, which have been or may be used for Tesla EVs.

Model	**Early NCR 18650 (before 2006)**	**NCR18650 A (2009)**	**New 18650 (2012)**	**Latest 18650 (2014)**	**21700 (2016)**
Cathode	$LiNiO_2$	NCA	High-density NCA	High-density NCA	High-density NCA
Anode	Graphite	Graphite	Graphite	Si/C composite	Si/C composite
Maximum capacity (Ah)	2.6	3.1	3.4	4.0	4.8
Mass (g)	~44	~44.5	~46	~54	~68.6
Discharging voltage (V)	3.6	3.6	3.6	3.4	3.6
Charging voltage (V)	4.2	4.2	4.2	4.2	4.2
Energy (Wh)	10.4	11.2	12.2	13.6	17.5
Volume energy density (Wh l^{-1})	620	675	730	800	722
Mass-energy density (Wh kg^{-1})	236	252	265	252	255

for i3, i5, and SUV models. Its latest EV midsize SUV model ix3 released in 2021 has a mileage range of 310 and 6.8 seconds 0–60 mph. Mercedes released the EQ series EVs beginning in 2018, among them the midsize SUV EQC model has a mileage range of 257 and 5.1 seconds 0–60 mph. Audi also released the E-tron series, including both EV and PHEV versions.

The aforementioned EV models above mainly employ $Li_x(Ni_xCo_yMn_z)\ O_2$-based Li-ion cells. With growing HEV/EV demand, the battery industry also experienced rapid development and competition over the past 10 years. Until 2021, China, Korea, and Japan dominated the Li-ion battery market, and the top 10 battery companies occupied over 90% of the market share. The biggest battery company is CATL from China, which supplies almost all car companies, and occupies around 24% of the global market. LG Chem is the second-largest company and holds a 23.5% market share. Other big battery companies include Panasonic, which mainly supplies Tesla with 18650/21700/46800 cylindrical cells, and SK innovation, which mainly supplies Hyundai & Kia, Daimler, and Ford. Another Chinese company EVE Energy Co., Ltd. supplies BMW, Daimler, Land Rover, and Jaguar, while Gotion & A123 supplies Volkswagen.

In order to beat the competition, car companies have taken the approach of setting up plants to produce in-house Li-ion batteries. Tesla was the first who began this initiative and now Volkswagen and GM are following, with the latter claiming to be fully electric by 2025.

A significant boost took place in the EV market between 2015 and 2021, which is reflected not only by the rapid growth of EV sales but also by the numerous new EV brands and models. Table 4.3 shows the market sales of different EVs from 2015 and forecasts this market to 2025. The global sales of EVs compared to conventional vehicles was 1% in 2015 and rose to 9% in 2021. The global HEV/EV production is predicted to grow at a compound annual growth rate (CAGR) of 21% and is expected to be 40 million units by 2030.

Aside from environmental awareness, a key factor that has allowed for this booming in the EV market is that battery technologies continuously improved in the past five years. Although the Tesla models, S and X, use cylindrical 18 650 or 21 700 Li-ion cells, they are not the best option. 18 650 and 21 700 refer to the cell

Table 4.3 Market sales of different EVs from 2015 and forecast of this market by 2025.

Product	PHEV	EV	E-Bus
LIB average capacity	11.7 kWh	33–100 kWh	80–200 kWh
Capacity ratio	1200	5000	15 000
2015 sales	220 000 ea 2.5 GWh	320 000 ea 10.8 GWh	120 000 ea 9.3 GWh
2019 sales	530 000 ea 6.6 GWh	1 600 000 ea 90.0 GWh	77 000 ea 15.0 GWh
2025 forcast sales	1 790 000 ea 26.2 GWh	14 700 000 ea 998.6 GWh	210 000 ea 39.0 GWh

dimensions, which are 18-mm diameter/65-mm length for the 18 650 and 21-mm diameter/70-mm length for the 21 700. Their electrode and electrolyte chemistry (and therefore capacity/energy density) can differ based on the manufacturer.

The auto industry has been long searching for the best type of Li-ion cells to solve the short-mileage range issue. According to the form of packing, Li-ion cells for cars are typically classified into three types, which are cylindrical cells, prismatic cells, and pouch cells. Cylindrical cells were widely used in the early Tesla models. The advantages are that the 18 650 and 21 700 have fixed dimensions, and therefore all vehicle models can use them without having to make changes in the cell production line, which dramatically reduces production costs. Tesla is currently developing a larger cylindrical cell of size 4680 (46 mm diameter/80 mm length).

Cylindrical cells have some significant disadvantages, with the biggest one being that a very limited volume of the cell pack is utilized due to the geometric characteristic of their cylindrical shape. Another disadvantage is that cylindrical cells have a very low unit-cell capacity; therefore, many cells are needed for a single vehicle. For the 18 650 cells, the highest capacity of a single cell is around 3.5 Ah, and for the 21 700 cells, it is around 5 Ah. Hence, to provide enough electricity for a single vehicle, it requires over 6000 cells, resulting among other things in a very challenging battery-management system (BMS). Therefore, prismatic and pouch cells are promising alternatives. Prismatic cells can be closely packed and efficiently utilize pack space to allow more electricity to be provided. The drawback of prismatic cells is that they have a low weight energy density, which leads to a high total weight of the battery pack, which increases the energy consumption rate of the vehicle. In contrast to cylindrical and prismatic cells, pouch cells use an aluminum-plastic film as their shell, instead of pure aluminum or steel. Aluminum-plastic films are a very light composite material for cell packing, since the several super thin plastic and aluminum layers have a total thickness 100–150 μm. Such a lightweight shell material helps enhance the energy density of pouch cells, i.e. decrease their weight. Prismatic and pouch cells can be made available in many sizes. First, the cell pack dimensions are determined according to the vehicle chassis, and then the module dimensions are designed according to pack dimensions, finally, the cell dimensions are determined according to the module dimensions. Such a design strategy offers prismatic and pouch cells of very high space efficiency, and they can provide more electricity within the limited space below the chassis; however, the drawback is that each chassis must match a specific cell dimension. This results in high production costs for cells, as each process of cell production requires the construction of multiple tools for fabricating the different cell dimensions. Some features of the three forms of cells are listed in Table 4.4.

In summary, each form of cell has its advantages and disadvantages, and the best solution for future EVs remains an open argument.

In order to develop superior EVs, industry turned towards using different material chemistries for the electrodes, than those employed in cells for smartphones and laptops. $LiCoO_2$ electrodes used in commercial Li-ion cells wear out after a couple of years, and they become unstable if overheated or overcharged. So, despite these cells being energy-storage efficient, their tendency to exhibit thermal runaway has

Table 4.4 Various forms of Li-ion cells and their features.

	Cylindrical	**Prismatic**	**Pouch**
Capacity	2.6–5.2 Ah	15–25 Ah for HEV 40–100 Ah for EV	8–25 Ah for HEV 40–80 Ah for EV
Highest weight energy density	240 Wh kg^{-1}	270 Wh kg^{-1}	310 Wh kg^{-1}
Cell costs	Low	Mid	High
Auto company selection	Tesla, Xpeng	Tesla, Volkswagen, BMW, General motor,	Volkswagen, Daimler, Nissan, Hyundai & Kia, General motor, Ford

led the industry to employ alternative cathodes for EV applications. In particular, Co-based cathodes experience oxidation, upon charging, of Co^{3+} to unstable Co^{4+}. Phosphate-based cathodes, however, undergo oxidation of Fe^{2+} to the stable Fe^{3+} resulting in a safer, fault-tolerant, chemistry. Therefore, $LiFePO_4$ cathodes are predicted as very promising for use in EVs, as they not only provide the required capacity and have a lifetime of over 3000 cycles, but they also improve the safety during charging. This is important for EVs where safety issues and power capacity are more important than energy capacity. As will be described in Chapter 5, $LiFePO_4$ was successfully employed as a cathode much later than $LiCoO_2$, and that is why it was not until after 2010 that electric vehicles became widely popular.

In this connection, it should be noted that Li-ion batteries operate best and safest within a charge-state window, which is typically 10–90%; hence, engineers can exploit this fact by charging to a percentage of the maximum capacity. It follows, therefore, that the technology for monitoring the charge status of the Li-ion cells is crucial for the safety and the operation of the EV. For each type of cell chemistry, the open-circuit voltage versus state of charge curve (OCV–SOC map) is known and in pouch cells of electric vehicles, the BMS can monitor the cell module voltage and match it with the OCV–SOC map, determining the charge status of cells and if there is any cell with an abnormal low voltage. This is significantly more difficult than monitoring the gas in the vehicle tank.

Despite the fact that EVs are zero-emission vehicles (the only emission is thermal energy), concerns have risen about the use of electricity for charging the batteries that power them. In the USA, 33% of the nation's electricity is generated by burning coal and 33% is generated by natural gas [13]. According to the Energy Information Administration, a gallon of gas produces 19.564 lbs of CO_2 from an internal combustion engine. A recent analysis by the Automotive Testing and Development Services found that for every 100 mi of travel, a Tesla Roadster [7] needs to be recharged with 31 kWh of electricity. Generating a kilowatt-hour of electricity produces an average of 1.55 lb of CO_2. These results will vary according to the use of coal as the main source of electricity per state. In high coal-dependent states, the carbon emissions for the Tesla Roadster recharge may be as high as the carbon emissions from an ICE car. In the technical analysis of the Pacific Northwest National Laboratory [14], it is pointed out that there is plenty of surplus electricity that goes unused at night and this might change with the commercialization of the electric cars without significantly affecting the carbon emissions. Nevertheless, a significant difference in CO_2 emissions between traditional ICE vehicles and EVs exists in states where renewable energy sources are the main electricity source as opposed to coal. Therefore, to take full advantage of the preferred environmental features of EVs, renewable energy source technologies, such as solar cells, must be advanced and reinforced. For example, optional solar cells may be installed on the rooftop of the vehicle to provide additional recharging for the batteries during sunlight.

Currently, a more generalized definition that is widely used is the PEV. This refers to any vehicle that plugs into the electric grid for all or part of its power source. PEVs include battery-electric vehicles (EVs), such as the Nissan Leaf and Tesla Model S, the Chevy Volt (also referred to as a PHEV; plug-in hybrid-electric vehicle), or extended-range electric vehicles (EREVs), such as the BMW i3.

A predicted trend for the future vehicle market is large-scale commercialization of electric vehicles, due to environmental concerns. According to reports based on historical vehicle sales, it was predicted that from 2013 to 2023 the cumulative sale of EVs in the United States would rise from 1.8 to 7.3 million vehicles [15]. Figure 4.2 shows actual sales of PEV from 2019 to 2020, and a prediction from 2021 to 2025. It can be seen that the road presence of PEVs is predicted to grow fast, and to reach 14.5% in 2025.

In conclusion, it should be noted that a recent study by the Electric Power Research Institute and the Natural Resources Defense Council suggests that electric vehicles could eliminate billions of tons of greenhouse-gas emissions between 2010 and 2050. A study by General Electric indicates that if half the vehicles on the road in 2030 are electric-powered, petroleum consumption in the United States will shrink by six million barrels a day.

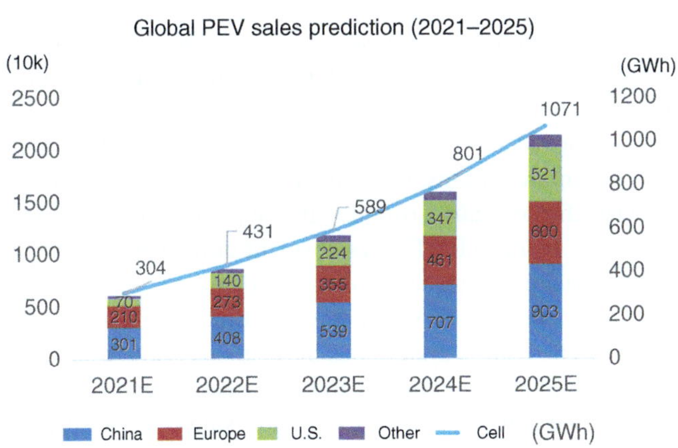

Figure 4.2 Global PEV sales prediction.

4.3 Aerospace Applications

When talking about aerospace engineering, we primarily refer to aircraft-related technologies. Aircrafts could be divided into two classes based on their working environment, which are endoatmospheric aircraft and exoatmosphere aircrafts, respectively. The latter are most commonly referred to as spacecrafts. The energy storage system (ESS) is a critical component for all modern aircrafts, and the requirements for its performance and reliability presented by the aircraft industry are quite strict. The ESS is responsible for providing the startup power for the engines and the working power for the main computer systems. It also provides emergency power when the engine is not in operation. It, therefore, follows that the ESS is particularly important for exoatmosphere aircrafts (such as satellites) as some of them are solely powered by solar energy obtained from photovoltaic modules. As satellites travel along their orbits and they enter dark regions where they are shadowed by the earth the power supply will be disrupted and unstable; hence, they must employ very reliable energy sources. Currently, most of endoatmospheric aircrafts still apply nickel–cadmium (Ni–Cd) and lead–acid batteries since they have been commercialized for over 40 years and fully developed. However, the aforementioned unique advantages of Li-ion batteries make them outstanding power source candidates for future aircrafts and spacecrafts. These advantages include high specific energy and power density, wide working temperature, long shelf life, and the most important one, lightweight. Although the ability to reduce the volume and mass of the energy source without reducing its output is highly desirable for both endoatmospheric and exoatmosphere aircrafts, the latter are especially sensitive to the weight of the power sources since launch costs increase significantly with liftoff load. With lighter power sources, more space and payload could be saved for instruments, allowing, therefore, for greater flexibility in designing the aircraft and making it more efficient.

Many research institutes and laboratories focus, hence, on the development of Li/Li-ion batteries for aviation engineering applications. Examples of these applications are given below.

One of the early aviation applications of Li-ion batteries was in the Lange Antares 20E; the glider built by Lange Aviation. The first prototype flew in 2003. It is a special aircraft as it applies a 42-kw (56 horsepower) electric motor, which is fully driven by the propulsion system consisting of 72 individual Li-ion cells in series made by Saft (Saft VL 41 M cells) [16]. The Antares is considered the first aircraft with an electric propulsion system to get a certificate of airworthiness. It is able to reach an altitude of up to 3000 m by using solely Li-ion cells. The total weight of the battery system is approximately 76 kg, the capacity of each cell is 39 Ah (with an energy density of 136 Wh kg^{-1}), and the operating voltage is 3.6 V. The charging time is eight hours and the charging voltage is 220 V. The battery system has up to 1500 charge/discharge cycles [17]. The difference between these batteries and those employed by Tesla is the cathode material. The Lange Antares employs a nickel oxide cathode instead of a NCA ternary cathode, resulting in a much lower energy density. A schematic diagram of a hybrid-electric propulsion plane is shown in Figure 4.3, where it is seen

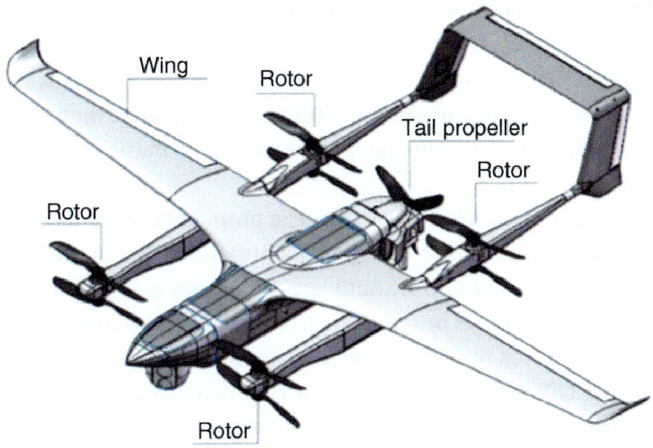

Figure 4.3 Hybrid-wing VTOL UAV; Source: From Zong et al. [18], MDPI.

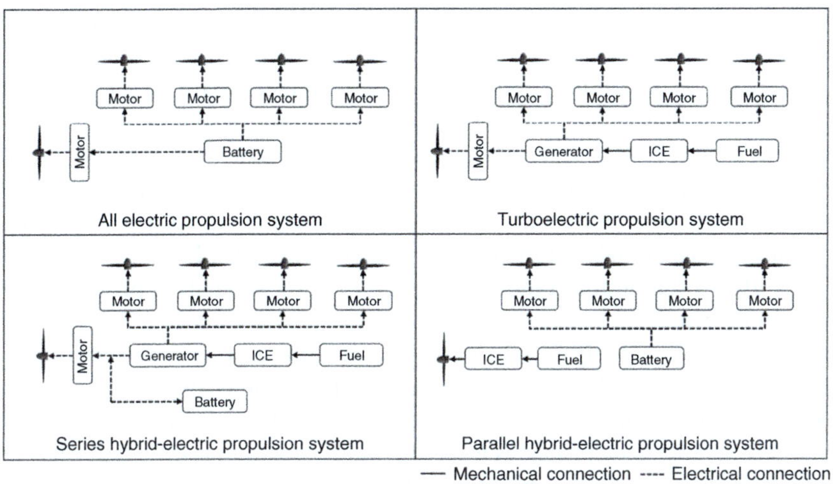

Figure 4.4 Schematic diagram of electric propulsion configurations; Source: Zong et al. [18]/MDPI.

that multiple rotors are used [18]. Figure 4.4 further illustrates how a hybrid eVTOL is configured using both electric rotors and an internal combustion engine [18].

In June 2005, Thales announced that they will incorporate lithium-ion batteries made by GS Yuasa Lithium Power (GYLP) into Boeing's next-generation commercial airliner, the 787 Dreamliner. Batteries play a significant role in planes as they not only supply electricity when the engine is not running but are also used in the auxiliary power unit (APU), which is the backup power system. GYLP is a commercial lithium-ion battery manufacturer that focuses on providing power source solutions for aerospace, military and undersea vessels. This contract is considered a historic first as it marked the first commercial aviation application of Li-ion batteries [17]. The chosen batteries, LVP-10 and LVP-65, not only offer nearly twice the energy that traditionally used Ni–Cd batteries do but also have a fast charging time of 75 minutes from 0% to 90% [19]. GYLP also claims compatibility with extreme operating conditions. In 2008, Saft announced that Airbus had chosen its Li-ion battery systems as the starting and emergency power supply for their A350 aircraft [16]. Since then, the largest two commercial aircraft manufacturers have joined the new battle initiated by upgrading their power systems.

Recently, there have been significant advances in electric propulsion systems and various fully electric aircrafts (with a small passenger capacity) have been manufactured that can fly short distances. Their applications are aimed towards providing transportation to small islands or airports in large cities, as they can achieve a limited flight time. Their design requires multiple propellers, and it has to be ensured that even if one fails due to a battery malfunction the aircraft can safely continue its journey. In 2019, MagniX teamed with the small commuter airline in British Columbia Harbour Air to develop an electric version of the DHC-2 Beaver, which is of the most popular small planes for reaching remote locations. A flight time of 15 minutes could be achieved. MagniX also teamed up with AeroTEC to manufacture the eCaravan and in 2020, it became the largest fully electric plane to fly. It employs Li-ion cells that weigh 2000 lb (560 kW) and is able to fly a distance of 160 km (100 mi) in 30 minutes. Eviation has also announced that its electric luxury plane Alice should be operational soon. It will operate using a ~8000 lb Li-ion battery (820 kWh), using a NMC (nickel–manganese–cobalt) cathode. It is anticipated that with nine passengers and two crew members, it can travel for ~650 mi (1046 km) at speeds of ~250 mph (402 km h^{-1}). It should be noted that, in 2020, a Li-ion battery used for Alice exploded at the Prescott Regional Airport, where the plane was located [20]. Rigorous testing, therefore, needs to be done before these planes can be airborne for testing.

Recently, the air taxi company Joby Aero was able to fly its prototype electric plane for 150 mi in 1 hour and 17 minutes. It used commercial Li-ion batteries that employ a NMC cathode and graphitic anode. This aircraft is different than the aforementioned electric propulsion planes, as it uses multiple rotors and vertical take-off and landing (eVTOL) takes place, while during flight time, it has a horizontal position. Its capacity will be four passengers and the pilot. Archer Aviation has also unveiled an eVTOL aircraft, called Maker, however, it has not demonstrated that it is in flight mode. Powered by a 75 kW Li-ion battery, it will supposedly have a range of 60 mi, cruising at a top speed of 150 mph (241 km h^{-1}). In addition to the environmental friendliness of such electric aircrafts, the noise reduction they offer is what makes them truly attractive for use in large cities. Archer claims that Maker is: "100X quieter than a helicopter, generating only 45 decibels of sound."

As previously mentioned, lithium-ion batteries have attractive applications for exoatmospheric aircrafts, especially satellites, due to their ability to be a stable power source in combination with solar energy harvesting and to reduce aircraft weight. As the life duration of satellites continuously increases, non-rechargeable batteries could no longer meet long-term power needs, and since the 2000s, an increasing shift from Ni–Cd and lead-acid-based power systems to Li-ion battery systems has occurred. Saft is the world's leading designer and manufacturer of

Li-ion batteries for satellites and began providing Li-ion batteries for satellites in 2002. In 2013, it announced that it will equip with its Li-ion batteries the Inmarsat 5 F-1 satellite. This satellite was the 100th satellite that applied Saft Li-ion batteries, which are the VES 140, VES 180, and VL 48E cells. The voltage of these modules ranges from 4 to 100 V with a capacity from 5.8 to 52 Ah [16]. Another large manufacturer of Li-ion batteries for satellite use is Quallion, which made its landmark in the satellite battery business in 2006 when it was awarded the US government contract for the Title III program [21], for producing the cathode and anode materials used for satellite applications. The large cell manufacturing line was set up in 2013 and Quallion was able to produce the latest 1.8 Ah Li-ion cell for aerospace applications. The cell's cycle life is over 20 000, its specific mass-energy is 105 Wh kg^{-1} and its specific volumetric energy is 215 Wh l^{-1}.

In addition to satellites, Li-ion batteries are also being considered for space stations and other spacecrafts, but there remain unresolved issues that must be addressed.

4.4 Medical Applications

4.4.1 Heart Pacemakers

Another area that promises many applications for next-generation rechargeable Li-ion batteries is that of biomedical implantable devices, such as pacemakers. Pacemakers comprise of a battery power source that creates electric impulses (periodic electric stimulations) that are carried to appropriate parts of the body through electrodes.

The initial application of pacemakers was to regulate the beating of the heart (treat cardiac arrhythmia disorders) by delivering the electrical impulses to the heart muscles. The first fully implanted pacemaker procedure took place in 1957 at the Karolinska University Hospital in Sweden by Åke Senning and the pacemaker was powered by a secondary Ni–Cd battery, designed by Rune Elmqvist. The Ni–Cd cell, however, was replaced by primary batteries, since the recharging process was not very efficient, as it was performed externally through an appropriate vest that provided electromagnetic waves [22]. The primary cells used, thereafter, were toxic, such as the ceramic plutonium oxide battery and the mercury battery.

The real breakthrough in implantable cardiac pacemakers took place in 1972 when the primary Li–I battery was invented by Wilson Greatbatch and his team. This type of battery has a life span of over 10 years and continues to this day to be the power source for many cardiac pacemakers. In addition to its long lifetime, the Li–I cell has a very low self-discharge rate, which results in a long shelf-life until the pacemaker is implanted; it also has a stable voltage and consumes its energy in a gradual and predictable manner. This is important for the follow-ups of patients, where the status of the battery has to be checked and the replacement needs to be programmed. Moreover, the battery produces no toxic waste or gas and is easily encased in a noncorrosive material. An additional significant benefit is that the size of these devices was reduced because of the higher capacity of the lithium batteries used. As a relatively new member in this family, Li carbon monofluoride (CF_x) cells have been commercialized for cardiac pacemakers since 1996 [23], and have recently been reported to offer higher energy density and pulses at currents above 20 mA. Moreover, the whole battery system is encased in a titanium casing, allowing a 50% reduction in weight over the same size Li–I battery, without any compromise in the safety. Nowadays, Li–CF_x batteries are commercially available in various packing sizes, including cylindrical batteries and cell/button batteries.

Modern heart pacemakers consist of a primary Li-battery power source, electrodes that are attached on the myocardium of the heart, a sensing amplifier that processes the electrical manifestation of naturally occurring heartbeats as sensed by the electrodes, and the computer logic that delivers the pacing impulse to the electrodes. Most commonly, the pacemaker (with the exception of the electrodes that are placed on the myocardium) is placed below the subcutaneous fat of the chest wall, above the muscles and bones.

Heart pacemakers have evolved to being able to treat other heart problems, in addition to arrhythmias, such as those consisting of multiple electrodes, that can stimulate different positions within the heart to improve synchronization of the lower chambers of the heart. Furthermore, modern pacemakers can be programmed externally, once they are implanted, allowing the cardiologist to adjust the pace according to the individual patient's needs. Some pacemakers are dual-functional as they can also act as a defibrillator.

4.4.2 Neurological Pacemakers

Since mid 2000, pacemakers have found significant applications in neurology as well. Neurological disorders result from chemical imbalances that cause parts of the brain to send out the wrong electric signals. By implanting a pacemaker in the upper chest, electrodes can carry out the proper electric impulses to the dysfunctional parts of the brain and interfere with disordered neural activity, treating, therefore, various, incurable otherwise, disorders such as epilepsy, severe depression, tremor, Parkinson's disease symptoms, and chronic pain.

This treatment is called deep-brain stimulation (DBS). The components of DBS are: the implanted pulse generator (IPG), the lead, and the extension. The IPG consists of the primary Li-battery, which is encased in titanium, and generates the electric pulses; it is implanted below the clavicle or in some cases at the abdomen. The lead is a coiled wire insulated in polyurethane and has four Pt–Ir electrodes for carrying impulses to the damaged region; they are placed, therefore, directly in the brain by drilling a hole in the skull. The extension is an insulated wire that connects the lead to the IPG; it, therefore, runs from the head, down the side of the neck, and behind the ear to the IPG. The advantages of DBS are that the electric pulses can be calibrated externally depending on the patient's needs and if side effects appear during operation, they completely disappear by turning off the battery. The underlying mechanisms by which DBS functions are not fully understood; however, it is documented that it can change brain activity in a controlled manner, and once the battery is switched off the symptoms resume. Complications during implantation, such as brain bleeding, can occur, however, this is not common.

The USA Food and Drug Administration (FDA) allowed the use of DBS for the first time in 1997 for the treatment of tremor, and has proven to be very efficient; only 40% of tremor patients can see improvements through conventional medicine. FDA approval was also given for the treatmet in 2002. Another area in which DBS has potential is depression. It has been observed that 10% of depression patients see no improvement by using antidepressant medicines, psychotherapy, and not even electroshock therapy. An experimental study in 2005 showed that four out of six such patients saw great improvement with DBS [24], however, FDA approval has not been given. DBS is also used in treating chronic pain, in which cases the pacemaker can be placed on the spine.

The most miraculous perhaps application of DBS was reported in 2007. A 38-year-old male had been in a vegetative state after his skull was severely damaged during a robbery and he was left for dead. After extensive brain surgery, he survived but the doctors gave him zero chances of recovery. After six years of being in a vegetative state, neurosurgeon Dr. Ali Rezai and his team at the Cleveland Clinic Foundation were able to activate parts of the patient's brain through DBS, by sending appropriate electrical impulses. As a result, the patient was able to regain his consciousness, recognize and talk with his family, and perform basic functions by himself [25].

Although all medical implantable devices make use of primary Li batteries, instead of secondary Li cells, the aforementioned medical applications are summarized as they illustrate the unique possibilities that Li-powered devices offer in treating incurable, otherwise, diseases. Nonetheless, the current primary Li-battery technology is considered to be deficient for DBS applications because of its short battery life. This deficiency could be alleviated if the primary battery is replaced with secondary batteries. Unfortunately, current commercial secondary Li batteries cannot be used in medical implantable devices due to safety issues. Therefore, it is of great significance that new material chemistries for rechargeable Li batteries are developed, not only for economic and environmental reasons but also for improving directly the quality of human life, as it will allow for great improvements in pacemaker applications in medicine and the replacement operations will be diminished.

4.5 Grid Energy Storage

Clean energy resources have attracted a lot of attention within the last decades, e.g. solar energy, wind energy, tidal energy, and biomass energy. However, these clean energy resources have not been widely commercialized, partially due to stability issues. A continuous and stable power supply from these energy resources highly depends on environmental conditions. For example, solar energy harvesting systems are able to provide a continuous and stable energy supply only when sunshine is available, that is, mostly during daytime; wind energy harvesting systems are operational only when wind is powerful and stable. What if customers do not need much power supply when these clean energy-harvesting systems are functional, and they do when the clean energy harvesting systems are not? Also, unstable power output will result in a pulse and negative impact on power grids. The solution to this issue is energy storage technologies. The demand for large-scale and highly efficient energy storage technologies arises as an auxiliary to applying clean energy technologies. Energy storage technologies are based either on physical or chemical methods. One representative physical technology is pumped hydro storage, which uses surplus electrical pumps to drain water and store it in a higher place. In this way, surplus electricity is physically changed into potential energy. However, such systems cannot be widely used, since sufficient hydro resources are necessary and constructions costs are very high. Batteries with various chemistries can also be used to transfer surplus electricity into chemical energy, with a historical example being vanadium flow batteries, but they are also very expensive and their chemistry is not environmentally friendly. As Li-ion batteries with higher energy density and reliability are being developed, grid energy storage systems based on them are the most promising. According to statistical information given by China, the total amount of electricity power supplied by chemical energy storage systems (EPSCESSs) from 2010 to 2014 was 840.3 MW, with an annual growth rate of 135% [26]. The U.S. and Japan, which ranked Nos. 1 and 2 in this market, respectively, occupied over 80% of the market. Sodium–sulfur battery systems used to be the most widely applied technology, however, with continuous technical breakthroughs, Li-ion batteries are growing fast and are believed to be the most promising technology in the next-generation energy storage systems. From 2013 to 2014, the percentage of the systems that applied sodium–sulfur batteries in the EPSCESS was 40% and it decreased by 6%, while that of lithium-ion battery systems was 35% and it increased by 15%. Table 4.5 illustrates some important investment projects of large-scale chemical energy storage systems. Na-ion batteries are also promising for grid storage, but a stable and economical system has yet to be developed.

Table 4.5 Several investment projects for large-scale chemical energy-storage systems initiated in 2014.

Project name	**S&C Electric Company, energy storage system(ESS) project, United Kingdom**	**BYD Auto Co., Ltd., China State Grid project, China**	**India energy-storage array project, India**	**Renewable Energy Systems Americans Inc., Elwood Energy-Storage Center project, Joliet, U.S.A.**
Battery type	Lithium-ion battery	Lithium-ion battery	Vanadium-flow battery	Lithium-ion battery
Installed capacity	10 MWh	20 MWh	10 MWh	20 MWh
Supplier	Samsung SDI	BYD Company Ltd.	Panasonic India Pvt. Ltd and AES India Private Limited	BYD Company Ltd.

One limitation of large-scale energy storage systems based on Li-ion cells is their high cost. Solar (as well as wind) power stations are highly sensitive to operational costs, and the industry has concluded that only if the price of the energy-storage system drops below \$0.23 Wh^{-1}, renewable resources can be beneficial.

An ideal cathode material that seems promising in solving this issue is $LiFePO_4$ since it allows the Li-ion battery system price to drop from around \$0.33 Wh^{-1} in 2015 to around \$0.1 Wh^{-1} in 2021, which is more than sufficient to fulfill the demand for renewable energy resource costs. Moreover, the cycle life of Li-ion cells keeps improving and it can reach around 8000 cycles, which dramatically reduces operational costs. As technical breakthroughs solve the economical bottlenecks, the presence of Li-ion based power energy storage systems increased rapidly from 2014 to 2020. It is reported that in 2020 around 1.9 GWh energy-storage systems were built globally, and this number is three times higher than it was in 2014. Industry predicts that this fast growth and high demand will continue for the next 10 years, and by 2025 Li-ion based global energy storage systems will reach around 33 GWh, with an annual growth rate of 95%.

In addition to large-scale chemical energy storage systems designed for industrial purposes, microscale energy storage systems combined with clean energy harvesting systems, have recently received much attention, as a promising candidate for stationary applications. The Tesla PowerWall is such an example, which is a home power supply system that charges using electricity generated from solar panels, or when utility rates are low, and powers the house in the evening and night. It also fortifies the house against power outages by providing a backup electricity supply. A 13.5 kWh model is available for daily cycle applications and it could provide 5 kW continuous and 7 kW peak power [27]. It is believed that the PowerWall will apply Li-ion batteries produced by Panasonic with a nickel–cobalt–aluminum (NCA) or nickel–manganese–cobalt oxide (NMC) cathode and a carbon anode. Orison also released, in 2016, the Orison Panel and Orison Tower energy storage systems for home use, both with a 2.2 kWh capacity. Microscale energy storage systems are also needed as the backup power supply for uninterruptible power supply for telecommunication towers and computer systems with a demand for high-data security.

4.6 Conclusions

This chapter gave an overview of current and future applications of Li based energy storage devices. Expanding the usage of Li-ion and other secondary battery systems will ensure a sustainable environment with limited CO_2 emissions. In order to achieve this, it is necessary to address the key issues that are presented in the following chapters.

Acknowledgments

KEA is grateful to the National Science Foundation for supporting this work through the CMMI grant [CMMI-1762602], and IRES Track II grant [1854528].

References

1 Nazri, G.-A. and Pistoia, G. (ed.) (2009). *Lithium Batteries, Science and Technology*. Berlin: Springer.

2 Croce, F., D'Epifanio, A., Hassoun, J. et al. (2003). Advanced electrolyte and electrode materials for lithium polymer batteries. *J. Power Sources* 119: 399.

3 Motorola Cell Phone http://www.motorola.com (accessed December 2021).

4 Samsung SDI Co., Ltd. (2021). www.samsungsdi.com/lithium-ion-battery/mobile-phone (accessed October 2021).

5 Apple (2021). www.apple.com/macbook-air/specs.html (accessed October 2021).

6 SNE research, global battery market outlook and key issues, 2020 KABC. http://www.sneresearch.com/_new/eng/sub/sub1/sub1_01.php?sub_cat=2 (accessed November 2021)

7 Tesla Motor (2021). www.tesla.com (accessed October 2021).

8 www.gg-lb.com (2021). www.gg-lb.com/asdisp2-65b095fb-13403-.html (accessed October 2021).

9 Chevrolet (2021). www.chevrolet.com (accessed October 2021).

10 Ford Moter Company (2021). www.ford.com (accessed October 2021).

11 Volkswagen (2021). www.vw.com (accessed October 2021).

12 Nissan (2021). www.nissanusa.com (accessed October 2021).

13 Independent statistics and analysis, https://www.eia.gov/electricity/monthly/epm_table_grapher.cfm?t=epmt_1_1 (accessed October 2021).

14 Kintner-Meyer, M. (2007). *Impacts Assessment of Plug in Hybrid Vehicles on Electric Utilities and Regional U.S. Power Grids*. Pacific Northwest National Laboratory.

15 Block D. and Harrison J. (2014). Electric Vehicle Transportation Center Report on Electric Vehicle Sales and Future Projections. EVTC Report Number: EVTC-RR-01-14.

16 www.saftbatteries.com. (accessed July 2015).

17 www.gsyuasa-lp.com/content/thales-selects-gs-yuasa-lithium-ion-battery-system-boeing%E2%80%99s-787-dreamliner (accessed July 2015)

18 Zong, J., Zhu, B., Hou, Z. et al. (2021). Evaluation and comparison of hybrid wing VTOL UAV with four different electric propulsion systems. *Aerospace* 2021 (8): 256.

19 Anonymous (2008). A350 XWB equipped with Saft lithium-ion battery systems. *Aircraft Engineering and Aerospace Technology*. 80: 4.

20 https://www.flightglobal.com/airframers/eviation-alice-fire-involved-lithium-ion-batteries-which-ignited-after-hours-of-powerplant-tests/141228.article (accessed July 2021).

21 Reyes, A. (2013). Quallion announces 1.8 Ah lithium-ion cell for aerospace/military applications. www.quallion.com (accessed July 2015).

22 Aquilina, O. (2006). A brief history of cardiac pacing. *Images Paediatr. Cardiol.* 8: 17.

23 Greatbatch, W., Holmes, C.F., Takeuchi, E.S., and Ebel, S.J. (2006). Lithium/carbon monofluoride (Li/CF*x*): a new pacemaker battery. *Pacing Clin. Electrophysiol.* 19: 1836.

24 Mayberg, H.S., Lozano, A.M., Voon. V. et. al. (2005). Deep brain stimulation for treatment resistent depression. *Neuron* 45: 651.

25 Schiff, N.D., Giacino, J.T., Kalmar, K. et. al. (2007). Behavioural improvements with thalamic stimulation after severe traumatic brain injury. *Nature* 448: 600.

26 CCid think tank. Ministry of Industry and Information Technology of the People's Republic of China.

27 Tesla Motor (2021). www.teslamotors.com/powerwall (accessed September 2021).

5

Cathode Materials for Lithium-Ion Batteries

Pu Hu[1], Lintao Dou[1], Tao Huang[2], and Aishui Yu[2]

[1] *Wuhan Institute of Technology, School of Materials Science and Engineering, LiuFang Campus, No.206, Guanggu 1st road, Donghu New & High Technology Development Zone, Wuhan 430205, China*
[2] *Fudan University, Department of Chemistry, No.220, Handan Road, Shanghai 200438, China*

In the first secondary Li batteries that were developed in the 1970s, pure Li was used as the anode. The cathode material, hence, was required to efficiently intercalate Li-ions. In 1965, Walter Rüdorff illustrated that TiS_2 could intercalate Li, and in 1976, due to the significant contributions of Stanley M. Whittingham, the first rechargeable Li-battery was developed at Exxon, using a Li anode, a TiS_2 cathode, and $LiClO_4$ in dioxolane as the electrolyte. However, significant safety issues resulted from the use of a pure Li anode, since dendrites initiated on the anode and could reach the cathode forming short circuits. As a result, researchers resorted to another configuration proposed again by Walter Rüdorff in 1938, that of ions being continuously intercalated between the anode and cathode. This system is known as the "rocking chair" battery and Rüdorff demonstrated it in 1938, by shuttling hydrogen sulfate ions between two graphitic electrodes. Therefore, instead of having the anode as the Li source, the cathode was to be a Li compound and the anode a material that could host Li. Since both electrodes must have the ability to host Li, these chemistries were termed Li-ion batteries. This chapter summarizes the development of the most promising cathode materials for current Li-ion batteries.

5.1 Layered Materials

5.1.1 $LiCoO_2$

$LiCoO_2$ was the first material that was discovered and applied as a commercial cathode for Li-ion batteries. Goodenough and coworkers designed the transition-metal lithium oxide $LiCoO_2$ with "ABCABC"-type stacking in the space group R3m [1]. As seen in Figure 5.1, during Li-ion electrochemical deintercalation, there are three distinct phases of $Li_{1-x}CoO_2$ detected by plateaus in the "voltage vs. Li content" profiles [2]. In the range of $0 < x < 0.25$, the coexistence of two hexagonal phases is

Rechargeable Ion Batteries: Materials, Design, and Applications of Li-Ion Cells and Beyond, First Edition. Edited by Katerina E. Aifantis, R. Vasant Kumar, and Pu Hu.

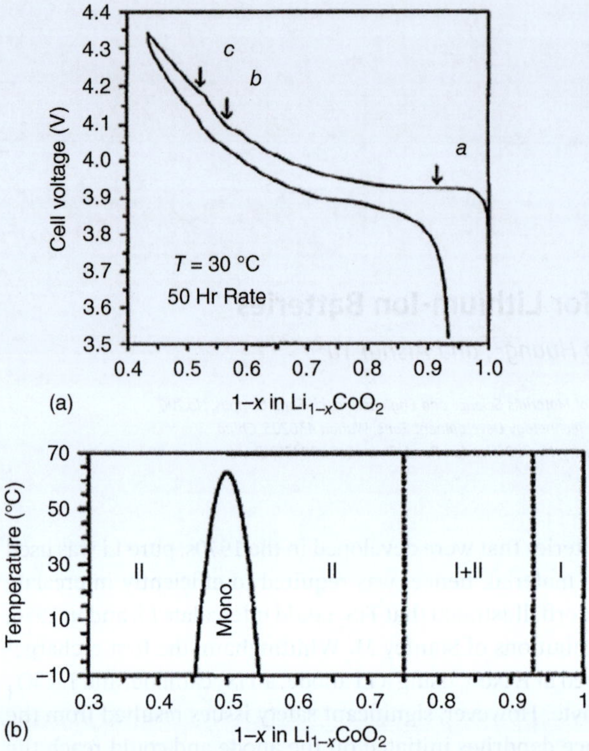

Figure 5.1 (a) Cell voltage as a function of $1-x$ in $Li_{1-x}CoO_2$ at the initial cycle. (b) A global phase diagram for $Li_{1-x}CoO_2$. Source: Reproduced with permission from Reimers and Dahn [2]/IOP Publishing.

observed indicating that a topotactic two-phase reaction takes place in this voltage region. In the range of $0.25 < x < 0.75$, a single-phase reaction takes place and a monoclinic phase is observed at $x = 0.5$ and this single-phase reaction is associated with phase changes, resulting in two plateaus in the voltage curves. In the range of $x > 0.75$, a two-phase reaction appears again, associated with the Jahn–Teller deformation of CoO_6 octahedra. When charged to more than 4.3 V, the oxygen loss and migration of Co cations to the Li layer lead to a faster capacity fade. Thus, the reversible capacity of $Li_{1-x}CoO_2$ is restricted to ~140 mAh g^{-1} in the region $0 < x < 0.5$, even though its theoretical capacity is 272 mAh g^{-1} [3, 4].

Significant work is being performed on improving these cathodes, and most of it focuses on the suppression of cobalt dissolution and the nonuniform distribution of local strains; both of which can be achieved by surface coatings. Coating, for example, $LiCoO_2$ with $AlPO_4$, allows for excellent electrochemical performance at a cut-off voltage higher than 4.6 V [5]. In the case of ZnO-coated $LiCoO_2$, the ZnO coating reduces the cobalt dissolution and prevents the inorganic surface films from the electrolyte, such as LiF, from covering the $LiCoO_2$ particles. The ZnO coating also mitigates the cycle-life degradation caused by inappropriate conductive carbon [6]. Other oxide coatings are, therefore, widely investigated and all of them show some promise. Coatings have also been promising to use with solid electrolytes by providing an interface between the cathode and electrolyte. It was shown that spraying thin layer of $Li_4Ti_5O_{12}$ with thickness of a few nanometers on a $LiCoO_2$ cathode, ensured the electronic conduction and increased the power density when a solid sulfide electrolyte was used [7].

There are several drawbacks with $LiCoO_2$, such as the toxicity and high price of Co; however, the most important issue is lack of safety. During de-lithiation, i.e. operation of the device, when the concentration of Li reaches 0.49, $Li_{0.49}CoO_2$ becomes unstable at relatively high temperatures, resulting in collapse of the layered crystal structure. Moreover, it leads to thermal runaway and ultimately to fire or explosion of the batteries due to release of O_2 from lattices. This is a particular issue for the automotive industry and that is why alternative cathodes had to be developed before Li-ion batteries were widely used in the automotive industry.

5.1.2 Nickel-Rich Materials

In searching for a less toxic and cheaper cathode material, $LiNiO_2$ was considered instead of $LiCoO_2$. However, the displacement of Ni and Li-ions takes place in the Li layers and the exothermic release of oxygen in the highly delithiated state results in irreversible phase transitions and safety issues. Thus, $LiNiO_2$ is not a promising cathode material for current Li-ion batteries. Recently, a new Ni-rich layered cathode $LiNi_{0.8}Co_{0.15}Al_{0.05}O_2$(NCA) was designed, in which a low amount of Al is incorporated to stabilize the layered structure during the electrochemical process, leading to safer batteries. However, the presence of lithium carbonate on the surfaces of $LiNi_{0.8}Co_{0.15}Al_{0.05}O_2$, which forms upon electrolyte decomposition, hinders the lithium-ion diffusion across the electrode–electrolyte interface. The formation of Li_2CO_3 is presumed to take place via the reaction:

$$LiNi_{0.8}Co_{0.15}Al_{0.05}O_2 + {}^1\!/\!_4\,x\,O_2 + {}^1\!/\!_2\,x\,O_2$$
$$\rightarrow Li_{1-x}Ni_{0.8}Co_{0.15}Al_{0.05}O_2 + {}^1\!/\!_2\,x\,Li_2CO_3$$

Active particle isolation due to Li_2CO_3 formation on the surface of the active material is proposed as the mechanism that is responsible for power and capacity losses relative to fresh electrodes. To prevent the formation of Li_2CO_3, most researchers have resorted to coating the surface with an electrochemically active layer. As an example Kim et.al. [8] constructed a $LiCoPO_4$ layer on the NCA particle surface. Electrochemical testing showed that this coating allowed for a 30% increase in the life cycle (compared to the uncoated samples) and prevented the capacity decrease on the first cycle. In addition to protection from the electrolyte, surface coatings can also protect the NCA materials from degradation it experiences by coming in contact with the atmosphere, which occurs during material fabrication and battery production. It has been shown that the surface coating of $LiCoO_2$ suppresses the reactions between $LiNi_{0.8}Co_{0.15}Al_{0.05}O_2$ and the atmosphere effectively, contributing to the enhancement of storage performance [9]. Specifically, it was shown that storing the $LiCoO_2$-coated NCA particles in various atmospheric conditions did not affect their electrochemical performance, whereas the performance of the uncoated particles was drastically reduced. A less expensive method for improving the storage of NCA is washing it in water. In Ref. [10] it was shown that washing NCA did not alter its structure, but reduced the formation of LiOH and Li_2CO_3 when the cells were being fabricated [10]. The storage of NCA materials, with exposure to air with certain humidity, can also be improved by reducing the nickel content or increasing the cobalt content in the compound. It also follows storing the material in an inert atmosphere without CO_2 and H_2O can also help the cathode to exhibit a larger discharge capacity.

An even more promising Ni-rich material is $LiNi_{1-x-y}Co_xMn_yO_2$. The Ni-rich ternary nickel–cobalt–manganese oxide layered materials (NCM), comes from the doping modification of $LiNiO_2$, $LiCoO_2$, and $LiMnO_2$ (which will be described in detail later). NCM has an α-$NaFeO_2$ layered rock-salt structure similar to $LiCoO_2$, belongs to the hexagonal crystal system, and the space point group is R3m; its crystal structure is shown in Figure 5.2 [11]. It can be seen from Figure 5.2 that the Li mainly occupies the 3a position, O occupies the 6c position, forming the MO_6 octahedral structure, and the disorder of Ni, Co, and Mn occupies the 3b position. The whole crystal can be regarded as alternately stacked [MO_6] octahedral layers and [LiO_6] octahedral layers, which is very suitable for the insertion and de-inertion of Li-ions.

In the layered NCM, nickel is the main redox element. Therefore, increasing the nickel content can effectively increase the specific capacity of the NCM. The increase in Ni content can increase the specific capacity but will reduce the cycle performance and stability. The increase in Co content can inhibit the phase change and improve the rate performance. The increase of Mn improves the structural stability but lowers the capacity. All the above are summarized in Figure 5.3 [12]. The high-nickel-content NCM material (Ni ≥ 0.6) has a high-specific capacity and low cost, but also has the drawbacks of lower capacity-retention rate and poor thermal stability, and thus fails to meet the commercial application requirements. It should be noted here that by rate performance, we mean the time allowed to fully charge and recharge the battery.

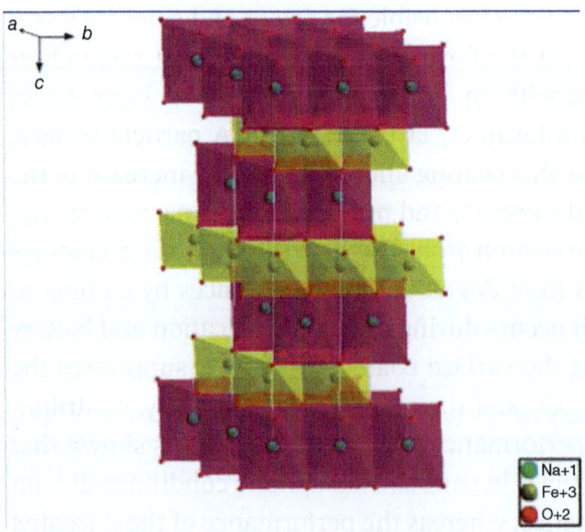

Figure 5.2 Structure of R-$NaFeO_2$ (R3m). Source: Reproduced with permission from Ellis et al. [11]/American Chemical Society.

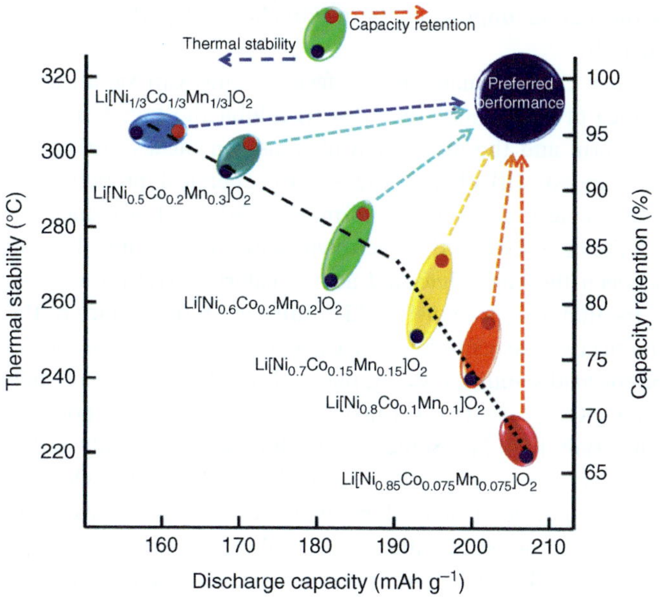

Figure 5.3 A map of relationship between discharge capacity, and thermal stability and capacity retention of $Li/Li[Ni_xCo_yMn_z]O_2$ ($x = 1/3, 0.5, 0.6, 0.7, 0.8$, and 0.85). Source: Reproduced with permission from Noh et al. [12]/Elsevier.

The properties and structure of high-nickel NCM materials are closely related to the preparation process. Different preparation processes and conditions directly affect the final structure and performance of the materials. The radius of Ni^{2+} (0.069 nm) is close to that of Li^+ (0.076 nm), and as a result Ni^{2+} easily enters the inter-chip and occupies the position 3a of Li^+, while Li^+ enters the main chip and occupies the position 3b, causing mixed cations and resulting in an enlarged the unit cell parameter a, causing the intensity of the (003) diffraction peak to weaken. The radius of Ni^{2+} in the Li layer is smaller than Li^+, which will reduce the thickness of the inter-chip and oxidize to Ni^{3+} or Ni^{4+} during charging, causing local collapse of the inter-chip space, increasing the difficulty of Li-ion insertion during discharge, and reducing the material's reversible capacity. The entry of Li into the transition-metal layer will increase the thickness of the main lattice spacing and make it difficult to de-intercalate, deteriorating the electrochemical performance of the material. The degree of ion mixing can be characterized by the value of c/a (the c/a represents the ratio of the unit cell parameter c value to a value) and I(003)/I(104). When $c/a > 4.9$ and I(003)/I(104)>1.2, the degree of mixing is low. In addition, the degree of splitting of the diffraction peaks of the (006)/(102) crystal plane and the (108)/(110) crystal plane reflects the integrity of the layered structure of the material and has a greater impact on the electrochemical performance of the material: The greater the splitting degree of the two pairs of diffraction peaks, the more complete the α-$NaFeO_2$ type layered structure will be, and the better the electrochemical performance. Therefore, during the preparation process, maintaining a proper Li^+/Ni^{2+} ratio, that is, a low degree of mixing and a complete layered structure, is the key to improving the electrochemical performance of high-nickel NCM materials.

As in the case of other cathode materials, surface coatings can significantly improve the performance of NCM since they can (i) inhibit the transformation of the material crystal form and the dissolution of transition metals during the charge and discharge process, (ii) change the surface chemical properties of the material to improve its electrochemical performance, (iii) avoid or reduce the direct contact between the material and the electrolyte, which can greatly alleviate the side reactions between the electrolyte and active materials, (iv) the coating layer as a conductive medium can promote the Li^+ diffusion on the surface of the particles, which is an effective means to improve capacity-retention performance, rate performance, and thermal stability. Usually, materials used for coating include metal oxides (ZrO_2 and TiO_2 et al.), [13–16] metal phosphates [17, 18], polymer [19–22], and other materials [23, 24]. Some coating layers are not only stable in their own structure, but also have certain electrochemical activity, which is beneficial to increasing the specific capacity of the material. Furthermore, as in the aforementioned case of NCA, the NCM cathode material requires special storage conditions, and coatings can reduce the contact between the cathode material and the air during mass production, and extend the storage life.

An example of SiO_2-coated NCM is shown in Figure 5.4, where a homogeneous precipitation method was employed to form a uniform silicon dioxide (SiO_2) coating on the NCM811 surface [26]; 811 denotes the content of Ni at 0.8, Co at 0.1, and Mn at 0.1. It is seen that the SiO_2 coating formed nanoparticles that fully covered the NCM surface, and therefore a double microstructure was observed. ~1-μm NCM particles formed a several micron NCM sphere, and when SiO_2 was precipitated, each NCM microparticle was covered with SiO_2 nanoparticles. Comparing the electrochemical properties of the coated and uncoated NCM showed that the SiO_2 coating was able to suppress the voltage decay and pulverization of the NCM811 particles during long-term cycling, and allowed for superior cycling stability and rate performance.

Figure 5.5, compares the cycling stability of the coated (with SiO_2) and uncoated pristine NCM811 at 0.2 C. The initial discharge-specific capacity for both cases was approximately the same at 208.4 and 204.3 mAh g^{-1}, for the NCM@SiO_2 and NCM, respectively. After 100 cycles, however, the capacity for NCM811@SiO_2 was significantly higher at 172.8 mAh g^{-1} (84.9% retention) than that for the pure NCM, which had dropped to 114.3 mAh g^{-1}. It can also be seen that the SiO_2 coating allowed for a higher coulombic efficiency and more stable cycling.

To study the effect of SiO_2 on the rate performance of the battery, the C rate varied from 0.2 C to 5 C. According to Figure 5.6, the SiO_2 coating allowed for an improved rate capability. Particularly at 5 C, the discharge capacity for NCM811 was 142.7 mAh g^{-1}, while for the coated case, it was 142.7 mAh g^{-1}, which is 91.3% of the capacity at 1 C.

The preferred electrochemical properties for the coated NCM could also be partly attributed to the more robust microstructure after cycling. Comparing the initial microstructure before cycling (Figure 5.4) and after 100 cycles (Figure 5.7), it is seen that no differences could be obtained for the NCM811@SiO_2 case, whereas the

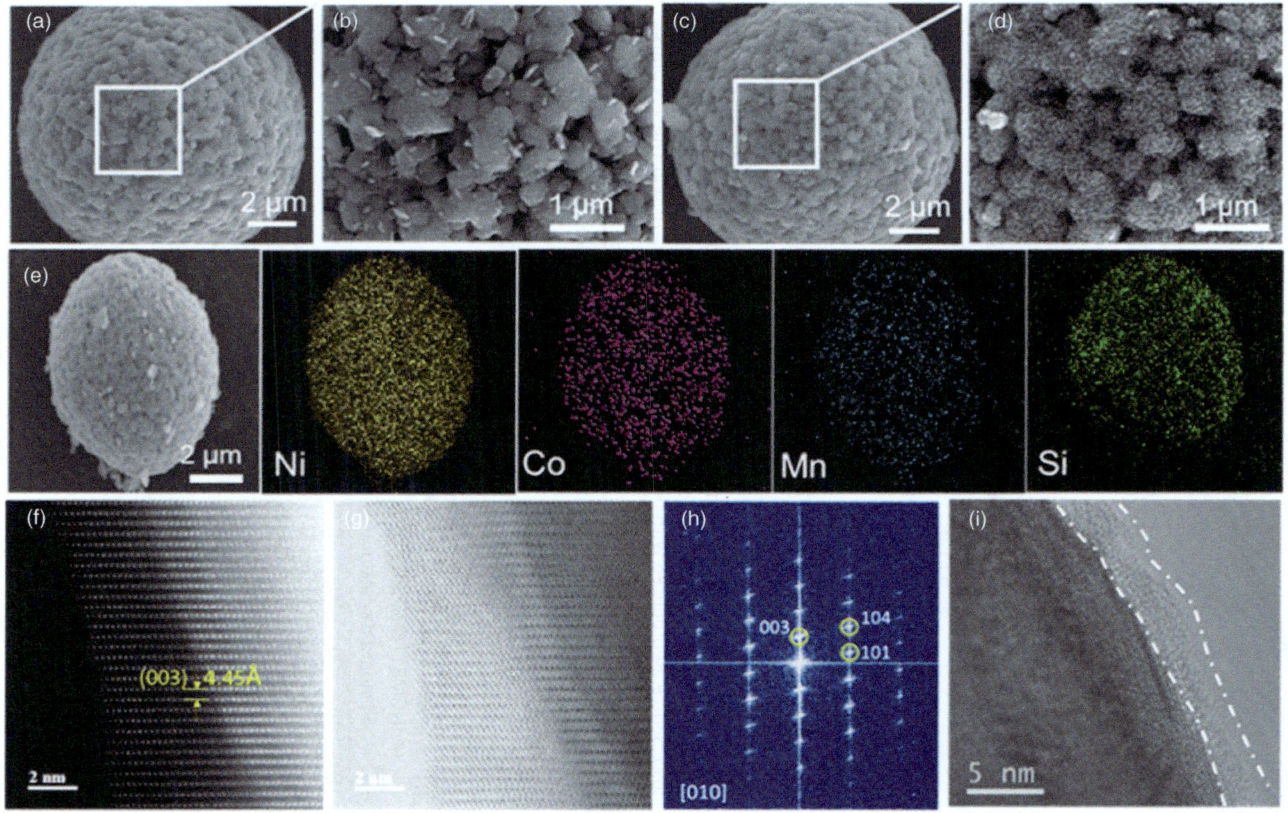

Figure 5.4 The SEM images of pristine NCM811 (a, b) and NCM811@SiO2 (c, d); EDS elemental mapping of NCM811@SiO2 (e); HAADF-STEM image (f) and HRTEM images (g) of pristine NCM811; The corresponding FFT of pristine NCM811 (h); HRTEM images of NCM811@SiO2 (i). Source: Reproduced with permission from Dou et al. [25]/John Wiley & Sons.

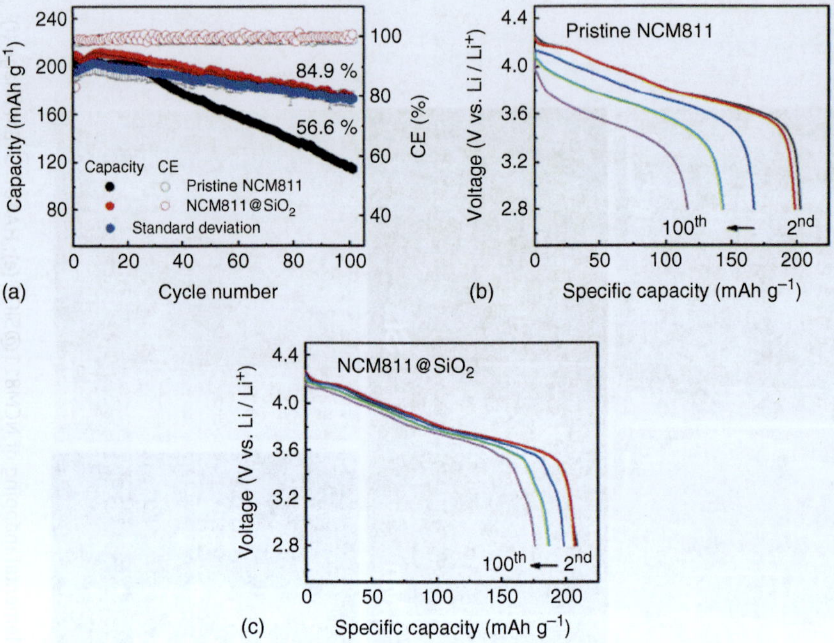

Figure 5.5 Electrochemical performance of pristine NCM811 and NCM811@SiO_2 Cycling stability (a); Corresponding discharge curves at different cycles (2nd, 25th, 50th, 75th, and 100th) (b,c). Source: Reproduced with permission from Dou et al. [25]/John Wiley & Sons.

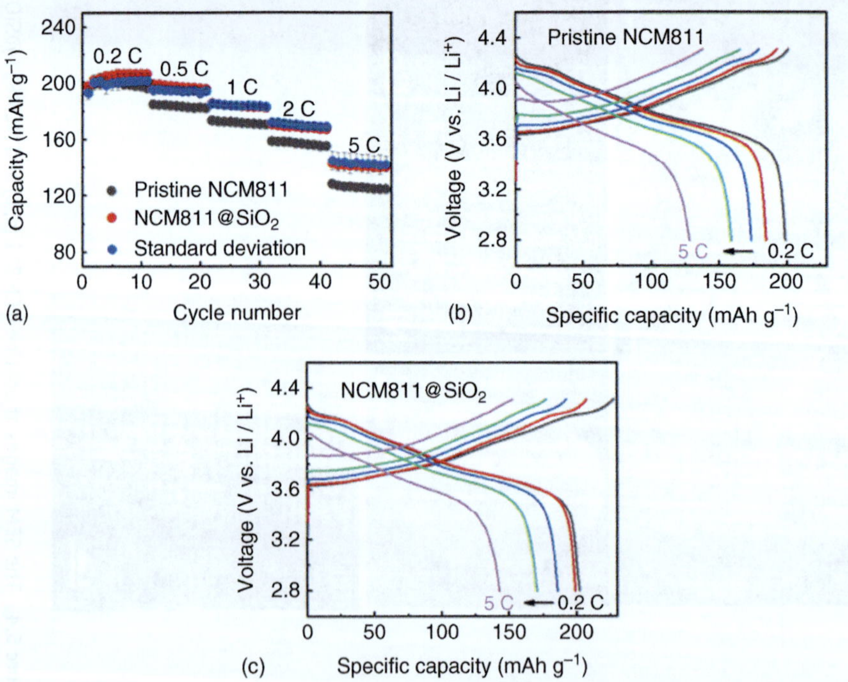

Figure 5.6 Rate performance (a); The charge–discharge curve of pristine NCM811 (b) and NCM811@SiO_2 (c) at different rate (0.2 C, 0.5 C, 1 C, 2 C, and 5 C). Source: Reproduced with permission from Dou et al. [25]/John Wiley & Sons.

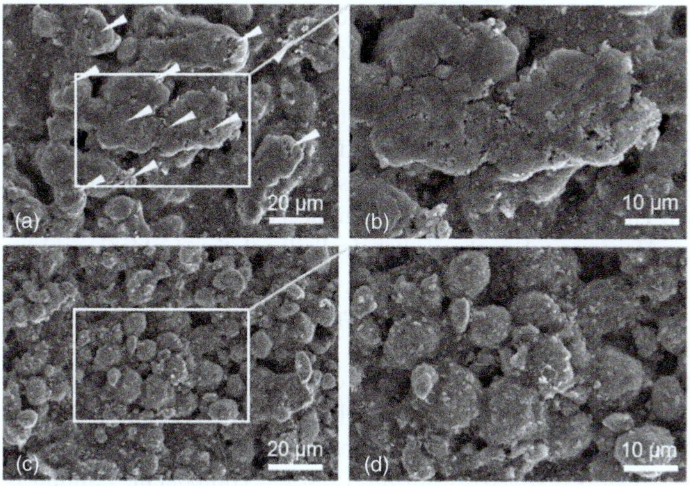

Figure 5.7 SEM images after 100 cycles of (a, b) pristine NCM811 and (c, d) NCM811@SiO_2. Source: Reproduced with permission from Dou et al. [25]/ John Wiley & Sons.

uncoated particles appeared to have lost their spherical shape, and formed pores and small cracks. This damage results from the volume expansions of NCM during Li-ion insertion and de-insertion and also the direct contact with the electrolyte. Surface coatings can constrain the volume changes and also protect from electrolyte attacks.

5.1.3 Excess Manganese Oxide Layered Cathode Materials

In addition to the previously mentioned lithium-rich transition-metal oxide cathodes, it is possible to fabricate excess manganese oxide Li-rich transition-metal oxide materials. These materials have alternating layers of $Li[Li_{1/3}Mn_{2/3}]O_2$ and $LiMO_2$ (M = Co, Ni, Mn, Fe, etc.). LMO can be $Li[Ni_{1/3}Mn_{1/3}Co_{1/3}]O_2$, in which case the cathode is denoted as $Li[Li_{0.2}Mn_{0.54}Ni_{0.13}Co_{0.13}]O_2$. Like all lithium-rich materials, they have rock-salt-type structures, in which the excess lithium is accommodated within the transition-metal layers at the expense of some metal (M) ions. Schematic illustrations of idealized Li_2MnO_3 and $LiMO_2$ (M = Co, Ni, Mn, Fe, etc.) layered structures are shown in Figure 5.8a,b, respectively [28].

Despite the difference in crystallographic space group symmetry, as well as atomic and electronic configurations (monoclinic C2/m for Li_2MnO_3 and trigonal R3m for

Figure 5.8 Layered structures of (a) Li_2MnO_3, (b) $LiMO_2$. Source: Reproduced with permission from Thackeray et al. [27]/Royal Society of Chemistry.

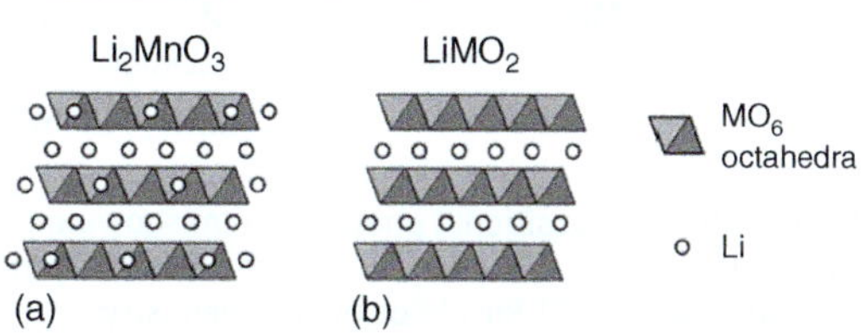

$LiMiO_2$), the close-packed layers in each of these compounds ((001) of monoclinic structure and (003) of trigonal structure) have an interlayer spacing close to 4.7 A.

Between 4.4 and 2.0 V, the $LiMO_2$ component is electrochemically active and operates as a common insertion electrode. By contrast, the Li_2MnO_3 component is electrochemically inactive over this potential window and it does not contribute to electrochemical capacity, because all the manganese ions are tetravalent and cannot be oxidized further. Thus, the Li_2MnO_3 component acts as a stabilizing unit in the electrode structure. However, when the electronically insulating Li_2MnO_3 regions are extremely small (nano-dimensional) and if they are distributed randomly throughout the composite structure, then these regions will function as solid electrolyte constituents in facilitating Li^+ ion transport between the capacity-generating $LiMO_2$ regions of the structure and the Li_2MnO_3 component acts as a reservoir of surplus lithium that can be used to stabilize the electrode structure at low lithium loadings. When the electrochemical potential is raised above 4.4 V during charge, then further lithium can be extracted from the Li_2MnO_3 component. Lithium is extracted from the Li_2MnO_3 component with the simultaneous release of oxygen and the net loss from the $xLi_2MnO_3{\cdot}(1{-}x)$ $LiMO_2$ is Li_2O. During such an electrochemical activation process, the extraction of two Li^+ per Li_2MnO_3 unit on the initial charge and the reintroduction of only one Li^+ on the subsequent discharge into the resulting MnO_2 unit necessarily means that there must be an irreversible capacity loss on the initial cycle. The compositional phase diagram and electrochemical plot of $xLi_2MnO_3{\cdot}(1{-}x)LiMO_2$ electrode are shown in Figure 5.9 [22].

However, the real electrochemical mechanism and structure of layered Li-rich cathode materials are still unclear. There has been an ongoing debate in the literature on whether these Li-rich cathode materials form homogeneous solid solutions or Li_2MnO_3 domains within a $LiMO_2$ matrix. By combining high-resolution

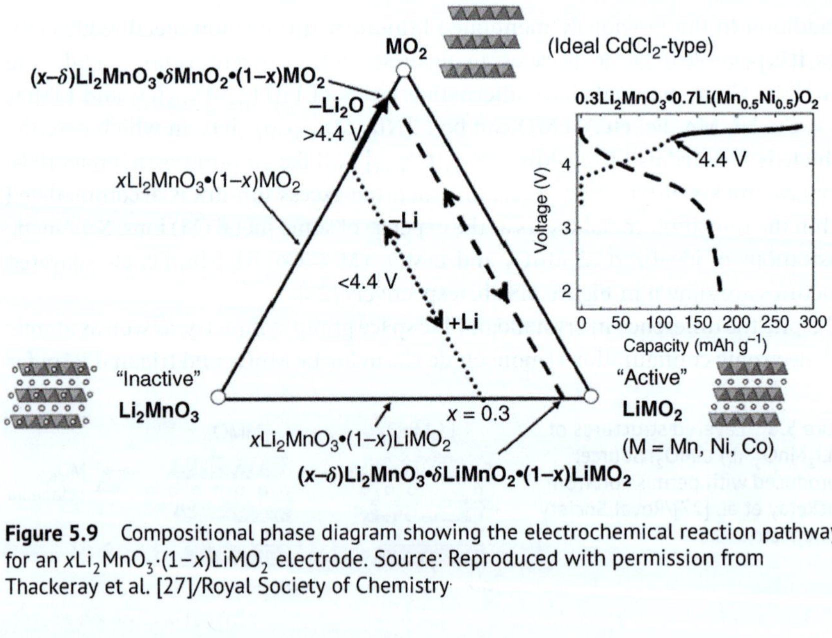

Figure 5.9 Compositional phase diagram showing the electrochemical reaction pathways for an $xLi_2MnO_3{\cdot}(1{-}x)LiMO_2$ electrode. Source: Reproduced with permission from Thackeray et al. [27]/Royal Society of Chemistry.

transmission electron microscopy (HRTEM) with electron energy-loss spectroscopy (EELS), the locally monoclinic (Li_2MnO_3-like) regions have been shown to exist in the parent rhombohedral structure of these Li-rich cathode materials.[26, 29, 30] High-angle annular dark-field scanning transmission electron microscopy (HAADF-STEM) has also revealed the coexistence of two phases inside the Li-rich cathode materials [31]. On the other hand, some researchers think that these Li-rich cathode materials are homogeneous solid solutions between the two components Li_2MnO_3 and $LiMO_2$ because their lattice parameters vary linearly with the composition of their end members [32]. In fact, most researchers think that the large difference between the atomic size and tendency for clusters of the elements inside these Li-rich cathode materials can induce large lattice distortions, although these Li-rich cathode materials can preserve the periodical long-range structure. The reaction mechanisms of these Li-rich cathode materials are also proved quite complicated. Besides the electrochemical reaction pathways in the bulk of the cathode shown in Figure 5.8, a new surface-reaction mechanism has been proposed, as shown in Figure 5.10. It is suggested that some of the extra-discharge capacity for these Li-rich cathode materials originated from the electrochemical-reduction reaction of the oxygen molecules at the electrode surface ($O_2 \rightarrow O_2^-$), but these contributions are suppressed by the accumulated lithium carbonate formation at the electrode surface [33].

In spite of the structural stability and large rechargeable capacity, the large first-cycle irreversible capacity, the inferior rate capability of these materials and their structural degradation are hindering the commercial application of these materials. A wide research effort has been devoted to overcoming these disadvantages, again focusing on coatings. Surface coatings of vanadium oxide and inactive oxide or fluorides, such as AlF_3 and Al_2O_3, are considered to be the most promising approach for reducing the irreversible capacity loss [34–39], prolonging the cycles and rate capability, although the exact mechanisms are still not clear. The coated materials always exhibit better rate capability, cyclic performance, and higher efficiency of the initial cycle. The improved performance is always attributed to the suppression of electrolyte decomposition by the coating, but it also enhances their

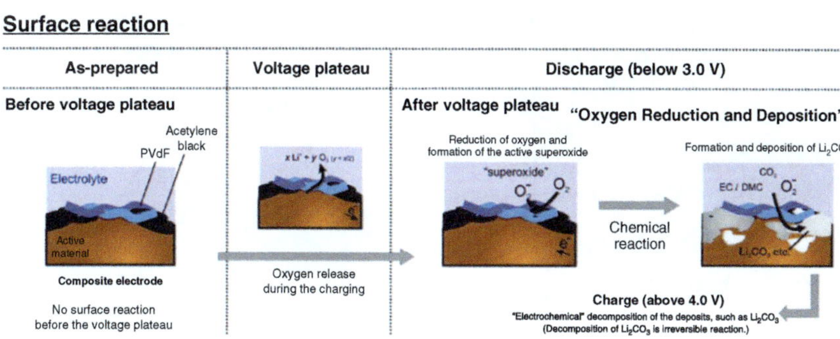

Figure 5.10 Schemes of the proposed surface reaction mechanisms in the $Li_{1.2}Ni_{0.13}Co_{0.13}Mn_{0.54}O_2$ composite electrodes. Source: Reproduced with permission from Yabuuchi et al. [33]/American Chemical Society.

structural stability. In Ref. [34], for example, Al_2O_3 and $AlPO_4$ were coated onto Li-excess Mn-based oxide particles ($Li[Li_{0.2}Mn_{0.54}Ni_{0.13}Co_{0.13}]O_2$) to successfully reduce the irreversible capacity loss. This reduction was ascribed to the retention of part of the oxygen ion vacancies in the lattice during first charge, due to the presence of the coating. Another technique to modify the $Li[Li_{0.2}Mn_{0.54}Ni_{0.13}Co_{0.13}]O_2$ by adding $Al_2O_3 + RuO_2$ since it was shown that surface-modified samples were more resistant to the formation of the solid-electrolyte interface (SEI) layer and had high-diffusivity paths for Li^+. It is interesting to note that the wt% of the Al_2O_3 and RuO_2 played a significant role since 1 wt% contents allowed for better performance than the 2 wt% contents [40]. An alternative method to coatings for improving the performance of excess manganese oxide cathodes is to pre-cycle them since this makes them more resistant to damage and structural rearrangements during cycling [41]. Another interesting approach, but more challenging, for improving the rate performance of these Li-rich cathode materials, is to use nanostructures that allow for a short Li^+ diffusion pathway. For example, $Li[Ni_{0.25}Li_{0.15}Mn_{0.6}]O_2$ nanowires with an aspect ratio of several hundred and a diameter of about 30 nm allowed excellent high-rate capability [42], while $Li[Li_{0.17}Ni_{0.25}\ Mn_{0.58}]O_2$; crystal habit-tuned nanoplate materials with significantly increased (010) planes allowed for a high-rate performance [43].

In conclusion, although there are many debates on these Li-rich cathode materials, they are the most widely used in the automotive industry.

5.2 Spinel Materials

Spinel lithium manganese oxide ($LiMn_2O_4$) is regarded as a promising cathode material for lithium-ion battery devices due to its low cost, lower toxicity, high theoretic capacity, and production ease [44]. As illustrated in Figure 5.11, $LiMn_2O_4$

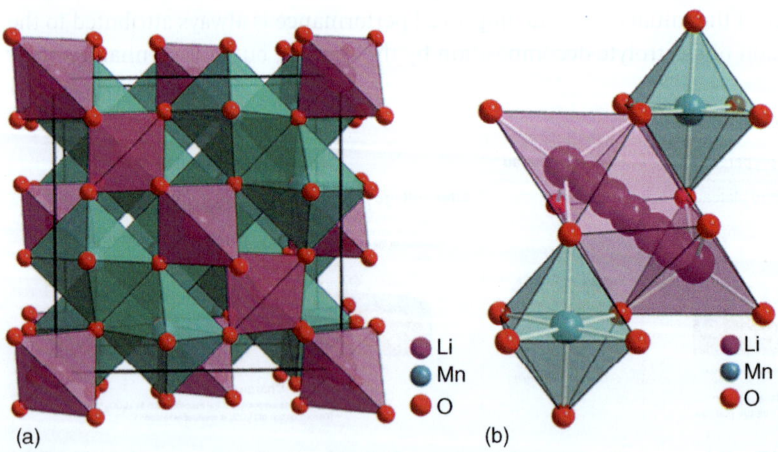

Figure 5.11 The crystal structure of spinel $LiMn_2O_4$. Source: Reproduced with permission from Zhang et al. [45]/Elsevier.

adopts the spinel structure with the space group Fd3m, and Li and Mn occupy the 8a tetrahedral and 16d octahedral sites of the cubic close-packed oxygen ions, respectively. The edge-shared dioctahedral Mn_2O_4 host structure is highly stable and possesses a series of intersecting tunnels formed by the face-sharing of tetrahedral lithium (8a) sites and empty-octahedral (16c) sites. Such tunnels allow the three-dimensional diffusion of Li^+. The lithium intercalation/deintercalation into/from the 8a tetrahedral sites occurs at about 4 V while maintaining the initial cubic spinel symmetry.

Surplus lithium can be intercalated into the spinel $LiMn_2O_4$ to reach a maximum composition of $Li_2Mn_2O_4$, which occurs at about 3 V against the lithium anode with a phase transformation from the cubic to tetragonal phase. The cubic to tetragonal-phase transition, often referred to as Jahn–Teller distortion, is accompanied by a 6.5% increase in the unit cell volume, which damages the structural integrity of the electrode during charge/discharge and results in rapid-capacity fading. To prevent Jahn–Teller distortion, $LiMn_2O_4$ is only used below 4 V vs. Li/Li^+ with a limited practical capacity of around 140 $mAh\,g^{-1}$ (the theoretical capacity of $LiMn_2O_4$ is 148 $mAh\,g^{-1}$ while the theoretical capacity of $Li_2Mn_2O_4$ is 296 $mAh\,g^{-1}$). However, $LiMn_2O_4$ tends to exhibit capacity fading even in the 4 V region, especially at elevated temperatures [46]. Several detrimental effects, including the Jahn–Teller distortion at the surface of the particles under conditions of non-equilibrium cycling [47], dissolution of Mn into the electrolyte[48], formation of two cubic phases in the 4 V region, loss of crystallinity [49], and development of microstrain during cycling [50], could be the source of the capacity fade.

Several strategies have been developed to overcome these effects, such as surface modification and cationic substitutions that result in $LiM_yMn_{2-y}O_4$ (M = Li,Cr,Co,Ni, and Cu). For surface modification, chemically stable oxides are usually chosen. Al_2O_3 and ZnO [51] are the most frequently used coating materials that could stabilize the structure, reduce the volume expansion of the cathode materials, and protect the electrode surface from direct contact with the electrolyte, thus impeding side-reactions. During the cycling, HF always emerges in the electrolyte, which leads to the dissolution of Mn into the electrolyte. Metal fluorides are chosen to protect the cathode material from HF attack; for example, AlF_3 [52] and MgF_2 [53]. Iron phosphates are effective coating materials since the strong P=O bond (bond energy = 5.64 eV) is very resistant to chemical attack, and the strong covalency of the $PO4^{3-}$ polyanions with the metal ions leads to high thermal stability. What is more important is that iron phosphates, $FePO_4$ [54] for example, have highly ordered three-dimensional macroporous channels and thereby are usually electrochemically-active. An ideal coating should be not only chemically stable but also electrochemically active on which the future work should focus.

As for cationic substitutions, Co [55], Ti [56], Cr [57], and Ni [45] are the most frequently used elements. Nickel-substituted "high-voltage" spinel cathodes $LiMn_{1.5}Ni_{0.5}O_4$ offer the advantages of both a higher capacity and higher operating voltage than the manganese-only $LiMn_2O_4$ spinel [45, 58]. The high-voltage spinel cathode has the benefit of a robust cubic close-packed crystal structure

Figure 5.12 Crystal structure of $LiMn_{1.5}Ni_{0.5}O_4$. The cation ordering in the octahedral sites is indicated with different colors for MnO_6 and NiO_6 octahedra. Source: Reproduced with permission from Manthiram et al. [59]/Royal Society of Chemistry.

(Figure 5.12) made up of strong edge-shared MO_6 octahedra (M = Mn or Ni). The large differences in the charges and ionic radius of Mn^{4+} and Ni^{2+} lead to a spectrum of cation-ordering behavior, which is influenced sensitively by the heating conditions [60]. Owing to the 3 : 1 ratio of Mn:Ni, it is theoretically possible to obtain a perfectly ordered structure in which all Ni^{2+} ions are fully coordinated by 6 Mn^{4+} nearest neighbors in the lattice. This would result in a space group of $P4_332$, with a distinction between the sites occupied by Mn^{4+} and Ni^{2+} [61]. In contrast, it is also theoretically possible to obtain a disordered structure in which Ni^{2+} and Mn^{4+} are distributed randomly among the 16d octahedral sites. This would result in a space group of Fd3m, which has a higher degree of symmetry than $P4_332$ [62].

The degree of cation ordering also profoundly affects the charge/discharge profiles depending on the amount of impurity phase, and Mn^{3+} ions present in the sample. Representative charge and discharge curves of the ordered and disordered samples can be seen in Figure 5.13. The long-flat plateau present at about 4.7 V represents the redox reaction of the $Ni^{2+/3+/4+}$ couple, and the activity near 4.0 V can be attributed to the $Mn^{3+/4+}$ couple [63]. It is clear that for the disordered Fd3m sample, a reasonable amount of Mn^{3+} is present, and there is a clear distinction between the $Ni^{2+/3+}$ plateau at 4.6 V and the $Ni^{3+/4+}$ plateau at 4.8 V. Examination of the ordered $P4_332$ electrochemical profile shows the absence of activity near 4.0 V, indicating that there is little or no Mn^{3+} in the spinel lattice. Another marked difference is that the Ni plateau at 4.7 V is continuous with no obvious break between the $Ni^{2+/3+}$ and $Ni^{3+/4+}$ reactions, supporting the premise that a two-phase mechanism dominates rather than a solid-solution regime [64].

As mentioned earlier, the high-redox voltage for the $Ni^{2+/3+}$ and $Ni^{3+/4+}$ couples (4.7 V) lies just above the upper limit of the stability window of commercial electrolytes. This may lead to side reactions, which eventually result in capacity fade [65]. To solve the problems associated with the unstable interface between $LiNi_{0.5}Mn_{1.5}O_4$ and conventional electrolytes, several approaches have been explored, with surface coatings being again the most facile method. Among the candidates, oxides (e.g. Bi_2O_3 [66], Al_2O_3 [67], and ZnO [68]), with good chemical stability, are often used to prevent the direct contact with electrolytes, thus impeding

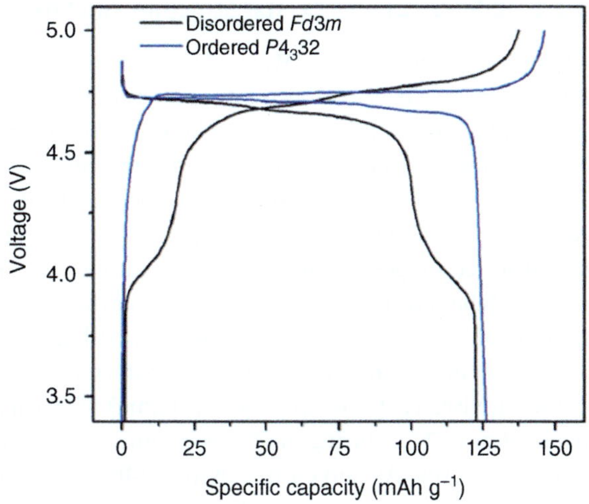

Figure 5.13 Typical electrochemical charge and discharge profiles of ordered and disordered $LiMn_{1.5}Ni_{0.5}O_4$ spinel cathodes. Source: Reproduced with permission from Manthiram et al. [59]/Royal Society of Chemistry.

the side reactions; however, most oxides are poor electronic conductors, which may increase the resistance. Carbon materials are widely employed as electronic conductors and among them, graphene is most frequently applied as an active-particle coating since it has superior electrical properties [69]. Alternatively, to protect from electrolyte side reactions, an artificial SEI was produced on $LiNi_{0.5}Mn_{1.5}O_4$ by using the solid-electrolyte lithium phosphorus oxynitride (Lipon) [70]. It is believed that with proper thickness, electrolyte decomposition can be largely suppressed by this artificial SEI film. Furthermore, alternative electrolytes with wider electrochemical potential windows [71] and additives suppressing the decomposition of the electrolyte and stabilizing the SEI [72–74] have also been developed.

5.3 Polyanion (Phosphate, Silicates) Framework Cathode Materials

Polyanion compounds are a class of materials in which tetrahedral polyanion structure units $(XO_4)^{n-}$ with strong covalent bonding combine with MO_x (M = transition metal). This type of material shows better structure and thermal stability due to the strong covalently bonded oxygen atoms [75]. Hence, polyanion compounds have attracted lots of attention and have been intensively investigated. The first class of such cathodes was proposed by Arumugam Manthiram and John B. Goodenough (Nobel Prize in Chemistry, 2019) in 1987 [76]. A common feature of all olivine materials, however, is their poor electronic conductivity, and therefore, they could not be electrochemically cycled initially. Hence, there are systematic works of many research groups worldwide to overcome the transport limitations of electrons and ions, aiming at evaluating their electrochemical performance. The first studies that were able to cycle $LiFePO_4$ were published in the early 2000 and this was achieved by coating the particles with a highly conductive carbon material[77] or doping them [78], to make them conductive. More details are given in Section 5.3.3.

5.3.1 $LiMPO_4$ Olivine Crystal Structure and Intercalation Mechanism

Olivine-structured-based $LiMPO_4$ (M = transition metal) cathode materials as hosts for lithium-ions have attracted lots of attention due to the advantages of flat a voltage profile, low material cost, abundant material supply, and better environmental compatibility [79–83]. They have an orthorhombic unit cell with the space group of Pnma, and consist of corner-shared FeO_6 octahedra and edge-shared LiO_6 octahedra running parallel to the *b*-axis [84–86]. As shown in Figure 5.14, the $LiFePO_4$ crystal has distorted oxygen hexagonal packing into which Li^+ and Fe^{2+} occupy one-half of the octahedral sites and P occupies 1/8 of the tetrahedral sites. In this structure, lithium diffusion proceeds along with the [010] direction [87, 89].

Upon delithiation, the Li-ions are extracted to yield heterosite MPO_4. Both phases are isostructural (space group Pnma) with slight difference in the unit cell parameters (<7%). During the Li insertion/de-insertion process, the $LMPO_4/MPO_4$ interface moves. As shown in Figure 5.14b, the flat-voltage profile indicates the $LiFePO_4/FePO_4$ two-phase reaction process [88]. The redox reaction of Fe^{2+}/Fe^{3+} and Mn^{2+}/Mn^{3+} for $LiFePO_4$ and $LiMnPO_4$ are 3.4 and 4.1 V vs. Li/Li^+, respectively, and the other $LiMPO_4$ (M = Co, Ni) shows 4.8 and 5.2 V that are outside the stability window of the electrolyte.

Even though $LiFePO_4$ has a lower capacity than other cathode materials, it is safer and allowed the wide use of Li-ion batteries in electric vehicles. The bond of oxygen with phosphorous is stronger than that with cobalt in the $(PO_4)^{3-}$ and the $(CoO_2)^-$ ions, respectively, and therefore, the oxygen atoms are released slower in $LiFePO_4$ if short-circuiting or overheating occurs.

5.3.2 $LMSiO_4$ Orthosilicate Crystal Structure and Intercalation Mechanism

Orthosilicate $LMSiO_4$ (M = Fe, Mn, Co, Ni) is another type of polyanion cathode material with high potential and capacity. $LMSiO_4$ are compounds related to the

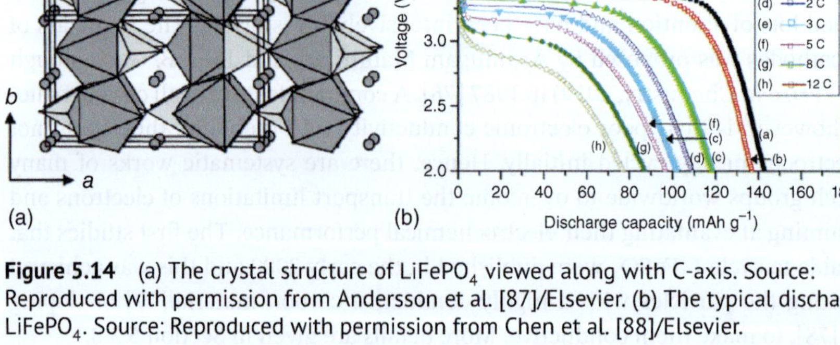

Figure 5.14 (a) The crystal structure of $LiFePO_4$ viewed along with C-axis. Source: Reproduced with permission from Andersson et al. [87]/Elsevier. (b) The typical discharge of $LiFePO_4$. Source: Reproduced with permission from Chen et al. [88]/Elsevier.

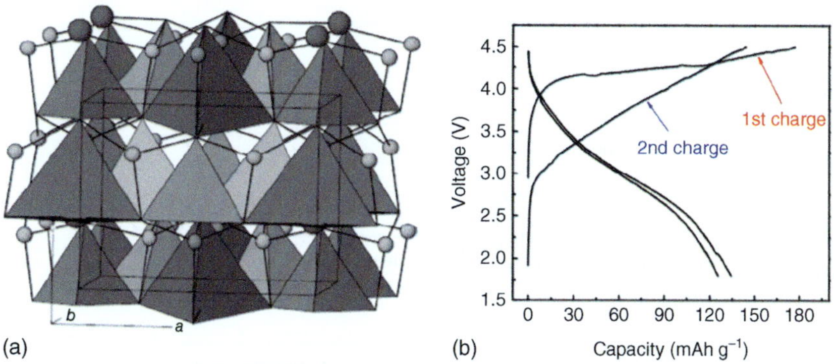

Figure 5.15 (a) crystal structure of Li_2MnSiO_4. Source: Reproduced with permission from Dominko et al. [90]/Elsevier. (b) The typical charge–discharge curves of Li_2MnSiO_4. Source: Reproduced with permission from Zhang et al. [91]/Elsevier.

Li_3PO_4 structure and belong to the large family of tetrahedral structure materials, which exhibit rich polymorphism. Three possible Li_2MnSiO_4 forms have been considered for crystallizing the Pmnb, $Pmn2_1$ (β-Li_3PO_4 derivatives), and $P2_1/n$ (γ-Li_3PO_4 derivatives) space group. Due to the higher stability at high temperatures of the $P2_1/n$ structure, $Li_2Mn/FeSiO_4$ is always described as $Pmn2_1$ space group, as shown in Figure 5.15a [90].

Li_2MSiO_4 provides the potential of insertion/extraction of two lithium-ions per formula unit. This indicates a high theoretical capacity of about $330\,mAh\,g^{-1}$, which is almost twice the capacity of olivine $LiMPO_4$ materials. However, this type of material shows extremely low electronic conductivity, which is several orders of magnitude lower than that of $LiMPO_4$, so the capacity of Li_2MnSiO_4 is not as high as predicted (Figure 5.15b).

5.3.3 Factors to Improve Electrochemical Performance of $LMXO_4$

For the intrinsic low ionic/electronic conductivity, preparation of nano-sized $LiMXO_4$ (M = transition-metal oxide, X = nonmetallic elements, such as P and Si) is of great importance. Under a high-current rate, the discharge capacity and cycling stability are almost dependent on the particle size. Hence, lots of research focuses on reducing the particle size to the nanoscale, which could increase the contact area of the active material and electrolyte.

The low electronic conductivity of the olivine particles leads to coating the particles with a thin carbon layer that is conducive for both electrons and Li-ions. Moreover, carbon coatings prevent particle agglomeration at high annealing temperatures. For the carbothermal-reduction process, the carbon source could also act as the reducing agent to transfer Fe^{3+} to Fe^{2+} for synthesizing $LiFeXO_4$.

In general, the carbon layer is always obtained from the thermal decomposition of carbon-containing precursors. However, those approaches based on the thermal decomposition of carbon-containing precursors can only produce $LiFePO_4$ particles with a partial coating of carbon. And to obtain uniform carbon layer, *in situ*

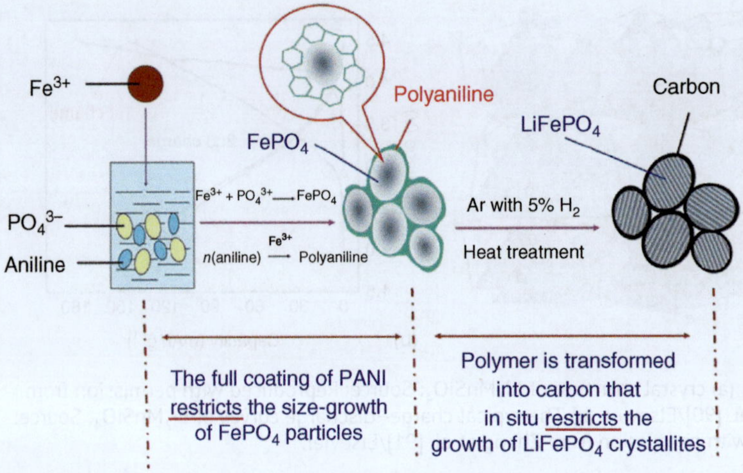

Figure 5.16 *In situ* carbon coating process of preparing $LiFePO_4$. Source: Reproduced with permission from Wang et al. [92]/John Wiley & Sons.

polymerization restriction method is used, as shown in Figure 5.16. Aniline was oxidized and polymerized on the outer surface of $FePO_4$ with the presence of Fe^{3+} and then, Fe^{3+} ions were reduced to Fe^{2+} during the process. With subsequent heat-treatment, the polymer shell transformed into a carbon layer.

In addition to carbon coating, doping with cations (as mentioned earlier) can increase the electronic conductivity of $LiMPO_4$ by a factor of ~10^8. It has been shown that aliovalent substitutions could reduce the lithium miscibility gap, increase phase-transformation kinetics during cycling, and expand Li-diffusion channels in the space [93, 94]. For Li_2MSiO_4, the electrochemical behavior could be improved by substitution of M, Li, Si, and O sites. It is found that the Na substitution of Li-ions in Li_2CoSiO_4 could lead to the expansion of the interlayer spaces, which benefits Li-ion diffusion.

After $LiFePO_4$, $LiMnPO_4$ has attracted lots of attention for its higher redox voltage. For $LiMSiO_4$, the electronic conductivity is much smaller than $LiMnPO_4$; so, much smaller particle size and good-conductive coatings are much more important. Until now, many methods have been developed to prepare polyanion nanoparticles, such as solid-state reactions, sol–gel method, co-precipitation, spray technology, and hydrothermal/solvothermal method.

Solid-state reaction is the most popular and conventional method to synthesize polyanion $LiMXO_4$ compounds. The mixtures of all starting materials are ground extensively and annealed at high temperatures. The starting mixture firstly decomposes at the temperature of 300–400 °C to expel the gases, and, after being reground, calcinated at temperatures ranging from 400 to 800 °C for 10–24 hours. The carbon source is always added to the precursor mixture, and after high-temperature calcination, an amorphous carbon layer forms on the particles. The purity and degree of crystallization for the final $LiMXO_4$ product is depended on the calcination procedure.

The motivation for sol–gel processing primarily lies in the fact that it ensures a potentially higher purity and homogeneity and lower processing temperatures than traditional ceramic powder methods [95]. The starting precursors are all dissolved in water with certain amounts of ascorbic acid or citric acid as the chelating agent. The gels are obtained after the solvent evaporates, which is followed by calcination at temperatures from 500 to 700 °C in inert (argon or nitrogen) or a reductive atmosphere.

The concept of the hydrothermal process for preparation of $LiFePO_4$ was first realized by Yang et. al in 2001 [96]. The process takes place in a stainless autoclave and always reacted at 150–250 °C. In the hydrothermal process, water is used as the main mineralizer, and in the solvothermal process, organic solvents are used as the mineralizer [96]. Compared to the conventional solid-state method, the hydrothermal/solvothermal methods occurred are relatively low temperatures. In addition, the solvothermal method can control the morphology of final product by using various kinds of surfactants.

There are many other methods to prepare $LiMXO_4$ polyanion cathode materials, such as carbothermal reduction, microwave processing, precipitation method, and spray pyrolysis. Critical to the good performance of polyanion $LiMXO_4$ cathode materials, all these methods are focused on nano-sized particle preparation and good-conductivity coating layers.

5.4 Conclusions

This chapter gives an overview of the main types of cathode materials that are currently used or will be used in the future in Li-ion batteries. There is continuous research to improve the performance of these materials, and the effect of the electrolyte is extensively examined. It should be emphasized that future Li solid-state batteries will also use similar cathodes to those described herein, and therefore, the improvement of the presented materials is necessary for the further advancement of all Li battery systems.

References

1. Mizushima, K., Jones, P.C., Wiseman, P.J., and Goodenough, J.B. (1980). *Mater. Res. Bull.* 15: 783.
2. Reimers, J.N. and Dahn, J.R. (2019). *J. Electrochem. Soc.* 139: 2091–2097.
3. Amatucci, G.G., Tarascon, J.M., and Klein, L.C. (1996). *Solid State Ionics* 83: 167.
4. Chebiam, R.V., Prado, F., and Manthiram, A. (2001). *Chem. Mater.* 13: 2951.
5. Cho, J., Kim, B., Lee, J.-G. et al. (2005). *J. Electrochem. Soc.* 152: A32.
6. Fang, T., Duh, J.-G., and Sheen, S.-R. (2005). *J. Electrochem. Soc.* 152: A1701.
7. Ohta, N., Takada, K., Zhang, L. et al. (2006). *Adv. Mater.* 18: 2226–2229.
8. Kim, Y. and Cho, J. (2007). *J. Electrochem. Soc.* 154: A495.
9. Liu, W., Hu, G., Du, K. et al. (2013). *J. Power Sources* 230: 201–206.

10 Kim, J., Hong, Y., Ryu, K.S. et al. (2006). *Electrochem. Solid-State Lett.* 9: A19.

11 Ellis, B.L., Lee, K.T., and Nazar, L.F. (2010). *Chem. Mater.* 22: 691–714.

12 Noh, H.-J., Youn, S., Yoon, C.S., and Sun, Y.-K. (2013). *J. Power Sources* 233: 121–130.

13 Pan, W., Peng, W., Yan, G. et al. (2018). *Energy Technol. Ger* 6: 2139–2145.

14 Zhang, H., Xu, J., and Zhang, J. (2019). *Front. Mater.* 6: 309.

15 Nisar, U., Al-Hail, S.A.J.A., Petla, R.K. et al. (2019). *ACS Appl. Energy Mater.* 2: 7263–7271.

16 Schipper, F., Bouzaglo, H., Dixit, M. et al. (2018). *Adv. Energy Mater.* 8: 1701682.

17 Zou, P., Lin, Z., Fan, M. et al. (2020). *Appl. Surf. Sci.* 504: 144506.

18 Lee, S.-W., Kim, M.-S., Jeong, J.H. et al. (2017). *J. Power Sources* 360: 206–214.

19 Yang, H., Wu, K., Hu, G. et al. (2019). *ACS Appl. Mater. Interfaces* 11: 8556–8566.

20 Li, B., Li, G., Zhang, D. et al. (2019). *ChemistrySelect* 4: 6354–6360.

21 Gan, Q., Qin, N., Zhu, Y. et al. (2019). *ACS Appl. Mater. Interfaces* 11: 12594–12604.

22 Paik, Y., Grey, C.P., Johnson, C.S. et al. (2002). *Chem. Mater.* 14: 5109–5115.

23 Yang, H., Wu, H.H., Ge, M. et al. (2019). *Adv. Funct. Mater.* 29: 1808825.

24 Liu, Y., Tang, L.-B., Wei, H.-X. et al. (2019). *Nano Energy* 65: 104043.

25 Dou, L., Hu, P., Shang, C. et al. (2021). *ChemElectroChem* 8: 4321–4327.

26 Bareño, J., Balasubramanian, M., Kang, S.H. et al. (2011). *Chem. Mater.* 23: 2039–2050.

27 Thackeray, M.M., Kang, S.-H., Johnson, C.S. et al. (2007). *J. Mater. Chem.* 17: 3112.

28 Johnson, C.S., Kim, J.-S., Kropf, A.J. et al. (2003). *Chem. Mater.* 15: 2313.

29 Kikkawa, J., Akita, T., Tabuchi, M. et al. (2008). *J. Appl. Phys.* 103: 104911.

30 Bareno, J., Lei, C.H., Wen, J.G. et al. (2010). *Adv. Mater.* 22: 1122–1127.

31 Boulineau, A., Simonin, L., Colin, J.-F. et al. (2012). *Chem. Mater.* 24: 3558–3566.

32 Ohzuku, T., Nagayama, M., Tsuji, K., and Ariyoshi, K. (2011). *J. Mater. Chem.* 21: 10179.

33 Yabuuchi, N., Yoshii, K., Myung, S.T. et al. (2011). *J. Am. Chem. Soc.* 133: 4404–4419.

34 Wang, Q.Y., Liu, J., Murugan, A.V., and Manthiram, A. (2009). *J. Mater. Chem.* 19: 4965.

35 Deng, H., Belharouak, I., Yoon, C.S. et al. (2010). *J. Electrochem. Soc.* 157: A1035.

36 Rosina, K.J., Jiang, M., Zeng, D. et al. (2012). *J. Mater. Chem.* 22: 20602.

37 Kim, J.-H., Park, M.-S., Song, J.-H. et al. (2012). *J. Alloys Compd.* 517: 20–25.

38 Li, G.R., Feng, X., Ding, Y. et al. (2012). *Electrochim. Acta* 78: 308–315.

39 Kulova, T.L. and Skundin, A.M. (2016). *Russ. J. Electrochem.* 52: 501–524.

40 Liu, J. and Manthiram, A. (2010). *J. Mater. Chem.* 20: 3961.

41 Ito, A., Li, D., Sato, Y. et al. (2010). *J. Power Sources* 195: 567–573.

42 Kim, M.G., Jo, M., Hong, Y.S., and Cho, J. (2009). *Chem. Commun.* 218–220. https://doi.org/10.1039/B815378G.

43 Wei, G.Z., Lu, X., Ke, F.S. et al. (2010). *Adv. Mater.* 22: 4364–4367.

44 Pitchai, R., Thavasi, V., Mhaisalkar, S.G., and Ramakrishna, S. (2011). *J. Mater. Chem.* 21: 11040.

45 Zhang, T., Li, D., Tao, Z., and Chen, J. (2013). *Prog. Nat. Sci.: Mater. Int.* 23: 256–272.

46 Thackeray, M.M. (1997). *Solid State Chem.* 25: 1–71.

47 Thackeray, M.M., Shao-Horn, Y., Kahaian, A.J. et al. (1998). *Solid-State Lett.* 1: 7–9.

48 Guan, D., Jeevarajan, J.A., and Wang, Y. (2011). *Nanoscale* 3: 1465–1469.

49 Yongyao, X. and Yoshio, M. (1996). *J. Electrochem. Soc.* 143: 825–833.

50 Huang, H., Vincent, C.A., and Bruce, P.G. (1999). *J. Electrochem. Soc.* 146: 3649–3456.

51 Gnanaraj, J.S., Pol, V.G., Gedanken, A., and Aurbach, D. (2003). *Electrochem. Commun.* 5: 940–945.

52 Sun, Y.K., Lee, M.J., Yoon, C.S. et al. (2012). *Adv. Mater.* 24: 1192–1196.

53 Wang, Y., Wang, X., Yang, S. et al. (2012). *J. Solid State Electrochem.* 16: 2913–2920.

54 Bai, Y., Chang, Q., Yu, Q. et al. (2013). *Electrochim. Acta* 112: 414–421.

55 Arora, P., Popov, B.N., and White, R.E. (1998). *J. Electrochem. Soc.* 145: 807–815.

56 Krins, N., Hatert, F., Traina, K. et al. (2006). *Solid State Ionics* 177: 1033–1040.

57 Robertson, A.D., Lu, S.H., and Averill, W.F. (1997). *J. Electrochem. Soc.* 144: 3505–3512.

58 Shin, Y. and Manthiram, A. (2002). *Electrochem. Solid-State Lett.* 5: A55.

59 Manthiram, A., Chemelewski, K., and Lee, E.-S. (2014). *Energy Environ. Sci.* 7: 1339.

60 Zhong, Q., Bonakdarpour, A., Zhang, M. et al. (1997). *J. Electrochem. Soc.* 144: 205.

61 Idemoto, Y., Narai, H., and Koura, N. (2003). *J. Power Sources* 119–121: 125–129.

62 Ariyoshi, K., Iwakoshi, Y., Nakayama, N., and Ohzuku, T. (2004). *J. Electrochem. Soc.* 151: A296.

63 Terada, Y., Yasaka, K., Nishikawa, F. et al. (2001). *J. Solid State Chem.* 156: 286–291.

64 Kunduraci, M. and Amatucci, G.G. (2008). *Electrochim. Acta* 53: 4193–4199.

65 Aurbach, D., Markovsky, B., Salitra, G., and Markevich, E. (2007). *J. Power Sources* 165: 491–499.

66 Fan, Y., Wang, J., Ye, X., and Zhang, J. (2007). *Mater. Chem. Phys.* 103: 19–23.

67 Noguchi, T., Yamazaki, I., Numata, T., and Shirakata, M. (2007). *J. Power Sources* 174: 359–365.

68 Liu, J. and Manthiram, A. (2009). *J. Electrochem. Soc.* 156: A833.

69 Liu, J. and Manthiram, A. (2009). *J. Electrochem. Soc.* 156: A66.

70 Fang, X., Ge, M., Rong, J., and Zhou, C. (2013). *J. Mater. Chem. A* 1: 4083.

71 Abouimrane, A., Belharouak, I., and Amine, K. (2009). *Electrochem. Commun.* 11: 1073–1076.

72 Lee, H., Choi, S., Choi, S. et al. (2007). *Electrochem. Commun.* 9: 801–806.

73 Zhu, Y., Li, Y., Bettge, M., and Abraham, D.P. (2012). *J. Electrochem. Soc.* 159: A2109–A2117.

74 Wu, Q., Lu, W., Miranda, M. et al. (2012). *Electrochem. Commun.* 24: 78–81.

75 Gong, Z. and Yang, Y. (2011). *Energy Environ. Sci.* 4: 3223.

76 Manthiram, A. and Goodenough, J.B. (1987). *J. Solid State Chem.* 7: 349–360.

77 Huang, H., Yin, S.C., and Nazar, L.F. (2001). *Electrochem. Solid-State Lett.* 4: A170.

78 Chung, S.Y., Bloking, J.T., and Chiang, Y.M. (2002). *Nat. Mater.* 1: 123–128.

79 Tarascon, J.M. and Armand, M. (2001). *Nature* 414: 359.

80 Ritchie, A. and Howard, W. (2006). *J. Power Sources* 162: 809–812.

81 Ohzuku, T. and Brodd, R.J. (2007). *J. Power Sources* 174: 449–456.

82 Whittingham, M.S. (2008). *MRS Bull.* 33: 411.

83 Scrosati, B. and Garche, J. (2010). *J. Power Sources* 195: 2419–2430.

84 Pahdi, A.K., Nanjundaswamy, K.S., and Goodenough, J.B. (1997). *J. Electrochem. Soc.* 144: 1188.

85 Fisher, C.A., Hart Prieto, V.M., and Islam, M.S. (2008). *Chem. Mater.* 20: 5907.

86 Li, J., Yao, W., Martin, S., and Vaknin, D. (2008). *Solid State Ionics* 179: 2016–2019.

87 Andersson, A.S. and Thomas, J.O. (2001). *J. Power Sources* 97: 498.

88 Chen, J.-M., Hsu, C.-H., Lin, Y.-R. et al. (2008). *J. Power Sources* 184: 498–502.

89 Bao, L., Gao, W., Su, Y. et al. (2012). *Chin. Sci. Bull.* 58: 575–584.

90 Dominko, R., Bele, M., Gaberšček, M. et al. (2006). *Electrochem. Commun.* 8: 217–222.

91 Zhang, Q., Zhuang, Q., Xu, S. et al. (2012). *Ionics* 18: 487–494.

92 Wang, Y., Wang, Y., Hosono, E. et al. (2008). *Angew. Chem. Int. Ed.* 47: 7461–7465.

93 Hench, L.L. and West, J.K. (1990). *Chem. Rev.* 90: 33–72.

94 Doeff, M.M., Hu, Y., McLarnon, F., and Kostecki, R. (2003). *Electrochem. Solid-State Lett.* 6: A207.

95 Yang, S., Zavalij, P.Y., and Whittingham, M.S. (2001). *Electrochem. Commun.* 3: 505.

96 Devaraju, M.K. and Honma, I. (2012). *Adv. Energy Mater.* 2: 284–297.

6

Next-Generation Anodes for Secondary Li-Ion Batteries

Katerina E. Aifantis[1], Utkarsh Ahuja[1], Yanhong Wang[2], and Hong Li[2]

[1] *University of Florida, Mechanical and Aerospace Engineering, 1064 Centre Drive, Gainesville, FL 32611, USA*
[2] *Renewable Energy Laboratory, Institute of Physics, Chinese Academy of Sciences, Beijing, 100190, China*

6.1 Introduction

This chapter will provide an overview of the materials chemistries and configurations that are being studied experimentally in order to develop the next-generation anodes for rechargeable Li-ion batteries.

The first laboratory Li-ion cells used Li metal and alloys as the anode material [1], and various solid solution Li materials (cobalt oxide) as the cathode and non-aqueous liquid electrolytes (propylene carbonate:diethyl carbonate [PC:DEC]). During the Li^+ transfer between the electrodes upon electrochemical cycling, Li was non-uniformly deposited on the anode leading to the formation of thin Li protrusions (dendrites) that could reach the cathode and therefore provide an alternate internal path for electron transfer. This was termed short-circuiting and led to severe safety issues, including explosions. As mentioned in Chapter 5, intercalation materials were proposed as the cathode, which contained Li, allowing the replacement of the Li anode with other materials. Akira Yoshino (Nobel Prize in Chemistry in 2019) first proposed the use of carbon as the anode [2]. The carbon structure, as well as the anode thickness, play a detrimental role in the number of Li ions it can intercalate [3]. Petroleum coke carbon, for example, which has a turbostratic structure, can intercalate a low amount of Li, forming Li_xC_6, where $0 < x < 0.5$ [4]. Graphite, however, which has an almost perfectly layered structure, is able to intercalate twice the amount of Li, forming Li_xC_6, where $0 < x < 1$, and giving a capacity of $372\,mAh\,g^{-1}$ [4]. During the first electrochemical cycles, the electrolyte decomposes on the carbon surface, forming a passivation layer; once this layer forms over the entire graphite surface, the electrolyte does not further decompose and the cell can be cycled continuously with no capacity loss [4]. Graphitic anodes are, therefore, used in all commercial Li-ion batteries. Despite the anode being graphitic, in fabricating the porous electrode, highly conductive carbon is added in order to increase the conductivity and allow for a better capacity

Rechargeable Ion Batteries: Materials, Design, and Applications of Li-Ion Cells and Beyond, First Edition. Edited by Katerina E. Aifantis, R. Vasant Kumar, and Pu Hu.

retention [3], despite resulting in a lower Li intercalation. It has been possible to exceed the Li intercalation in graphite beyond LiC_6 [5], but at the expense of safety issues, such as Li dendrite formation on the anode surface [3, 5].

At the time that research was being performed to establish graphitic anodes, concurrent studies were being performed on metals that could host Li-ions, without the occurrence of dendritic deposition of metallic lithium on their surface upon cycling [6]. These early papers suggested that Li intermetallics with various metals, such as Sn, Si, Al, Bi, Pb, and In, exhibited promising electrochemical properties [7–10]. The commercialization of these high-capacity anodic materials has not been possible, however, due to the large volume expansions, over 100% [6], that they experience upon lithiation. As a result, during electrochemical cycling, the anodes expand (lithiation) and contract (delithiation) leading to fracture. This fracture produces active material that is no longer in electrical contact with the remainder of the electrode, which therefore becomes unable to respond to the applied voltages necessary to recharge or control the discharge of the battery. Furthermore, fracture increases the surface area available to chemical attack by corrosive agents present in the cell (HF and H_2O). Significant capacity drops are, therefore, observed after the first few cycles. Depending on the stability of the system, the surface cracks may lead to complete fracture of the electrode, thus resulting in safety issues as well.

In particular, the most promising materials for next-generation Li-ion batteries are Si and Sn, as they both allow for a maximum Li insertion that is over four times greater than that of graphite, $Li_{4.4}Si$ and $Li_{4.4}Sn$; while the respective specific capacities given off upon the formation of these Li-rich alloys are 4200 and 996 mAh g^{-1}, respectively [11], and their corresponding volume expansions can be over 300%. In this connection, it should be mentioned that the commercially used graphite has a very high mechanical stability as the volume expansions it undergoes are significantly smaller, ~6–10%. In order to minimize the expansion of Si and Sn, nanostructured Sn and Si anodes are being considered as nanomaterials not only have unique deformation mechanisms [12, 13], but also preferred electrochemical properties [14].

It is well known that during charging of the battery, a voltage is applied such that electrons flow from the cathode to the anode; as a result, the anode attains a negative charge and the positive Li^+ from the cathode are attracted to it. The exchange of Li ions during cycling affects anode materials significantly, both mechanically and chemically. This chapter will give an overview of the progress that has been made thus far in examining promising high-capacity, metal-based anodes. In addition to various materials chemistries, different configurations will be considered, as they also affect significantly the capacity of anodes.

Although studies existed since the 1970s [8, 9], suggesting that various metals such as Sn, Si, and Bi could intercalate Li, it was in the mid-1990s that industry began producing patents on such promising metal-based anodes; the proposed anodes would comprise of about 80% of the active metal-base powder, 10% carbon (for conductivity), and 10% PVDF (for connectivity). These patents by Fuji [15] concerned anodes comprised of 5–10 μm powder particles of SnO, also known as tin composite oxide (TCO). It provided a capacity of about 600 mAh g^{-1} [16].

The Sn lithiation takes place in two steps. During the first electrochemical cycle, the Li ions react with the oxygen to form Li_2O, resulting in Sn nanoparticles dispersed in a Li_2O glass matrix [16]. Once this two-phase composite forms, the additional Li ions that diffuse into the anode react with Sn to form Li-rich Sn phases. During discharging, Li is attracted back to the cathode, and therefore extracted from the tin-rich phases, and then re-inserted upon recharging. The formation of Li_2O during the first cycle results in the high irreversible capacity loss observed in TCO materials, upon the first cycle. The aforementioned reactions can be written as [15–18]:

$$SnO_2 + 4Li^+ + 4e^- \rightarrow 2Li_2O + Sn \text{ (irreversible reaction)}$$

$$xe^- + xLi^+ + Sn \leftrightarrow Li_xSn \ (0 \leq x \leq 4.4) \text{ (during continuous cycling)}$$

The first reaction (irreversible) is known as the conversion reaction, while the second one (reversible) as the insertion reaction. It was previously mentioned that electrochemical cycling of graphitic anodes resulted in the formation of a passivation layer on graphite; this protective layer inhibits co-intercalation of electrolyte solvents that result in exfoliation. In graphitic anodes, this layer forms once on the graphite surface during the first cycle (at voltages below 0.8 V Li/L$^+$) and its drawback is that it introduces the irreversible capacity. Such a layer also forms in these new TCO materials, termed solid electrolyte interface (SEI) and the sequence of reactions is not well clarified. There are two possibilities:

1. The SEI layer forms first on the surface of the TCO, then Li ions diffuse through it, and decompose the TCO to Sn particles and the Li_2O matrix; finally Li reacts with Sn to form rich Li–Sn alloys.
2. The Li reacts with the TCO to form the Li_2O and Sn particles, then the SEI layer forms on the surface of the particles, and finally the Li–Sn reaction takes place.

Impedance studies of TCO anodes [19] concluded that prior to the reaction of Li with O_2, Li first reacts with the surface of the material forming a SEI shell, which is distorted; then Li diffuses through this shell to distort the crystal structure of TCO and form the Li_2O matrix in which Sn is distributed, and then the reversible Li_xSn reaction takes place. The reaction during SEI formation is written as

$$Li^+ + (EC, DEC, \ldots) + e^- \rightarrow SEI\ (Li_2CO_3, ROCO_2Li, \ldots),$$

where (EC, DEC, …) denote the electrolyte components.

The composition of the SEI layer has been identified for SnO anodes as Li_2CO_3, which forms at 1.2 to 0.9 V Li/Li$^+$, and $ROCOLi_2$, which forms later at voltages below 0.9 V Li/Li$^+$; it follows, therefore, that the Li_2CO_3 forms first [19]. The SEI layer on TCO is a greater problem than the passivation layer is in graphitic anodes, since as the Sn undergoes large volume expansions, during cycling, that lead to fracture more Sn surface becomes available to chemical attack by the electrolyte. Furthermore, as the sites expand and contract, upon Li insertion and de-insertion, the SEI layer forms and disintegrates continuously, and therefore new SEI layers form on aged layers, and eventually a thick layer forms that inhibits Li diffusion in the active material, and reduces the battery capacity.

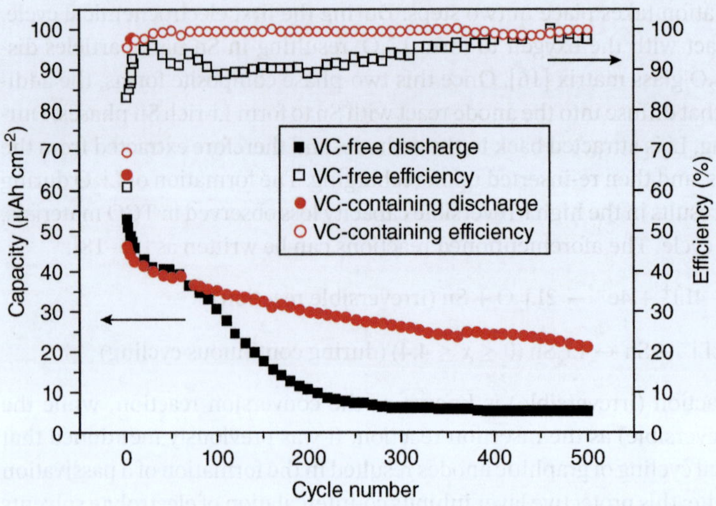

Figure 6.1 Si thin films cycled in a VC-free and VC-containing electrolyte. Source: Reproduced with permission from Chen et al. [22]/Elsevier.

Li intercalation and SEI layer formation are similar in other metal-based anodes, and studies are being performed to fully understand this process. In trying to minimize the SEI thickness, studies are being performed to estimate the operation voltages that inhibit such electrolyte decomposition; in [20], for example, it is indicated that the formation of this layer on the surface of AlSb anodes is reduced at voltages below 0.5 V.

Another way to face this problem is by considering alternate electrolytes. It was shown that adding vinylene carbonate (VC) in electrolytes enhances the performance of graphitic anodes, by reducing their irreversible capacity and improving their cyclability [21]. This led to further studies [22] in which the same thin-film Si anodes were cycled using an electrolyte that contained VC and one that did not. In the VC-free electrolyte case, the SEI thickness kept increasing with cycling, and it contained a higher content of compounds, such as LiF, that decrease the capacity. As a result, the VC-free electrode had an increased anode polarization, which resulted in the degradation of cycle performance. The VC-containing electrolyte, however, allowed for an SEI layer to form that had a stable thickness and, therefore, a higher capacity was retained for 500 cycles for the VC-containing electrolyte [22], as shown in Figure 6.1. New electrolytes that are being investigated for Li batteries will be described in detail in Chapter 7.

6.2 Mechanical Instabilities During Electrochemical Cycling

In addition to the formation of the SEI, fracture of the electrodes also plays a key role in capacity decay. In Figure 6.1, it was illustrated that controlling the formation of the SEI layer, through the electrolyte, resulted in a higher capacity. It can be seen, however, in Figure 6.1 that for both types of electrolytes, a significant capacity drop took place, regardless of the structure of the SEI layer. This capacity drop resulted from the over 300% expansion of the Si thin film, which leads to fracture. In continuing the study of these next-generation anodes, it is essential to describe the mechanical instability phenomena that arise upon electrochemical cycling (a brief description of how candidate electrode materials are cycled was given in Chapter 3).

During charging of the battery the Li^+-ions enter the anode, they diffuse through the inactive regions that may be present and they enter the active, with respect to Li, metal inclusions (Si, Sn, Al, Sb, Bi can act as active sites). As Li enters the active sites, the metal expands over 200%. During discharge of the battery, that is, during operation, electrons flow into the cathode and hence the Li^+-ions are attracted back to their original positions. As a result, the active sites attain the initial volume they had – prior Li insertion – and a contraction takes place, leading to fracture.

This cracking forms has been documented very nicely in [11]. In [11], a 4-μm Cu film, which is inactive with respect to Li, was sputtered on a stainless steel substrate, and then a 1-μm SiSn thin film was sputtered on the Cu. As a complete battery system was not examined in [11], but instead the behavior of the SiSn film upon Li insertion and de-insertion was considered, the authors of [11] refer to the insertion of Li in SiSn as the discharge process, while the removal of Li is referred to as the charging process; pure Li metal was used as the counter electrode. [When Li is the counter electrode it is a half cell.] Upon Li insertion, Si and Sn experienced an expansion of 310% and 260%, respectively. To determine the direction of this expansion, razor marks were carved on the SiSn film prior to the initiation of electrochemical cycling (Figure 6.2a). Figure 6.2b was taken after the first discharge

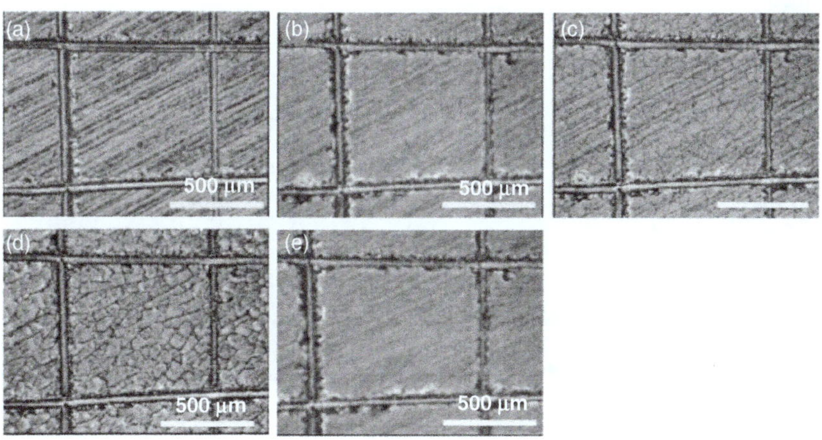

Figure 6.2 Experimental evidence taken from [11] on electrochemical cycling of a SiSn thin film. (a) Before cycling begins, (b) after first Li-insertion, (c) one hour during first Li de-insertion, (d) after complete Li de-insertion, (e) after second Li-insertion. Source: Reproduced with permission from Beaulieu et al. [11]/IOP Publishing.

(i.e. after SiSn was fully lithiated); no lateral expansion is observed and, therefore, the film expanded in the out-of-plane direction. In Figure 6.2c, it can be seen that as Li was removed from SiSn, cracks began to form, which became more apparent as Li de-insertion progressed (Figure 6.2d). The flake-like active particles produced from these cracks are 100 μm. These particles have the ability to continuously expand and contract upon cycling, and it can be seen that once the film became fully lithiated (Figure 6.2e), the cracks were not visible and the film looked as it did during the first discharge (Figure 6.2b). Therefore, even though SiSn fractured, it is possible to maintain electrical connectivity and control the charging and discharging processes; this electrical contact results from the fact that the center of the cracked particles remains pined to the Cu film and, therefore, electrical contact is sustained. After further cycling, however, the cracked SnSi particles can detach from the Cu, and therefore the capacity drops to zero.

This submicroscopic fracture of the SiSn anode that results from electrochemical cycling resembles the macroscopic phenomenon of drying mud, as seen in Figure 6.3; it is very interesting that the pattern in both the cracked active site and cracked mud are similar, regardless of the great difference in the scale. In this case, Li acts in a similar way as the evaporating water; in fact, just as dry mud, the edges of the cracked SiSn particles curl at the edges, where they detach from the Cu film.

Similar cracking behavior has been observed when electrochemically cycling a SiO_2 thin film [23]. The cracked particles produced, however, had a much smaller diameter of 100 nm. It should be noted here that it has been proposed that these cracks might act as high diffusivity paths for the Li ions, allowing for better electrochemical properties of the anode but this has not been examined in-depth.

In the remainder of this chapter, various methods that have been developed to deal with the aforementioned instability phenomena of these promising anodes will be described.

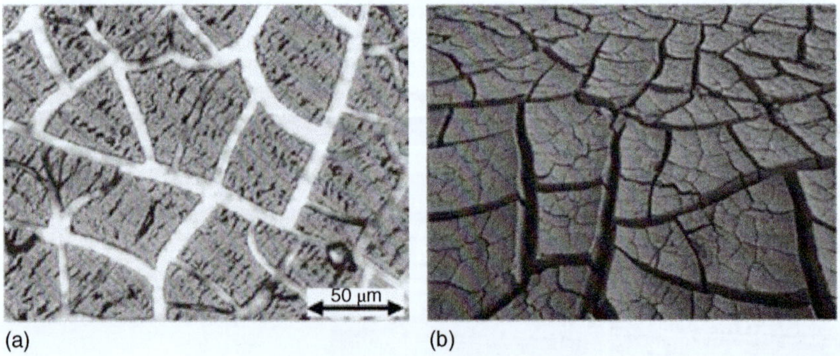

(a) (b)

Figure 6.3 (a) an optical micrograph of a Li alloy film after expansion and contraction due to electrochemical cycling; (b) cracked mud in a dry lake bed bottom. Source: Reproduced with permission from Beaulieu et al. [11]/IOP Publishing.

6.3 Nanostructured Anodes

Extensive experimental research has yielded that the most effective way to minimize the large expansions of metals that are active with respect to Li is to minimize their dimensions since deformation and fracture mechanisms are less severe at the nanoscale.

In addition to the mechanical stability, the electrochemical properties are also enhanced at the nanoscale for two main reasons: (i) the large surface area present in nanomaterials allows for higher charge/discharge rates, (ii) nanodimensions result in shorter path lengths for Li-ion transport, meaning higher Li-diffusivities, which allow for an increase in power capabilities; for example, bulk Si reacts with Li at 400 °C, whereas nanoscale Si reacts with Li at room temperature [24]. This discovery by Li and co-workers was key in paving the way to commercialize Si, and since then the majority of articles focusing on anodes are concerned with Si.

There exist numerous experimental studies that propose promising Si and Sn-based anodes that allow for capacities nearly double that of current graphitic anodes. They exist in various configurations such as (i) thin films, (ii) fiber-like/nanowires, and (iii) active/less-active nanocomposites. In the sequel, the various next-generation anodes will be presented in the aforementioned three categories. In Section 6.8, Sb, Al, and Bi anodic materials will also be summarized, although they appear to be the furthest away from commercialization.

6.4 Sn-Based Materials

Despite Sn not being the material that gives the highest capacity upon its reaction with Li, it will be examined first since it was the metal that popularized the use of high-capacity metal-base anodes, after the patent by Fuji Film [15]. The volume expansion that Sn undergoes upon the formation of $Li_{4.4}Sn$ is ~298%. As previously mentioned, these acute volume expansions cause pulverization of the active material, loss of electronic conductivity, formation of a non-uniform SEI layer, delamination of the electrode from the current collector, and other issues, which negatively impact the performance of the anode [25, 26]. This volume expansion is unavoidable since it is a result of the high Li uptake of Sn, and various strategies have been employed to mitigate its effect on capacity retention. As mentioned, nanoscale structures limit fracture, but also reduces diffusion path lengths, and increases the surface area to facilitate electron transfer, thereby increasing high rate cycling stability [11, 27–29].

The reaction of Li with SnO_2 exhibits faster reaction kinetics but a lower capacity than pure Sn anodes [30]. Moreover, the lower discharge potential of SnO_2 over Sn makes it a viable alternative [31]. As mentioned earlier, SnO_2 reacts with Li in two steps and the theoretical capacity is 1497 mAh g^{-1} [29].

As in the case of Sn, shortcomings of SnO_2 anodes include the formation of unstable and excessive SEI layers due to severe volume expansions, and the agglomeration of electrochemically active pulverized tin clusters, which subsequently lead to an increase in capacity fade. Nevertheless, use of nanostructured active particles or hollow structures to buffer the expansion can be employed to sustain the capacity during cycling [32–36].

6.4.1 Sn-Based Conversion Reaction Materials

In [37], 550 nm films were fabricated by electroplating SnO_2 on a Pt current collector. The resulting electrodes had an initial capacity of 560 mAh g^{-1}, which was reduced to 440 mAh g^{-1} after 40 cycles.

More efficient Sn-based thin film anodes can be achieved by using a different matrix material. It was observed in [38] that Li could form a buffering matrix with S, similarly as it does with O. Studies performed in [39] propose that the Li intercalation takes place again by a conversion reaction followed by insertion

$$xLi^+ + SnS + xe^- \rightarrow Li_xS + Sn \text{ (irreversible reaction)}$$
$$xe^- + xLi^+ + Sn \leftrightarrow Li_xSn\ (0 \leq x \leq 4.4) \text{ (during continuous cycling)}$$

In [40], SnS/C thin film anodes were fabricated by homogeneously dispersing 5–6 nm SnS particles into carbon aerogel and then coating this thin film on copper foil through the spin-coating method. Carbonizing the mixture resulted in a SnS/C nanocomposite thin film that was 600 nm. Therefore, C provided additional constraints to the expansion of Sn, and also since C is active with respect to Li, it could store additional Li$^+$. The initial capacity of this thin film was nearly 1000 mAh g^{-1}, while after a few cycles it stabilized at 530 mAh g^{-1} for 40 cycles. The significant capacity drop of about 500 mAh g^{-1} that occurred during the first cycle (Figure 6.4)

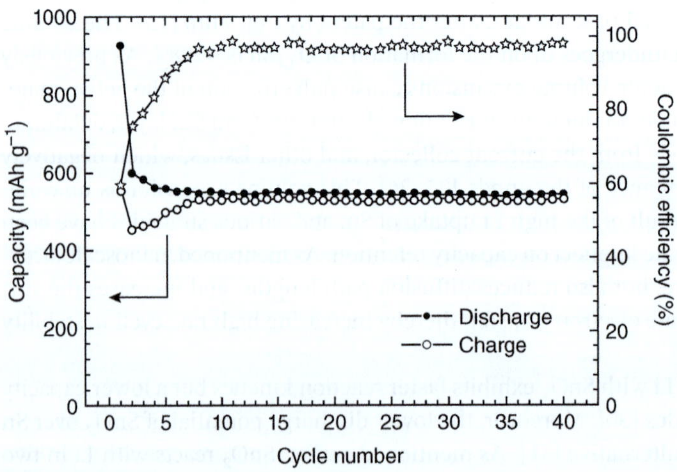

Figure 6.4 Capacity retention of SnS/C nanocomposite thin film. Source: Reproduced with permission from Li et al. [40]/Elsevier.

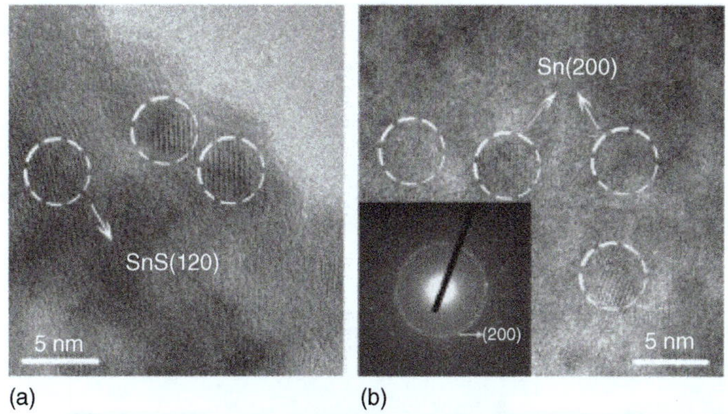

Figure 6.5 SnS/C nanocomposite film. The left high resolution transmission electron microscopy (HRTEM) micrograph was taken prior electrochemical cycling, while the left was taken after the fourth electrochemical cycle. It can be seen that no significant damage took place. Source: Reproduced with permission from Li et al. [40]/Elsevier.

was due to the irreversible bond of Li ions to form the Li_xS matrix. Once this matrix formed the capacity was stable, and the presence of C and S prevented aggregation and fracture of the Sn particles (Figure 6.5), and minimized the formation of the SEI layer; the small dimensions of the Sn particles also play a detrimental role in the minimal damage observed.

There exist additional Sn-based thin films, to those described above; however, they give capacities below 400 $mAg\,h^{-1}$, which are comparable to those of commercial carbonaceous anodes and therefore will not be discussed.

Another promising Sn-based anode that undergoes a conversion reaction is that of layered SnS_2 nanoplates. The fabrication of these unique nanostructures is achieved by thermal decomposition of the precursor, $Sn(S_2CNEt_2)_4$, in an organic solvent at an elevated temperature [41]. These nanoplates are 120 nm in the lateral direction, and they are stacked on each other along the [001] direction; their stacking configuration is enforced by van der Waals interactions [42]. Each nanoplate is a single crystal comprised by 30 layers according to Figure 6.6, while their shape is hexagonal as depicted in Figure 6.7.

During electrochemical cycling, lithiation takes place through the conversion mechanism as follows:

$$SnS_2 + 4Li^+ \rightarrow 2Li_2S + Sn \text{ (irreversible reaction)}$$

$$xLi^+ + Sn \leftrightarrow Li_xSn \ (0 \leq x \leq 4.4) \text{ (during continuous cycling)}.$$

The initial capacity of these materials was 645 $mAh\,g^{-1}$, which stabilized to 583 $mAh\,g^{-1}$, as seen in Figure 6.8. The high capacity retention of these SnS_2 nanoplates was attributed to the following factors: (i) the Li_2S matrix buffers the expansion of Sn, and (ii) their 2D layered structure results in a finite lateral size with enhanced open-edges. As a result, the Li-ion diffusion through the active materials is increased, while the overvoltage associated with the formation of Li–Sn is decreased [43–45].

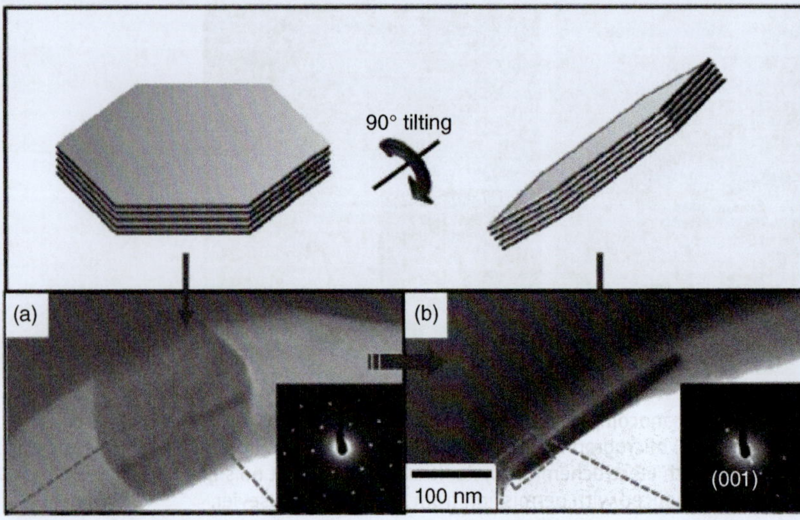

Figure 6.6 TEM images, obtained by rotating the TEM holder stage 90°, indicating the layered structure of the SnS_2 nanoplates. Source: Reproduced with permission from Seo et al. [42], John Wiley & Sons.

Figure 6.7 (a) TEM and (b) field emission scanning electron microscopy (FESEM) images of SnS_2 nanoplates. Source: Reproduced with permission from Seo et al. [42], John Wiley & Sons.

6.4.2 Sn-Based Alloys

In addition to binding Sn to O, intermetallic compounds (two-component or multi-component), such as Sn–Se and Sn–Sb–Co, Sn–Co–C, have shown better cycling stability than pure Sn anodes at the expense of capacity due to inactivity of some metals towards Li [46, 47]. A ternary Sn–Co–C electrode in a proper composition was developed by SONY in 2005. When cycled against Li, its capacity was as high as 500 mAh g^{-1} and it could be cycled at rates exceeding 1 C, while

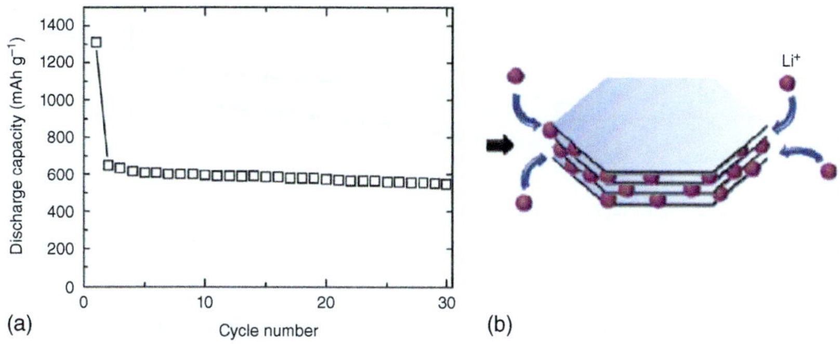

Figure 6.8 (a) Capacity retention of SnS_2 nanoplates. Source: Reproduced with permission from Seo et al. [42]. (b) Li-insertion in $SnSb_2$ nanoplates. Source: Reproduced with permission from Seo et al. [42]/John Wiley & Sons.

still recovering 50% of its total capacity [48–50]. A complete cell was assembled where the $Sn_{0.31}Co_{0.28}C_{0.41}$ anode was coupled with a high-voltage $LiNi_{0.5}Mn_{1.5}O_4$ cathode in a 1M $LiPF_6$ EC:DMC electrolyte solution. As shown in Figure 6.9b, this corresponds to a capacity of the order of 130 mAh g^{-1} for the cathode (i.e. approaching 85% of the theoretical value) and to about 380 mAh g^{-1} for the anode. The performance was quite good, suggesting that the $Sn_{0.31}Co_{0.28}C_{0.41}$ ternary compound is a promising electrode material for the development of new generation Li-ion batteries. The Nexelion battery launched by Sony in 2005 made use of such an amorphous Sn–Co–C anode allowing for an energy density of 226 Wh kg^{-1} [51–53]. In addition to higher capacities and energy densities, Sn-based anodes offer an order of magnitude lower electrical resistivity than graphite.

Use of electrochemically active metals in Sn intermetallic compounds can still boost the cycling stability without severely impacting the capacity of anode [54]. In the case of Sn–Sb, for example, Sb is used to buffer the volume expansion of Sn, but it is also active with respect to Li, providing a capacity of 660 mAh g^{-1}, upon maximum lithiation (Li_3Sb). To show the enhanced electrochemical properties that result from alloying Sn with Sb, various Sn–Sb alloys were considered as anodes in [55]. It was noted that the capacity of pure Sn nanoparticles (diameters between 50 and 200 nm, Figure 6.10a) was initially 990 mAh g^{-1} but droped rapidly to 210 mAh g^{-1} after 20 cycles. Adding, however, 46.5 at% Sb (Sn–Sb particle diameters between 50 and 200 nm, Figure 6.10b) resulted in anodes that gave an initial capacity of 701 mAh g^{-1}, but reduceds to 650 mAh g^{-1} after 20 cycles. Other at% Sb compositions (Sn–30.7%Sb, Sn–46.5%Sb, Sn–47.2%Sb, Sn–58.5%Sb, Sn–80.8%Sb) also result in a better cyclability than pure Sn, but 46.5 at% Sb gave the best capacity retention [55]. It follows that since Sb provides a lower capacity than Sn, increasing the amount of Sb in the Sn–Sb alloy lowers the initial capacity; this side effect, however, is greatly compensated by the significantly increased capacity retention of Sn–Sb alloys.

To further enhance the cyclability of Sn-based anodes, there have been studies that coated the Sn–Sb alloy with a C shell. In [56], the cyclability of a crystalline Sn_2Sb

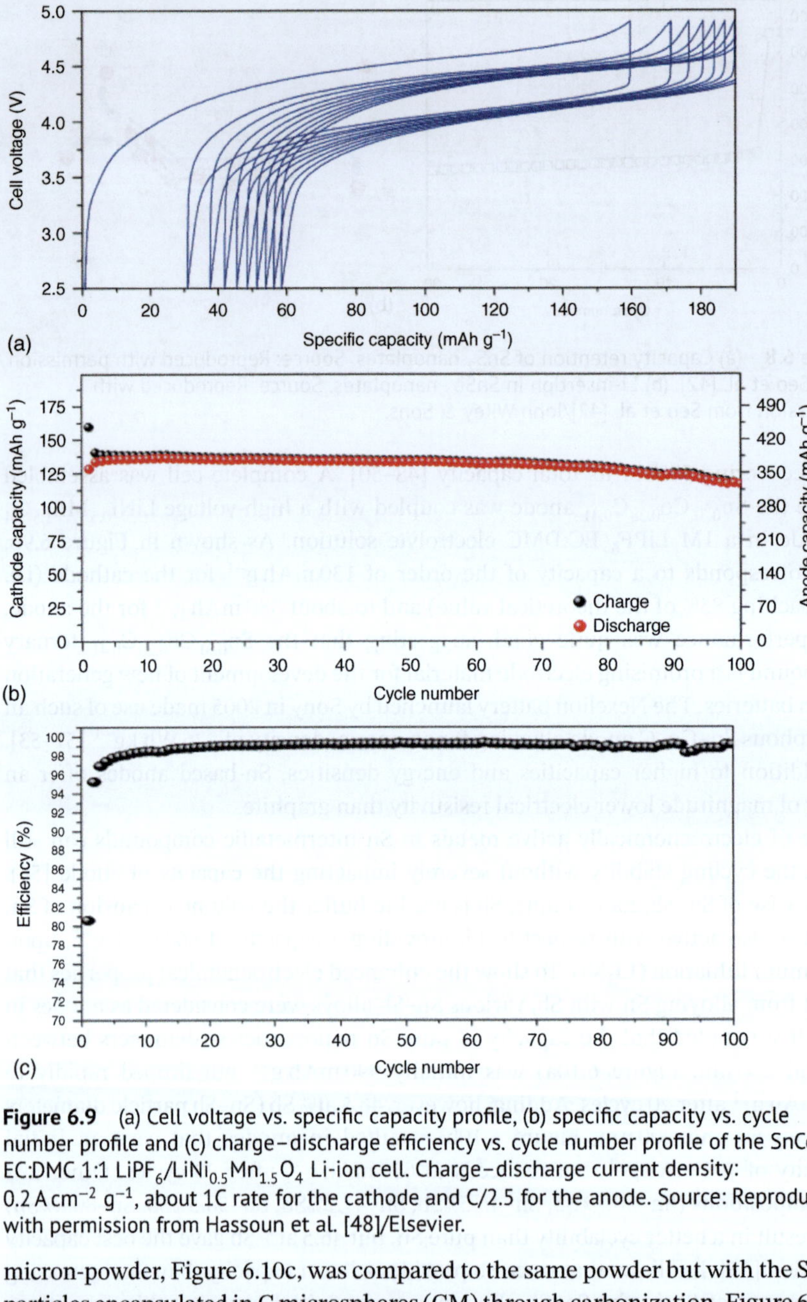

Figure 6.9 (a) Cell voltage vs. specific capacity profile, (b) specific capacity vs. cycle number profile and (c) charge–discharge efficiency vs. cycle number profile of the SnCoC−4/EC:DMC 1:1 $LiPF_6/LiNi_{0.5}Mn_{1.5}O_4$ Li-ion cell. Charge–discharge current density: 0.2 A cm^{-2} g^{-1}, about 1C rate for the cathode and C/2.5 for the anode. Source: Reproduced with permission from Hassoun et al. [48]/Elsevier.

micron-powder, Figure 6.10c, was compared to the same powder but with the Sn_2Sb particles encapsulated in C microspheres (CM) through carbonization, Figure 6.10d. The starting reversible capacity of the 5–10 μm Sn_2Sb particles was 689 mAh g^{-1}, but a significant capacity fade was observed after the 13th cycle (approximately), and after the 60th cycle the capacity had reduced to 150 mAh g^{-1}; even lower than that of graphite. Once, however, these powders were encapsulated in C, the capacity was retained to 600 mAh g^{-1} after 60 cycles, Figure 6.11.

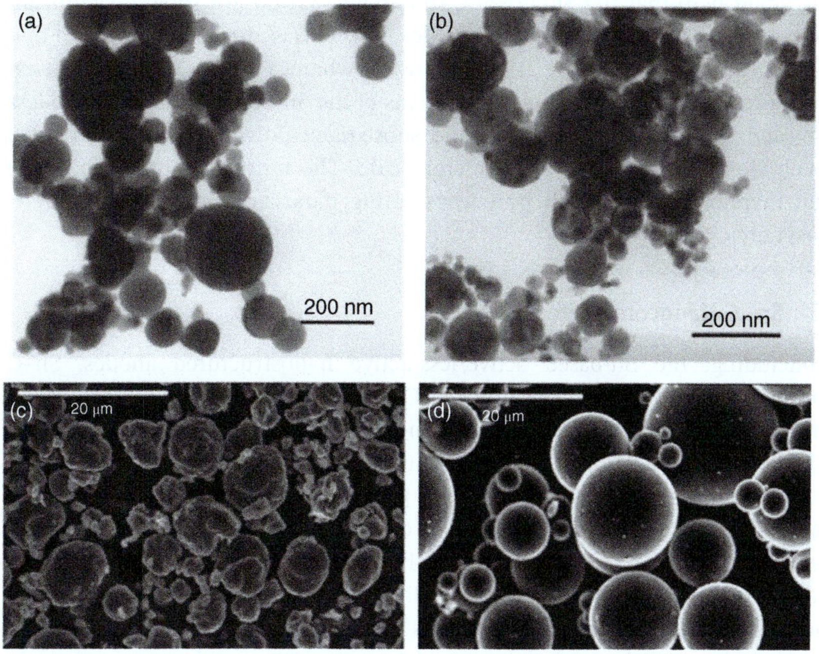

Figure 6.10 (a) Sn nanoparticles. Source: Reproduced with permission from Wang et al. [55]/ELSEVIER (b) Sn-46.5 at% Sb Source: Reproduced with permission from Wang et al. [55]/ELSEVIER; (c) Sn_2Sb micron-powder particles of multiangular shape. Source: Reproduced with permission from Wang et al. [56]/Elsevier. (d) Sn_2Sb particles encapsulated in C microspheres. Source: Reproduced with permission from K. Wang et al. [56]/Elsevier.

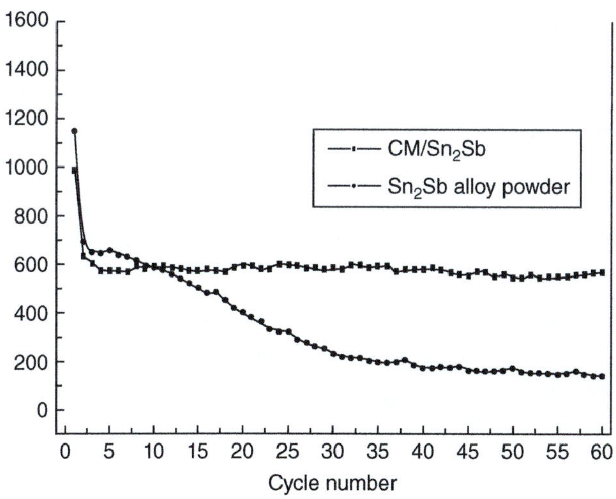

Figure 6.11 The cyclability of Sn_2Sb alloy powder and CM/Sn_2Sb composite electrodes. Source: Reproduced from Wang et al. [56]/with permission of Elsevier.

It is noted here that the Sn-46.5 at% Sb anodes [55] do not show a significant capacity loss at 13 cycles as the Sn_2Sb anodes [56]. This is attributed to the fact that the Sn-46.5 at% Sn particles are at the nanoscale, whereas the Sn_2Sb have a micron size. As a result, the deformation mechanisms of the Sn-46.5 at% Sb anode are less severe, and therefore agglomeration and subsequent pulverization are observed at much slower rates than in the micron-scale. The most effective SnSb anode would, therefore, could result by encapsulating nanosized Sn_2Sb powder by C microspheres.

6.4.3 Sn–C Nanocomposites

In concluding, the Sn-based active/less-active nanostructured anodes, Sn–C nanocomposites will be described, where C can act as a buffering matrix for the volume expansions. In [57], a fabrication method by which Sn nanoparticles were attached to the surface of a synthetic carbon was proposed; however, the resulting capacities were not above 400 mAh g^{-1}. In [58], a similar fabrication method was presented but the resulting Sn/C composite contained Cl, which is corrosive and its presence should be avoided in battery cells. In [59], a stable capacity of 500 mAh g^{-1} was attained for 25 cycles; however, in addition to Sn particles that were attached to the C surface, free-standing Sn particles were also present; these free-standing particles agglomerated and fractured, and hence were not able to contribute in the capacity of the anode. In [60], however, a fabrication method was developed for binding all the Sn particles on the C surface; in some cases, Sn was oxidized and hence SnO_2 was bound on the C surface. As seen in Figure 6.12, the smallest Sn or SnO_2 particles were 4 nm, while in some cases they aligned adjacent to each other forming larger diameters of Sn or SnO_2 islands (this microstructure is also referred to as Sn/SnO_2 island microstructure). In Figure 6.13, it can be seen that a stable capacity of 380 mAh g^{-1} was attained for over 400 cycles for such an 8 wt% Sn–C island composite. Although this capacity is slightly higher than that of currently used graphitic anodes, it should be noted that the carbon matrix was Vulcan-C, which has a capacity of 180 mA g^{-1}; therefore, the 200 mAh g^{-1}

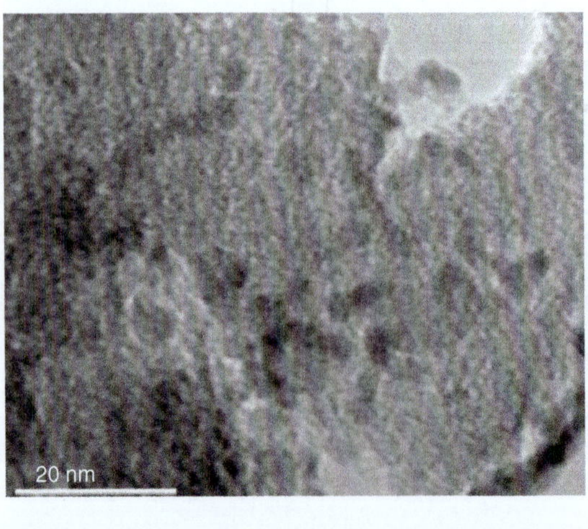

Figure 6.12 TEM image of Sn or SnO_2 islands on Vulcan C. The Sn or SnO_2 islands are shown as dark spots such as those pointed to by the arrows. The Sn content is 8 wt%. Source: Reproduced with permission from Aifantis et al. [60]/Elsevier.

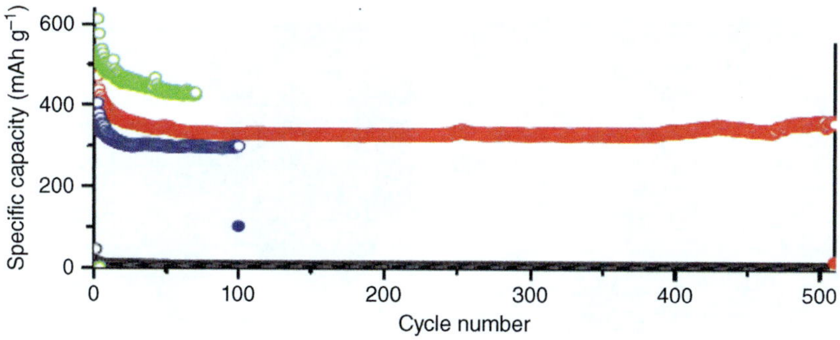

Figure 6.13 Capacity retention at various current densities for 8 wt% Sn C nanopowder (again Vulcan C was used). Green: 0.2 A (cm^2 g$_{AM}$)$^{-1}$, Red: 1.025 A (cm^{2}g$_{AM}$)$^{-1}$, Blue: 1.011 A (cm^{2}g$_{AM}$)$^{-1}$, Black: 1.018 A (cm^{2}g$_{AM}$)$^{-1}$. Source: Reproduced with permission from Aifantis et al. [60]/Elsevier.

capacity increase observed was solely due to the addition of 8 wt% Sn. In attempts to increase the capacity further a graphitic matrix was used, since it had a capacity of over 300 mAh g^{-1} and the Sn content was increased to over 25 wt%. The capacity retention for these materials, however, was very low and dropped below that of graphite before reaching 100 cycles. Although most studies assume such capacity fade is due to fracture, not many document it through scanning and transmission electron microscopy (SEM and TEM). Performing electron microscopy before and after cycling showed that the initial Sn particle size was 100 nm, but agglomeration took place forming Sn micron-sized islands on the C [51]. These large particles underwent fracture, as seen in Figure. 6.14, and detached into the electrolyte leading to the severe capacity loss. The image after cycling (also termed post-mortem) was obtained by opening the coin cell in the glove box, removing the Sn-based porous electrode and dissolving the SEI layer, and then scraping some of the electrode material on a TEM grid.

To better document the microstructural changes during cycling, various half cells were prepared and cycled, with each cell being opened at different stages of cycling to perform post-mortem TEM [51]. It is seen in Figure 6.15 that fracture occurred upon Li-ion insertion and the volume expansions were anisotropic.

It should be mentioned that Sn–C nanocomposites have also been fabricated in which the Sn nanoparticles were embedded in C (e.g. [53]). When the Sn particles were over 100 nm [53], the capacities were not as high; if the Sn nanoparticles, however, had an average diameter of 35 nm, a capacity of 500 mAh g^{-1} was retained for 100 cycles. To achieve this capacity, the Sn content had to be 50 wt%. It is therefore seen that particle size and Sn content play a crucial role in increasing the capacity.

Sn–C composites prepared by aerosol spray pyrolysis were also able to attain an 88% discharge capacity retention at 470 mAh g^{-1} after 200 cycles [26]. Further advancements have been reported in Sn–C composites when carbon nanotubes (CNTs) and graphene were used to increase the conductivity and surface stability of the electrode [61].

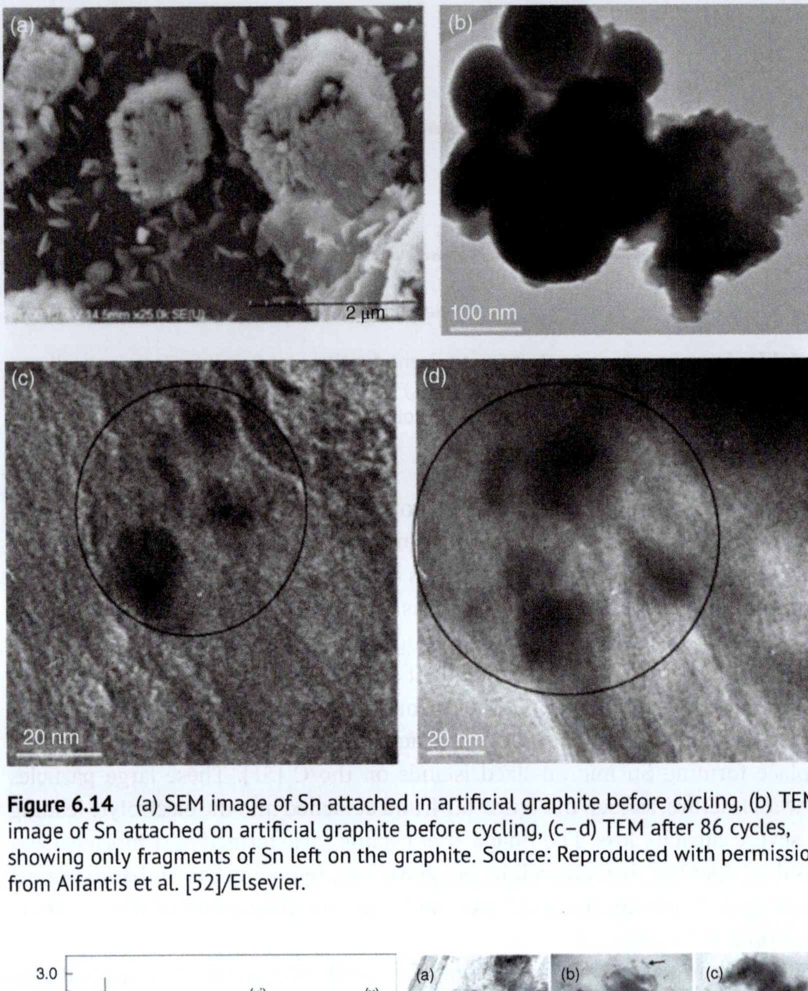

Figure 6.14 (a) SEM image of Sn attached in artificial graphite before cycling, (b) TEM image of Sn attached on artificial graphite before cycling, (c–d) TEM after 86 cycles, showing only fragments of Sn left on the graphite. Source: Reproduced with permission from Aifantis et al. [52]/Elsevier.

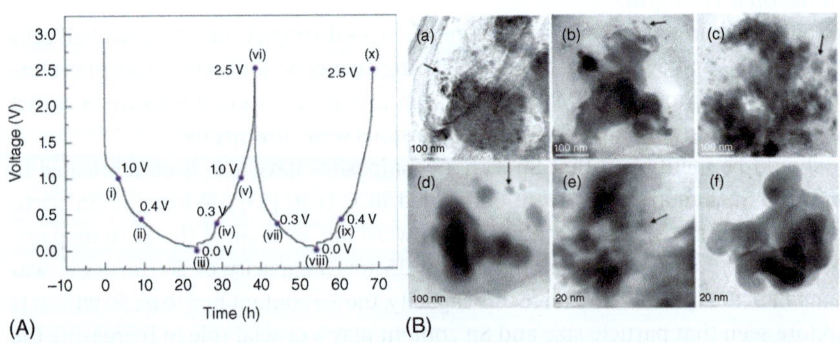

Figure 6.15 (A) Voltage-time curve for the first two cycles of Sn attached on artificial graphite. (B) TEM images of Sn-based anode at stages indicated in the voltage-time curve: (a) at stage i, (b and c) at stage iii, (d) at stage v, (e and f) at stage viii. It is seen at stage f that the smaller Sn fractured particles had delaminated in the electrolyte. Source: Reproduced with permission from Aifantis et al. [51]/Elsevier.

6.4.4 Sn-Based Nanofiber/Nanowire Anodes

In Section 6.4.1, the results for the SnO_2 thin film described in [37] were reported, and as seen they did not have a high capacity retention. In addition, however, to a thin film SnO_2 anode, the authors in [37] examined a fiber-like SnO_2 anode. These fiber-like materials were produced similarly as the thin film, but on performing the template method a microporous membrane was placed on the Pt current collector, and therefore the SnO_2 precipitated in the pores and formed fibers that protruded from the current collector as seen in Figure 6.16. The nanofibers had diameters of ~110 nm. The initial capacity of these anodes was 700 mAh g^{-1}, which increased after the first few electrochemical cycles and stabilized to 770 mAh g^{-1}. It was possible to cycle such electrodes up to 800 times without a decrease in capacity. The unique electrochemical performance of these anodes was attributed to their nanodimensions and also to the fact that the brush-like fibers accommodate the expansion during Li^+ insertion. For illustration purposes, the capacity of the SnO_2 thin film is also shown in Figure 6.17.

Another confirmation of the preferred electrochemical properties of nanofibers was presented in [62]. SnO_2 nanorods that can be used as anodes were fabricated

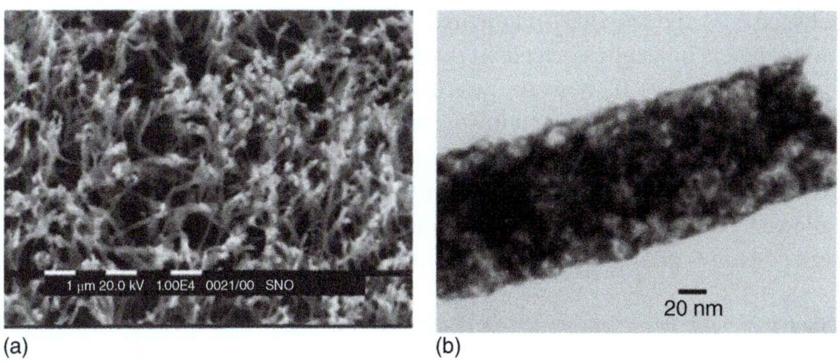

Figure 6.16 (a) Scanning electron micrograph of the nanostructured SnO_2 electrode. The nanofibers are protruding from an underlying Pt current-collector surface; (b) Transmission electron micrograph of a single SnO_2 nanofiber before electrochemical cycling. Source: Reproduced with permission from Li et al. [37]/Elsevier.

Figure 6.17 Capacity vs. cycle number for the nanostructured and thin-film control electrodes at a charge/discharge rate of 8C over the potential window of 0.2–0.9 V. Source: Reproduced with permission from Li et al. [37]/Elsevier.

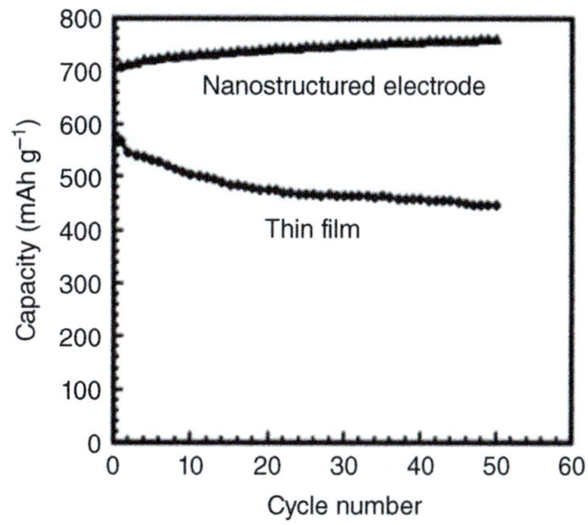

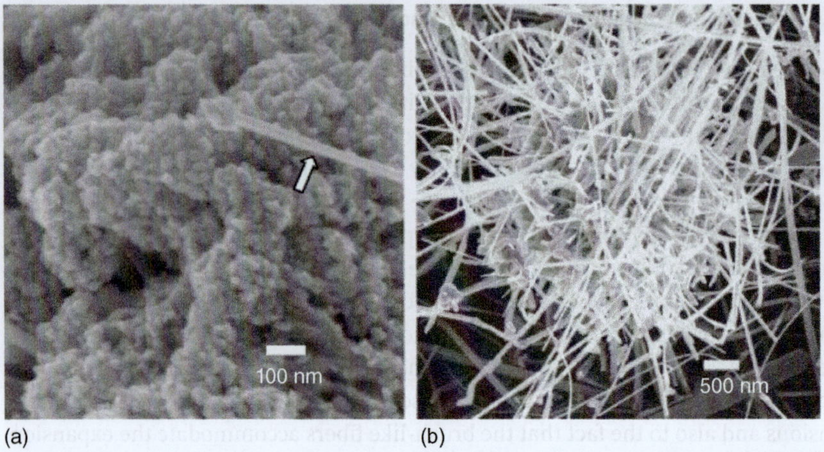

Figure 6.18 SEM images of SnO_2 produced by heat treatment of Sn–Ag alloys: (a) at 800 °C for 30 minutes in air and (b) at 1000 °C for 30 minutes in a gas of Ar–1% O_2. Source: Reproduced with permission from Ahn et al. [62]/Elsevier.

by heat-treating Sn–20%Ag alloys in gas containing oxygen. If the oxygen volume was between 0.5% and 1%, SnO_2, nanowires (30–70 nm in diameter and several hundred micrometers in length) formed on the Sn–Ag surface as shown in Figure 6.18. If the oxygen volume level, however, was above 2%, nanoparticles formed that had a mean diameter of 30 nm; nanowires were still sparingly observed in this case. As Figure 6.18 indicates the SnO_2 powder particles agglomerated, not leaving much room to accommodate the volume expansion upon cycling; the SnO_2 nanorods, however, allowed for a more spaced structure and therefore a higher capacity retention was seen (Figure 6.19).

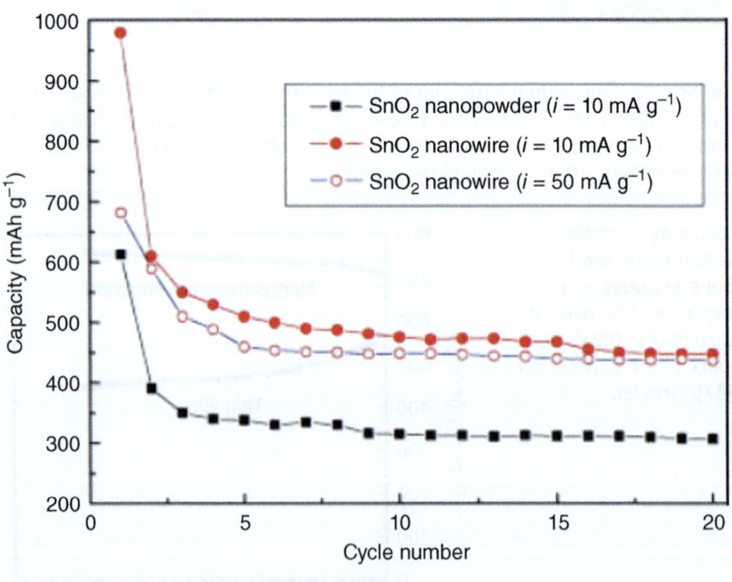

Figure 6.19 Superior capacity retention of SnO_2 nanorods over SnO_2 nanoparticles. Source: Reproduced with permission from Ahn et al. [62]/Elsevier.

6.5 Si-Based Materials

The reversible alloying of Si with Li was first reported in 1976 at elevated temperatures [63] in partnership with General Motor Corporation, but as mentioned it was not until 1999 (with the discovery that nano Si can react with Li at room temperature) that Si became a popular anode candidate [24]. As opposed to Sn, which is rarely used in its pure form, pure Si has been used as an experimental anode base material. The abundance of Si along with its low cost, non-toxic nature, and appropriate discharge potential of ~50 mV makes it an ideal candidate for use as an anode in LIBs. Theory predicts that the formation of $Li_{4.4}Si$ and $Li_{3.75}Si$ produces a capacity of 4200 and 3579 mAh g^{-1}, respectively; according to phase diagrams $Li_{3.75}Si$ forms at room temperature. Due to the high Li uptake capacity Si even at room temperature, fully lithiated alloys can expand up to ~400% (320% at room temperature). Volume expansions of this magnitude generate stresses and repeated volume changes during lithiation and de-lithiation, causing irreversible damage to Si active particles [64]. Delamination of the electrode from the current collector due to repeated expansion/contraction can lead to a loss of electronic pathways. Furthermore, continuous reformation of the SEI layer consumes Li^+, which decreases the Coulombic efficiency and causes capacity fade, since a thick and unstable SEI layer can hinder the Li^+ diffusion. Figure 6.20 depicts the schematic effects of these volume changes, which are similar to the experimental images for Sn-based anodes in Figure 6.15.

Due to the continuous volume changes, the capacity of bulk Si is reduced by 90% after the fifth electrochemical cycle, such that it approaches the capacity of C. For

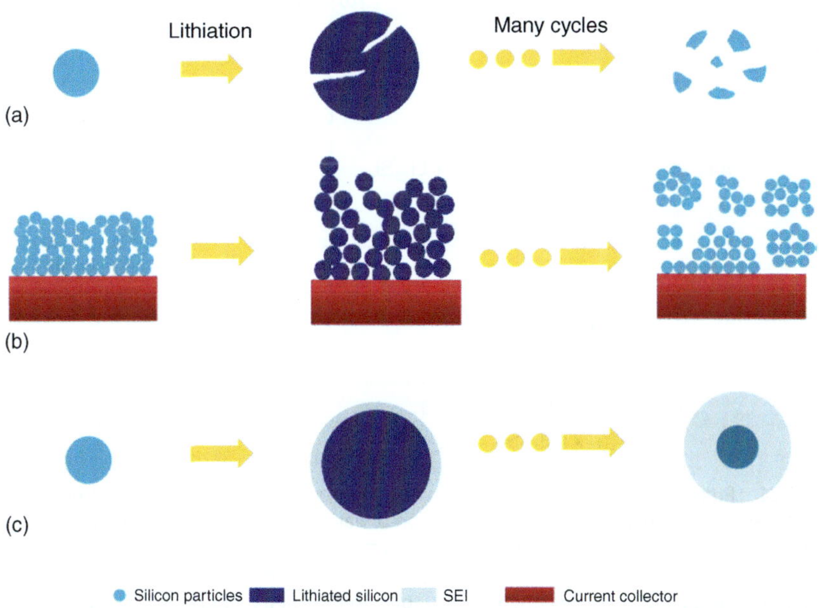

Figure 6.20 Electrode failure mechanisms for silicon active material: (a) material cracking and pulverization. (b) Electrode expansion and impingement leading to detachment from current collector. (c) Continuous formation of SEI. Source: Reproduced with permission from Li and Dahn [65]/IOP Publishing.

nanoscale Si, however, it has been possible to maintain capacities close to the theoretical value for tens of cycles.

In situ X-ray diffraction (XRD) of crystalline Si shows that Si goes through multiple phase changes during initial lithiation [65]. Initial lithiation of crystalline Si (c-Si) results in an amorphous (a-) partially lithiated phase along with c-Si. Below 70 mV, this phase rapidly transforms to c-$Li_{15}Si_4$. Delithiation of the electrode yields (amorphous) a-Li_xSi, and a-Si with remnants of the c-$Li_{15}Si_4$ phase. The formation and residual existence of the permanent c-$Li_{15}Si_4$ phase minutely contributes to lower initial coulombic efficiency and capacity fade. The corresponding reactions are shown below and the phase diagram in Figure 6.21.

$$\textit{Step}\ 1(a): \text{c-Si} + x\text{Li}^+ + xe^- \rightarrow \text{a-Li}_x\text{Si}$$

$$\textit{Step}\ 1(b): \text{a-4Li}_x\text{Si} + 4*(3.75 - x)\text{Li}^+ + 4*(3.75 - x)e^- \rightarrow \text{c-Li}_{15}\text{Si}_4$$

$$\textit{Step}\ 2: \text{c-Li}_{15}\text{Si}_4 \rightarrow \text{a-Si} + y\text{Li}^+ + ye^- + \text{Li}_{15}\text{Si}_4\ (\text{residual})$$

To prevent rapid capacity fade due to volume expansion upon alloying and subsequent volume shrinkage upon de-alloying, numerous strategies have been applied with varying degrees of complexity and benefits. Theoreticians and experimentalists alike have conducted studies to ascertain ways to maximize capacity retention, with the most accepted method being the reduction of the Si dimensions, as it has

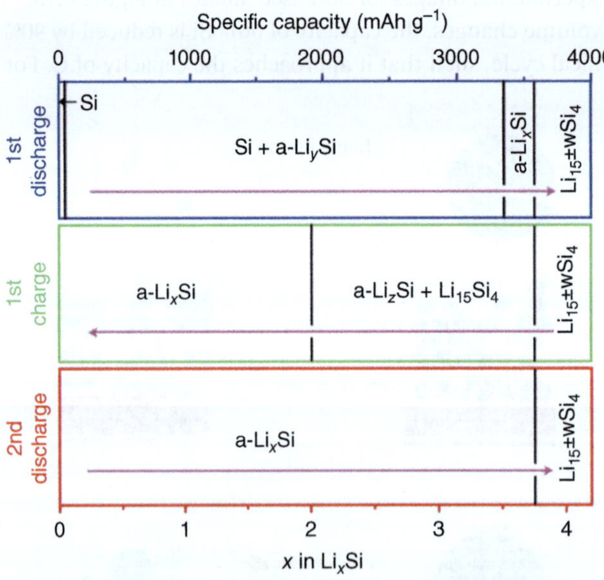

Figure 6.21 Phase diagram describing the phases that form during charge-discharge of a Li/Si cell. a-Li_xSi designates a single-phase region of variable composition, Li_ySi designates a fixed composition that coexists with c-Si during the first lithiation but not during the subsequent lithiations, Li_zSi designates the fixed composition that co-exists with $Li_{15}Si_4$ during de-lithiation. $Li_{15\pm w}Si_4$ designates a variable stoichiometry single-phase region. Source: Reproduced with permission from Li and Dahn [65]/IOP Publishing.

been noted that fracture was less severe and capacity retention improved. This is consistent with the observation that the fracture toughness of Si increases at small scales, especially in nanospheres [66]. In the following sub-sections, various promising Si nano-configurations will be summarized.

6.5.1 Si-Films Anodes

The highest capacities have been obtained for Si in the form of films with nanodimensions. The focus of researchers has been to increase the film thickness while maintaining high capacities over 1500 mAh g^{-1}. It has been observed that the capacity not only depends on the film thickness, but also on the fabrication method. Therefore, an overview will be given here on the various capacity retentions that have been achieved in various studies. It should be noted that the common factor in all the proceeding studies is that Si films have an amorphous structure instead of a crystalline, as it has been documented that amorphous Si has a more open structure as compared to crystalline Si, and can therefore accommodate more efficiently the Li insertion and de-insertion by allowing for a homogenized expansion–contraction upon cycling [67].

In [67] it was shown that a capacity of 2000 mAh g^{-1} for over 50 cycles can be achieved for an amorphous 100 nm Si thin film fabricated by physical vapor deposition. In [68] it was observed that both the film thickness and capacity can be increased by fabricating the films through radio frequency magnetic sputtering. This allowed the fabrication of 250 nm amorphous Si thin films that gave capacities of 3500 mAh g^{-1} for 30 cycles [68]. Increasing the film thickness to 500 nm, while maintaining this fabrication method (but using a rough Cu collector), gave capacities of 1500 mAh g^{-1} after 30 cycles [69]. In [70] a capacity of 3000 mAh g^{-1} for 70 cycles was achieved for 200 nm Si thin films. This radio frequency magnetic sputtering technique, however, significantly reduces the favorable properties of Si when used for the production of microscale films [68]. Therefore, despite the significantly high capacities obtained, it is not likely that this method will allow for the development of commercial Si anodes.

In [71] it was possible to increase the film thickness, while maintaining a high capacity by roughening the surface of the Ni substrate onto which n-type Si was deposited in a vacuum. Roughening of the Ni surface substrate was performed by etching. Letting the thickness of the Si film be 500 nm allowed capacities of 3600 mAh g^{-1} to be retained for 200 cycles [71].

Aside from the cost and difficulty of fabricating thick films for commercialization, fracture is another limiting factor for films beyond a few hundred nanometers. Figure 6.22 depicts fracture and delamination from the current collector from the first electrochemical cycle [72].

In attempts to find alternative fabrication methods that allow the production of thicker films that retain the favorable properties of Si, while maintaining the mechanical stability, electron-beam deposition was used to form 2-μm Si films [73]. These films were able to deliver a 3400 mAh g^{-1} capacity for 50 cycles, which decreased to 2550 mAh g^{-1} after 200 cycles [73]. Motivated by these results, a

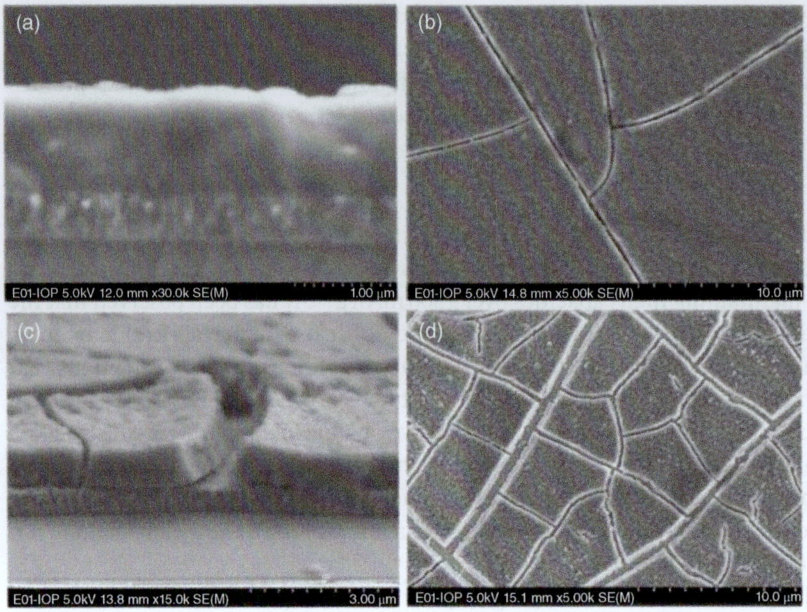

Figure 6.22 SEM images of Si-films deposited on Ti. (a) Side view of the film upon the first lithiation, showing that the thickness of the Si film increases from 500 to 1080 nm due to volume expansions. (b) Initiation of cracks upon onset of de-lithiation. (c) Side view of the film after de-lithiation, note that delamination from Ti occurs. (d) Crack pattern after de-lithiation. Source: Reproduced with permission from Réthoré et al. [72]/Elsevier.

battery cell was developed [73] in which the anode was a Si thin film with 4–6 μm thickness, while the cathode consisted of $LiCoO_2$; this cell retained a capacity of 1.5 mAh cm^{-2} for 200 cycles, with an efficiency of 97%. It is interesting to note the microstructure effects of Li-insertion in the Si film. Figure 6.23 illustrates the as-deposited Si film on the roughened Cu substrate. Due to the roughness of Cu, the Si film has a columnar structure. These columns in the Si film resulted in a discontinuous structure, which as a result contains additional Si interfacial surface, perpendicular to Cu. The discontinuous structure accommodated the Si expansion upon Li-insertion, while the perpendicular interfaces allowed for the fracture to form along the interface, and therefore the microcolumn structure was accentuated upon cracking during discharge. Planar fracture, as that for the continuous films of Figure 6.22 was, therefore, avoided. Figure 6.24 depicts the aforementioned microcolumn structure after 250 cycles. This fracture morphology most likely occurred during the initial electrochemical cycle. Just as the flake-like particles of Figure 6.2d, the Si columns remained in contact with the Cu current collector (allowing cycling to continuously take place), and ample space existed to accommodate the volume expansions. This good adhesion is achieved due to the roughness of Cu [73].

Aside from the above Si film with a discontinuous columnar structure, nano-textured silicon thin films (NTSTFs) with a thickness of 600 nm also showed interesting behavior [74]. In [74] the NTSTF was prepared using a wet

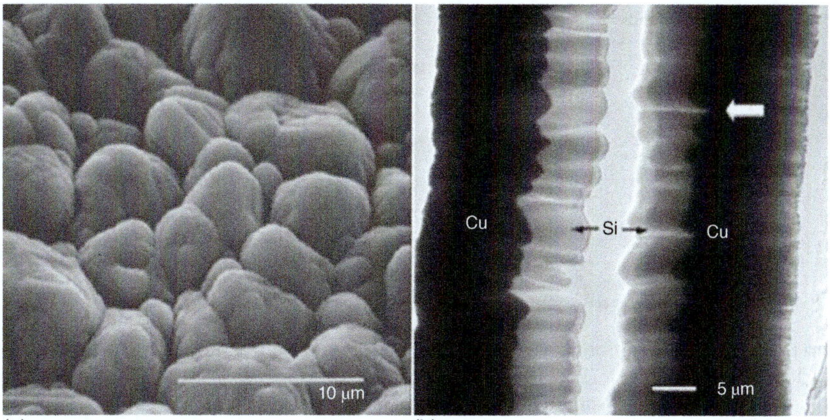

Figure 6.23 Morphologies of the 5-μm thick Si thin film deposited on Cu substrate: surface SEM image left and cross-sectional TEM image right. Source: Reproduced with permission from Yin et al. [73]/IOP Publishing.

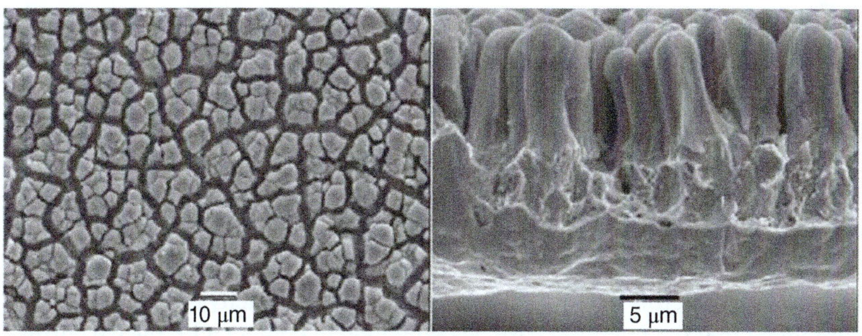

Figure 6.24 SEM images of a 6-μm Si film anode after 250 cycles (produced capacities of 1.8 mAh cm^{-2}): (a) surface, (b) cross-section. It can be seen that Li insertion and de-insertion in the first cycles resulted in Si-micron fibers/columns. Source: Reproduced with permission from Yin et al. [73]/IOP Publishing.

chemical etching method from commercial silicon wafers, and a side view of the films showed a columnar structure. To examine the effect of coatings NTSTF samples were coated with 5 nm Al_2O_3, and 5 or 10 nm Cu thin films. The pristine NTSTF revealed a very stable capacity of 2300 mAh cm^{-2} for the first 10 cycles and improved rate performance as shown in Figure 6.25. Despite the nano-texture with a high ratio of porosity, cracking was unavoidable due to the electrochemical agglomeration after the first cycle. The 5 nm Al_2O_3 coating resulted in poorer performance while the best performance was obtained for the 5 nm Cu coating. An interesting observation was that during cycling, the height of the NTSTF was unchanged, and the volume changes occurred only in the lateral direction.

Photolithography and reactive-ion etching have also been used to produce Si patterns [75], and coat them with a 20 nm Al_2O_3 film. The initial charge capacity was

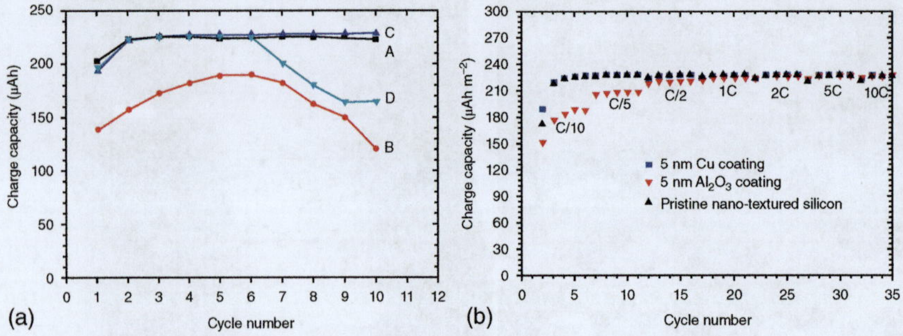

Figure 6.25 (a) Cyclic performance and (b) rate performance of the four samples in the first 10 cycles: (A) pristine NTSTF; (B) NTSTF coated with 5 nm Al_2O_3; (C) NTSTF coated with 5 nm Cu; (D) NTSTF coated with 10 nm Cu. Source: Reproduced with permission from Wang et al. [74]/IOP Publishing.

1125 mAh g^{-1}, and remained practically constant (at 1100 mAh g^{-1}) after 100 cycles. The alumina acted as a protective layer for preventing side reactions between the electrode and electrolyte. At the same time, the fracture resistance of the silicon columns was enhanced by atomic layer deposition.

In addition to the aforementioned $LiCoO_2$/Si thin-film cells, another type of Si thin films were cycled with an $LiCoO_2$ cathode [76]. The Si film was sputtered on rough Cu foils by dc magnetron sputtering; the resulting films were 2 μm thick, and when cycled against a $LiCoO_2$ cathode. A capacity of 0.54 mAh cm^{-2} for 300 cycles was achieved, with only a 0.01% capacity decay per cycle. Figure 6.26 depicts the amorphous structure of Si (which allowed for a better accommodation of the volume expansions) prior to cycling, and the fractured Si particles after 300 cycles. Apparently, however, this fracture was not severe, as the capacity retention was very good, implying that the fractured particles remained in contact with the Cu current

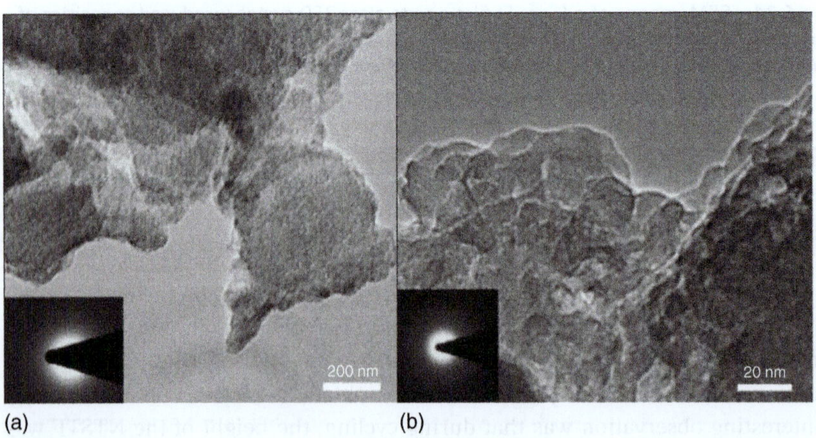

Figure 6.26 (a) HRTEM micrograph of an amorphous 2 μm Si thin film deposited on Cu. (b) HRTEM micrograph of the Si thin film after 300 cycles. Source: Reproduced with permission from Yang et al. [76]/Elsevier.

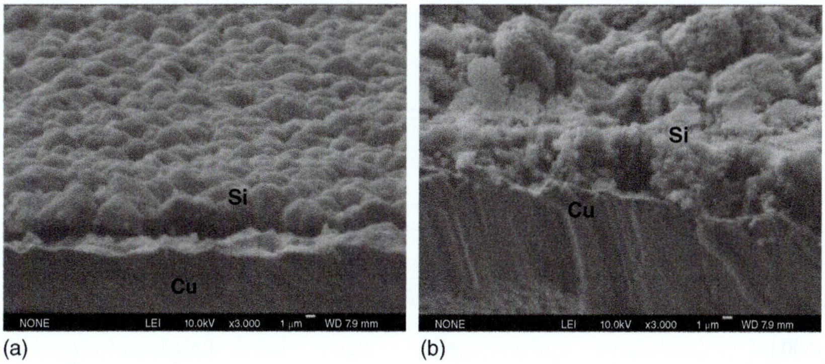

Figure 6.27 (a) SEM micrograph of Si thin film (2 µm thick) deposited on Cu before cycling. (b) SEM micrograph of Si thin film after 300 cycles (the thickness has increased to 6µm). Source: Reproduced with permission from Yang et al. [76]/Elsevier.

collector, and could respond to electrochemical. It is of interest to point out that the formation of the SEI layer is significant on such Si films, as shown in Figure 6.27. Despite the thick SEI layer that formed, the capacity was not reported to be affected [76]; in [73], however, it was stated that Li-ions were captured in this layer and resulted in an increase in the irreversible capacity.

In addition to obtaining insight into the electrochemical performance of Si, the thin film configuration can help understand the volume changes during Li^+ insertion and extraction [77, 78]. Unless specific patterning is done, it has been concluded that the volume variation along the vertical and horizontal directions is anisotropic and asymmetric. The maximum volume change has been shown to be about 310% under the fully lithiated state (at 10 mV) for amorphous Si and ~300% for crystalline Si (at room temperature), which is consistent with the expected value for crystalline Li_xSi alloys.

In addition to patterning, another method for improving the performance of Si films is by alloying with another element. One such composition is a Ti–Si nanocomposite thin film with Ti_xSi_y alloy grains dispersed within amorphous Si, which allowed to improve the cyclability and reduce the volume changes and fracture [79]. The purely amorphous Si thin film showed a poor capacity retention, exhibiting large directional cracks and fractured particles, after 10 cycles, while the Ti–Si nanocomposite films had finer and wiggled cracks and a smaller fractured particle size.

Adding N in amorphous Si thin films, to form $Si_{1-x}N_x$ thin films also allows for improved properties [80]. Once Li enters these films three phases form: Li_3N, Si_3N_4, and Si; with the Si active sites being distributed throughout. In addition to offering mechanical stability, the matrix formation enhances the electrochemical properties, as Li_3N is one of the best Li-ion conducting compounds. To study the effect of N content, three different thin films were examined: a pure Si thin film, a $Si_{0.73}N_{0.24}$ film, and a $Si_{0.59}N_{0.41}$ film [80]. Each anode exhibited a very different electrochemical performance, which is seen in Figure 6.28. The pure Si anode, in this case, exhibited a high initial capacity of 3000 mAh g^{-1}, which decreased to 700 mAh g^{-1} at the 200th cycle. The $Si_{0.73}N_{0.24}$ anode exhibited practically no capacity for the first several

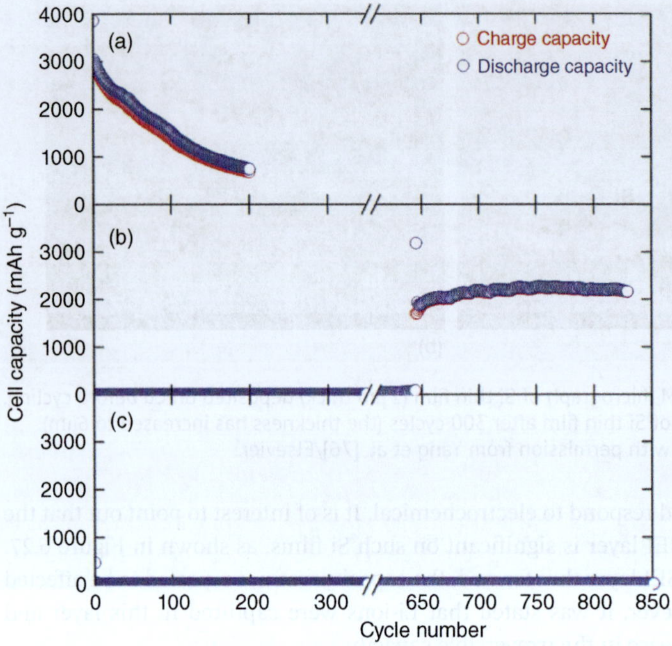

Figure 6.28 Cycle-life performance of amorphous, 200 nm-thick (a) Si, (b) $Si_{0.76}N_{0.24}$, and (c) $Si_{0.59}N_{0.41}$ thin films at 0.5C rate (1C = 4200 mA g^{-1}) at 25 °C. Source: Reproduced with permission from Ahn et al. [80]/Elsevier.

hundred cycles, while an abrupt capacity increase of nearly 2000 mAh g^{-1} was recorded at the 646th cycle, which increased to 2300 mAh g^{-1} at the 750th cycle, and was retained until cycling ceased at 850 cycles. The $Si_{0.59}N_{0.41}$ anode did not give off any capacity throughout the whole cycling process. The unique electrochemical behavior of $Si_{0.73}N_{0.24}$ was attributed to the formation of Li_3N, which began forming at the 646th cycle and, therefore, pure Si particles were also formed that could react with Li; further cycling formed additional Li_3N, which enhances the kinetics of Li insertion and de-insertion, and therefore the capacity increased as shown in Figure 6.28. Furthermore, it was observed that performing cycling at higher temperatures allowed the abrupt capacity increase for the $Si_{0.73}N_{0.24}$ anode to occur around the 30th cycle, since high temperatures aid the formation of Li_3N.

Another film nanocomposite was that of Si nanoparticles (30–50 nm) distributed homogenously in $LiTi_2O_4$ [81] using the sol–gel method. The resulting nanocomposite film was porous and had a thickness of 600 nm. The porosity and a capacity of 1100 mAh g^{-1} were retained for 50 cycles. Through appropriate impedance measurements, it was possible to measure the Li-ion diffusion coefficient as 10^{-13} cm^2 s^{-1}.

The deficiency of using thin-film electrodes is that their thickness does not provide sufficient active material to make them commercially viable. Hence, alternative nano-configurations are sought after. It is noted that in the case of thin film anodes the active material is coated directly on the current collector, rather than creating a slurry (from the powder active material, a binder and carbon) and then coating on the collector.

6.5.2 Si-Nanowire Anodes

It was seen in Section 6.5.1 that the most promising thin-film configuration was that in which the Si thin film had a columnar structure, since it could accommodate fracture through the formation of micron columns (Figures 6.23 and 6.24). In this section, it will be further illustrated that such type of a vertical/columnar-like microstructure results in higher capacities.

Si nanowires have stimulated intensive interest in their physical properties and potential application because of their low dimensionality and quantum-confinement effect. Si nanowires with different Li^+ doping levels were obtained by controlling the discharging/charging of Si/Li half cells [82]. The microstructure of the Si NWs with different contents of Li^+ was investigated by high-resolution transmission electron microscopy, which showed that the crystalline structure of the Si NW was destroyed gradually as the Li-ion concentration increased. When the Li-ions were extracted from the amorphous Li-doped Si NWs by the same electrochemical method, local ordering of atoms occurred and recrystallization was observed.

As mentioned in Chapter 5, carbon is often used to coat active materials so as to protect the surface and accommodate the volume expansion. Therefore, as Li_2O acted as the buffering matrix in the aforementioned Sn-based anodes, C is used [83] to protect the surface of Si nanowires. Since carbon nanotubes have been given considerable attention during the past decade, Cr-doped Si nanowires were coated by C nanotubes as seen in Figure 6.29. Chromium was added in order to increase the conductivity of Si (and hence assist Li^+ diffusion). The nanowires were coated with

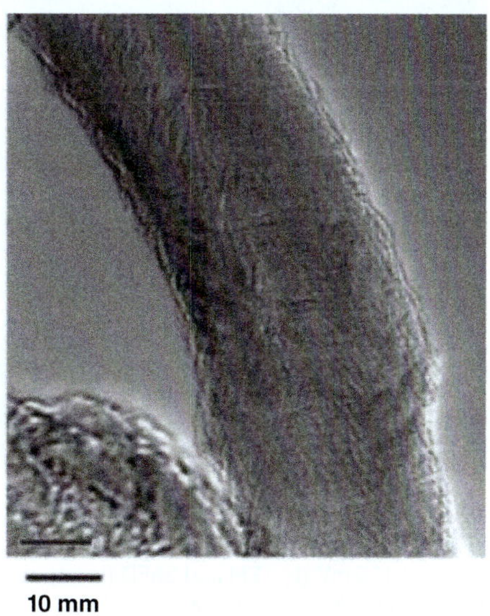

Figure 6.29 Cr-doped Si, coated with carbon nanotube. Source: Reproduced with permission from Ishihara et al. [83]/Elsevier.

C nanotubes through decomposition of CH_4 by using Ni as the catalyst at a temperature of 1073 K for six hours. The initial capacity was 1700 mAh g^{-1}, which was reduced to 1250 mAh g^{-1} after the tenth cycle.

Promising capacities have also been obtained for pure Si nanowires with a crystalline Si core and an amorphous Si shell [84]. The amorphous Si has a lower activation voltage than the crystalline Si; hence by electrochemical cycling at particular voltages, it is possible to activate only the amorphous Si shell so as to accept Li-ions. The crystalline Si core allows for high electrical conductivity and mechanical support of the nanowire, while the amorphous shell can store Li-ions and accommodate the resulting volume expansions. The capacity attained for these anodes was as high as 1000 mAh g^{-1} for 100 cycles, while the Coulombic efficiency was 90% [84]. Figure 6.30 depicts the crystalline/amorphous core/shell Si nanowire, after electrochemical cycling. The dark and light contrast regions of the core, after the 15th cycle, indicate the straining that took place upon cycling due to the volume expansions.

Amprius, a spin-off from Stanford University developed a 100% Si anode that uses Si nanowires. These cells were tested by Airbus in their Zephyr S pseudo-satellite [85].

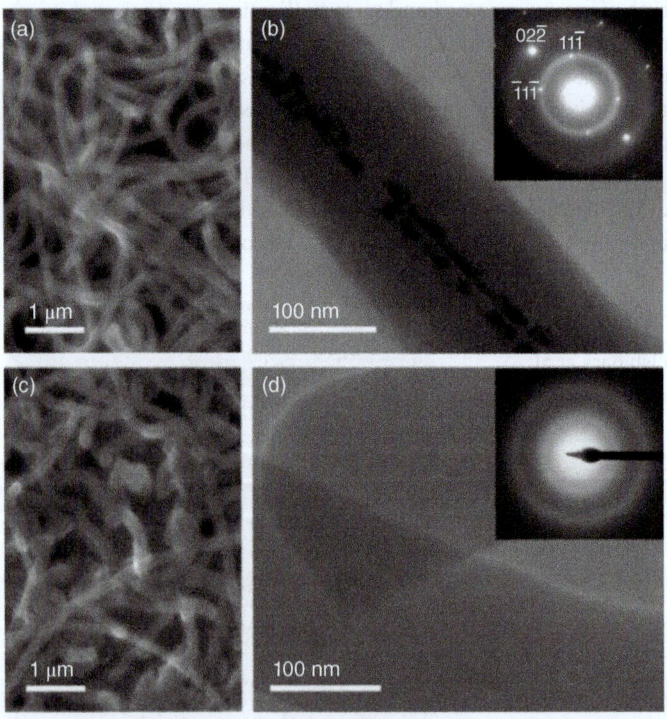

Figure 6.30 SEM, TEM, and SAED images of the c-a core–shell NWs after cycling at 0.2C (0.85 A g^{-1}) and different low voltage cutoffs. (a) SEM image of NWs after 15 cycles with cutoff of 150 mV. (b) TEM and SAED (inset) images of a NW after 15 cycles with cutoff of 150 mV. (c) SEM image of NWs after 3 cycles with cutoff of 10 mV. (d) TEM and SAED (inset) images of NWs after 3 cycles with cutoff of 10 mV. Source: Reproduced with permission from Cui et al. [84]/American Chemical Society.

6.5.3 Si Microparticle Based Porous Electrodes

The most commercially viable Si anodes would be those comprised of Si micron particles, since they are easier to mass produce. Such anodes would have a porous electrode structure since the Si powder is mixed with binder and conductive carbon to form a slurry that is then coated on the current collector. However, they have not been explored to a significant extent as they exhibit a faster capacity decay than Si nanoparticles. Initial studies assumed that this was due to more severe fracture that the larger particles experienced, however, it was not until recently that extensive studies began for porous anodes that used Si micro rather than nano particles. Figure 6.31a shows the voltage profile of a porous electrode with microscale Si particles at 0.2C (~715 mA g^{-1}) [86]. The initial discharge profile demonstrates a flat voltage plateau corresponding to the formation of Li–Si alloy during lithiation. Specifically, the initial discharge and charge capacities of ~3390 mAh g^{-1} ($Li_{3.54}Si$) and ~3050 mAh g^{-1}, respectively, correspond to an initial Coulombic efficiency (C.E.) of 90%. Subsequent cycling yielded a C.E. of 98.2% and 98.4%, respectively. Figure 6.31b shows that the discharge capacity decayed with continuous cycling. An initial discharge capacity of 2996 mAh g^{-1} is observed along with an initial C.E. of 86.6%. It should be noted that small differences in the initial capacity and C.E. between Figure 6.31(a) and (b) are due to a different mass loading of the electrode. The initial C.E. of micron-scale Si electrodes at 86.6% is higher than some of the nanocomposite Si anodes that will be described in the sequel. This is because, microparticles have a smaller surface area and therefore less SEI forms, consuming less Li^+. After 100 cycles, the micro-Si electrode exhibited a specific discharge capacity of 1932 mAh g^{-1}. This amounts to a capacity retention of 64.4% over 100 cycles. After 200 cycles, the electrode had a discharge capacity of 1792 mAh g^{-1}. Less discharge capacity fade was observed in the last 100 cycles than in the first 100 cycles. Overall, the micronscale Si electrode boasts a discharge capacity retention of 59.8% over 200 cycles.

In order to understand the mechanisms that give rise to this capacity decay, cells were opened after one electrochemical cycle, and electron microscopy was performed, revealing that the micron-size Si particle fractured in half after the first electrochemical cycle and the crack began to cleave (Figure 6.32b) after two cycles. It is interesting to note that even though Si was studied extensively for the past 20 years, the documentation of such fracture in microparticles was obtained very recently [86], and is consistent with the predictions of multiphysics simulations, as will be shown in Chapter 10.

To minimize damage in Si microparticles of ~1–3 μm, graphene has been used to fully encapsulate them [87]. The half-cell showed a capacity retention of 1500 mAh g^{-1} over ~350 cycles and a C.E. of 99% in a full cell configuration using traditional $LiCoO_2$ as the cathode.

6.5.4 Si/C Nanocomposites and other Si Nanoconfigurations

The first electrochemical studies on Si nanoparticles (80 nm in diameter) gave a capacity of 1700 mAh g^{-1} on the 10th cycle [14], which was significantly less from the initial capacity. As in the case of thin films, this poor retention was attributed to

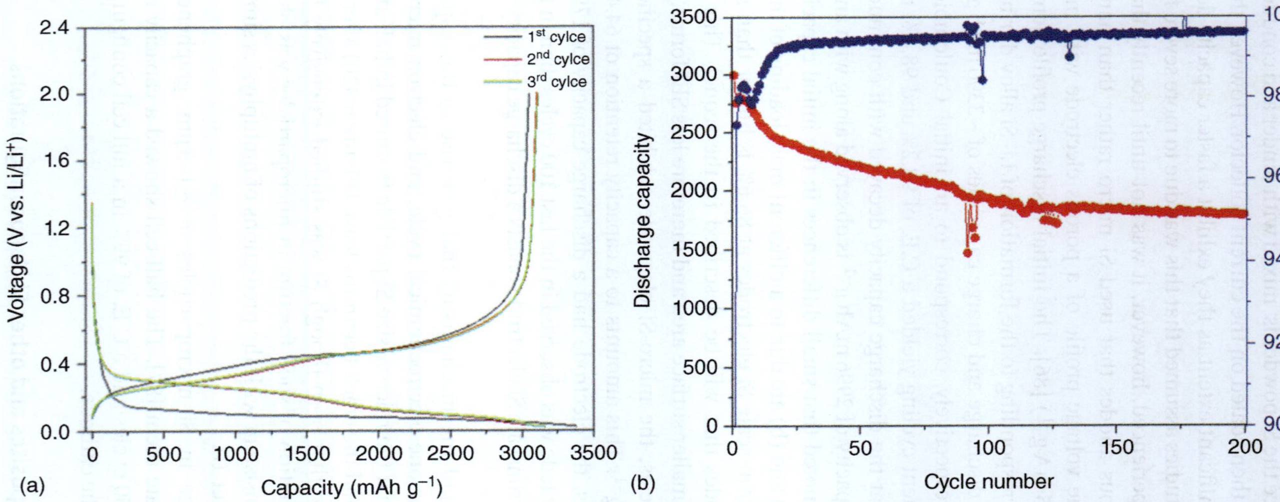

Figure 6.31 Micron-scale Si electrode. (a) Initial galvanostatic charge/discharge profile (reproduced with permission from Ahuja et al. [87]/Eslevier) and (b) extended cycling performance at 0.2C.

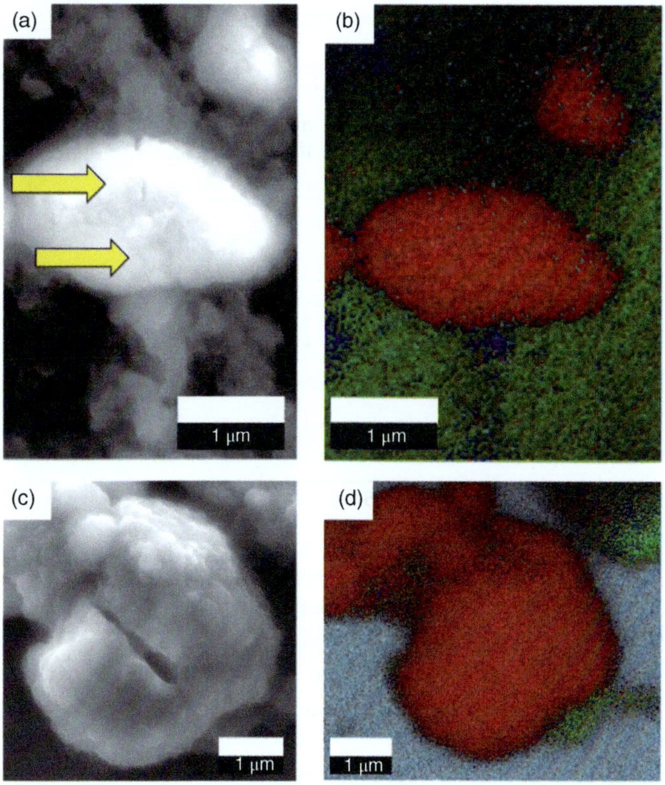

Figure 6.32 SEM image capturing the fracture in micron-scale Si (a) SEM image after 1 cycle, (b) EDS map after 1 cycle, (c) SEM image after 2 cycles, (d) EDS map after 2 cycles (scale bar 1 μm). Source: Reproduced with permission from Ahuja et al. [86]/Elsevier.

fracture and agglomeration of Si [14] that takes during cycling. Subsequent studies on pure nano Si-based electrodes showed a similar capacity decay. Cycling Si nanoclusters 50 times resulted in a capacity reduction from 2400 to 525 mAh g^{-1} [67], while cycling Si nanoparticles reduced their capacity from 2400 to 600 mAh g^{-1} after 20 cycles, implying a capacity retention of only 25% [88].

Just as in the case of Sn, carbon has been used to buffer the volume expansions and protect the Si surface from the SEI, to improve the electrochemical performance. One such early configuration proposed was that of coiled CNTs with different lengths grown on superfine silicon particles via a chemical vapor deposition method [89]. When cycled against Li, a reversible specific capacity of 940 mAh g^{-1}, and a C.E. of 80%, were obtained for the first-cycle. It was observed that the longer CNTs were able to cover more efficiently the Si particles, forming a better conductive network and continuous SEI film, and leading to a more stable capacity during cycling, as shown in Figure 6.33. Another early Si–C composite (20 wt%Si–80 wt%C) was fabricated through a dehydration reaction using concentrated H_2SO_4 as the dehydration agent and the thermal decomposition of sucrose [90]. The Si particle sizes were below 25 nm, and they were embedded in C such that the metal active surfaces were protected and agglomeration was prevented. Since the C content was significantly higher than the Si content, the initial capacity was 1000 mAh g^{-1},

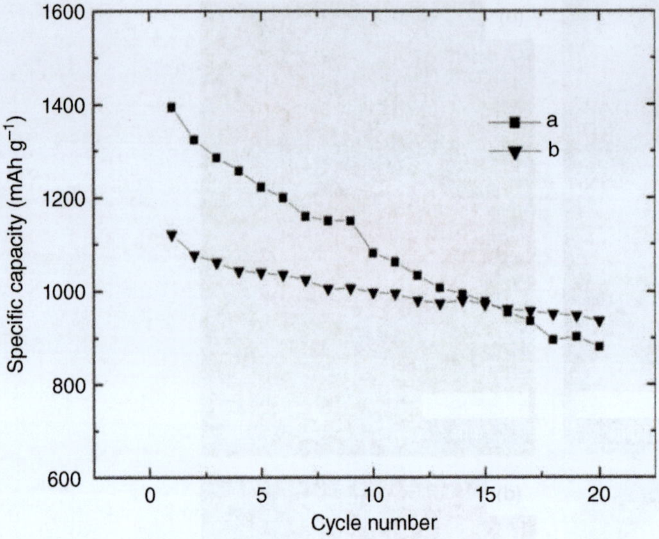

Figure 6.33 Cyclic performance of the CNTs/Si composites cycled between 0.0 and 2.0 V at a current density of 0.156 mA cm^{-2}. The CNT length different in each sample. Source: Reproduced with permission from Shu et al. [89]/Elsevier.

and a 100% retention was obtained for 30 cycles, which is far from the 1000 cycles needed for commercialization.

Si/C composites have also been fabricated [91] using the Rochow reaction method under mild reaction conditions (325–450 °C, atmospheric pressure, and various Cu-based catalysts). This allowed for interconnected porous Si coated by a conductive carbon-coated layer, giving a capacity of 1036 mAh g^{-1} after 100 cycles.

Porous C coatings have also been used. A template method was used to embed the Si in the carbon shell with voids for accommodating the expansion during lithiation. This method realized a stable SEI layer on the porous shell, without damaging it. Stable capacities of ~1000 mAh g^{-1} were obtained for these structures [92].

Another configuration that has been proposed to further accommodate the Si volume expansions, is that of a yolk-shell structure, which allows for a void between the Si and C shell, and gave a capacity of over 1300 mAh g^{-1} at 1 C for 1000 cycles [93]. The authors indicated that their results were relatively independent of the binder used (binders will be elaborated in the sequel), thereby proving the efficacy of the structural design and fabrication.

Hollow nanospheres also accommodate the volume expansions and have therefore been explored. Further coating the hollow Si with Ag nanoparticles showed an excellent rate capability and a high capacity of 3400 mAh g^{-1} after 100 cycles. Similarly, monodisperse porous Si spheres synthesized via surface protected magnesiothermic reduction along with hydrolysis can extend the cycle life of Si [94], allowing a high reversible capacity retention of over 1500 mAh g^{-1} after 500 cycles at a rate of 0.5C. Other configurations, such as a dual-yolk structure, pomegranate structure, and solid core–shell structures, have yielded similar cycling stabilities [95–97].

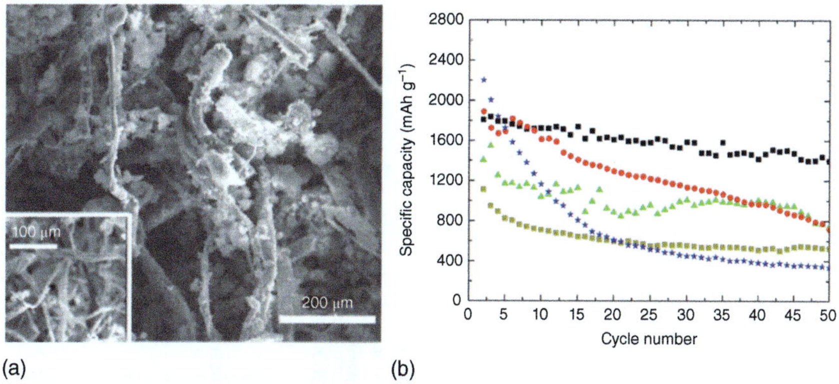

Figure 6.34 (a) SEM image of Si nanoparticles attached on cellulose fibers; inset indicates fibers prior Si adhesion, (b) Capacity retention of various wt% Si/cellulose composites cycled at 200 mA. The most promising is the 34% Si/cellulose composite (black squares), while pure Si underwent the greatest capacity fade (stars). Source: Reproduced with permission from Gómez Cámer et al. [98]/IOP Publishing.

Nonetheless, fabrication of such hierarchical intricate geometries is not possible in large amounts. Even at pilot scales, batch uniformity and control over quality become complicated.

In concluding this subsection, it is interesting to note that nano-Si/cellulose nanocomposites have been used as anodes [98], allowing for capacities of 1400 mAh g^{-1} to be attained for 50 cycles, as seen in Figure 6.34. The Si nanoparticles (40–50 nm) were mixed with the cellulose in isopropanol and stirred, adhering to the cellulose fibers, and, thus, forming a type of island structure, as seen in Figure 6.34. The high capacity obtained from these materials was attributed to the mechanical support that the cellulose fibers offered to the Si. It should be mentioned that this microstructure was initially attained for Sn particles on cellulose [99]; however, the resulting electrochemical properties were not comparable to that of the Sn-based anodes in Section 6.4.

6.5.5 Si/Polymer Nanocomposites

In addition to using carbon to constrain the volume expansion of Si, polymer coatings have also been employed. Such coatings also help improve the electronic conductivity, Li-ion diffusivity and adhesion with the binder. Figure 6.35 depicts polydopamine (DPA) coated Si nanoparticles (50 nm) [87]. Thermogravimetry indicated that there was about 10.4 wt% of DPA in the composites. The TEM image shown in Figure 6.35b shows that the DPA thickness was ~1.5–3 nm (the coating is denoted by red lines/arrows).

The voltage profiles of the pure nano Si and Si/DPA electrodes at 0.2C are shown in Figure 6.36a,b. Both exhibited a flat voltage plateau at 50 mV, which is consistent with the formation of Li–Si [65] alloy during lithiation. The flat portion of the plot describes an interface-limited step wherein crystalline Si converts to amorphous Li_xSi [100]. Subsequent cycling did not show the same plateau, instead, lithiation

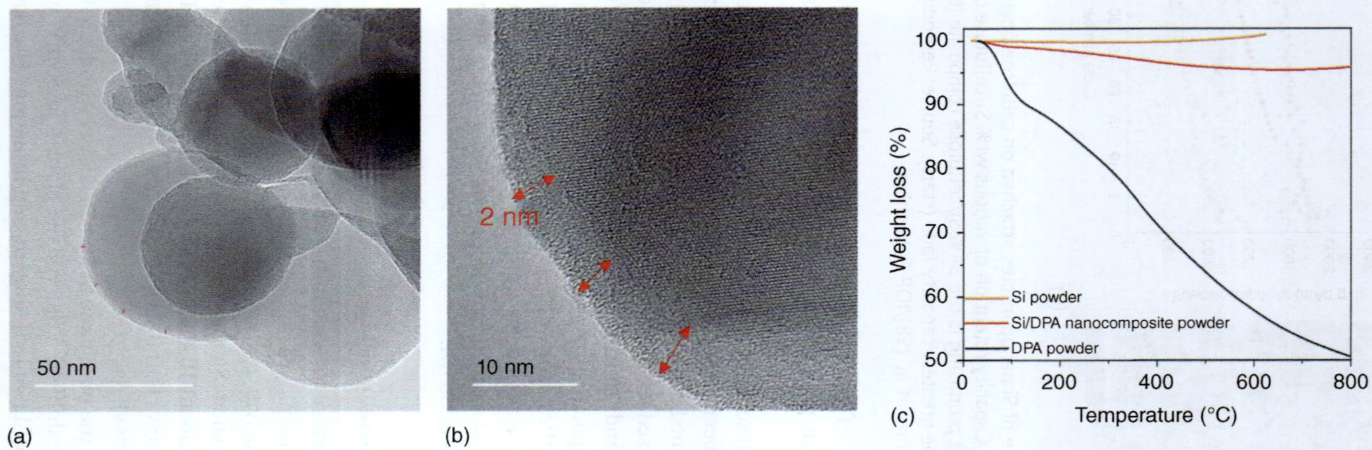

Figure 6.35 (a, b) TEM images of Si/DPA nanocomposite (c) TGA of Si/DPA nanocomposite. Reproduced with permission from Ahuja et al. [87]/Elsevier.

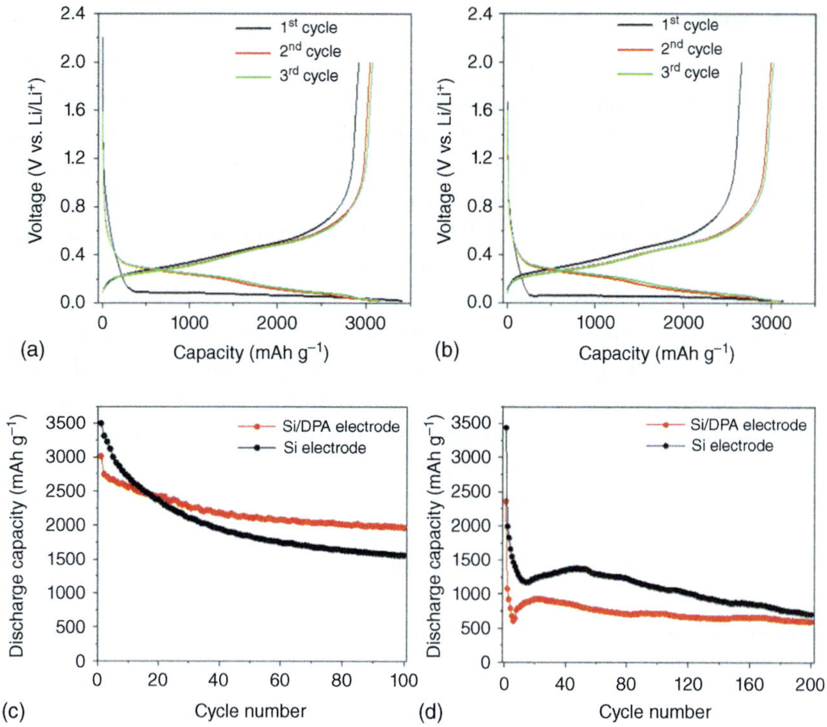

Figure 6.36 Initial voltage profile of (a) nano Si electrode, (b) Si/DPA nanocomposite electrode. Cycling performance of nano Si and Si/DPA electrode (c) at 0.2C and (d) 1C. Reproduced with permission from Ahuja et al. [87]/Elsevier.

and delithiation occurred at higher potentials at a finite voltage slope [100]. Since the Si/DPA electrode contained less wt% Si than the pure Si, it exhibited a lower initial capacity. The lower capacity of Si/DPA can also be due to the fact that DPA does not host Li^+.

Long-term cycling is shown in Figure 6.36c,d. The initial discharge capacity and C.E. for the electrode containing uncoated nano Si were 3500 mAh g^{-1} and 89.8%, respectively. After 100 cycles, the discharge capacity had decayed to 1550 mAh g^{-1}, and the C.E. had increased to 99.3%. The Si/DPA nanocomposite exhibited an initial discharge capacity and C.E. that were lower at 3000 mAh g^{-1} and 84%, respectively. Upon continuous cycling, however, the discharge capacity was 1971 mAh g^{-1} and the C.E. approached 100%. It is, therefore, seen that a significantly better long-term performance was obtained when a thin polymeric layer covered the Si active particles.

Figure 6.36d compares the cycling performance of the electrodes at 1C. It is observed that the Si/DPA electrode shows a rapid initial capacity fade but is able to retain an almost constant capacity of 600 mAh g^{-1} for 200 cycles. The uncoated nano Si electrode initially showed a high capacity, but it continuously decayed over the entire cycling period. Other polymers that have been used to increase the capacity of Si are polypyrrole, and polyaniline, which can also increase the cycling stability [101, 102].

6.5.6 Binders

In addition to coating the Si with polymers or C, the binder in porous electrodes also provides mechanical support to the Si [103, 104]. Polymeric binders are needed in porous electrodes even if the Si did not expand, since along with the ability to contain the volume changes, they allow the active material to disperse well during the electrode fabrication process, preventing aggregation and forming a slurry that uniformly coats the current collector [105]. Binders, therefore, need to have a good adhesion with both the active material and current collector, while providing the least resistance to the flow of electrons and Li^+. Binders are expected to withstand wide temperature and potential changes during cycling of LIBs, and they should be chemically inert to the electrolyte while remaining electrically and ionically conductive. The commercially successful binder for graphitic anodes polyvinylidene fluoride (PVDF) shows good electrochemical stability and adhesion to the current collector [106]. However, when applied to high-capacity anodes, such as Si and Sn, its ability to sustain electrochemical stability is highly compromised [107]. This stems from weak van der Waals forces and decreased stability over time, and also the poor mechanical support that PVDF provides to the Si during its volume changes. Therefore, alternative binders are being explored. Conductive polymers, such as carboxymethyl cellulose (CMC), alginate, guar gum, and xanthan gum [108–110], are also considered since they allow for enhanced electronic and ionic conductivities.

Figure 6.37 depicts the capacity retention for Si-based anodes using different binders. It is seen that cross-linkage between polyacrylic acid (PAA) and CMC binders allowed for the best response and the authors attributed this to the ability of PAA–CMC to accommodate the strains produced upon cycling [112].

Furthermore, in situ polymerization of conductive binders ensures complete wetting of the Si, thereby boosting the electrode conductivity. Polyaniline (PANi) coated

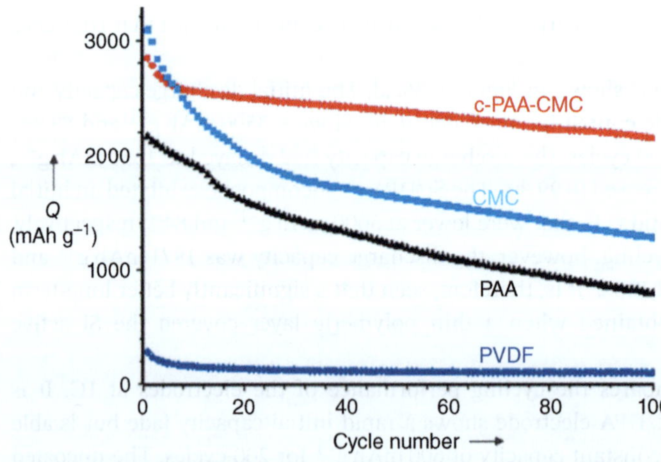

Figure 6.37 Discharge capacity of Si anode with different binders. *X*-axis denotes cycle number and *Y*-axis denotes discharge capacity. Source: Reproduced with permission from Wu et al. [111]/Springer Nature.

on Si active particles via a facile in-situ polymerization method [111] allowed for a high capacity of ~1200 mAh g^{-1} after 1000 cycles at a discharge rate of 1 A g^{-1}.

More recently rationally constructed designer binders and cocktail electrolytes with additives have emerged as a potent method to decrease capacity fade [113–115]. It should be noted that the aforementioned polymer-coated Si particles (Section 6.6.5) have an improved performance since the polymer coating can allow for better adhesion between the Si and binder, ensuring, hence, a better connectivity within the electrode.

6.5.7 Si–SiO_2–C Composites

SiO, SiO_2, and SiO_x represent an alluring choice apart from Si. However, silicon oxides experience similar challenges as those that exist for Si-base anodes, which include lack of electronic conductivity, large volume expansions, and low initial coulombic efficiency. In 2012, a low yield method was used to synthesize a 3D nanoporous SiO system [116]. The multicomponent system of SiO and Si showed promise as it provided a high reversible capacity of 1600 mAh g^{-1} after 100 cycles. Prelithiation of stoichiometric crystalline SiO provided a high initial C.E. (more than 99%) and good cycling stability (1600 mAh g^{-1}) over 100 cycles at 1C.

Nonstoichiometric SiO_x base anodes undergo lithiation via a conversion reaction followed by the alloying reaction, similarly as tin oxide. In order to increase conductivity and facilitate the multistep lithiation reaction, reduced graphene oxide (rGO) has been used [117]. Employing a facile self-assembly process, a cage of rGO formed around the SiO_x nanoparticles, which allowed the electrode to maintain a capacity of 780 mAh g^{-1} after 1000 cycles, which was 80% of the original capacity.

Rice husk has also been used to fabricate a composite of SiO_x porous microparticles in C [118]. A reversible capacity of 1230 mAh g^{-11} was observed with less than 0.5% capacity fade after 200 cycles. Such a stable cycling performance was attributed to the lack of fracture of the electrode material, shorter Li^+ pathways, and constrained expansions during cycling.

SiO_2 has also been used as the matrix, into which Si particles were embedded. In order to make the deformation mechanisms of the Si active sites less severe, the Si nanocrystals were taken to be 2–10 nm in diameter [119]. Using solely SiO_2 as a matrix material is not efficient since SiO_2 is nonconductive and traps the Li-ions within it; hence Si–SiO_2–C composite anodes were proposed instead [119]. To fabricate the aforementioned composite, a mixture of SiO powder milled with graphite powder was dispersed into a monomer solution and then polymerized. With heat treatment, Si nanocrystals and amorphous clusters were dispersed in SiO_2, which in turn were embedded in graphite. It should be noted that Si reacts with C to form SiC above 1000 °C, so it is necessary to maintain appropriate temperatures in order to inhibit the formation of this additional nonconductive phase. Upon electrochemical cycling, Li enters the pure Si active sites and once they are fully lithiated, it is intercalated with C. Li, though, is also trapped in the SiO_x phases by forming Li_2SiO_3 and Li_2SiO_4; these phases are formed irreversibly during the first Li-insertion process, and therefore both the amounts of Si and Li that can control the capacity of the

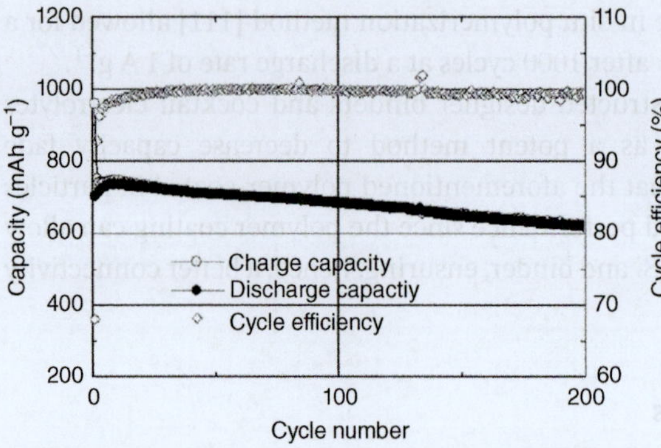

Figure 6.38 Capacity retention of Si–SiO$_2$–C nanocomposite. Source: Reproduced with permission from Morita and Takami [119]/IOP Publishing.

battery are reduced. The initial capacity of this Si–SiO$_2$–C anode was 700 mAh g^{-1}, and after 200 cycles it had reduced to 620 mAh g^{-1} [119]. It was noted in [119] that the smaller the active sites, the higher the capacity retention. In Figure 6.38, it is seen that the C.E. was 100% after the first few cycles. The lower C.E. observed initially is due to the SEI layer formation.

In order to optimize the performance of these promising anode materials, the wt% of the oxide phase must be reduced. A step toward this direction was taken in [88], where the performance of pure Si and Si–SiO$_2$ electrodes, with 12 wt% O, was compared directly. The pure Si nanocrystals had an initial capacity of 2400 mAh g^{-1} and a 25% capacity retention after 20 cycles, while the oxide-covered Si (Figure 6.39) not only had a lower starting capacity of 1500 mAh g^{-1} but also a lower capacity retention of 21%. This suggested that the SiO$_2$ limits the diffusivity of Li into the Si particles, and therefore C was used as the protective matrix, as opposed to SiO$_2$ [88]. The Si–C composites were fabricated by mixing Si powder with a resorcinol-formaldehye (RF) gel and then carbonizing the mixture to produce Si particles coated by hard C, as shown in Figure 6.40. The disadvantage of this technique is that Si reacted with the oxygen in the RF gel to produce SiO that covered a certain surface area of the Si particles and during the carbonization this SiO decomposed to Si and SiO$_2$. Due to the presence of SiO$_2$, the anticipated Si–C core–shell composite gave initial capacities of 950 mAh g^{-1}, which reduced to 350 mAh g^{-1} after 20 cycles.

The fact that the Si-based composite of [119] had a significantly higher capacity retention than that of [88], despite the existence of higher amounts of nonconductive SiO$_x$ phases in [119], is attributed to the fact that the nanocrystalline Si particles were significantly smaller in [119]; in [119], the Si diameter was 2–10 nm, whereas in [88] it was 30–50 nm. These larger particles resulted in more severe cracking that

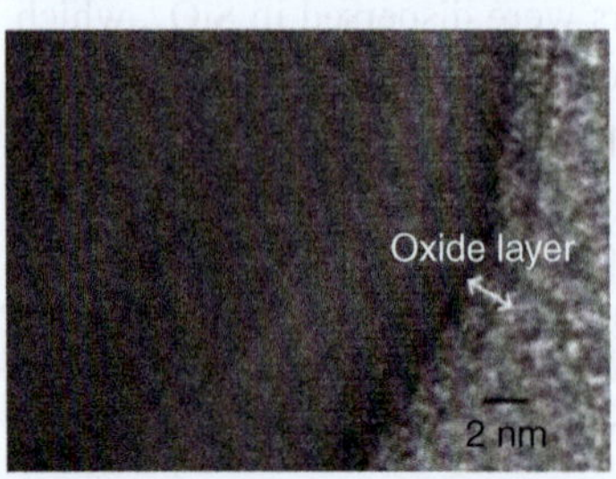

Figure 6.39 Si nanoparticle covered by a SiO layer. Source: Reproduced with permission from Jung et al. [88]/Elsevier.

Figure 6.40 Carbon coated Si; (a) SEM and (b) TEM images. Source: Reproduced with permission from Jung et al. [88]/Elsevier.

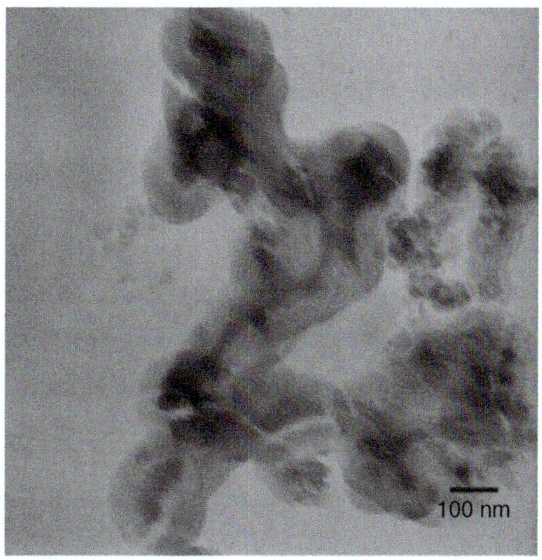

Figure 6.41 TEM image of a silicon–carbon core-shell composite after 30 cycles. Source: Reproduced with permission from Jung et al. [88]/Elsevier.

disrupted the connectivity of the anode, as seen in Figure 6.41, and hence gave a poor capacity retention.

BTR China has commercialized high energy density LIBs with SiO as the anode [120, 121]; however, they are not cost-efficient preventing the replacement of graphitic-based systems.

6.6 Other Anode Materials

Although all the research in metal-based anodes is focused on having Sn and Si as the active sites due to their very high capacities, 996 and 4200 mAh g^{-1}, respectively, there exist other metals that are active with respect to Li as mentioned in the introduction. Although the resulting capacities from using such metal-based anodes are much lower, it is of interest to include some examples herein, even though will are not considered canditates for commercilization.

6.6.1 MXene Electrodes

MXenes are 2D materials with a large interlayer spacing, which facilitates intercalation and deintercalation of Li^+ [122, 123], and they have, therefore, been investigated as anodes [124]. MXenes are made from precursors known as MAX phases, and their generic formula is $M_{n+1}X_nT_x$, wherein n+1 layers of early transition metals (such as Ti, V, W) are interwoven with n layers of either C or N. Furthermore, T_X are surface termination moieties, which may include H, O, F [123]. The surface area of MXenes is found to be considerably higher than other 2D materials, such as graphene or its precursor MAX phases (Ti_2AlC). Figure 6.42a,b shows the exfoliation of the 2D layers of Ti_3C_2, whereas Figure 6.42d,e shows the scroll formation. Both the exfoliation and the scroll formation of MXene materials showcase their similarity to graphene. Figure 6.42f is the schematic representation of the scrolls that are formed upon sonication of nanosheets.

Intercalation of Li in Ti_3C_2 results in $Ti_3C_2Li_2$ which has a theoretical capacity of 447.8 mAh g^{-1}

$$Ti_3C_2 + 2Li \rightarrow Ti_3C_2Li_2.$$

Figure 6.42c shows the formation of the intercalation compound via density functional theory (DFT).

The high conductivity in combination with the 2D structure of MXenes enhances the diffusion of Li^+ and hence a high-rate capability is observed [125]. This was also

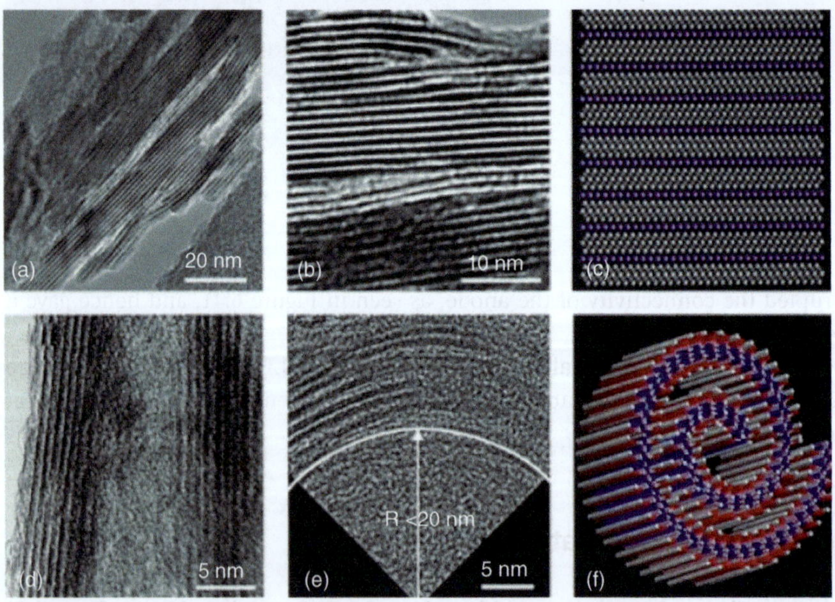

Figure 6.42 TEM images and simulated structures of multilayer MXene. (a) TEM images for stacked layers of Ti–C–O–F. Those are similar to multilayer graphene or exfoliated graphite that finds use in electrochemical storage. (b) The same as (a) but at a higher magnification. (c) Model of the Li-intercalated structure of Ti_3C_2 ($Ti_3C_2Li_2$). (d) Conical scroll of about 20 nm in outer diameter. (e) Cross-sectional TEM image of a scroll with an inner radius of less than 20 nm. (f) Schematic for MXene scroll (OH-terminated). Source: Reproduced with permission from Naguib et al. [124]/John Wiley & Sons.

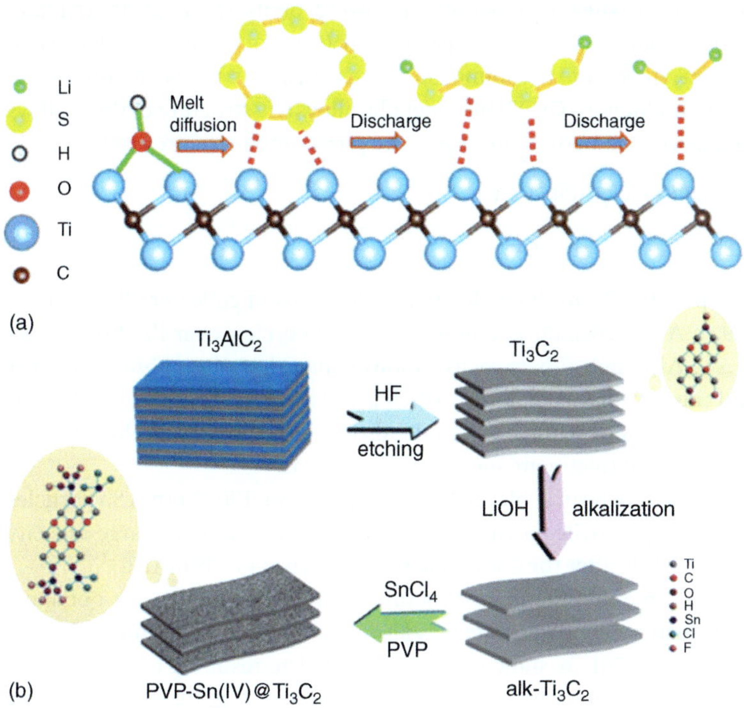

Figure 6.43 (a) Schematic illustration of replacement of the Ti-OH bond on the MXene surface with a S–Ti–C bond by heat treatment or by contact with polysulfides. Source: Reproduced with permission from Kannan et al. [127]/MDPI. (b) Schematic illustration of the fabrication process of PVP-Sn(IV)@Ti_3C_2 nanocomposites. Source: Reproduced with permission from Liang et al. [128]/John Wiley & Sons.

predicted by theoretical studies [126]. MXene-based composites, such as 70 wt% S/Ti_2C (Figure. 6.43a) and Sn^{4+} ion-decorated Ti_3C_2 (Figure 6.43b), have shown enhanced stability upon extended cycling, since the MXenes help maintain the electrical pathway and structural integrity while the matrix constraints the volume change of the particles [128, 129]. A superior volumetric capacity of 1375 mAh cm^{-3} was obtained for such MXene composites, which is two times higher than that of a graphic electrode (550 mAh cm^{-3}).

In connection to the above study, it is emphasized that the lithium storage capacity of MXene and MXene-based materials heavily depends on surface decoration, functional group grafting, variation of composition, interlayer distance, and surface terminations [127]. In addition to Sn, their performance can be significantly enhanced by adding Si, Ag, and SiO_x [130]; for example, Si-Mxene electrodes can retain a specific capacity of 881 mAh g^{-1} over 90 cycles [131].

6.6.2 Sb-Based Anodes

In Section 6.5.2, it was shown that SnSb alloys can be used as anodes. Sn is the most active material with respect to Li, and therefore such anodes are considered as Sn-based; the main role of the Sb is to constrain the expansion of the Sn, and inhibit agglomeration of the Sn sites. Sb, however, can also act on its own as an active materials since upon lithiation, it gives a capacity of 660 mAh g^{-1}, and experiences an expansion of 150%. In order to accommodate this expansion, it has been proposed to alloy Sb with Cu producing a Cu_2Sb intermetallic alloy. The structure of these alloys are alternating layers and upon Li insertion, the phases produced are [132]

$$Cu_2Sb + 2Li \leftrightarrow Li_2CuSb + Cu$$

$$Li_2CuSb + Li \leftrightarrow Li_3Sb + Cu.$$

Depositing a powdered Cu_2Sb anode on a Cu foil current collector allowed for a capacity of 290 mAh g^{-1}, which was retained after 25 cycles, over the full voltage range of 0–1.2 V. Although this capacity is lower than that of graphite, it is near the theoretical capacity of this material, which is 323 mAh g^{-1} [133]. In [134] it was shown that a capacity of 290 mAh g^{-1} could also be attained for Cu_2Sb thin films fabricated through heat treatment and electrodeposition.

Higher capacities have been achieved by embedding pure 100–200 nm Sb particles in a pyrolytic polyacrylonitrile polymer matrix [135]. The second discharge capacity was 485 mAh g^{-1}, while after the 20th cycle it had reduced to 408 mAh g^{-1}.

About the same time that the microstructure of Figure 6.34 for Si was obtained, the potential of such a nanocomposite was also illustrated for Sb particles that are adhered to cellulose [136], as shown in Figure 6.44. The resulting capacity of the Sb/cellulose nanocomposite was 470 mAh g^{-1} for 40 cycles (Figure 6.45), while the capacity of pure Sb nanoparticles degraded below 70 mAh g^{-1} after the 15th cycle. Hence, this island-type structure seemed to be very promising for Sb as well.

In [137], another promising Sb-anode was proposed. It was shown that cycling $LiSbO_3$ against Li metal gave starting capacities that were nearly 600 mAh g^{-1}, but a significant drop to 500 mAh g^{-1} was observed at the 15th cycle. In particular, upon lithiation the reversible electrochemical reaction that takes place is

$$LiSbO_3 + 4Li^+ + 4e^- \leftrightarrow Li_5SbO_3.$$

This indicates that other materials of the same family, such as $LiSb_3O_8$ and Li_3SbO_4, could also act as anode candidates.

(a) (b)

Figure 6.44 (a) Sb nanoparticles adhered on the surface of cellulose fibers, (b) unattached Sb nanoparticles. Source: Reproduced with permission from Han et al. [136]/Elsevier.

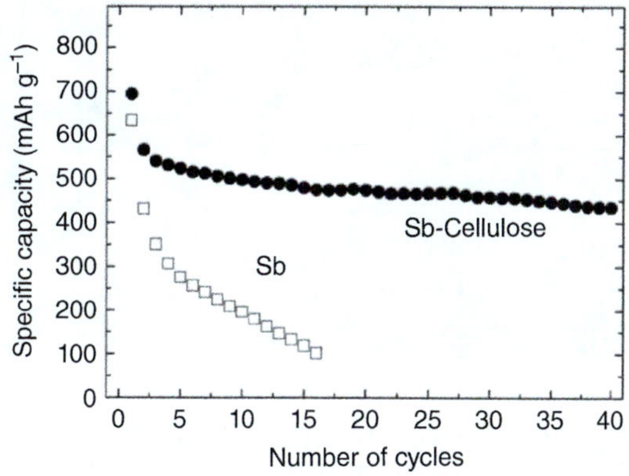

Figure 6.45 Comparison between capacity retention in pure Sb nanoparticles and Sb/Cellulose nanocomposite. Source: Reproduced with permision from Han et al. [136]/Elsevier.

6.6.3 Al-Based Anodes

Al–Li alloys give a wide range of theoretical capacities depending on the Li content; AlLi 993 mAh g^{-1} (similar to that of $Li_{4.4}Sn$), Al_2Li_3 1490 mAh g^{-1}, and Al_4Li_9 2235 mAh g^{-1}. These high capacities are partly due to the lower atomic weight of Al, as compared to that of Sn. It is very difficult, however, to maintain such high capacities after the first cycle. This may be partly attributed to the fact that the diffusion coefficient in Al (6×10^{-12} cm^2 s^{-1} at 293 K) [138] is two orders of magnitude lower than that in Sn [139].

Just as in most metal-based anodes, the first Al-anode configuration that was examined had a thin film form. The capacity increased as the Al thin film thickness decreased. The thinnest film examined was 0.1 μm and it gave an initial capacity of 800 mAh g^{-1}; however, it was not possible to retain it [139].

Powder Al particles have also been examined but again it was not possible to retain the initial capacity. Various particle sizes in the micron-scale were examined, but a coherent trend relating the reduction in the particle size with better capacities was not obtained [141], and therefore further studies need to be performed. In [140], it was documented that pure Al particles gave starting capacities of 789 mAh g^{-1}, which dropped to 325 mAh g^{-1} after 10 cycles. This significant capacity decrease was attributed to severe cracking and pulverization that occurred from the first cycle, as seen in Figure 6.46. In order to protect the Al particles from cracking and from contact with the electrolyte, SnO nanoparticles were attached, through chemical means on the Al surface, as seen in Figure 6.47. After one cycle, it was observed that no damage was apparent for such composite particles due to an increase in fracture strength, tensile strength, and elastic modulus of the reinforced Al composites [142]. On initial Li insertion, SnO decomposed to Li_2O and Sn, and therefore the conductivity also increased; the possibility of Sn-lithiation was not examined in [140]. Three different wt% for SnO were considered: 5 wt% SnO, 10 wt% SnO, and 20 wt% SnO. In Figure 6.48 it is seen that the Al sample covered with 10 wt% SnO showed the best electrochemical properties, giving a starting capacity of 810 mAh g^{-1}, which reduced to 582 mAh g^{-1} after 10 cycles. Figure 6.48, however, suggests that further

Figure 6.46 (a) Pure Al particles before cycling; (b) pure Al particles after the first cycle. Source: Reproduced with permission from Lei et.al. [141]/Elsevier.

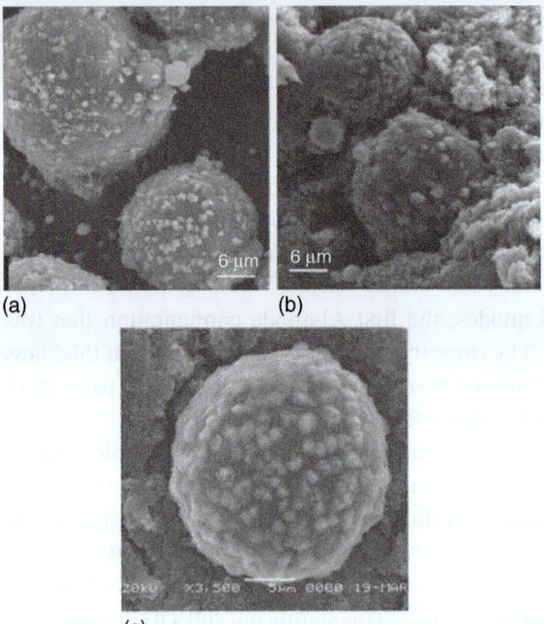

Figure 6.47 (a) Al particles covered with 10 wt% SnO nanoparticles before cycling; (b) and (c) Al particles covered with 10 wt% SnO after first cycle. Source: Reproduced with permission from Lei et al. [140]/Elsevier.

cycling will result in additional capacity reduction. Finally, it is noted that increasing the SnO content to 20 wt% or reducing it to 5 wt% gave lower capacities. This shows that the amount of SnO must be such so as to protect the Al from fracture and pulverization, but not so much so as to bind a significant amount of Li, during the decomposition of SnO.

6.6.4 Bi-Based Anodes

Another Li-active material is Bi. The theoretical capacity upon the formation of Li_3Bi is 385 mAh g^{-1} [143], which is practically near that of currently used graphic anodes, and therefore from a commercial point of view, Bi is not promising. From a scientific viewpoint, however, it is of interest to look at the electrochemical cycling

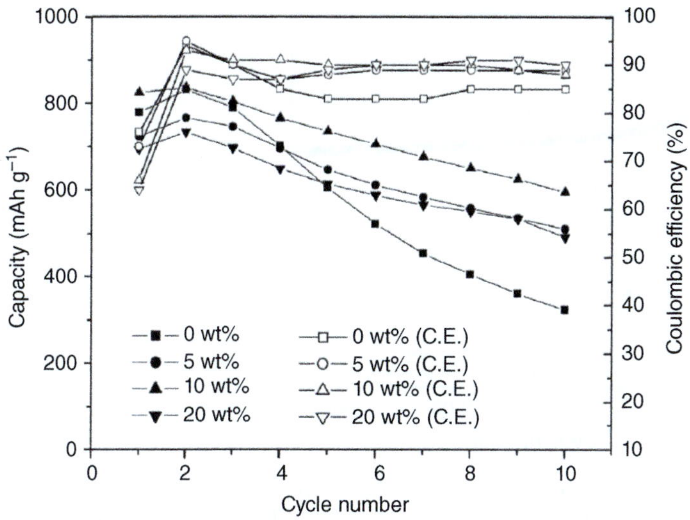

Figure 6.48 Capacity vs. cycle for various Al-wt% SnO particles. Source: Reproduced with permission from Lei et al. [140]/Elsevier.

properties of Bi since it is consistent with previous observations on Sn- and Si-based anodes.

Upon maximum Li insertion, Bi expands 210%, and as a result, it also has a low cyclability in the bulk and microscale due to severe deformation. In order to decrease the volume expansion, nanoscale Bi particles with an average diameter of 300 nm were embedded inhomogeneously in a graphite-PVDF/HFP SOLEF copolymer (Solvay), which acted as the binder. A 50% capacity fade occurred after ten cycles. The Bi content of the anode was varied to various wt%, and it was noted that lower Bi content allowed for a better capacity retention. The low capacity retention was attributed to the fact that the Bi nanoparticles were distributed randomly, favoring, therefore, agglomeration and pulverization (Figure 6.49) that resulted from the 210% volume expansion upon Li insertion.

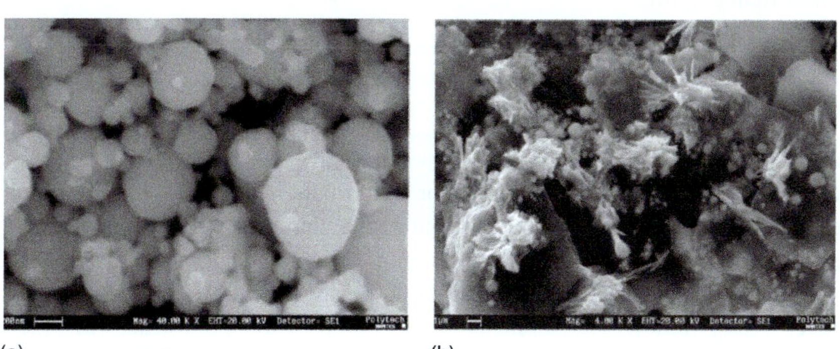

Figure 6.49 Micrographs showing Bi nanoparticles of 300 nm diameter (a) before cycling, (b) damage formation after 10 cycles.

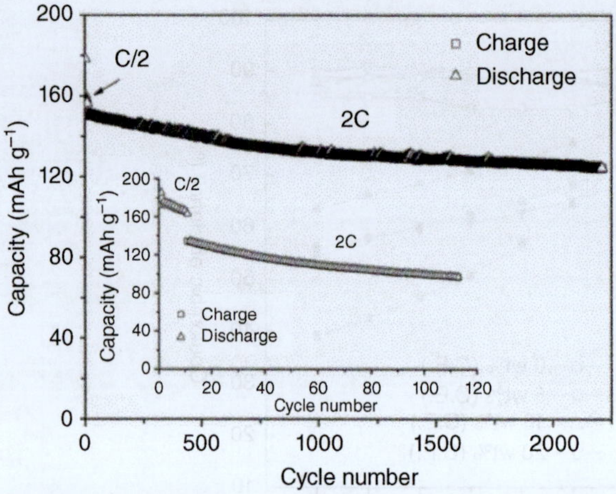

Figure 6.50 Cycling performance of coated $Li_4Ti_5O_{12}$. Inset: The cycling performance of pristine $Li_4Ti_5O_{12}$. Source: Reproduced with permission from Zhao et al. [148]/John Wiley & Sons.

6.6.5 LiTiO-Based Anodes

Spinel $Li_4Ti_5O_{12}$ has a relatively high lithiation voltage plateau at 1.54 V versus Li/Li^+, which can prevent the formation of SEI, is very safe [144, 145] and does not undergo volume expansions. In particular, as a zero-strain insertion material [146, 147], it has excellent cycling performance, particularly when it is coated with C. Figure 6.50 depicts the performance with and without the C coating [149]. For $Li_4Ti_5O_{12}$-C, the initial capacity was 150 mAh g^{-1} at 2C, which decreased to 124 mAh g^{-1} after 2200 cycles at 2C, indicating a very good capacity retention of 83%. However, for the pristine (uncoated) sample, the initial capacity was 135 mAh g^{-1} at 2C, which decreased to 98 mAh g^{-1} after only 100 cycles (73% capacity retention). These features make it a promising anode material for large-scale long-life energy storage batteries. The drawback of these zero strain materials is that their capacity is lower than graphite.

6.6.6 Metal Oxide-Based Anodes

Transition-metal oxides (TMOs, here TM = Mn, Co, Fe, Ni, Cu, etc.) have also been investigated widely as anode materials for Li-ion batteries because of their much higher lithium-storage capacities than that of the commercially used graphite, as well as their safer nature, being environmental benignity, and low cost [150, 151]. Many nanostructured TMOs, such as Co_3O_4 nanowires [152], hierarchical NiO spheres [153], hollow TiO_2 spheres [154], porous CuO nanostructures [155], mesoporous MnO_2 [156], porous Fe_3O_4 [157], porous Fe_2O_3 [158], and hollow Fe_2O_3 spheres [159], show good electrochemical performance. However, it is not anticipated that they will be commercialized as anodes, and therefore are not elaborated herein.

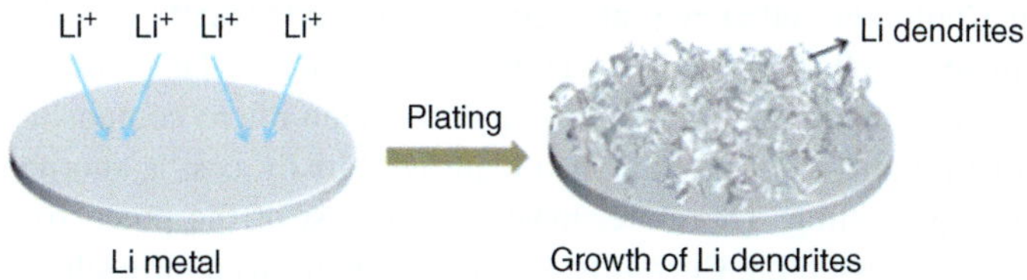

Figure 6.51 Dendrites formed on Li metal in solid-state batteries. Source: Reproduced with permission from Chen et al. [160]/American Chemical Society.

6.7 Solid-State Batteries

As mentioned in the introduction of this chapter, pure Li metal cannot be utilized as the anode due to the severe dendrite formation that occurs in liquid electrolytes.

However, pure Li remains the most promising anode from an energy density point of view, as it allows the complete anode to be active material, without requiring the use of "dead" weight, such as binder and conductive agents. To limit and dispense with dendrite formation and growth, solid electrolytes have been proposed. To date, however, an effective system that can achieve this goal has yet to be proposed. Figure 6.51 depicts dendrite formation in a solid-state Li battery, as these cells are referred to, to distinguish them from Li-ion that employ non-Li anodes. Chapter 7 gives an overview of the different types of electrolytes in Li-ion cells, and then focuses on recent advances in solid-state systems.

The energy density of solid-state Li systems can reach double that of Li-ion (~280 Wh kg^{-1}); however, Li-air batteries can provide 1700 Wh kg^{-1} which is close to that of gasoline. The challenges that are needed to be overcome for commercializing Li-air are greater than any other Li-based chemistry and therefore will not be covered in the present book.

6.8 Conclusions

In concluding this overview, it should be mentioned that the research articles published in this area, every year, are plethora and it would be impossible to summarize all of them. The purpose was to describe some of the most promising configurations and fabrication methods for the next-generation anodes.

As was seen Si-based anodes allow for the highest capacities. Although thin film and nanowire configurations allow for very good electrochemical properties, more applicable configurations for next-generation anodes are nanostructured composites, in which a more active material is alloyed or embedded in a less-active material with respect to Li.

From a practical viewpoint, the volume variation for batteries (especially tablets) has been limited strictly to less than 10–20%. Therefore, the capacity of Si or Sn anodes has to be limited to satisfy this requirement. Furthermore, in order to improve the volumetric energy density, anode materials and electrode sheets with a high pressing density are desirable. Achieving a high initial Columbic efficiency

(>90%) and a high Coulombic efficiency at successive cycles (>99.5%) are also required for realizing an excellent cyclic performance in the full cell. This requires a low surface area of the anode materials. Consequently, nano-micro composites with a surface area less than $1-3\,m^2\,g^{-1}$ are also required. Finally, cost is another important issue for large scaling. Therefore, lowering the cost of the precursors and fabrication methods is essential. Currently, the price of nanosized pure silicon particles is still too expensive ($850 for 50 gm), and should become less than $50-100\,\$\,kg^{-1}$. This is feasible but still quite a challenge.

Acknowledgments

The authors are grateful to the National Science Foundation for supporting this work through the CMMI grant [CMMI-1762602].

References

1 Murphy, D.W. and Christian, P.A. (1979). Solid state electrodes for high energy batteries. *Science* 205: 651–656. https://doi.org/10.1126/SCIENCE.205.4407.651.

2 Akira Yoshino, Kenichi Sanechika, Takayuki Nakajima (1986). Secondary battery. United States Patent (19), filed 10 May 1985 and issued 26 May 1987.

3 Shu, Z.X., McMillan, R.S., and Murray, J.J. (1993). Electrochemical intercalation of lithium into graphite. *J. Electrochem. Soc.* 140: 922–927. https://doi.org/10.1149/1.2056228/XML.

4 Fong, R., von Sacken, U., and Dahn, J.R. (1990). Studies of lithium intercalation into carbons using nonaqueous electrochemical cells. *J. Electrochem. Soc.* 137: 2009–2013. https://doi.org/10.1149/1.2086855/XML.

5 Sato, K., Noguchi, M., Demachi, A. et al. (1994). A mechanism of lithium storage in disordered carbons. *Science* 264: 556–558. https://doi.org/10.1126/SCIENCE.264.5158.556.

6 Yang, J., Winter, M., and Besenhard, J.O. (1996). Small particle size multiphase Li-alloy anodes for lithium-ion-batteries. *Solid State Ionics* 90: 281–287. https://doi.org/10.1016/S0167-2738(96)00389-X.

7 Rao, B.M.L., Francis, R.W., and Christopher, H.A. (1977). Lithium-aluminum electrode. *J. Electrochem. Soc.* 124: 1490–1492. https://doi.org/10.1149/1.2133098/XML.

8 Besenhard, J.O. (1978). Cycling behaviour and corrosion of Li–Al electrodes in organic electrolytes. *J. Electroanal. Chem. Interfacial Electrochem.* 94: 77–81. https://doi.org/10.1016/S0022-0728(78)80400-8.

9 Besenhard, J.O. and Fritz, H.P. (1975). Reversibles elektrochemisches legieren von metallen der V. hauptgruppe in organischen Li+—Lösungen. *Electrochim. Acta* 20: 513–517. https://doi.org/10.1016/0013-4686(75)90042-0.

10 Wang, J. (1986). Behavior of some binary lithium alloys as negative electrodes in organic solvent-based electrolytes. *J. Electrochem. Soc.* 133: 457. https://doi.org/10.1149/1.2108601.

11 Beaulieu, L.Y., Eberman, K.W., Turner, R.L. et al. (2001). Colossal reversible volume changes in lithium alloys. *Electrochem. Solid-State Lett.* 4: A137. https://doi.org/10.1149/1.1388178.

12 Konstantinidis, D.A. and Aifantis, E.C. (1998). On the "Anomalous" hardness of nanocrystalline materials. *Nanostruct. Mater.* 10: 1111–1118. https://doi.org/10.1016/S0965-9773(98)00145-7.

13 Aifantis, K.E. and Konstantinidis, A.A. (2009). Yielding and tensile behavior of nanocrystalline copper. *Mater. Sci. Eng., A* 503: 198–201. https://doi.org/10.1016/J.MSEA.2008.04.084.

14 Li, H., Huang, X., Chen, L. et al. (1999). A high capacity nano-Si composite anode material for lithium rechargeable batteries. *Electrochem. Solid-State Lett.* 2 (11): 547. https://doi.org/10.1149/1.1390899.

15 Idota, Y., Kubota, T., Matsufuji, A. et al. (1997). Tin-based amorphous oxide: a high-capacity lithium-ion-storage material. *Science (80-.)* 276: 1395–1397. https://doi.org/10.1126/science.276.5317.1395.

16 Y. Idota, M. Mishima, Y. Miyiaki, T. Kubota, T. Miyasaka, Nonaqueous secondary battery, 1994. https://patents.google.com/patent/US5618640A/en (accessed 16 January 2022).

17 Crosnier, O., Brousse, T., Devaux, X. et al. (2001). New anode systems for lithium ion cells. *J. Power Sources* 94: 169–174. https://doi.org/10.1016/S0378-7753(00)00599-1.

18 Courtney, I.A. and Dahn, J.R. (1997). Electrochemical and in situ X-ray diffraction studies of the reaction of lithium with tin oxide composites. *J. Electrochem. Soc.* 144: 2045–2052. https://doi.org/10.1149/1.1837740/XML.

19 Li, H., Huang, X., and Chen, L. (1999). Electrochemical impedance spectroscopy study of SnO and nano-SnO anodes in lithium rechargeable batteries. *J. Power Sources* 81–82: 340–345. https://doi.org/10.1016/S0378-7753(99)00214-1.

20 Stjerndahl, M., Bryngelsson, H., Gustafsson, T. et al. (2007). Surface chemistry of intermetallic AlSb-anodes for Li-ion batteries. *Electrochim. Acta* 52: 4947–4955. https://doi.org/10.1016/J.ELECTACTA.2007.01.064.

21 Aurbach, D., Gamolsky, K., Markovsky, B. et al. (2002). On the use of vinylene carbonate (VC) as an additive to electrolyte solutions for Li-ion batteries. *Electrochim. Acta* 47: 1423–1439. https://doi.org/10.1016/S0013-4686(01)00858-1.

22 Chen, L., Wang, K., Xie, X., and Xie, J. (2007). Effect of vinylene carbonate (VC) as electrolyte additive on electrochemical performance of Si film anode for lithium ion batteries. *J. Power Sources* 174: 538–543. https://doi.org/10.1016/J.JPOWSOUR.2007.06.149.

23 Brousse, T., Retoux, R., Herterich, U., and Schleich, D.M. (1998). Thin-film crystalline SnO_2-lithium electrodes. *J. Electrochem. Soc.* 145: 1–4. https://doi.org/10.1149/1.1838201/XML.

24 Gao, B., Sinha, S., Fleming, L., and Zhou, O. (2001). Alloy formation in nanostructured silicon. *Adv. Mater.* 13: 816–819.

25 Winter, M., Besenhard, J.O., Spahr, M.E., and Novák, P. (1998). Insertion electrode materials for rechargeable lithium batteries. *Adv. Mater.* 10: 725–763. https://doi.org/10.1002/(SICI)1521-4095(199807)10:10<725::AID-ADMA725>3.0.CO;2-Z.

26 Guo, J., Yang, Z., and Archer, L.A. (2013). Aerosol assisted synthesis of hierarchical tin-carbon composites and their application as lithium battery anode materials. *J. Mater. Chem. A* 1: 8710–8715. https://doi.org/10.1039/c3ta11802a.

27 Yang, Z., Meng, Q., Yan, W. et al. (2015). Novel three-dimensional tin/carbon hybrid core/shell architecture with large amount of solid cross-linked micro/nanochannels for lithium ion battery application. *Energy* 82: 960–967. https://doi.org/10.1016/j.energy.2015.01.105.

28 Schmuelling, G., Oehl, N., Knipper, M. et al. (2014). Synthesis and electrochemical performance of surface-modified nano-sized core/shell tin particles for lithium ion batteries. *Nanotechnology* 25: 355401. https://doi.org/10.1088/0957-4484/25/35/355401.

29 Chen, J.S., Archer, L.A., and Wen Lou, X. (2011). SnO_2 hollow structures and TiO_2 nanosheets for lithium-ion batteries. *J. Mater. Chem.* 21: 9912–9924. https://doi.org/10.1039/c0jm04163g.

30 Cho, Y.J., Kim, C.H., Im, H.S. et al. (2013). Germanium-tin alloy nanocrystals for high-performance lithium ion batteries. *Phys. Chem. Chem. Phys.* 15: 11691–11695. https://doi.org/10.1039/c3cp51366a.

31 Tang, X., Yan, F., Wei, Y. et al. (2015). Encapsulating SnxSb nanoparticles in multichannel graphene-carbon fibers as flexible anodes to store lithium ions with high capacities. *ACS Appl. Mater. Interfaces* 7: 21890–21897. https://doi.org/10.1021/acsami.5b06377.

32 Ferraresi, G., Villevieille, C., Czekaj, I. et al. (2018). SnO_2 model electrode cycled in Li-ion battery reveals the formation of Li_2SnO_3 and Li_8SnO_6 phases through conversion reactions. *ACS Appl. Mater. Interfaces* 10: 8712–8720. https://doi.org/10.1021/acsami.7b19481.

33 Cheng, Y., Nie, A., Gan, L.Y. et al. (2015). A global view of the phase transitions of SnO_2 in rechargeable batteries based on results of high throughput calculations. *J. Mater. Chem. A* 3: 19483–19489. https://doi.org/10.1039/c5ta03521j.

34 Jiang, D., Wang, C., Sun, L. et al. (2017). Facile hydrothermal synthesis of SnO_2 nanoparticles with enhanced lithium storage performance. *Chem. Lett.* 46: 1639–1642. https://doi.org/10.1246/cl.170757.

35 Zhang, D.-F., Sun, L.-D., Yin, J.-L., and Yan, C.-H. (2003). Low-temperature fabrication of highly crystalline SnO_2 nanorods. *Adv. Mater.* 15: 1022–1025. https://doi.org/10.1002/adma.200304899.

36 Weng, W., Lin, J., Du, Y. et al. (2018). Template-free synthesis of metal oxide hollow micro-/nanospheres: via Ostwald ripening for lithium-ion batteries. *J. Mater. Chem. A* 6: 10168–10175. https://doi.org/10.1039/c8ta03161d.

37 Li, N., Martin, C.R., and Scrosati, B. (2001). Nanomaterial-based Li-ion battery electrodes. *J. Power Sources* 97–98: 240–243. https://doi.org/10.1016/S0378-7753(01)00760-1.

38 Brousse, T., Lee, S.M., Pasquereau, L. et al. (1998). Composite negative electrodes for lithium ion cells. *Solid State Ionics* 113–115: 51–56. https://doi.org/10.1016/S0167-2738(98)00366-X.

39 Li, Y., Tu, J.P., Huang, X.H. et al. (2006). Nanoscale SnS with and without carbon-coatings as an anode material for lithium ion batteries. *Electrochim. Acta* 52: 1383–1389. https://doi.org/10.1016/J.ELECTACTA.2006.07.041.

40 Li, Y., Tu, J.P., Huang, X.H. et al. (2007). Net-like SnS/carbon nanocomposite film anode material for lithium ion batteries. *Electrochem. Commun.* 9: 49–53. https://doi.org/10.1016/J.ELECOM.2006.08.019.

41 Harreld, C.S. and Schlemper, E.O. (1971). The crystal and molecular structure of tetrakis-(*N*,*N*-diethyldithiocarbamato)-tin(IV). *Acta Crystallogr., Sect. B: Struct. Sci.* 27: 1964–1969. https://doi.org/10.1107/S056774087100517X/FULL.

42 Seo, J.W., Jang, J.T., Park, S.W. et al. (2008). Two-dimensional SnS_2 nanoplates with extraordinary high discharge capacity for lithium ion batteries. *Adv. Mater.* 20: 4269–4273. https://doi.org/10.1002/ADMA.200703122.

43 Momma, T., Shiraishi, N., Yoshizawa, A. et al. (2001). SnS_2 anode for rechargeable lithium battery. *J. Power Sources* 97–98: 198–200. https://doi.org/10.1016/S0378-7753(01)00723-6.

44 Mukaibo, H., Yoshizawa, A., Momma, T., and Osaka, T. (2003). Particle size and performance of SnS_2 anodes for rechargeable lithium batteries. *J. Power Sources* 119–121: 60–63. https://doi.org/10.1016/S0378-7753(03)00125-3.

45 Elidrissi Moubtassim, M.L., Bousquet, C., Olivier-Fourcade, J. et al. (1998). Local effects of the lithium insertion into cation-deficient tin–indium thiospinels revealed by 119Sn Mössbauer spectroscopy. *Chem. Mater.* 10: 968–973. https://doi.org/10.1021/CM970100C.

46 Chen, J.S. and Lou, X.W.D. (2013). SnO_2-based nanomaterials: synthesis and application in lithium-ion batteries. *Small* 9: 1877–1893. https://doi.org/10.1002/smll.201202601.

47 Deng, J., Yan, C., Yang, L. et al. (2013). Sandwich-stacked SnO_2/Cu hybrid nanosheets as multichannel anodes for lithium ion batteries. *ACS Nano* 7: 6948–6954. https://doi.org/10.1021/nn402164q.

48 Hassoun, J., Mulas, G., Panero, S., and Scrosati, B. (2007). Ternary Sn–Co–C Li-ion battery electrode material prepared by high energy ball milling. *Electrochem. Commun.* 9: 2075–2081. https://doi.org/10.1016/J.ELECOM.2007.05.033.

49 Hassoun, J., Panero, S., Mulas, G., and Scrosati, B. (2007). An electrochemical investigation of a Sn–Co–C ternary alloy as a negative electrode in Li-ion batteries. *J. Power Sources* 171: 928–931. https://doi.org/10.1016/J.JPOWSOUR.2007.06.067.

50 H. Seino, SONY NEXELION, 2005. https://www.sony.com/en/SonyInfo/News/Press/200502/05-006E/ (accessed in June 2022).

51 Aifantis, K.E., Huang, T., Hackney, S.A. et al. (2012). Capacity fade in Sn–C nanopowder anodes due to fracture. *J. Power Sources* 197: 246–252. https://doi.org/10.1016/J.JPOWSOUR.2011.09.025.

52 Aifantis, K.E., Haycock, M., Sanders, P., and Hackney, S.A. (2011). Fracture of nanostructured Sn/C anodes during Li-insertion. *Mater. Sci. Eng., A* 529: 55–61. https://doi.org/10.1016/J.MSEA.2011.08.060.

53 Kim, I.S., Blomgren, G.E., and Kumta, P.N. (2004). Sn/C composite anodes for Li-ion batteries. *Electrochem. Solid-State Lett.* 7: A44. https://doi.org/10.1149/1.1643792/XML.

54 Cheng, D., Yang, L., Hu, R. et al. (2019). Sn–C and Se–C Co-bonding SnSe/few-layered graphene micro-nano structure: route to a densely compacted and durable anode for lithium/sodium-ion batteries. *ACS Appl. Mater. Interfaces* 11: 36685–36696. https://doi.org/10.1021/ACSAMI.9B12204/SUPPL_FILE/AM9B12204_SI_001.PDF.

55 Wang, Z., Tian, W., and Li, X. (2007). Synthesis and electrochemistry properties of Sn–Sb ultrafine particles as anode of lithium-ion batteries. *J. Alloys Compd.* 439: 350–354. https://doi.org/10.1016/J.JALLCOM.2006.08.247.

56 Wang, K., He, X., Ren, J. et al. (2006). Preparation of Sn2Sb alloy encapsulated carbon microsphere anode materials for Li-ion batteries by carbothermal reduction of the oxides. *Electrochim. Acta* 52: 1221–1225. https://doi.org/10.1016/J.ELECTACTA.2006.07.020.

57 Lee, J.Y., Zhang, R., and Liu, Z. (2000). Dispersion of Sn and SnO on carbon anodes. *J. Power Sources* 90: 70–75. https://doi.org/10.1016/S0378-7753(00)00450-X.

58 Santos-Peña, J., Brousse, T., and Schleich, D.M. (2000). Search for suitable matrix for the use of tin-based anodes in lithium ion batteries. *Solid State Ionics* 135: 87–93. https://doi.org/10.1016/S0167-2738(00)00337-4.

59 Balan, L., Schneider, R., Ghanbaja, J. et al. (2006). Electrochemical lithiation of new graphite–nanosized tin particle materials obtained by $SnCl_2$ reduction in organic medium. *Electrochim. Acta* 51: 3385–3390. https://doi.org/10.1016/J.ELECTACTA.2005.09.045.

60 Aifantis, K.E., Brutti, S., Hackney, S.A. et al. (2010). SnO_2/C nanocomposites as anodes in secondary Li-ion batteries. *Electrochim. Acta* 55: 5071–5076. https://doi.org/10.1016/J.ELECTACTA.2010.03.083.

61 Landi, B.J., Ganter, M.J., Cress, C.D. et al. (2009). Carbon nanotubes for lithium ion batteries. *Energy Environ. Sci.* 2: 638–654. https://doi.org/10.1039/b904116h.

62 Ahn, J.H., Kim, Y.J., and Wang, G. (2009). Electrochemical properties of SnO_2 nanowires prepared by a simple heat treatment of Sn–Ag alloys. *J. Alloys Compd.* 483 (1–2): 422–424. https://doi.org/10.1016/J.JALLCOM.2008.07.227.

63 Sharma, R.A. (1976). Thermodynamic properties of the lithium-silicon system. *J. Electrochem. Soc.* 123: 1763. https://doi.org/10.1149/1.2132692.

64 Abraham, K.M. (2015). Prospects and limits of energy storage in batteries. *J. Phys. Chem. Lett.* 6: 830–844. https://doi.org/10.1021/jz5026273.

65 Li, J. and Dahn, J.R. (2007). An in situ X-ray diffraction study of the reaction of Li with crystalline Si. *J. Electrochem. Soc.* 154: A156. https://doi.org/10.1149/1.2409862.

66 Gerberich, W.W., Mook, W.M., Perrey, C.R. et al. (2003). Superhard silicon nanospheres. *J. Mech. Phys. Solids* 51: 979–992. https://doi.org/10.1016/S0022-5096(03)00018-8.

67 Graetz, J., Ahn, C.C., Yazami, R., and Fultz, B. (2003). Highly reversible lithium storage in nanostructured silicon. *Electrochem. Solid-State Lett.* 6: A194. https://doi.org/10.1149/1.1596917/XML.

68 Maranchi, J.P., Hepp, A.F., and Kumta, P.N. (2003). High capacity, reversible silicon thin-film anodes for lithium-ion batteries. *Electrochem. Solid-State Lett.* 6: A198. https://doi.org/10.1149/1.1596918.

69 Lee, K.L., Jung, J.Y., Lee, S.W. et al. (2004). Electrochemical characteristics of a-Si thin film anode for Li-ion rechargeable batteries. *J. Power Sources* 129: 270–274. https://doi.org/10.1016/J.JPOWSOUR.2003.10.013.

70 Moon, T., Kim, C., and Park, B. (2006). Electrochemical performance of amorphous-silicon thin films for lithium rechargeable batteries. *J. Power Sources* 155: 391–394. https://doi.org/10.1016/J.JPOWSOUR.2005.05.012.

71 Takamura, T., Ohara, S., Uehara, M. et al. (2004). A vacuum deposited Si film having a Li extraction capacity over 2000 mAh/g with a long cycle life. *J. Power Sources* 129: 96–100. https://doi.org/10.1016/j.jpowsour.2003.11.014.

72 Réthoré, J., Zheng, H., Li, H. et al. (2018). A multiphysics model that can capture crack patterns in Si thin films based on their microstructure. *J. Power Sources* 400: 383–391. https://doi.org/10.1016/j.jpowsour.2018.07.106.

73 Yin, J., Wada, M., Yamamoto, K. et al. (2006). Micrometer-scale amorphous Si thin-film electrodes fabricated by electron-beam deposition for Li-ion batteries. *J. Electrochem. Soc.* 153: A472. https://doi.org/10.1149/1.2160429/XML.

74 Wang, Y., Liu, Y., Zheng, J. et al. (2013). Electrochemical performances and volume variation of nano-textured silicon thin films as anodes for lithium-ion batteries. *Nanotechnology* 24: 424011. https://doi.org/10.1088/0957-4484/24/42/424011.

75 He, Y., Yu, X., Wang, Y. et al. (2011). Alumina-coated patterned amorphous silicon as the anode for a lithium-ion battery with high coulombic efficiency. *Adv. Mater.* 23: 4938–4941. https://doi.org/10.1002/ADMA.201102568.

76 Yang, H., Fu, P., Zhang, H. et al. (2007). Amorphous Si film anode coupled with $LiCoO_2$ cathode in Li-ion cell. *J. Power Sources* 174: 533–537. https://doi.org/10.1016/J.JPOWSOUR.2007.06.249.

77 He, Y., Yu, X., Li, G. et al. (2012). Shape evolution of patterned amorphous and polycrystalline silicon microarray thin film electrodes caused by lithium insertion and extraction. *J. Power Sources* 216: 131–138. https://doi.org/10.1016/J.JPOWSOUR.2012.04.105.

78 Wei Wang, J., He, Y., Fan, F. et al. (2013). Two-phase electrochemical lithiation in amorphous silicon. *Nano Lett.* 13: 709–715. https://doi.org/10.1021/nl304379k.

79 Wang, Y.H., He, Y., Xiao, R.J. et al. (2012). Investigation of crack patterns and cyclic performance of Ti–Si nanocomposite thin film anodes for lithium ion batteries. *J. Power Sources* 202: 236–245. https://doi.org/10.1016/J.JPOWSOUR.2011.11.027.

80 Ahn, D., Kim, C., Lee, J.G., and Park, B. (2008). The effect of nitrogen on the cycling performance in thin-film Si1-*x*N*x* anode. *J. Solid State Chem.* 181: 2139–2142. https://doi.org/10.1016/J.JSSC.2008.04.040.

81 Zeng, Z.Y., Tu, J.P., Wang, X.L., and Zhao, X.B. (2008). Electrochemical properties of Si/$LiTi_2O_4$ nanocomposite as anode materials for Li-ion secondary batteries. *J. Electroanal. Chem.* 616: 7–13. https://doi.org/10.1016/J.JELECHEM.2007.12.017.

82 Zhou, G.W., Li, H., Sun, H.P. et al. (1999). Controlled Li doping of Si nanowires by electrochemical insertion method. *Appl. Phys. Lett.* 75: 2447–2449. https://doi.org/10.1063/1.125043.

83 Ishihara, T., Nakasu, M., Yoshio, M. et al. (2005). Carbon nanotube coating silicon doped with Cr as a high capacity anode. *J. Power Sources* 146: 161–165. https://doi.org/10.1016/J.JPOWSOUR.2005.03.110.

84 Cui, L.F., Ruffo, R., Chan, C.K. et al. (2009). Crystalline-amorphous core-shell silicon nanowires for high capacity and high current battery electrodes. *Nano Lett.* 9: 491–495. https://doi.org/10.1021/nl8036323.

85 100% Silicon Nanowire Batteries from Amprius Technologies | amprius.com, (n.d.). https://www.amprius.com/technology/ (accessed 26 May 2020).

86 Ahuja, U., Wang, B., Hu, P. et al. (2021). Polydopamine coated Si nanoparticles allow for improved mechanical and electrochemical stability. *Electrochim. Acta* 392: 138993. https://doi.org/10.1016/J.ELECTACTA.2021.138993.

87 Li, Y., Yan, K., Lee, H.W. et al. (2016). Growth of conformal graphene cages on micrometre-sized silicon particles as stable battery anodes. *Nat. Energy* 1: 1–9. https://doi.org/10.1038/nenergy.2015.29.

88 Jung, Y.S., Lee, K.T., and Oh, S.M. (2007). Si–carbon core–shell composite anode in lithium secondary batteries. *Electrochim. Acta* 52: 7061–7067. https://doi.org/10.1016/J.ELECTACTA.2007.05.031.

89 Shu, J., Li, H., Yang, R. et al. (2006). Cage-like carbon nanotubes/Si composite as anode material for lithium ion batteries. *Electrochem. Commun.* 8: 51–54. https://doi.org/10.1016/J.ELECOM.2005.08.024.

90 Wen, Z., Yang, X., and Huang, S. (2007). Composite anode materials for Li-ion batteries. *J. Power Sources* 174: 1041–1045. https://doi.org/10.1016/J.JPOWSOUR.2007.06.041.

91 Zhang, Z., Wang, Y., Ren, W. et al. (2014). Scalable synthesis of interconnected porous silicon/carbon composites by the rochow reaction as high-performance anodes of lithium ion batteries. *Angew. Chem. Int. Ed.* 53: 5165–5169. https://doi.org/10.1002/ANIE.201310412.

92 Yue, L., Zhang, W., Yang, J., and Zhang, L. (2014). Designing Si/porous-C composite with buffering voids as high capacity anode for lithium-ion batteries. *Electrochim. Acta* 125: 206–217. https://doi.org/10.1016/j.electacta.2014.01.094.

93 Liu, N., Wu, H., McDowell, M.T. et al. (2012). A yolk-shell design for stabilized and scalable Li-ion battery alloy anodes. *Nano Lett.* 12: 3315–3321. https://doi.org/10.1021/nl3014814.

94 Zuo, X., Zhu, J., Müller-Buschbaum, P., and Cheng, Y.J. (2017). Silicon based lithium-ion battery anodes: a chronicle perspective review. *Nano Energy* 31: 113–143. https://doi.org/10.1016/j.nanoen.2016.11.013.

95 Yang, L.Y., Li, H.Z., Liu, J. et al. (2015). Dual yolk-shell structure of carbon and silica-coated silicon for high-performance lithium-ion batteries. *Sci. Rep.* 5: 10908. https://doi.org/10.1038/srep10908.

96 Liu, N., Lu, Z., Zhao, J. et al. (2014). A pomegranate-inspired nanoscale design for large-volume-change lithium battery anodes. *Nat. Nanotechnol.* 9: 187–192. https://doi.org/10.1038/nnano.2014.6.

97 Hwa, Y., Kim, W.S., Hong, S.H., and Sohn, H.J. (2012). High capacity and rate capability of core–shell structured nano-Si/C anode for Li-ion batteries. *Electrochim. Acta* 71: 201–205. https://doi.org/10.1016/j.electacta.2012.03.138.

98 Gómez Cámer, J.L., Morales, J., and Sánchez, L. (2008). Nano-Si/cellulose composites as anode materials for lithium-ion batteries. *Electrochem. Solid-State Lett.* 11: A101. https://doi.org/10.1149/1.2902305/XML.

99 Caballero, Á., Morales, J., and Sánchez, L. (2005). Tin nanoparticles formed in the presence of cellulose fibers exhibit excellent electrochemical performance as anode materials in lithium-ion batteries. *Electrochem. Solid-State Lett.* 8: A464. https://doi.org/10.1149/1.1993388/XML.

100 Schnabel, M., Arca, E., Ha, Y. et al. (2020). Enhanced interfacial stability of Si anodes for Li-ion batteries via surface SiO_2 coating. *ACS Appl. Energy Mater.* 3: 8842–8849. https://doi.org/10.1021/acsaem.0c01337.

101 Li, Z.F., Zhang, H., Liu, Q. et al. (2014). Novel pyrolyzed polyaniline-grafted silicon nanoparticles encapsulated in graphene sheets as Li-ion battery anodes. *ACS Appl. Mater. Interfaces* 6: 5996–6002. https://doi.org/10.1021/am501239r.

102 Tian, M. and Wu, P. (2019). Nature plant polyphenol coating silicon submicroparticle conjugated with polyacrylic acid for achieving a high-performance anode of lithium-ion battery. *ACS Appl. Energy Mater.* 2: 5066–5073. https://doi.org/10.1021/acsaem.9b00734.

103 Huang, S., Ren, J., Liu, R. et al. (2018). The progress of novel binder as a non-ignorable part to improve the performance of Si-based anodes for Li-ion batteries. *Int. J. Energy Res.* 42: 919–935. https://doi.org/10.1002/er.3826.

104 Mazouzi, D., Karkar, Z., Hernandez, C.R. et al. (2015). Critical roles of binders and formulation at multiscales of silicon-based composite electrodes. *J. Power Sources* 280: 533–549. https://doi.org/10.1016/j.jpowsour.2015.01.140.

105 Su, H., Barragan, A.A., Geng, L. et al. (2017). Colloidal synthesis of silicon-carbon composite material for lithium-ion batteries. *Angew. Chem. Int. Ed.* 56: 10780–10785. https://doi.org/10.1002/anie.201705200.

106 Chen, H., Ling, M., Hencz, L. et al. (2018). Exploring chemical, mechanical, and electrical functionalities of binders for advanced energy-storage devices. *Chem. Rev.* 118 (18): 8936–8982. https://doi.org/10.1021/acs.chemrev.8b00241.

107 Li, T., Yang, J.Y., and Lu, S.G. (2012). Effect of modified elastomeric binders on the electrochemical properties of silicon anodes for lithium-ion batteries. *Int. J. Miner. Metall. Mater.* 19: 752–756. https://doi.org/10.1007/s12613-012-0623-1.

108 Karkar, Z., Guyomard, D., Roué, L., and Lestriez, B. (2017). A comparative study of polyacrylic acid (PAA) and carboxymethyl cellulose (CMC) binders for Si-based electrodes. *Electrochim. Acta* 258: 453–466. https://doi.org/10.1016/j.electacta.2017.11.082.

109 Yoon, J., Oh, D.X., Jo, C. et al. (2014). Improvement of desolvation and resilience of alginate binders for Si-based anodes in a lithium ion battery by calcium-mediated cross-linking. *Phys. Chem. Chem. Phys.* 16: 25628–25635. https://doi.org/10.1039/c4cp03499f.

110 Chen, D., Yi, R., Chen, S. et al. (2014). Facile synthesis of graphene-silicon nanocomposites with an advanced binder for high-performance lithium-ion battery anodes. *Solid State Ionics* 254: 65–71. https://doi.org/10.1016/j.ssi.2013.11.020.

111 Wu, H., Yu, G., Pan, L. et al. (2013). Stable Li-ion battery anodes by in-situ polymerization of conducting hydrogel to conformally coat silicon nanoparticles. *Nat. Commun.* 4: 1–6. https://doi.org/10.1038/ncomms2941.

112 Koo, B., Kim, H., Cho, Y. et al. (2012). A highly cross-linked polymeric binder for high-performance silicon negative electrodes in lithium ion batteries. *Angew. Chem. Int. Ed.* 51: 8762–8767. https://doi.org/10.1002/anie.201201568.

113 Eshetu, G.G. and Figgemeier, E. (2019). Confronting the challenges of next-generation silicon anode-based lithium-ion batteries: role of designer electrolyte additives and polymeric binders. *ChemSusChem* 12: 2515–2539. https://doi.org/10.1002/cssc.201900209.

114 Nguyen, C.C., Yoon, T., Seo, D.M. et al. (2016). Systematic investigation of binders for silicon anodes: interactions of binder with silicon particles and electrolytes and effects of binders on solid electrolyte interphase formation. *ACS Appl. Mater. Interfaces* 8: 12211–12220. https://doi.org/10.1021/acsami.6b03357.

115 Designer Binders Protect Silicon Battery Electrodes | Department of Energy, (n.d.). https://www.energy.gov/science/bes/articles/designer-binders-protect-silicon-battery-electrodes (accessed 17 November 2020).

116 Lee, J.-I., Lee, K.T., Cho, J. et al. (2012). Chemical-assisted thermal disproportionation of porous silicon monoxide into silicon-based multicomponent systems. *Angew. Chem. Int. Ed.* 51: 2767–2771. https://doi.org/10.1002/anie.201108915.

117 Xu, Q., Sun, J.K., Yu, Z.L. et al. (2018). https://doi.org/10.1002/adma.201707430). SiO_x encapsulated in graphene bubble film: an ultrastable Li-ion battery anode. *Adv. Mater.* 30: 1707430. (1-6).

118 Cui, J., Cui, Y., Li, S. et al. (2016). Microsized porous SiO_x@C composites synthesized through aluminothermic reduction from rice husks and used as anode for lithium-ion batteries. *ACS Appl. Mater. Interfaces* 8: 30239–30247. https://doi.org/10.1021/acsami.6b10260.

119 Morita, T. and Takami, N. (2006). Nano Si cluster-SiO_x-C composite material as high-capacity anode material for rechargeable lithium batteries. *J. Electrochem. Soc.* 153: A425. https://doi.org/10.1149/1.2142295/XML.

120 Scott, A. (2019). In the battery materials world, the anode's time has come. *Chem. Eng. News C&EN* 97: 18–21. https://doi.org/10.1021/cen-09714-feature1.

121 New Hitachi Maxell Li-Ion battery increases battery power by 60% without taking up extra space, (n.d.). https://www.companyowl.com/i/n/new-hitachi-maxell-liion-battery-increases-battery-power-60-taking-extra-space/201397 (accessed 26 May 2020).

122 Ali, G., Iqbal, M.Z., and Jan Iftikhar, F. (2021). MXene. *Adv. Supercapacitor Supercapattery* 1: 255–269. https://doi.org/10.1016/B978-0-12-819897-1.00005-7.

123 Gogotsi, Y. and Anasori, B. (2019). The rise of MXenes. *ACS Nano* 13: 8491–8494. https://doi.org/10.1021/ACSNANO.9B06394.

124 Naguib, M., Kurtoglu, M., Presser, V. et al. (2011). Two-dimensional nanocrystals produced by exfoliation of Ti_3AlC_2. *Adv. Mater.* 23: 4248–4253. https://doi.org/10.1002/ADMA.201102306.

125 Mashtalir, O., Naguib, M., Mochalin, V.N. et al. (2013). Intercalation and delamination of layered carbides and carbonitrides. *Nat. Commun.* 4 (1): 1–7. https://doi.org/10.1038/ncomms2664.

126 Yoo, E., Kim, J., Hosono, E. et al. (2008). Large reversible Li storage of graphene nanosheet families for use in rechargeable lithium ion batteries. *Nano Lett.* 8: 2277–2282. https://doi.org/10.1021/NL800957B.

127 Kannan, K., Sadasivuni, K.K., Abdullah, A.M., and Kumar, B. (2020). Current trends in MXene-based nanomaterials for energy storage and conversion system: a mini review. *Catalysts* 10: 495. https://doi.org/10.3390/CATAL10050495.

128 Liang, X., Garsuch, A., and Nazar, L.F. (2015). Sulfur cathodes based on conductive MXene nanosheets for high-performance lithium–sulfur batteries. *Angew. Chem. Int. Ed.* 54: 3907–3911. https://doi.org/10.1002/ANIE.201410174.

129 Luo, J., Tao, X., Zhang, J. et al. (2016). Sn^{4+} ion decorated highly conductive Ti_3C_2 MXene: promising lithium-ion anodes with enhanced volumetric capacity and cyclic performance. *ACS Nano* 10: 2491–2499. https://doi.org/10.1021/ACSNANO.5B07333.

130 Zou, G., Zhang, Z., Guo, J. et al. (2016). Synthesis of MXene/Ag composites for extraordinary long cycle lifetime lithium storage at high rates. *ACS Appl. Mater. Interfaces* 8: 22280–22286. https://doi.org/10.1021/ACSAMI.6B08089.

131 Han, X., Zhou, W., Chen, M. et al. (2022). Interfacial nitrogen engineering of robust silicon/MXene anode toward high energy solid-state lithium-ion batteries. *J. Energy Chem.* 67: 727–735. https://doi.org/10.1016/J.JECHEM.2021.11.021.

132 Thackeray, M.M., Vaughey, J.T., Johnson, C.S. et al. (2003). Structural considerations of intermetallic electrodes for lithium batteries. *J. Power Sources* 113: 124–130. https://doi.org/10.1016/S0378-7753(02)00538-4.

133 Fransson, L.M.L., Vaughey, J.T., Benedek, R. et al. (2001). Phase transitions in lithiated Cu_2Sb anodes for lithium batteries: an in situ X-ray diffraction study. *Electrochem. Commun.* 3: 317–323. https://doi.org/10.1016/S1388-2481(01)00140-0.

134 Bryngelsson, H., Eskhult, J., Nyholm, L., and Edström, K. (2008). Thin films of Cu_2Sb and Cu_9Sb_2 as anode materials in Li-ion batteries. *Electrochim. Acta* 53: 7226–7234. https://doi.org/10.1016/J.ELECTACTA.2008.05.005.

135 He, X., Pu, W., Wang, L. et al. (2007). Synthesis of nano Sb-encapsulated pyrolytic polyacrylonitrile composite for anode material in lithium secondary batteries. *Electrochim. Acta* 52: 3651–3653. https://doi.org/10.1016/J.ELECTACTA.2006.10.029.

136 Caballero, Á., Morales, J., and Sánchez, L. (2008). A simple route to high performance nanometric metallic materials for Li-ion batteries involving the use of cellulose: the case of Sb. *J. Power Sources* 175: 553–557. https://doi.org/10.1016/J.JPOWSOUR.2007.09.100.

137 Kundu, M., Mahanty, S., and Basu, R.N. (2009). Lithium antimonite: a new class of anode material for lithium-ion battery. *Electrochem. Commun.* 11: 1389–1392. https://doi.org/10.1016/J.ELECOM.2009.05.013.

138 Hamon, Y., Brousse, T., Jousse, F. et al. (2001). Aluminum negative electrode in lithium ion batteries. *J. Power Sources* 97–98: 185–187. https://doi.org/10.1016/S0378-7753(01)00616-4.

139 Huggins, R.A. (1999). Lithium alloy negative electrodes. *J. Power Sources* 81–82: 13–19. https://doi.org/10.1016/S0378-7753(99)00124-X.

140 Lei, X., Xiang, J., Ma, X. et al. (2007). Surface modification of aluminum with tin oxide coating. *J. Power Sources* 166: 509–513. https://doi.org/10.1016/J.JPOWSOUR.2006.12.105.

141 Lei, X., Wang, C., Yi, Z. et al. (2007). Effects of particle size on the electrochemical properties of aluminum powders as anode materials for lithium ion batteries. *J. Alloys Compd.* 429: 311–315. https://doi.org/10.1016/J.JALLCOM.2006.04.019.

142 Rao, K.T.V. and Ritchie, R.O. (2013). Fatigue of aluminium—lithium alloys. *Int. Mater. Rev.* 37: 153–186. https://doi.org/10.1179/IMR.1992.37.1.153.

143 Crosnier, O., Devaux, X., Brousse, T. et al. (2001). Influence of particle size and matrix in “metal” anodes for Li-ion cells. *J. Power Sources* 97–98: 188–190. https://doi.org/10.1016/S0378-7753(01)00617-6.

144 Aricò, A.S., Bruce, P., Scrosati, B. et al. (2005). Nanostructured materials for advanced energy conversion and storage devices. *Nat. Mater.* 4: 366–377. https://doi.org/10.1038/NMAT1368.

145 Bruce, P.G., Scrosati, B., and Tarascon, J.M. (2008). Nanomaterials for rechargeable lithium batteries. *Angew. Chem. Int. Ed.* 47: 2930–2946. https://doi.org/10.1002/ANIE.200702505.

146 Colbow, K.M., Dahn, J.R., and Haering, R.R. (1989). Structure and electrochemistry of the spinel oxides $LiTi_2O_4$ and $Li_{43}Ti_{53}O_4$. *J. Power Sources* 26: 397–402. https://doi.org/10.1016/0378-7753(89)80152-1.

147 Ferg, E., Gummow, R.J., de Kock, A., and Thackeray, M.M. (1994). Spinel anodes for lithium-ion batteries. *J. Electrochem. Soc.* 141: L147–L150. https://doi.org/10.1149/1.2059324/XML.

148 Zhao, L., Hu, Y.-S., Li, H. et al. (2011). Porous $Li_4Ti_5O_{12}$ coated with N-doped carbon from ionic liquids for Li-ion batteries. *Adv. Mater.* 23: 1385–1388. https://doi.org/10.1002/ADMA.201003294.

149 Ohzuku, T., Ueda, A., and Yamamoto, N. (1995). Zero-strain insertion material of Li [Li1/3Ti5/3] O4 for rechargeable lithium cells. *J. Electrochem. Soc.* 142: 1431–1435. https://doi.org/10.1149/1.2048592/XML.

150 Armand, M. and Tarascon, J.M. (2008, 2008). Building better batteries. *Nature* 451: 652–657. https://doi.org/10.1038/451652a.

151 Wagemaker, M., Borghols, W.J.H., and Mulder, F.M. (2007). Large impact of particle size on insertion reactions. a case for anatase Li_xTiO_2. *J. Am. Chem. Soc.* 129: 4323–4327. https://doi.org/10.1021/JA067733P.

152 Wang, B., Zhu, T., Bin Wu, H. et al. (2012). Porous Co_3O_4 nanowires derived from long $Co(CO_3)0.5(OH)·0.11H_2O$ nanowires with improved supercapacitive properties. *Nanoscale* 4: 2145–2149. https://doi.org/10.1039/C2NR11897A.

153 Zhu, T., Chen, J.S., and Lou, X.W.D. (2012). Highly efficient removal of organic dyes from waste water using hierarchical NiO spheres with high surface area. *J. Phys. Chem. C* 116: 6873–6878. https://doi.org/10.1021/JP300224S/SUPPL_FILE/JP300224S_SI_001.PDF.

154 Ding, S., Chen, J.S., Wang, Z. et al. (2011). TiO_2 hollow spheres with large amount of exposed (001) facets for fast reversible lithium storage. *J. Mater. Chem.* 21: 1677–1680. https://doi.org/10.1039/C0JM03650A.

155 Wang, Z., Su, F., Madhavi, S., and Lou, X.W. (2011). CuO nanostructures supported on Cu substrate as integrated electrodes for highly reversible lithium storage. *Nanoscale* 3: 1618–1623. https://doi.org/10.1039/C0NR00827C.

156 Ren, Y., Armstrong, A.R., Jiao, F., and Bruce, P.G. (2010). Influence of size on the rate of mesoporous electrodes for lithium batteries. *J. Am. Chem. Soc.* 132: 996–1004. https://doi.org/10.1021/JA905488X/SUPPL_FILE/JA905488X_SI_001.PDF.

157 Xiong, Q.Q., Tu, J.P., Lu, Y. et al. (2012). Synthesis of hierarchical hollow-structured single-crystalline magnetite (Fe_3O_4) microspheres: The highly powerful storage versus lithium as an anode for lithium ion batteries. *J. Phys. Chem. C* 116: 6495–6502. https://doi.org/10.1021/JP3002178/SUPPL_FILE/JP3002178_SI_001.PDF.

158 Yao, X., Tang, C., Yuan, G. et al. (2011). Porous hematite (α-Fe_2O_3) nanorods as an anode material with enhanced rate capability in lithium-ion batteries. *Electrochem. Commun.* 13: 1439–1442. https://doi.org/10.1016/J.ELECOM.2011.09.015.

159 Wang, B., Chen, J.S., Bin Wu, H. et al. (2011). Quasiemulsion-templated formation of α-Fe_2O_3 hollow spheres with enhanced lithium storage properties. *J. Am. Chem. Soc.* 133: 17146–17148. https://doi.org/10.1021/JA208346S/SUPPL_FILE/JA208346S_SI_001.PDF.

160 Chen, T., Kong, W., Zhao, P. et al. (2019). Dendrite-free and stable lithium metal anodes enabled by an antimony-based lithiophilic interphase. *Chem. Mater.* 31: 7565–7573. https://doi.org/10.1021/ACS.CHEMMATER.9B02356/SUPPL_FILE/CM9B02356_SI_001.PDF.

7

Electrolytes for Lithium Batteries: The Quest for Improving Lithium Battery Performance and Safety

Claudio Capiglia

The Institute of Frontier Materials, Deakin University, 221 Burwood Highway, Burwood VIC 3125, Australia

7.1 Introduction

How can Li-ion batteries become the dream technology to power our cars in the twenty-first century?

The development of new electrolytes is the key. This chapter provides an overview of the electrolytes that are used today in lithium-ion batteries and gives a summary of solid-state electrolytes that are currently under development.

As mentioned in the previous chapters, the current battery systems for electronic applications, such as smartphones and tablets utilize, in the most cases, graphitic carbons as the anode, and NMC, LCA, LCO, and LMO, as the cathode, while the electrolyte, they use a polymeric separator imbibed with a mixture of alkyl carbonates with dissolved lithium salt. The electrolytes are indispensable for the fabrication of advanced lithium-ion batteries since they serve as the medium for the Li-ion transfer between the cathode and the anode. The electrolytes that are based on alkyl carbonate have an acceptable window of electrochemical stability, up to 4 V, to satisfy the needs for the cathodes used nowadays in Li-ion batteries. These electrolytes show a high polarity, with a reasonable temperature range between the boiling point and the freezing point, and have sufficiently low toxicity. The most commonly used solvents are ethylene carbonate (EC), ethyl methyl carbonates (EMC), dimethyl carbonate (DMC), and diethyl carbonate (DEC). $LiPF_6$ is the salt of choice for lithium-ion batteries although it undergoes decomposition into LiF and PF_5. The latter hydrolyzes with any water contamination inside the cell to form HF and PF_3O. These two hydrolysis products are highly reactive on both the negative and positive electrodes, and their unavoidable presence contributes to the degradation of lithium-ion batteries during the long-term charge–discharge process. In particular, HF can lead to further electrolyte decomposition and gas generation. PF_5 generated from the decomposition of $LiPF_6$ is also a strong Lewis acid, which can result in the opening of the cyclic carbonate of the solvent and attack the carbonyl oxygen atoms of the linear carbonates while producing side reaction products. The decomposition of the electrolyte can produce a number of different gas species, including CO_2, CH_4, C_2H_4, C_2H_5F, and other subsequent reaction products.

Rechargeable Ion Batteries: Materials, Design, and Applications of Li-Ion Cells and Beyond, First Edition.
Edited by Katerina E. Aifantis, R. Vasant Kumar, and Pu Hu.

Table 7.1 General classification of the electrolytes types for secondary lithium batteries.

Batteries	Lithium-ion	Lithium polymer		Solid state	
Electrolytes	Nonaqueous	Gel polymer	Solid oxide	Polymer	Sulfide-Type
Salt Solvent	Lithium salt ($LiPF_6$, LiBOB…)	Lithium Salt ($LiPF_6$, LiBETI…)	i.e. NASICON, LISICON, LiPON	Lithium salt ($LiPF_6$, imide…)	i.e. LGPS
Polymer	Organic solvent PE	Organic solvent (PVDF, PMMA)		(PEO, *Hyper-branched*)	

The electrolyte decomposition reaction can be complex; in particular, the gas generation under abuse conditions has a detrimental effect on the safety of the lithium-ion cells because of the venting of flammable solvent vapors into the surrounding environment. In fact, these vapors can force the lithium-ion cell to explode either due to the frequent sparks produced by the battery or due to the self-ignition generated by thermal runaway. The decomposition of the electrolyte that uses alkyl carbonates as the solvent begins at above 150 °C and the thermal runway begins at temperatures above 240 °C.

The decomposition of the electrolyte and the flammability of vented vapors are still open issues for Li-ion batteries. Enormous efforts have been carried out worldwide to solve the problem while improving the electrochemical performance of the electrolytes. Surely, the quest for improving the Li-ion battery performance and their safety targets the development of new electrolytes, and in fact, the name of the electrolyte gives the name to the lithium-ion battery technology. For the sake of simplicity, electrolytes are divided into three main categories: nonaqueous electrolytes, gel polymer electrolytes, and solid-state electrolytes. The nonaqueous electrolytes are composed of an organic solvent and lithium salt. The gel polymer electrolytes are composed of a polymer swollen with a nonaqueous electrolyte. Finally, the solid-state electrolytes that, as the name indicates, comprise inorganic solid materials or glass or polymer that conduct lithium-ions (Table 7.1).

In the following sections, we will describe the three groups of electrolytes, highlighting their advantages and disadvantages.

7.2 Nonaqueous Electrolytes

The vast majority of electrolytic solution types for lithium-ion batteries consist of salts dissolved in organic solvents, usually named nonaqueous electrolytes [1, 2]. In fact, in the 1950s lithium metal in primary batteries was found to be stable in nonaqueous solvents, thanks to the formation of a passivation film on the lithium surface, which prevented further electrolyte decomposition. Passivation is a phenomenon of all lithium primary cells (not rechargeable) related to the interaction of the metallic lithium anode and the electrolyte. A thin passivation layer forms on the surface of the anode, at the instant, the electrolyte is introduced into the cell. This layer is important because it protects the anode from reaction while the cell is dormant – resulting in a long shelf-life.

The research activities that followed this discovery resulted in the commercialization of lithium primary cells in the 1960s and 1970s. At this time, the electrolyte solvents that were used ranged from propylene carbonate (PC) to thionyl chloride and sulfur dioxide. The initial efforts to expand the lithium chemistry from a primary cell into a rechargeable technology encountered severe difficulties in terms of cycle life and safety. The main problem was the morphology of the lithium crystals (dendrites) growing on the anode upon charge and discharge. In Figure 7.1 lithium dendrite formation is schematically represented. In fact, the continuous growth of dendrites caused the perforation of the separator resulting in an internal short circuit with the consequent thermal runaway and explosion of the cell. Hence, the work on rechargeable lithium batteries from the 1970s to 1980s concentrated mostly on the development of electrolytes that would eliminate the dendritic formation of lithium. Experimental images of dendrites are found in Chapter 3.

Since 1989, because of accidents related to the use of pure lithium as the anode in rechargeable batteries, intercalation anodes began to gain attention. This concept of reversible chemistry was already in use in cathode materials for lithium batteries. Metal oxides with stable crystal lattices and layered structures were providing the pathways for lithium-ions to diffuse. In the 1990s a breakthrough was made in Japan where they used carbonaceous materials as the intercalation anode and proposed the name of lithium-ion battery for this type of chemistry. The advantage of this new kind of anode was the low cost for production and the high lithium-ion activity in the intercalation compound.

7.2.1 The Solid-Electrolyte Interface (SEI)

In 1991, lithium-ion cells based on an anodic carbon, nonaqueous electrolyte, and $LiCoO_2$ as cathode were commercialized. The electrolyte and graphitic anode were the key components that enabled the rechargeable lithium-ion battery to work reversibly; in fact, the electrolyte solvents decomposed on the carbonaceous anode while forming a protective film to prevent further reaction with the electrolyte components. It was discovered that the passive film was an ionic conductor (hence allowing Li-ions to diffuse through) and that the decomposition process occurred only during the first charge. The electrolyte solvents were discovered to influence

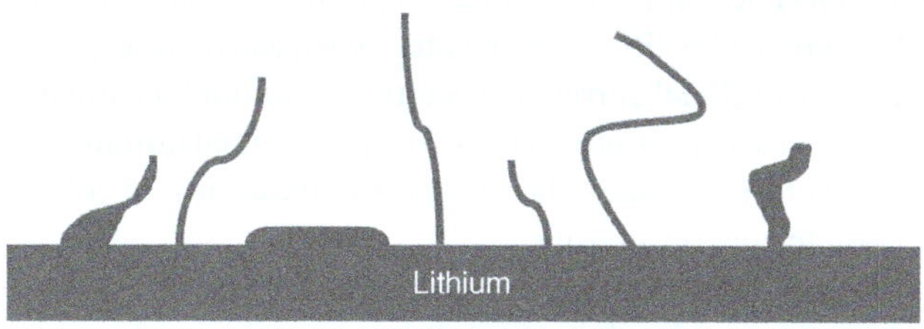

Figure 7.1 Schematic representation of lithium dendrite formation.

the nature of the protective film, and ethylene carbonate was found to be an essential component of the solvents to protect the structure of graphite. The film was named solid-electrolyte interface (SEI). The SEI formation was related to the reduction products of the electrolyte and it was demonstrated to influence the electrochemical properties of the electrode and prevent exfoliation of the graphitic anode during cycling. Controlling the SEI thickness was significant because thicker SEI increased the cell's electrical resistance with a detrimental effect on the lithium cell performance. SEI was also discovered to be formed at the cathode side and influence the battery performance as well but to a less extent.

Extensive research has evidenced that SEIs are formed by decomposed products from solvents and salts. SEIs are composed of Li_2CO_3, lithium alkyl carbonate, lithium alkyl oxide, and LiF when $LiPF_6$ salt type is present. The two mechanisms that describe SEI formation on carbonaceous anode materials are: the first one consists of the reduction of solvent-generated gaseous products, resulting in an SEI that is Li_2CO_3 based (unstable SEI is formed); in the second one, more polymer-based materials are formed and precipitated on the anode (with stable SEI formation). Generally, SEIs are formed in two steps: the first step occurs before Li^+ intercalation resulting in an SEI that is unstable and rich in inorganic compounds; then the second step occurs after Li^+ intercalation in the graphite, resulting in a stable SEI and rich in organic compounds. The two mechanisms for SEI formation are associated with the catalytic activity of the fresh graphite and are influenced by the morphology of graphite surface [3].

The following decade witnessed explosive growth in the research and development of lithium-ion batteries. Universities, government research institutes, and industries around the world were developing new carbon-based materials, metal oxide cathode materials, and novel nonaqueous electrolytes compatible with those electrodes. The result was that in 2000 the market value of Li-ion batteries was over a billion dollars, mainly for portable electronic applications. Today, after three decades of research, development, production, and commercialization of lithium-ion batteries, the market is expected to reach the value of more than 73 billion dollars in 2025 owing to automotive applications, such as electric vehicles (EV), hybrid electric vehicles (HEV), and industrial applications, such as stationary power.

7.2.2 Current Collector Corrosion

Current collectors are indispensable components bridging lithium-ion batteries and external circuits, greatly influencing the lithium-ion batteries' capacity, rate capability, and long-term stability. Conventional current collectors, aluminum for cathode, and copper foils for the anode, have been used since the first commercial lithium-ion battery. However, over the past three decades, the thickness of these current collectors has decreased to increase the energy density.

The primary current collector materials used in lithium-ion cells are susceptible to environmental degradation: aluminium to pitting corrosion and copper to environmentally assisted cracking. The electrolyte contact with the current collectors is the origin of corrosion, and this can occur once the electrodes fracture/delaminate/degrade. The highly oxidizing potential associated with the positive electrode-charge condition is the primary factor. The corrosion resistance of aluminum may be due to its native oxide. At the same time, for Cu, environmental cracking occurs at or near the lithium potential when specific metallurgical conditions exist (work hardening and large grain size). The corrosion of the current collector is also influenced by the salt used in the electrolyte. Therefore, the engineering of a suitable electrolyte is significant to reduce the corrosion of the current collectors.

7.2.3 Solvents for Nonaqueous Electrolytes

To be used in the production of lithium-ion batteries, the solvents for nonaqueous electrolytes must satisfy several requirements. In general, solvents need to have a high dielectric constant to dissolve lithium salts together with low viscosity for ion transport. Their melting point should be low so as to allow the charge/discharge at low temperature. In addition, solvents need to have, as essential safety requisite, a high enough boiling point to avoid solvent evaporation. Finally, the solvents should not allow dangerous side reactions with cell components and electrodes, they must be nontoxic for safe handling, and of course economically advantageous to reduce costs of cell fabrication. Solvents with active protons are excluded because the reduction of such protons occurs at 2.0 V vs. Li, which is exactly in the range of operation of lithium-ion batteries. Nonaqueous compounds qualified as solvents for lithium-ion batteries need polar groups. The most commonly used solvents for nonaqueous electrolytes are listed in Table 7.2.

Propylene carbonate (PC): The early interest among the community of researchers was focused on this solvent with a high dielectric constant, and therefore the ability to dissolve a wide variety of lithium salts. Its wide liquid window (i.e. low-melting point and high-boiling temperature), high dielectric constant, and stability vs. lithium, thanks to the SEI formed by PC, made it the solvent of choice for electrolytes in batteries that used anodes based on lithium metal. However, it was observed that batteries with PC electrolytes experienced a huge-capacity fade, and it was realized that the poor cycling efficiency was intrinsic to the solvent reaction between PC and the deposited lithium particles

Table 7.2 Solvent most commonly used in today's lithium-ion batteries.

Solvent	Potential value for solvent oxidation V vs. Li/Li^+ (purity 99%, 98%)
Propylene carbonate (PC)	5.40
Ethylene carbonate (EC)	5.50
Dimethyl carbonate (DMC)	5.30
Diethyl carbonate (DEC)	5.15

during the charge/discharge. This behavior was motivated by the nonuniform morphological change of the lithium surface during cycling which resulted in dendrite formation.

Ethers: Due to the aforementioned PC safety concerns, ethers were preferred as alternative solvents because of their low viscosity and the better lithium morphology during the charge/discharge process. In particular, the solvents of choice were tetrahydrofuran (THF), polymethoxy ethers, dimethoxy propane, diethyl ether, and a few others. The formation of dendritic lithium seemed to be suppressed even during high-charge rates. However, ether-based nonaqueous electrolytes in lithium-ion batteries showed again poor capacity retention and poor cycle life. The origin of this behavior was again related to dendrite depositions that were found during long-term charge–discharge tests. Furthermore, the oxidative decomposition of ether-based compounds at the cathode surface was discovered to be a new factor contributing to the capacity fade of the battery.

Ethylene carbonate (EC): The interest in EC was renewed when pure lithium was abandoned in favor of a carbon-based anode. EC has a good viscosity and high dielectric constant, which is comparable with water. However, because of its high-melting point (approximately 36 °C), it is solid at ambient temperature. EC was considered an electrolyte cosolvent in lithium-ion batteries because the addition of EC to the nonaqueous electrolyte solutions favored the ion conductivity. The presence of salt was found to enhance the EC melting point at room temperature, which is further enhanced by the addition of PC. The employment of EC as a cosolvent in various mixtures of electrolytes was found to improve the interfacial properties of various cathode surfaces. EC was found to form a protective film (SEI) on graphitic anodes that prevented electrolyte decomposition and graphite exfoliation, as observed with PC. The first commercialized rechargeable lithium batteries used $LiCoO_2$ as the cathode, an EC/PC mixture as nonaqueous electrolyte, and carbonaceous material as anode.

Linear carbonate After the lithium-ion batteries were commercialized, numerous manufacturers tried to develop new high-energy-density lithium-ion batteries. Highly crystalline graphitic carbon became the standard anode in the industry together with EC as the solvent for the electrolyte owing to its capability of avoiding graphite exfoliation. Efforts were made to expand the limited liquid range of EC-based nonaqueous electrolytes by using different cosolvents, including PC and ether-based solvents. However, the attempts were not successful. Therefore, the interest switched to linear carbonates as EC cosolvents. At this time several linear carbonates were investigated and developed: DMC, DEC, EMC, and propylmethyl carbonate (PMC), as well as the mixture of two or more solvents [4]. It was found that nonaqueous electrolytes based on linear carbonates had a wide electrochemical stability window up to 5.0 V, and they formed homogeneous mixtures with EC while enhancing both the melting point and the ion conductivity of the electrolyte. At the present, electrolyte solvents used in the over one billion lithium-ion cells manufactured each year are almost exclusively based on the mixture of EC with one or more of these linear carbonates.

7.2.4 Salts for Nonaqueous Electrolytes

The conduction mechanism in nonaqueous electrolytes is related to the amount of ions originating from the dissolution of lithium salt in the solvent [5]. When electrolytes are imbibed in the separator, they conduct electricity by means of the transport of ions, which are the charge carriers. The nonaqueous electrolyte should be inert with the electrodes during the Faradaic reaction and should be a poor electron conductor, but a good ionic conductor.

Even though the lithium-ion battery panorama offers a high number of electrodes and combinations of solvents, the lithium salts do not yet show the same amount of desirable options and properties. The $LiPF_6$ is today the best compromise to be employed, even if it poses several safety concerns, which might result in the explosion of the cell. Hence, the improvement of lithium salts is a crucial aspect and it is expected that this improvement will advance the electrochemical performances of lithium-ion batteries. Lithium salt candidates need to meet several fundamental characteristics: (i) the salts should have a high thermal and chemical stability and must be able to passivate the aluminum current collector to prevent its anodic dissolution. The salts also need to exhibit high ionic conductivity in various nonaqueous solvent systems. (ii) The anion must be stable vs. oxidative decomposition at the cathode. (iii) The salts should form less resistive solid-electrolyte interfaces on the electrodes, and they should be able to completely dissolve and dissociate in the wide variety of linear and cyclic carbonate solvents. (iv) The solvated lithium cations should move inside the nonaqueous electrolytes with high mobility. Both anion and cation salts must remain inert toward the other components of the cell. Particularly, the anion must be inert to the electrolyte solvents, nontoxic, and it must remain stable against thermally induced reactions with electrolyte solvents and other cell components [6]. In Table 7.3, the most common salts used in lithium-ion batteries are listed.

Table 7.3 Salts most commonly used in lithium-ion batteries with the product derived from the salt decomposition or side reaction.

Formula	Side reaction products
$LiClO_4$	$LiClO_3$ (hygroscopic and explosive), $LiClO_2$, LiCl
$LiBF_4$	BF_3 (highly reactive), LiF, HF
$LiAsF_6$	HF, AsF_5, AsF_3 (react with SEI)
$LiPF_6$	PF5 (highly reactive), HF
$LiN(CF_3SO_3)$	Aluminum corrosion (current collector)
$LiN[(CF_3SO_2)_2]$ or (LiTFSI)	Aluminum corrosion (current collector)
$LiN[(C_2F_5SO_2)_2]$ or (LiBETI)	Cathode decomposition (oxygen release with thermal runway)
LiBOB	High resistivity at the solid-electrolyte interface

7.2.4.1 Lithium Perchlorate ($LiClO_4$)

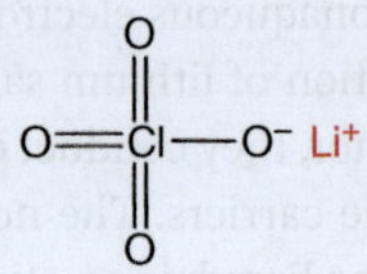

$LiClO_4$ exists in salts with high solubility in linear carbonate solvents that show good ionic conductivity and high anodic stability (up to 5.1 V). $LiClO_4$ is stable toward hydrolysis and forms a passivation film on anode materials both in carbonaceous and metallic lithium form. $LiClO_4$-based electrolytes have low interfacial resistance because they do not form thin resistive LiF species. The ClO_4^- decomposes, above 4.5 V (vs. Li/Li^+), into ClO_2 and HCl. $LiClO_4$ decomposes during the operation of the cell and the byproduct from this reaction, $LiClO_3$, is highly hygroscopic. The high oxidation state of chlorine (VII) makes the perchlorate a strong oxidant, which reacts with organic species at higher temperatures, and it may be explosive. For these reasons, $LiClO_4$ cannot be used in everyday batteries; thus, it is widely employed in research laboratories and academia as a reference salt.

7.2.4.2 Lithium Tetrafluoroborate ($LiBF_4$)

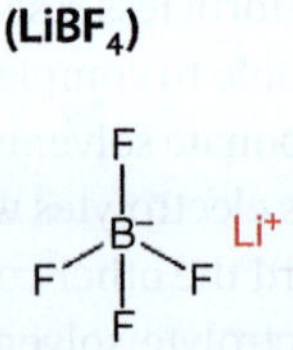

$LiBF_4$ is characterized by small anions with a high negative charge density that results in a high association with Li^+ ions. $LiBF_4$ reacts with trace amounts of moisture-producing HF, which degrades the transition-metal elements existing in the cathode. Another byproduct of hydrolysis, which reacts with solvents, is BF_3. LiF is the main component of the SEI, onto which polymeric species formed from alkyl carbonate reduction components are also observed. $LiBF_4$ passivates the Al current collector and the passivation phenomenon prevents its dissolution. However, $LiBF_4$ has restricted application in the electrolytes because SEI is not stable for lithium-ion batteries.

7.2.4.3 Lithium Hexafluoroarsenate ($LiAsF_6$)

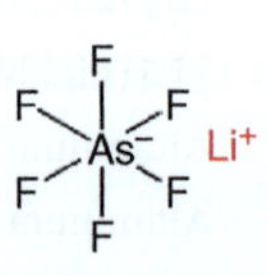

$LiAsF_6$ has good ionic conductivity, it is soluble in low-dielectric-constant solvents and has a good cycling efficiency. $LiAsF_6$ passivates the Al current collector and forms a stable SEI on both electrodes, which consists of alkyl carbonates rather than LiF. The electrochemical window stability of $LiAsF_6$-based solutions is found to be up to 6.3 V vs. Li/Li^+. Moisture may lead to the production of dangerous products as well as HF. Dangerous products, such as HF, AsF_5, or AsF_3, can react with the SEI and lead the battery to a thermal runaway. As(V) is nontoxic, but As(III) and As(0) are highly toxic for humans; the use of As is hindered in commercial batteries.

7.2.4.4 Lithium Hexafluorophosphate ($LiPF_6$)

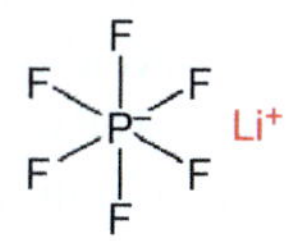

$LiPF_6$ is widely used in commercial lithium-ion batteries because of its low-melting point. It shows good solubility in low-dielectric solvents and forms a stable SEI on both carbonaceous anodes and cathodes. $LiPF_6$ is very reactive to trace amounts of moisture in nonaqueous solvents and at elevated temperatures has poor thermal stability. PF_5 is a gaseous byproduct of $LiPF_6$ decomposition that, by absorption of moisture, leads to the formation of HF. $LiPF_6$ forms SEI with both graphitizable (soft) and non-graphitizable (hard) carbon anodes. The SEI consists of LiF, polymeric species with a small amount of Li_2O, and $Li_xPF_yO_z$ components. $LiPF_6$ also effectively passivates the Al current collector and prevents the dissolution of Al.

Furthermore, $LiPF_6$ can efficiently suppress the decomposition of the cathode by producing polymeric species during solvent reduction, which hinders both the release of oxygen and the diffusion of transition-metal atoms from the cathode.

7.2.4.5 Lithium Trifluoromethanesulfonate ($Li(CF_3SO_3)$)

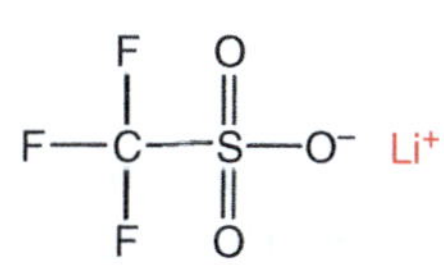

The attempt to use these salts originated from the hope that their dissociation constants would be higher even in low dielectric media. However, due to the formation of ion pairs, $Li(CF_3SO_3)$ exhibits poor ionic conductivity. The groups ($[CF_3SO_3]$) resist hydrolysis. $Li(CF_3SO_3)$ has good electrochemical windows of stability up to 4.8 V vs. platinum and good thermal stability; it is nontoxic, resistant to oxidation, and insensitive to ambient moisture conditions. The main drawback is the severe corrosion of the Al current collector, and therefore, this salt is not used in practical lithium-ion cells.

7.2.4.6 Lithium Bis(trifluoromethanesulfonyl)imide ($Li[N(CF_3SO_2)_2]$ or LiTFSI):

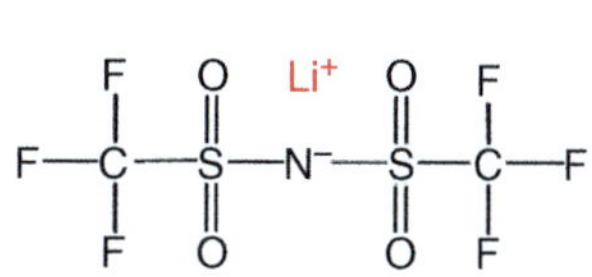

LiTFSI salts were commercialized by 3 M Corporation for research and industrial purposes, as it was found thermally stable and it forms high-conducting electrolytes. Despite the formation of ion pairs, LiTFSI dissociates well in linear carbonate solvents. The electrochemical window stability was found to be up to 4.3 V vs. Li/Li^+. The TFSI anion actively participates in the formation of surface films and the NSO_2 group is involved in these processes. LiTFSI does not generate HF and forms very stable SEI on carbonaceous materials. However, this salt is not able to form a protective film on the Al surface and it cannot avoid the dissolution of Al in nonaqueous electrolyte. The corrosion mechanism of LiTFSI on the Al current collector hinders the application of LiTFSI in commercial lithium-ion batteries.

7.2.4.7 Lithium Bis(perfluoroethylsulfonyl)imide ($Li[NC(C_2F_5SO_2)_2]$ or LiBETI)

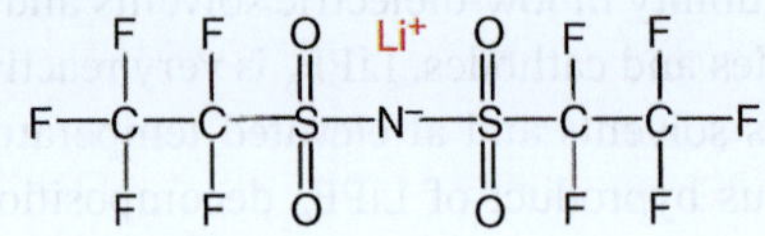

LiBETI was commercialized by 3M Corporation to overcome LiTFSI limitations for research and industrial applications. LiBETI exhibits superior behavior to LiTFSI, due to its anionic size. The largest BETI anions do not form ion pairs with small cations that provide enhanced ionic conductivity. The large anion size favors the high stability of the SEI passivation film. Similar to LiTFSI, LiBETI does not form HF as a reaction byproduct; therefore, there is no formation of LiF on carbonaceous anodes. This fact is detrimental to SEI stability. It was noted that LiBETI also has the ability to form a passivation film on the Al current collector preventing thus Al corrosion. LiBETI shows a wide electrochemical stability window up to 5.5 V vs. Li/Li^+ and exhibits good cycling behavior and better thermal stability than LiTFSI. However, LiBETI anions are not involved in the production of polymeric species during solvent reduction, which hinders the release of oxygen and transition-metal atoms diffusion from the cathode side. The absence of this mechanism reduces the battery safety causing the thermal runaway of the battery. Therefore, LiBETI cannot be used in commercial lithium-ion batteries.

7.2.4.8 Lithium Tris(trifluoromethanesulfonyl)methide ($Li[C(CF_3SO_2)_3]$ or LiTFSM)

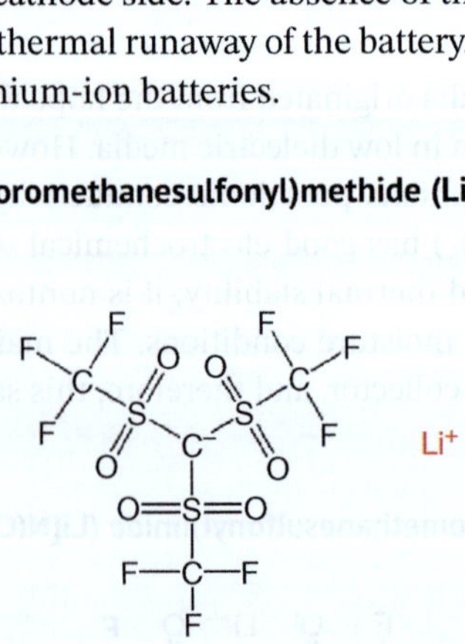

LiTFSM is a lithium salt based on a carbanion stabilized by three perfluorinated trifluoromethanesulfonyl groups. The anion is very stable and exhibits better ionic conductivity than LiTFSI. The CF bonds and the CF_3 groups strongly react with Li^+ to form LiF. Similarly, SO_2 reacts with Li^+ to form compounds, such as $Li_2S_2O_4$ and Li_2S. LiTFSM is stable up to 340 °C and has a good electrochemical window of stability up to 6.1 V vs. Li/Li^+. LiTFSM reacts with the Al current collector; therefore, it cannot be used in commercial lithium-ion batteries.

7.2.4.9 Lithium Tris(perfluoroethyl)trifluorophosphate ($Li[PF_3(CF_3CF_2)_3]$ or LiFAP)

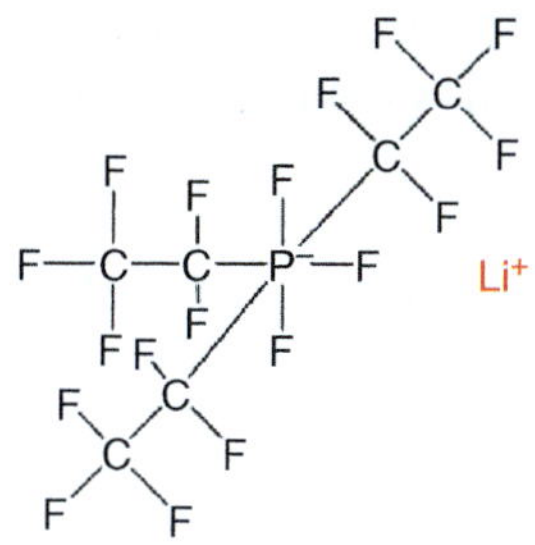

LiFAP has been proposed as the potential replacement in commercial cells containing $LiPF_6$. LiFAP has superior stability for hydrolysis than $LiPF_6$, improved thermal stability, and it exhibits a large electrochemical window up to 5.0 V vs. Li/Li^+. LiFAP participates in the formation of polymeric species at the cathode side and also in the formation of a stable SEI passivation film on carbonaceous anodes. The SEI contains organic and inorganic carbonates and trace amounts of LiF. LiFAP shows excellent capacity retention and is thermally stable up to 220 °C.

7.2.4.10 Lithium Fluoroalkylborate ($Li[BF_3(CCF_3CF_2)]$ or LiFAB)

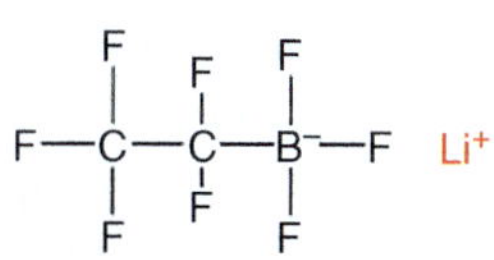

LiFAB salt has one fluorine atom replaced by a perfluoroalkyl group (CF_3CF_2), which hinders the strong coordination with Li^+ and reduces the ion-pair formation observed in $LiBF_4$. When compared with BF_4, LiFAB has the advantage that it forms a passivation layer on the Al current collector. LiFAB also appears to be thermally stable and forms a stable SEI layer on both anodes and cathodes. The electrochemical stability of LiFAB salt was found to be up to 4.25 V. LiFAB demonstrated superior performance with graphitic anodes over $LiPF_6$ and $LiBF_4$-based electrolytes. LiFAB electrolytes were found to contain low HF contamination.

7.2.4.11 Lithium Nonafluorobutylsulfonyltrifluoromethylsulfonylimide ($Li[N(C_4F_9SO_2)(CF_3SO_2)]$ or LiFBMSI): LiFBMSI:

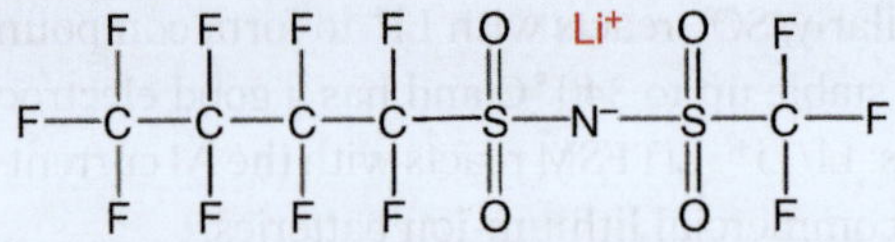

LiFBMSI is very resistant to the hydrolysis reaction in the presence of moisture; it shows good cycling efficiency in lithium-ion batteries and the electrolyte solution is electrochemically stable up to 5.0 V vs. Li/Li^+. The FBMSI anion does not form ion pairs with Li^+. LiFBMSI allows for a good passivation film to form on the Al current collector, indicating that it could be a suitable salt to replace $LiPF_6$ in commercially available lithium-ion batteries.

7.2.4.12 Lithium B(oxalato)borate ($Li[B(C_2O_4)_2]$ or LiBOB)

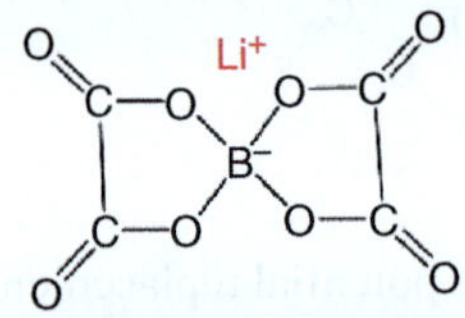

Recently, LiBOB has attracted much attention because of its peculiar properties. It was discovered that BOB anions are involved with the SEI formation on carbonaceous anodes, as well as cathodes. The SEI in carbonaceous materials contains semi-carbonates, which prevent solvent co-intercalation. Furthermore, LiBOB passivates the Al current collector. The synthesis process of this salt is quite simple; thus its employment allows to reduce the production costs. LiBOB does not hydrolyze to form HF and has good stability in a wide potential window up to 4.5 V, and it is soluble in alkyl carbonate organic solvents, such as ethylene carbonate, dimethyl carbonate, and DEC. LiBOB has a high conductivity in various nonaqueous electrolytes; good cycling behavior is nontoxic and has good thermal stability. During overcharge, LiBOB does not ignite sparks or fire. BOB anion implements a high degree of dissociation of the salt. It was discovered that LiBOB prevents dendrite formation on the Li metal anode. LiBOB can be used in combination with other salts to result in a better electrochemical performance of lithium-ion cells [7, 8].

7.2.5 Additives for Nonaqueous Electrolytes

Salts and solvents are the main components of nonaqueous electrolytes. Additives are not the main component but their addition of low amount in weight or volume can drastically improve the electrochemical performance and the safety of lithium-ion cells. The most commonly used additive is vinylene carbonate (VC), although several other options are available, which include sulfates, phosphates, and active gases, such as SO_2 and CO_2 aromatic compounds as overcharge protection agents (by a shuttle mechanism) and other possibilities. Generally, the amount in the electrolyte does not exceed 5% by weight or by volume. Additives help lithium-ion batteries to facilitate the formation of SEI on carbonaceous anodes; they enhance thermal stability of the electrolyte solution and they reduce parasite reactions and gas generation from the SEI during long-term cycling. Furthermore, additives protect cathode materials from dissolution and overcharge and improve the physical proprieties of the electrolyte solution, such as viscosity, wettability on the separator, and ionic conductivities. The additives can reduce the flammability of the electrolyte, provide protection and terminate the lithium-ion batteries operation during overcharge abuse conditions. They operate by chemical radical formation, which terminates radical chain reactions responsible for the combustion reaction. Tables 7.4a and 7.4b summarize the most common additives developed and used in the industry and academia.

7.2.5.1 Reductive Additives

These types of additives react electrochemically with the surface of graphitic anodes to form an insoluble product that forms the SEI. A list of reductive additives includes: Vinylene carbonate, vinyl ethylene carbonate, allyl ethyl carbonate, vinyl acetate, divinyl adipate, acrylic acid nitrile, 2-vinyl pyridine, maleic anhydride, methyl cinnamate, phosphonate, and vinyl-containing silane-based compounds. Other types of reductive additives include N_2O, nitrate, nitrite, halogenated ethylene carbonate, and halogenated lactone. Their mechanism for SEI formation is related to the bonds formed between the halogen species, which result from the reduced electrolyte and halogen-based additives. The action of those additives takes place at the initial stage of SEI formation and results in reducing gas formation, initial irreversible capacity, and SEI stabilization for long-term cycling.

7.2.5.2 Polymerizable Additives for Graphite-based Anode

These types of additives facilitate SEI formation by electrochemically induced polymerization, where obtained radicals can be terminated with the solvent to form an insoluble and stable product. On the other hand, oxidation may occur at the cathode side, deteriorating thus the lithium-ion battery performance. Therefore, the amount in the electrolyte should be low and not greater than 2%. Most of these types of additives are sulfur-based compounds and include SO_2, CS_2, polysulfide (Sx_2-), cyclic alkyl sulfites, such as ethylene sulfite and propylene sulfite, and aryl sulfites. The action mechanism of these additives is probably related to the poisoning effect that the sulfur has on the catalysts. Similarly, as reductive additives, the action of polymerizable additives takes place at the initial stage of SEI formation and results in reducing gas formation, initial irreversible capacity, and SEI stabilization for long cycling.

7.2.5.3 Reaction Additives for Graphite-based Anode

These types of additives are not reduced through electrochemical reactions in lithium-ion batteries but are able to form radical ions, which react with anions

Table 7.4a Most common type of additives for Li-ion cells using graphite-based anodes.

	Actions
Reductive additives Vinylene carbonate Vinyl ethylene carbonate Allyl ethyl carbonate Vinyl acetate Divinyl adipate Acrylic acid nitrile 2-Vinyl pyridine Maleic anhydride Methyl cinnamate Phosphonate Vinyl-containing silane-based compounds N_2O Nitrate Halogenated ethylene carbonate Halogenated lactone	This type of additive acts on the surface modification of graphite through electrochemical reactions to form an insoluble product that forms the SEI
Polymerizable additives for graphite-based anode SO_2 CS_2 Polysulfide Cyclic alkyl sulfites Ethylene sulfite Propylene sulfite Aryl sulfites	This type of additive facilitates SEI formation by electrochemical induced polymerization, where the radical formed can be terminated with the solvent to form an insoluble and stable product
Reaction additives for graphite-based anode CO_2 Carboxyl phenol Aromatic esters Anhydride Succinimide *N*-benzyloxy carbonyloxy succinimide.	This type of additive is able to form radical ions, which react with anions or with Lithium alkyl dicarbonate and lithium alkyloxide to form more stable SEI
Absorption additives for graphite-based anode Halogenated compound	This type of additive is absorbed into the SEI
Surface modifier additives for graphite-based anode Tris(pentafluorophenyl) borane (TPFPB)	This type of additive acts on the surface modification of the SEI on the anode

Table 7.4b Most common additives used in nonaqueous electrolytes with different effects on the battery electrochemical performance.

	Actions
Protective additives for cathode	This type of additive eliminates the trace of HF and H_2O responsible for cathode dissolution. Amine reduces the reactivity of PF_5.
N,N-dicyclohexylcarbodiimide	
N,N-diethylamino trimethylsilane	
Butylamine carbodiimide	
$LiPF_6$ additives to stabilize salt decomposition	They can form a complex with PF_5 while reducing its reactivity
Tris(2,2,2-trifluoroethyl) phosphite (TTFP)	
1-Methyl-2-pyrrolidinone	
Fluorinated carbamate	
Hexamethyl-phosphoramide	
Shuttle additives	This type of additive should be oxidized reversible and work as shuttle, when overcharged these additives should be oxidized at the cathode and then diffuse to the anode to be reduced again to neutrality
Anisole-family	
Hexaethylbenzene	
Bipyridyl or biphenyl carbonates	
Difluoroanisoles	
Phenothiazine-based compounds	
2,7-Diacetyl thianthrene	
Shutdown additives	These additives polymerize when the Lithium-ion battery is overcharged. They release gas that vents from the cell to activate the security device to shut down the cell.
Xylene	
Cyclohexylbenzene	
Biphenyl	
Phenyl-*tert*-butyl carbonate	
3-Thiopheneacetonitrile	
Fire-retardant additives	The working principle is to form an isolating layer between the condensed and gas phase to stop combustion or to terminate the radical chain reaction responsible for combustion
Tris(2,2,2-trifluoroethyl) phosphite	
Additives to reduce lithium plating	
Tetraalkylammonium chlorides	
Cetyltrimethylammonium chlorides	
Perfluoropolyethers	
Fluoro ethylene carbonate (FEC)	
Additives to increase the transport number	Those additives may coordinate the anion of the lithium salt
Tris(pentafluorophenyl) borane (TPFPB)	
Additive for hindering Aluminum current collector corrosion	It passivates Aluminum
LiBOB	
Additive to improve the wetting of separator	Surfactants that may improve wetting of the separator
Cyclohexane	

or with lithium alkyl dicarbonate and lithium alkyloxide to form a more stable SEI. A short list of these additives includes CO_2, carboxyl phenol, aromatic esters, anhydride, succinimide, and *N*-benzyloxy carbonyloxy succinimide. The effect of these types of additives is to reduce the irreversible capacity at the first cycle and to increase the cycling performance. Boron compounds have been investigated for controlling the SEI formation and they were found to reduce capacity fading, as well as to increase rate capability, thanks to the formation of more stable SEIs. Similar behavior was confirmed for LiBOB salts. It was found that, with LiBOB salt, the main SEI ingredients come from two competing sources – the reduction products of cyclic carbonate and those of BOB-anion.

7.2.5.4 Absorption Additives for Graphite-based Anode

These types of additives are absorbed on the SEI and they consist of halogenated compounds. The advantages of these additives lie in the fact that, since they are not involved in electrochemical or chemical reactions, they can be used either by addition to the electrolyte or by previous graphite pretreatment. The effect of this class of additives is to reduce the first irreversible capacity and to increase the cycling performance.

7.2.5.5 Surface Modifier Additives for Graphite-based Anode

These types of additives act on the surface of the SEI on the anode, which is more unstable when inorganic compounds, such as LiF, are present. They, therefore, modify the surface of the SEI. One example is tris(pentafluorophenyl) borane (TPFPB), which dissolves LiF. The use of potassium ions was found to work as a surface modifier, since K^+ can decrease the solvation of Li^+ with EC while mitigating solvent decomposition, improving SEI stability and therefore long-cycling performance. Alkali metal salts that form more uniform SEIs have also been investigated. Graphite pretreatment with Na_2CO_3 and LiCl results in better SEIs, as well as a small elevation of intercalation and de-intercalation potentials, which reduces lithium plating (i.e. lithium deposition on graphite). Hindering the lithium-plating phenomenon is a very important feature of obtaining high-rate lithium-ion battery performance.

7.2.5.6 Protective Additives for Cathode

Cathodes in lithium-ion batteries deteriorate during overcharging due to water, acidic impurities, and irreversible oxidation of electrolyte solvents. HF is the main compound responsible for cathode dissolution, especially for $Li_2Mn_2O_4$, and most efforts have been focused on the elimination of HF and H_2O traces (the latter is inevitably produced by the oxidation of solvents during charging). A list of these additives includes butylamine carbodiimide-based compounds, such as *N,N*-dicyclohexylcarbodiimide and bifunctional additives, such as *N,N*-diethylamino trimethylsilane. Furthermore, amine compounds can reduce the reactivity of PF_5, which is formed when $LiPF_6$ is used and is responsible for the dissolution of cathode active materials. Another approach is to form a protective film on the cathode surface: for instance, LiBOB when used as an additive seems to have these properties. The effect of this type of additive is to improve long-term cycling performance.

7.2.5.7 $LiPF_6$ Additives to Stabilize Salt Decomposition

$LiPF_6$ can decompose into LiF and PF_5; the former is very reactive with the solvent and deteriorates the SEI. Tris(2,2,2-trifluoroethyl) phosphite (TTFP) is capable of stabilizing $LiPF_6$, thanks to the formation of a complex between TTFP and PF_5. 1-methyl-2pyrrolidinone, fluorinated carbamate, and hexamethyl-phosphoramide can also react as a base to reduce the reactivity of PF_5.

7.2.5.8 Shuttle Additives

The function of these additives is to protect the cell against the overcharging irreversible passive reaction. These types of additives can undergo the reversible oxidation process and they can work as a shuttle, meaning that when overcharged, these additives undergo oxidation at the cathode side and then they diffuse to the anode to be reduced again to neutrality. These additives should have an oxidation potential slightly higher than the positive electrode and lower than the decomposition potential of electrolytes, they should have good electrochemical stability and, when oxidized and reduced, they have to show high solubility and mobility in the electrolyte solution.

Anisole-family compounds are one of the best candidates to be used as shuttle additives because of their relatively high-redox potential, which in most cases ranges from 3.8 to 4.0 V vs. Li^+/Li, and high solubility in lithium-ion battery electrolytes. 2,5-ditertbutyl-1,4-dimethoxybenzene also shows excellent redox shuttle behavior at 3.85–3.92 V vs. Li^+/Li. Other aromatic compounds are monomethoxy benzene class compounds, hexaethylbenzene, dipyridyl or biphenyl carbonates, difluoroanisoles, and some S- or N-containing heterocyclic aromatic compounds, such as thianthrene and 2,7-diacetyl thianthrene, as well as phenothiazine-based compounds.

7.2.5.9 Shutdown Additives

The working principle of these additives is based on their polymerization when the lithium-ion battery is overcharged. While the additive polymerizes it and releases gas that vents from the cell to activate the security currents interrupting the device, which, in turn, shutdown the cell. The polymer is produced during overcharging and then covers and isolates the cathode. These additives include xylene, cyclohexylbenzene, biphenyl, 2,2-diphenylpropane, phenyl-*tert*-butyl carbonate, phenyl-R-phenyl compounds (R = aliphatic hydrocarbon, fluorine substituted), and 3-thiopheneacetonitrile.

7.2.5.10 Fire-Retardant Additives

As their name suggests fire-retardant additives target to reduce the flammability of the liquid electrolyte and therefore improve the safety of lithium-ion batteries. The working principle is to form an isolating layer between the condensed and gas phases for stopping combustion or to terminate the radical chain reaction responsible for combustion. Most of these additives are based on organic phosphorus compounds and halogenate derivatives. For instance, tris(2,2,2-trifluoroethyl) phosphite was found to reduce flammability of the solvent and improve the cycling performance of lithium-ion batteries.

7.2.5.11 Additives to Reduce Lithium Plating

Lithium plating often occurs at high rates and low temperatures. This phenomenon is detrimental to safety and performance due to dendrite formation. The reactivity of lithium is correlated with the specific surface area and the formation of an ion conductive layer or lithium metal alloy to facilitate the uniform deposition of lithium. Surfactant molecules can be absorbed on the surface of lithium to suppress dendrite formation. Tetraalkylammonium chlorides, cetyltrimethylammonium chlorides, and perfluoropolyethers have been investigated as surfactant additives and their action is to facilitate a uniform lithium deposition on the anode. Another example of additive is the fluoro ethylene carbonate (FEC), which releases HF in a controlled manner and small quantity. In this case, HF is believed to improve the lithium morphology deposition. It is very interesting to notice that the previously mentioned vinylene carbonate (VC) additive, which is commonly used in industry, may have a detrimental effect on lithium deposition. This effect is associated with the concentration of VC contained in the solutions (generally for concentrations higher than 2% in weight), and it clearly indicates the importance of the selection of an appropriate concentration of the electrolyte additives.

7.2.5.12 Additives to Increase the Transport Number

The transport number indicates the fraction of total charge carried by cations. Generally, it is referred to as lithium and varied from 0 to 1. The closer the transport number of lithium to 1, the better the performance of the electrolyte. To increase the Li^+ transfer number, a series of Boron-based salts have been investigated. The principle behind those additives is that they have an anion electron deficiency and therefore may coordinate the anion of the lithium salt. Several borates and boranes have been investigated in the literature. Among them, particularly, interesting is the TPFPB, which was also found to increase the electrochemical window stability up to 5 V and the thermal stability of electrolyte solution.

7.2.5.13 Additives for Hindering Aluminum Current Collector Corrosion

It has been widely accepted in the battery community that lithium salts are responsible for aluminum corrosion. For example, as we discussed before, Li-imide is highly corrosive with the aluminum current collector. To overcome this problem, new Li-salts have been developed. LiBOB is one of the best candidates because it passivates and hinders aluminum dissolution.

7.2.5.14 Additives to Improve the Wetting of Separator

The problem of wetting is related to the inability of the electrolyte to properly wet the separator. Nonionic, ionic surfactant, cyclic alkyls, and aromatic compounds are the best candidates as wetting additives. Wetting additives should not bring side effects to the electrochemical performance. An example of wetting agent is cyclohexane.

7.3 Gel Polymer Electrolytes

In 1999, Panasonic (at the time was called Matsushita) introduced to the worldwide market Li-ion polymer batteries, which were comprised a graphitic anode, a gel polymer electrolyte (generally based on the copolymer poly(vinylidene fluoride) [VdF]/hexafluoropropylene [HFP]) (PVDF-HFP) technology covered by the Belcore patents [9, 10], and a $LiCoO_2$ cathode. A separator was not included since the gel polymer electrolyte works as a self-standing separator. The author of this chapter contributed to the development of the gel polymer electrolytes that were used in the first mass production with his research work at the Osaka National Research Institute.

The most common preparation process for gel polymer electrolytes is by embedding nonaqueous liquid electrolytes into polymer structures, or by mixing (solving) nonaqueous liquid electrolytes with polymers. Poly(ethylene oxide) (PEO), polyacrylonitrile (PAN), poly(vinylidene fluoride) (PVDF), and poly(methyl-methacrylate) (PMMA) were among the most investigated materials. Unlike nonaqueous liquid-based Li-ion batteries, which are limited to rigid metal cases, batteries using gel polymer electrolytes can have flexible and laminated cases of different shapes. This important characteristic allows for high freedom in the cell design; thus a number of specific applications can be addressed. The absence of the heavy casing increases the energy density of the battery cell as a whole; thus, lithium-ion polymer batteries are today preferred for portable electronics, vehicles, and other high-energy-density applications. Especially, polymer batteries are today popular in mobile devices, power banks, very thin laptop computers, portable media players, wireless controllers for video game consoles, wireless PC peripherals, electronic cigarettes, and other applications where small-form factors are sought and the high-energy density outweighs cost considerations.

Although very efficient and widely used, lithium-ion polymer batteries are associated with high costs and safety concerns with strict charging/discharging guidelines to prevent damage to the cell. Expansion/contraction while charging and discharging results in the delamination of the electrodes with polymer electrolytes. Long cycling can cause damage to the laminate case (pouch cell) resulting in exposure to the ambient. Lithium is highly reactive and unstable in ambient and in fact when punctured, Li-ion polymer batteries react quickly with the ambient moisture and may catch on fire.

Gel polymer electrolytes need to satisfy several requirements for applications in a wide range of temperatures. These electrolytes need to have a good lithium-ion diffusion and not decompose at the electrode due to polarization during charging/discharging. They need to have good thermal stability, a wide electrochemical window, good mechanical properties, processability, enhanced safety, and low cost; finally, they have to be nontoxic and environment-friendly. The major advantages of using gel polymers instead of nonaqueous electrolytes are the ability to overcome the constraints of the cell configuration and the possibility to make thin-film lithium-ion polymer batteries. In fact, both polymer and gel polymer electrolytes use a polymer membrane as a host matrix; however, they differ in the solvent content. PEO-containing $LiPF_6$ and polycarboxylate/PC is an example of gel polymer electrolyte. The PEO-based electrolyte system was first studied because of the possibility of easily casting the blend as a thin membrane. However, given PEO's high degree of crystallinity, the ionic conductivity of PEO-based electrolytes was found to be low, ranging from 10^{-8} to 10^{-4} $S\,cm^{-1}$ at temperatures between 40 and 100 °C. It was also observed that the ionic conductivity increases when a plasticizer is used; for example, polycarboxylate/PC, to form the gel polymer electrolyte. In the latest studies, PAN was instead used as the electrolyte host matrix material because of its small thermal resistance and flame-retardant properties. The ionic conductivity of PAN-based gel electrolytes was found to be between 10^{-5} and 10^{-3} $S\,cm^{-1}$ and the amount of transferred lithium-ions was larger than PEO-based gel polymer electrolytes. The negative side of PAN-based gel polymer electrolytes is the overall increased internal impedance of the lithium-ion polymer battery. A blend of these polymers (PEO/PAN) with electrolytes can further improve mechanical flexibility, ionic conductivity, and interfacial properties. PMMA-based gel electrolytes were also studied among other host matrix materials owing to their enhanced interface stability and lower cost, originating from the abundance of raw materials and simple synthesis process. However, PMMA's poor mechanical flexibility narrowed down the applications. Copolymerization between PMMA and other polymers provides better performance for lithium-ion gel polymer batteries. For example, copolymerizing polyvinyl chloride/PVC with PMMA leads to increased conductivity.

7.3.1 Gel Polymer Electrolyte Based on Copolymer PVDF–HFP

The most common polymer used in present lithium-ion gel polymer batteries is PVDF. PVDF-based gel polymer electrolytes are obtained by imbibing or by mixing for instance PVDF–HFP polymer with nonaqueous liquid electrolytes. Gel polymer electrolytes show good conductivity of the order of 10^{-3} $S\,cm^{-1}$ at room temperature, a value comparable with nonaqueous electrolytes. Furthermore, these polymers are highly processable by the industry and it is easy to make thin films. PVDF is a semicrystalline polymer while HFP, with its nonpolar character, affects the PVDF backbone and reduces the overall chain polarity. The CF_3 group in HFP inhibits chain packing by increasing the amorphous phase of the system. In the late 1990s and early 2000s, great research efforts were put into the study of the PVDF–HFP gel polymer electrolyte ion conduction mechanism to improve performance. It was necessary to understand the mechanism and to estimate independently the conduction factors of each ionic species (carrier concentration, ionic mobility, and transport number). The dissociation degree of the salts, the solvation effect of the solvent on the ions, and the interactions between the ions and the polymer were investigated. Powerful techniques, such as pulsed-field gradient-nuclear magnetic resonance (PFG-NMR), were used to measure the self-diffusion coefficient of ^{7}Li and various anions to obtain the ionic conduction properties and ionic conditions of the

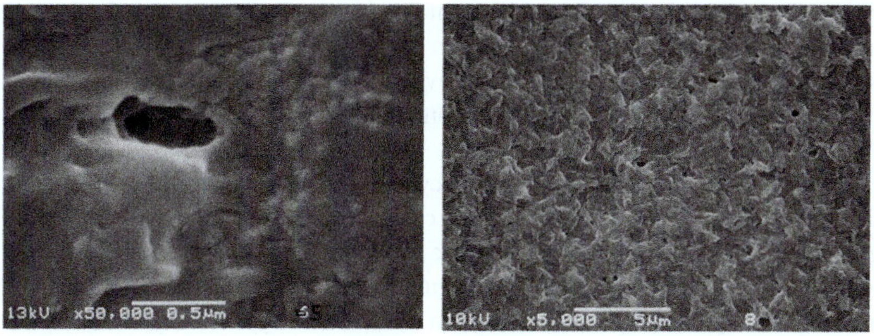

Figure 7.2 SEM micrograph of a sample with 50 wt% of EC:DEC 2:3 $LiN(CF_3SO_2)2$ 1 M solution, which shows interconnected pores of about 500 nm in diameter. Source: Reproduced with permission from Capiglia [11]/Elsevier.

gel electrolytes [11]. By using scanning electron microscopy (SEM), it was shown that the PVDF-based gel polymer electrolyte membrane is fully interconnected and contains open micropores (see Figure 7.2).

In this way, ion storage and mobility are enhanced. Another aspect evidenced by this work was the query to improve the transport number. This value, both for lithium-polymer and lithium-liquid electrolytes systems, is generally less than 1 (in most cases, less than 0.5). The problem of the transport number approaching 1 is nowadays an open issue that limits the performance of Li-ion and Li-ion polymer batteries. This process leads to an accumulation of lithium salt at the interface between the lithium electrode and the electrolyte, which, in turn, can result in interfacial polarization of the lithium electrode, thus interfering with the charge migration at the interface. Several approaches aimed to solve the transport number issue were proposed. An example consisted of adding several acid materials (i.e. trimethylene borate, trimethylboroxine) to the nonaqueous electrolytes to trap the anions of the Li salt and hopefully increase the Li-ion transport number. These nonaqueous electrolytes were then embedded into the polymer PVDF–HFP matrix.

The dynamics of the swelling of the gel polymer electrolyte membrane were found to be strongly influenced by the reduction of the polymer content [12], the overall membrane porosity, as well as by the pore dimensions. Figure 7.5 shows the conductivity behavior at different temperatures of gel polymer electrolytes activated with (a) 1 M and (b) 0.5 M solution of $LiN(CF_3SO_2)_2$ prepared using a different polymer amount. Reducing the amount of polymer from 80 to 20 wt% increases the conductivity of the gel polymer electrolyte. We can note that the VTF (Vogel–Tamman–Fulcher equation) behavior is more pronounced for the films with a low polymer content, whereas in the case of the samples with 20 wt% of electrolyte solution, the conductivity behavior is fairly Arrhenian in nature. This means that gel polymer electrolytes shift from a gel-like conduction mechanism for high polymer contents to a liquid-like conduction mechanism for low polymer contents. Furthermore, in the case of 0.5 M solution, an abrupt change of the slope appeared near 10 °C for the gels, which contained more than 60% of 0.5 M solution. The change is related to the crystallization of part of the solution. Many systems based on PVDF homo- and copolymers activated by several types of carbonates and lithium salts were tested. Generally, the conductivity values were high enough for technological applications.

Good capacity and coulombic efficiency were reported for lithium-ion batteries in the presence of new generation cobalt and nickel-based cathodes with high-performance graphitic anodes. In spite of these promising results, gel polymer electrolytes still suffer from some relevant drawbacks. The presence of a large amount of nonaqueous liquid electrolytes in PVDF-based systems negatively affects their mechanical proprieties, therefore, inducing electrolyte leakage, which in turn results in a reduced conductivity.

Another approach to improve gel electrolyte characteristics was the addition of fillers. It was proved that ceramic fillers in PVDF-based gel polymer electrolytes improve interfacial stability between the electrode and the electrolyte, especially in the case of Al_2O_3 fillers. To improve the mechanical stability of gel polymer electrolytes, membranes with nanosponge structures were also prepared. However, the most relevant problem of high polymer contents of nonaqueous electrolytes still remains an open issue. In fact, such electrolytes lead to high flammability, which may, in turn, lead to hazardous explosions in the case of thermal runaway (Figure 7.3).

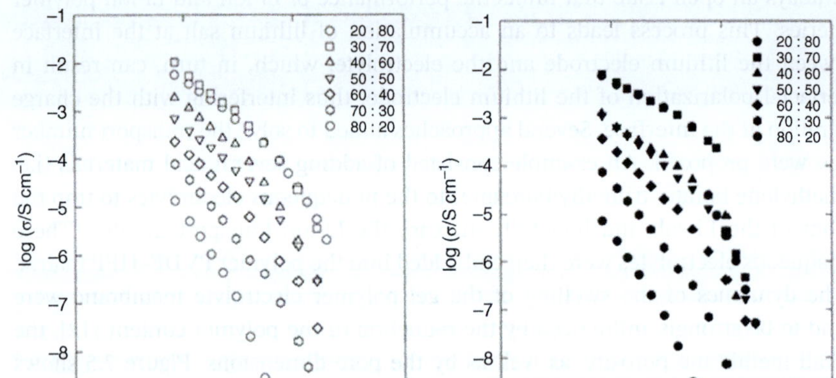

Figure 7.3 Arrhenius plots of the conductivity for the gel electrolytes prepared with different percentages of PVDF–HFP (20 : 80; 30 : 70; 40 : 60; 50 : 50; 60 : 40; 70 : 30; 80 : 20) and activated with (a) 1 M solution of $LiN(CF_3SO_2)_2$ and (b) 0.5 M solution. All the samples display a Vogel–Tamman–Fulcher (VTF) behavior. We can note that the VTF behavior is more pronounced for the films with high polymer content, whereas in the case of the samples with 20 wt% of the electrolyte solution, the conductivity behavior is fairly Arrhenian in nature. Source: Reproduced with permission from [12]/Elsevier.

7.3.2 Gel Polymer Electrolyte with Ionic Liquid

One approach to improve the safety of gel polymer electrolytes is by incorporating lithium-doped ionic liquids in the polymer host. Ionic liquids are nonflammable materials, therefore, when mixed with various polymers; the resulting gel polymer electrolyte is safer than gel polymer electrolytes obtained from swollen nonaqueous electrolytes. These gel polymer electrolytes are thermally stable; they can operate at high temperatures without polymer degradation and/or decomposition processes.

In 1997, (PVDF–HFP) gel electrolytes activated by ethyl methyl imidazolinium salts of CF_3SO_3 and BF_4 were investigated. Room temperature ionic conductivities up to $5.8\,mS\,cm^{-1}$ were obtained without any system degradation. Later, new promising flexible gel electrolytes were prepared by incorporating alkylpyrrolidinium–TFSI doped with LiTFSI inside PVDF-based polymers. Other attempts were carried out with a lithium solution of ethylbutyl pyrrolidinium salt to swell (PVDF–HFP) copolymers. The resulting membranes showed conductivity values in the range 0.34–0.94 $mS\,cm^{-1}$ without liquid leakage.

Another advantage of ionic liquids-based gel polymer electrolytes is the formation of a protective and quite stable SEI in the host matrix, which reduces the reactivity with Li. The interfacial properties were further improved by adding small amounts of EC or PC to the ionic liquid solutions or by dispersing mesoporous fillers in the gel polymers. Similar research was carried out also with PEO-based gel electrolytes. Several works describe the electrochemical behavior of PEO-based gel electrolytes with pyrrolidinium, imidazolinium, and piperidinium ionic liquids. Among these attempts, particularly, good results were obtained with the system PEO–LiTFSI gelled with different amounts of *N*-alkyl-*N*-methylpyrrolidinium. The addition of ionic liquid to the PEO–salt complex increased the conductivity and decreased the interfacial resistance with the Li anode. Conductivity values higher than $10^{-4}\,S\,cm^{-1}$ were observed at room temperature, about two orders of magnitude higher than those measured for the dry system PEO–LiTFSI. To further improve the mechanical stability of PEO-based systems, some attempts to cross-link PEO chains were made. The UV cross-linking of PEO in the presence of the lithium salt and pyrrolidinium ionic liquids results in a membrane with higher conductivity and improved mechanical properties. By comparing several types of gel polymer electrolytes, it was concluded that the PVDF-based system works better than the PEO-based systems.

To summarize, the ion mobility of gel polymer electrolytes containing ionic liquids depends on the miscibility of the polymer component in the ionic liquid. Ionic liquids can stabilize the reduction of the Li metal and the reaction occurring on the electrode surface. Overall, it is recognized that ionic liquids can drastically improve the performance of gel polymer electrolyte membranes and therefore of Li-ion polymer batteries [13–15].

7.4 Solid-State Batteries

7.4.1 First-Generation Solid-State Batteries

The first generation(s) of the solid-state Li battery will have solid electrolytes instead of liquid electrolytes. They will initially contain similar active electrode materials (i.e. LCO or NMC cathode and graphite as the anode). Since they will not employ Li-metal as the anode, they will not yet feature a higher energy density at the cell level than a Li-ion battery containing a liquid electrolyte.

Why, then, develop solid-state batteries at all, one may ask? Well, for one, solid-state batteries allow for different, more compact arrangements in the battery pack. They can be assembled in bipolar arrangements providing higher voltages at the cell level. This, in turn, simplifies connecting the cells and creates extra space in the battery pack to include more cells.

A solid-state battery will also be more intrinsically safe. Thus, less safety monitoring electronics are needed in the periphery of the battery module. Finally, solid-state batteries can have a larger voltage window than Li-ion batteries, which means that the risk of damaging the cell during charge and discharge is reduced, and a larger part of the cell energy becomes available for use.

Combining all of those factors, the amount of energy available in the battery pack will be higher for solid-state even if the energy density of the first generation(s) of solid-state cells will be the same, or perhaps slightly lower than the energy density of conventional Li-ion cells. The first electric cars equipped with these first-generation solid-state batteries are expected to launch somewhere in the middle of this decade.

7.4.2 Second-Generation Solid-State Batteries

As described above, simply replacing a liquid electrolyte in a Li-ion cell with a solid electrolyte does not make for an increase in energy density. On the contrary, ceramic solid electrolytes in powder form are likely to take up more space – and weigh more – than their liquid equivalents. Hence, solid electrolytes would somewhat lower the energy density with the same active electrode materials.

The benefit lies in the fact that some solid electrolytes can provide a larger electrochemical window. In other words, they remain stable also at very high voltages (where current liquid electrolytes do not). This means that higher-voltage cathode materials can be used than those employed in today's 3.6- to 3.8-V Li-ion battery, and a higher cell voltage leads to an increase in the cell energy, provided the Li-ion capacity remains the same.

For example, $LiMn_{1.5}Ni_{0.5}O_2$, or LMNO, has an electrode potential of 4.7 V and cannot be used with liquid electrolytes that react at these high voltages. However, solid electrolytes, such as lithium lanthanum zirconium oxide (LLZO) are stable up to 5 V and are thus compatible with these high-voltage cathodes. As such, a solid-state battery can break the 800-Wh l^{-1} barrier.

The real game-changer is the use of metallic lithium as an anode. Lithium metal has the highest energy density as an active anode material and provides the highest cell voltage from the anode contribution in the cell. Therefore, reaching the desired 1000-Wh l^{-1} energy density and even going beyond all-solid-state lithium–metal battery cells is the target.

So far, the use of lithium metal did not work because charging the battery leads to the formation of lithium–metal needles (or dendrites). These short the battery, internally, resulting in thermal runaway and explosions. Dr. Akira Yoshino, one of the Nobel prize winners in chemistry (along with Dr. John Goodenough and Dr. M. Stanley Whittingham) in 2019, came up with graphite as a safe solution, enabling the commercialization of the current wet Lithium-ion batteries in the early 1990s.

Inserting lithium-ions in between the layers of graphite was a safe compromise, but it came at the expense of the energy density and cell voltage. Several solid-state electrolytes are stable against metallic lithium; thus, thin lithium as an anode becomes a possibility. Lithium-metal batteries with solid polymer electrolytes are already commercially available, as mentioned in the introduction of this chapter (i.e. Bluecar in Paris). However, these batteries only operate at temperatures of 70 °C and therefore are not suitable for an electric family car. The reason for this is that the lithium-ion conductivity is too low for these solid polymer electrolytes. As a result, solid-state battery research has focused on finding solid electrolytes with a high enough ionic conductivity for many years.

Another option is to use high-energy-density silicon-based anode materials instead of lithium. In this case, the swelling of the silicon during Li insertion and deintercalation is an even more significant challenge than in wet lithium-ion cells and is described in the respective anode chapters of this book.

In the last few years, several good options have emerged with lithium-ion conductivities matching, or even surpassing, the ion conductivity of liquid electrolytes that are currently used for lithium-ion batteries. In 2011, Kanno and his team, working in collaboration with Toyota Motor Corporation and Japan's High-Energy Accelerator Research Organization (KEK), published a landmark paper in Nature Materials that introduced a solid electrolyte with the structure $Li_{10}GeP_2S_{12}$ (LGPS). This material became an important forerunner in the race to develop viable all-solid-state batteries. It exhibited an ionic conductivity of 1.2×10^{-2} S cm^{-1} at room temperature, a level comparable with – and even exceeding some – liquid electrolytes used in existing batteries. The team went on to design other solid electrolytes based on the same LGPS crystal structure, with promising results. In their latest study, the researchers kept the same framework structure of LGPS, and finely adjusted the ratio and positioning of the Sn, Si, and other constituent atoms. The resulting material LSSPS (composition: $Li_{10.35}[Sn_{0.27}Si_{1.08}]P_{1.65}S_{12}$ ($Li_{3.45}[Sn_{0.09}Si_{0.36}]P_{0.55}S_4$)) achieved an ionic conductivity of 1.1×10^{-2} S cm^{-1} at room temperature, almost reaching that of the original LGPS structure [16]. Another contender is LLZO, a garnet-type oxidic inorganic electrolyte, which even though it has a somewhat lower Li-ion conductivity than liquid electrolytes, is of interest because of its sizeable electrochemical window. On the downside, these inorganic electrolytes are extremely sensitive to moisture, making the solid-state cells' assembly difficult (and potentially costly). Now that several options are available, solid-state battery research has shifted its focus toward the assembly of the cells and the integration of all components into a functional cell. One issue is the reactivity of the materials forming resistive or blocking layers between the active components, severely inhibiting the cell's operation. Special thin-film coatings (or so-called "artificial interphase coatings") are being developed to prevent these reactions from happening.

7.5 Solid-Polymer Electrolytes

Since Wright reported in 1973 [17] that fast ionic conduction in high-molecular-weight PEO (CH_2CH_2O), and Armand et al. [18] noted the ability to use those materials as separators in storage devices, numerous studies have been devoted to polymer electrolytes. In 2000, the author of this chapter proposed the utilization of a solid-polymer electrolyte for solid-state batteries in EV and HEV and produced the first solid-state batteries prototype for Toyota, which was based on an alternative composite anode based on $Li_{2.6}Co_{0.4}N$ and SnSb and $LiNi_{0.8}Co_{0.2}O_2$ cathodes. As an alternative to the solid polymer electrolyte, he investigated also the sulfide glass-based solid-state electrolyte. In 2011, the first 250 production versions Bluecars (produced by Bollore') that used solid-state batteries based on solid polymers and instead lithium anodes were deployed in a car-sharing program in Paris which is the only commercial example up to date of electric vehicles that used solid-polymer electrolyte.

The interest in this field is strongly linked to the possible technological applications and has spread all around the world, including several branches of basic and applied sciences related to chemistry, physics, and engineering. Polymer electrolytes, unlike liquid electrolytes, allow the fabrication of flexible lithium-ion batteries of any shape. This is a tremendous advantage for practical applications. Polymer electrolyte-based LIBs are safer than the corresponding cells based on nonaqueous liquid electrolytes, and they are expected to permit the utilization of lithium as the anode because of their hampering of dendrite formation, and results in a higher energy density. Lithium has a theoretical energy capacity of $3860\,mAh\,g^{-1}$ (which is almost one order of magnitude higher than graphite that has a capacity of $372\,mAh\,g^{-1}$). However, the utilization of a lithium metal anode with polymer electrolyte is still far from being safe for the automotive industry.

The multidisciplinary approach of scientists working in both academia and industry has resulted in a better understanding of these new polymers, as well as in the improvement of their unique characteristics. The large amounts of articles, reviews, and books concerning this research area are used as the source for writing the present section.

For several years polymer electrolytes were generally comprised a mixture of a polymer (PEO) and a lithium salt. Most of the research efforts were directed to the improvement of their ionic conductivity and the understanding of fundamental problems related to ion-polymer, ion–ion interaction, ion molecular dynamics, and conduction mechanisms. Among these several issues, the starting point was the understanding of the molecular basis of PEO's ability to dissolve lithium salts. PEO is a semicrystalline polymer sic (is the word sic needed here?); it consists of amorphous and crystalline regions. The crystalline regions are roughly spherical and consist of a number of platelets and lamellae formed from folded polymer chains. The melting point of the crystal phases inside the PEO is approximately 65 °C and above this point, the polymer is in the molten state. A preliminary answer to the reason for such PEO affinity with salts can be found in the dielectric constant of both molten and solid PEO. The dielectric constant of molten PEO is about 8, which would explain partly the fair affinity of PEO for different kinds of salts. However successive works demonstrated other relevant aspects of such behavior. It was shown that the ion pair electrons on the oxygen of polymers coordinated with the cations in a manner similar to crown ethers. Surprisingly, poly(methylene oxide) CH_2–O and poly(trimethylene oxide) CH_2–CH_2–CH_2–O did not act as solvents for ions. This fact implied that not only the dielectric properties, or the presence of the group C–O, which gives the polarity to the material, but also the conformational flexibility provided by the PEO unit were optimal for cation coordination and therefore both effects explained the salt's solubility.

In continuing, the diffusion of ions, cations, and anions inside PEO (i.e. PEO mass-transfer processes) was investigated. The first original data presented by Armand on PEO-based polymer electrolytes illustrated that the ion conduction took place inside the crystalline region. However, following work showed that the ion transport occurred in the amorphous phase. This finding was very important and influenced all the research that came later, which focused on increasing the amount of the polymer amorphous phase inside the polymer electrolytes, with the aim of achieving a higher ionic conductivity. Inside the polymer electrolyte, cations were found to be coordinated to several oxygen atoms, and the conductivity was accepted to be confined in the amorphous phases above the glass transition temperature region in which the polymer chain motion facilitates or permits the movement of cations by breaking old oxygen links and allowing new attachments. This model is a unique feature of polymer electrolytes and an analogous process is not present in either liquid or solid electrolytes.

On the other hand, the anion place and function were less understood. Extensive works and studies revealed the presence of nonconductive ion pairs, which tell us where part of the anions can be found. Successive investigations made further clarification on the conduction mechanism and introduced a new model for the ion conduction in PEO-based polymer electrolytes: this approach considers them as solutions of an ionic salt in a polymer matrix that acts as an immobile solvent. Another intuitive and effective way to describe polymer electrolyte behavior is by assimilating them with peas in a bowl of cooked spaghetti, where the peas are the cations, which are changing positions while the spaghetti is stirred. Figure 7.2 illustrates a schematic representation of the morphology of PEO-based electrolytes; lines represent the PEO polymer chain while circles represent lithium cations. Lithium cations are bound with oxygen in PEO chains.

The conduction mechanism of PEO-based polymer electrolytes was clarified by Bruce [19] who observed that many highly conductive PEO polymer electrolyte systems form crystalline 6 : 1 complex whereas those with lower conductivities do not. Based on this, he determined the structure of the 6 : 1 PEO-based system with an

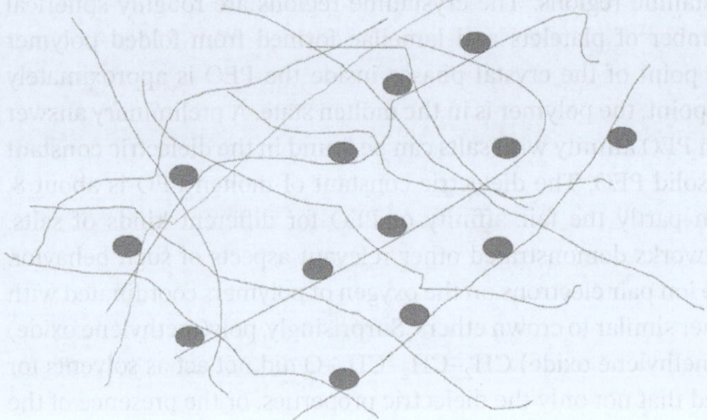

Figure 7.4 Schematic representation of PEO-based polymer electrolyte morphology, the circle represents the Li^+.

ab-initio method for flexible molecular structures, involving full-profile fitting to the X-ray powder diffraction data by simulated annealing. He found that the 6 : 1 complex formed double non-helical chains that interlocked to form a cylinder. The lithium ions reside inside these cylinders and, in contrast with other complexes, are not coordinated by the anions (Figure 7.4).

Finally, another critical aspect that hinders the mass production of solid-state polymer-based batteries is the mechanical stability and resistance to the compressibility of the solid-state polymer electrolytes. It is well known that ion transport in polymer electrolytes is governed by the VTF relationship. The VTF equation dictates that a lower T_g in a polymer electrolyte leads to higher ionic conductivity. T_g (the glass transition) is the gradual and reversible transition in amorphous materials (in this case polymer) from a hard and relatively brittle "glassy" state into a viscous or rubbery state as the temperature is increased. Because of this, a preponderance of polymer electrolyte research in the past 40 years has focused on reducing the T_g of polymer electrolytes to improve ionic conductivity. But, unfortunately, lowering the T_g of a polymer is harmful to its mechanical strength. Therefore, a polymer electrolyte with a low T_g could lead to hazards, such as short-circuiting via an external puncture or from dendrite formation. As such, developing polymer electrolytes with good mechanical strength and good ionic conductivity remains a big challenge.

7.5.1 The Advantage and the Query for the New Polymeric Materials for Polymer Electrolytes

Polymer electrolytes were developed from the perspective of their application in lithium batteries. The use of polymers produces several improvements in battery production processes with respect to the commonly used electrolytes. Flexible battery shaping, lighter weight, easy processing of thin films, and safety are major benefits of polymer electrolytes. In addition, the wide industrial background that exists in polymer processing ensures mass-production capability. The presence of such a variety of technological advantages justifies worldwide interest in the research centers located in Industries, Institutes, and Universities. Large efforts have been dedicated to increasing the low ionic conductivity at room temperature (between 10^{-6} and 10^{-3} S cm^{-1}) of early developed polymer electrolytes to a more practical 10^{-3} approximately 10^{-2} S cm^{-1} conductivity at room temperature, which is close to the conductivity of current nonaqueous liquid electrolytes.

Another important electrode property to be considered is the transport number. Generally, the transport number refers to lithium and varies from 0 to 1. It indicates the fraction of the total charges carried by cations. Electrolytes with transport number of lithium less than one imply that the migration of anions to the lithium electrode occurs until a steady-state discharge current is reached. The closer the transport number of lithium is to 1 the better the polymer electrolyte performance in a lithium battery. Large efforts have, therefore, been done to enhance the polymer electrolyte-transport number. The most common approach to preparing (PEO)/LiX systems consisted of mixing PEO with different types and amounts of LiX salts. The proper amounts of PEO and salt were mixed under magnetic stirring in solvents, such as acetone, acetonitrile, chloroform, ethanol, and tetrahydrofuran (THF). The solution was cast on a plate and the solvent was let slowly evaporated to obtain a self-supporting membrane. The electrical properties, electrochemical windows, and thermal and mechanical stabilities of the films were characterized by various techniques. The original choice of the anion was based on the hard–soft-acid-based (HSAB) principle following the nonaqueous electrolyte methodology. The selected anions were: trifluoromethanesulfonate ($CF_3SO_3^-$), tetrafluoroborate (BF_4^-), hexafluoroarsenate (AsF_6^-), thiocyanate (CSN^-), and perchlorate (ClO_4^-). The latter held the disadvantage to be potentially explosive. More recently, perfluorosulfonimide (CF_3SO_2)2N anion has attracted much interest by following the concept that in lithium salt, a larger anion enhances the formation of polymer amorphous phase and therefore facilitates the ion conduction. Following this concept, new fluorinated sulfonimide salt was synthesized: lithium bis(perfluoroethylsulfonyl)imide $LiN(C_2F_5SO_2)_2$. Even, more recently, analogs of $LiPF_6$, such as lithium fluoroalkylphosphates $Li(C_6F_{15}PF_3)$ and $Li(C_6F_{15}PF_4)$, have been synthesized. The development of novel lithium salt with bis(oxalato)borate (BOB) anion has also been reported. Li-BOB when used in lithium batteries with an organic liquid electrolyte has the characteristic of being thermally stable with an electrochemical performance similar to $LiPF_6$. Therefore, it represents a valid substitute for $LiPF_6$. Moreover, LiBOB and, in general, Boroxine ring can be easily synthesized in the laboratory and it can undergo polymerization with PEO chains: this feature is definitely an interesting perspective that may lead to a new polymer electrolyte using the salt structure as the starting monomer with the transport number of lithium equal to 1.

Other approaches are addressed to the synthesis and characterization of novel polymeric materials with features similar to PEO or based on new concepts. In the following, these polymers are listed and described.

Comb polymers: They consist of a backbone chain with short-side chains attached. An example is MEEP-type polyphoshazenes in which pairs of short PEO-like structures are attached to a N=P backbone at the phosphorous position. Generally, the flexible backbone represses the crystallinity and enhances conductivity; however, mechanical properties are reduced. An alternative and recent approach is to use poly(*p*-phenylene) backbone elements furbished with oligo(oxyethylene) side chains. Mechanical stability of the polymer electrolyte is improved and a good ability of dissolving salt together with good ion conductivity is attained by these polymers.

Ladder polymers: Two polymer chains are connected at intervals through cross-linking side chains to form a ladder-like structure.

Network polymers: Their structure resembles a fisherman's net in which the knots are generally siloxane or isocyanate and the grid is made by PEO-like polymer chain.

Cross-linked linear polymers: They are obtained by irradiation of, for example, PEO polymer.

Interpenetrating networks: Generally, they consist of a network polymer that imports the mechanical properties and a second interpenetrated polymer, which provides the matrix for ionic conduction. Examples are given by the systems based on PEO–salt with styrene or alternatively with poly(butadiene-*co*-acrylonitrile).

Copolymers: In this case, the structure consists of polymer chains, such as PEO in which the ethylene oxide (EO) sequence alternates with nonuniform distributions of sequence lengths of different polymers, for example, PMO/PEO (where PMO is for polyoxometalates).

Polymer blends: Generally, they are a mixture of two polymers. In this case, one can provide mechanical stability, while the other can provide pathways for ionic conductivity. Examples are PEO/PMMA mixtures and a more recently developed blend – a low-molecular-weight PEO dimethyl ethers mixed with polyvinylidene fluoride (PVDF)– hexafluoropropylene (HFP). This polymer blend showed conductivity as high as 10^{-4} S cm^{-1} at room temperature.

Polymers with anion incorporated: In these polymers the anions are attached to the polymer backbone to form a polyelectrolyte. However, the desired conductivity is not always obtained because the fixed anions hinder the cations movement.

Polymers with anion trapper: Here an anion trapper, for example, boroxine, is fixed on the backbone chain and traps the anion while the lithium is more free to move inside. A similar idea, but with an opposite approach was previously proposed for using crown ether groups that coordinate the cations very effectively.

Block copolymers: In this class of materials, chemically dissimilar polymers are bonded end-to-end. Under certain circumstances, a net repulsion between the polymer blocks induces a micro-phase separation. By choosing one polymer with dissolving salt ability, it is possible to obtain an ion conductive polymer while solid-like mechanical properties are provided by the nanoscale morphology.

Hyperbranched polymers: They consist of poly[bis(triethylene glycol) benzoate] capped with an acetyl group (HBP). The monomer, which contains nine oxygen atoms, has the form of a Y that is repeated to form a hyperbranched structure. This polymer can dissolve lithium salt and shows good ionic conductivity.

7.5.2 Polymer Composite Electrolytes

Polymer-filler composite electrolytes are one of the most promising ways to obtain components that satisfy the technological requirements for applications in polymer rechargeable batteries. The first developed system of this kind was based on PEO–NaI and alumina powders. The conductivity was found to be improved by about one order of magnitude compared to the pristine PEO–NaI electrolyte. This exciting result generated a deeper interest in these materials: several researchers tested different fillers in PEO-based polymer electrolytes and investigated their effects on ionic conductivity. Later, PEO–NaI–NASICON electrolytes were NASICON ($Na_{1+x}Zr_2Si_xP_{3-x}O_{12}$, $0<x<3$) that was added as the filler, showed one order of magnitude improved conductivity [20–27]. It was concluded that this enhancement was correlated to the increase of the amorphous phase. PEO:NaSCN:y-Al_2O_3 system presented similar results. The system PEO:NaI and PEO:$LiClO_4$ with B″-Al_2O_3 resulted in a reduced conductivity. However, a further study on $LiAlO_2(PEO)_8LiClO_4$ showed that this system has better mechanical properties and higher conductivities compared with pure PEO_8LiClO_4. It was also found that the maximum conductivity was obtained for composites with 10 wt% of filler. This fact was ascribed to the hindrance of polymer chain crystallization that is introduced by the filler and it was concluded that the amorphous phases allow for a superior ion-transport capability in the polymer. Further, investigation showed that also the particle size embedded in the polymer composite electrolyte is a critical factor: incorporating nanosized particles of TiO_2 and Al_2O_3 (with particle sizes of 5.8 and 13 nm, respectively) into PEO_8LiClO_4 allowed for conductivities of the order of 10^{-4} S cm^{-1} at 50 °C. This approach was proposed with the aim to increase the ion conductivity of PEO-based electrolytes at low temperatures. The advantage of the use of a filler instead of a commonly used liquid plasticizer is that normally, the latter promotes the deterioration of the electrolyte mechanical properties despite the increase of conductivity, while the former increases both mechanical and conductivity properties. Interesting results were obtained for LiI:PEO:Al_2O_3. In particular, the study of the salt-rich region of the system PEO:LiI evidenced that the major contribution to the conductivity in the composite polymer electrolyte was given by the complex $(PEO)_3$-LiI.

Polymer-filler interface chemistry in the system PEO:$LiCF_3SO_3$ with nano-SiO_2 was also investigated. In particular, the effect of the presence of surfactants (lithium dodecylsulfate, LDS), which were added to the system, was analyzed. Conductivities as high as 10^{-5} S cm^{-1} at room temperature were found. Therefore, it was concluded that the increased conductivity observed is related to the absorption of the surfactant molecules on the surface of the SiO_2 particles, which increase the amorphous phase of the polymer electrolyte. Several additional papers extended significantly the study of composite polymer electrolytes. The exciting results generated a lot of interest in the academic community. Later, it was found that the addition of finely dispersed particles of $BaTiO_3$ into PEO_8LiClO_4 resulted in the enhancement of the transport number as well as ionic conductivity. The highest conductivity enhancement was observed when 1.4 wt% $BaTiO_3$ with a particle size of 0.6–1.2 μm and 5 wt% $BaTiO_3$ with particle size of 1.8 μm were used as fillers. The conductivity increase was observed mainly at high temperatures and it was suggested that the spontaneous polarization of the ferroelectric material filler facilitates the salt dissociation while at lower temperatures only the amorphous phase fraction was increased. The subsequent works further considered the filler's dielectric properties and their influence on salt dissociation. Several PEO-based systems with lithium salts having different anions $LiClO_4$, $LiBF_4$, $LiPF_6$, $LiCF_3SO_3$, $Li[(CF_3SO_2)N]$, and LiBETI, and different fillers having good ferroelectric properties ($BaTiO_3$, $PbTiO_3$, and $LiNbO_3$) were investigated. It was found that in case of low-association tendency of the lithium salts, permanent dipoles of ferroelectric ceramic powders as a filler did not work effectively to enhance the dissociation of ion–ion pairs and ion clusters due to the already high-dissociation level of the salts. Furthermore, the "dielectric power" of the fillers with salts with high-association tendency was found to not be strong enough to promote the dissociation of the PEO complex with salts [28]. The size and the structure of the filler were discovered to be important parameters for controlling the formation of charge carriers by preventing ion–ion and ion-cluster associations

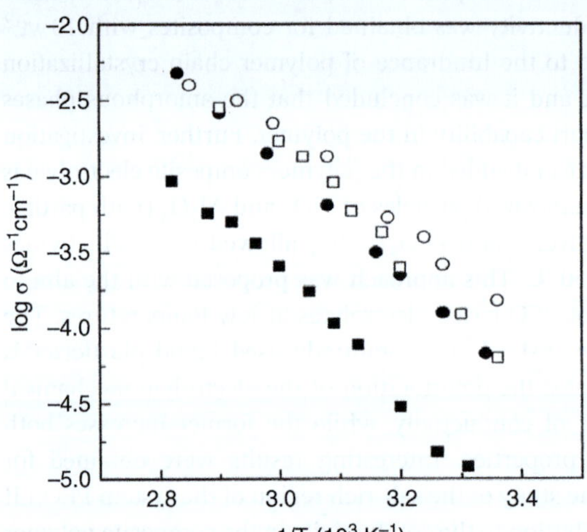

Figure 7.5 The Arrhenius plots of the conductivity for $(PEO)_8$–$LiClO_4$ with 5 wt% of nanoscale SiO_2 (open squares); $(PEO)_8$–$LiN(CF_3SO_2)_2$ with 5 wt% of nanoscale SiO_2 (open circles); $(PEO)_8$–$LiClO_4$ with 33 wt% of Li_2O-$3B_2O_2$ (dark square); $(PEO)_8$–$LiN(CF_3SO_2)_2$ with 33 wt% of Li_2O-$3B_2O_2$ (dark circles). Particle size matter and nanoscale SiO_2 allow reaching conductivity values generally higher than those offered by the tribotate glass. Source: Reproduced with permission from Quartarone and Mustarelli [29]/with permission of Royal Society of Chemistry.

in composite polymer electrolytes. In particular, fillers with high tetragonality structures hold a strong dipole moment, which would facilitate the stiffening of the PEO chain and therefore reduce the total conductivity. The filler size effect shows its relevance, especially at the nanoscale. PEO-based nanocomposite electrolytes prepared with nanoscale SiO_2 and compared with triborated glass were investigated, and their conductivity was found to increase drastically.

Figure 7.5 shows the conductivities of PEO-based nano-composite electrolytes with fumed SiO_2 (20–60 nm) and triborate glass at different temperatures. It is seen that particle size matters and nanoscale SiO_2 allowed for conductivity values, generally higher than those obtained by the triborate glass [29]. PEO-based nanocomposite electrolytes with nanoscale TiO_2 and Al_2O_3 fillers (5, 8–13 nm) were investigated with similar results, demonstrating the positive effect of the nanosize filler on the polymer electrolyte conductivity [30].

7.6 Solid Electrolytes

At the end of the nineteenth century, various solid ceramic oxides, with the peculiarity of being ion conducting, were discovered. In fact, later on and precisely, in the early decades of the twentieth century, fast ionic conductivity was observed for a variety of metal halides and sulfides. However, the main results in understanding and developing novel materials for practical applications were achieved during the last few decades. Solid electrolytes have a high ionic conductivity based on a single predominantly conducting anion or cation species, which means that the transport number is close to 1 for Li^+, furthermore they have a negligible or small electronic conductivity. Solid electrolytes exhibit ionic conductivities in the range from 10–1 to 10^{-5} S cm^{-1} at room temperature, namely the same order of magnitude for liquid and polymer electrolytes. Solid-state Li-ion batteries that use ceramic, glass, and glass–ceramic lithium-ion conductors show several positive aspects, such as better thermal stability, no solvent content, and therefore improved safety, and wider electrochemical windows. Solid electrolytes can be grouped in:

- Perovskite-type oxides, (e.g. (Li,La)TiO_3)
- NASICON-type lithium electrolytes, [e.g. LiMIV2$(PO_4)_3$ (MIV = Ti, Zr, Ge)]
- Garnet-type structures containing transition-metal oxides, [e.g. $Li_5La_3M2O_{12}$ (M = transition metal)]
- Glassy and glass–ceramic electrolytes, based on lithium nitrides, sulfides, borates, and phosphates.

Generally, these materials show a transport number close to unity for lithium, high stability of the electrode/electrolyte interface, and electrochemical windows up to 6 V. Among several types of solid electrolytes, sulfates and nitrates (Li_3N, Li_2S, $Li_2S–SiS_2–P_2S_5$) are considered good ones for Li-ion batteries owing to their high ionic conductivity (>10–5 S cm^{-1}), although they have the drawback to be highly hygroscopic. Li oxides are instead more attractive for their stability to moisture, but unluckily they have a low ionic conductivity. Among the inorganic materials explored during the last years, glassy lithium phosphorus oxynitride (LiPON) still represents the best choice to fabricate thin-film solid electrolytes for Li-ion micro-batteries. Rechargeable micro-batteries based on solid-state electrolytes will be the size of a grain of sand, produced for example by a 3D printer, or lithographic techniques, and could power biomedical implants, coin-size sensors, and other tiny electronics. LiPON does not react with the Li anode and its conductivity ranges between 10^{-7} and 10^{-6} S cm^{-1} upon modulation of the nitrogen concentration in the glass. To reduce the thickness to few microns of LIPON, solid electrolyte is the viable solution for application in solid-state batteries. For LiPON, the sputtering processes seem to offer the best compromise for solid-state lithium micro-batteries production.

7.6.1 Solid-State Electrolyte Issues

Solid electrolytes need to possess the following characteristics for their large-scale application in solid-state batteries:

(a) high ionic conductivity at room temperature;
(b) negligible electronic conductivity, interfacial impedance, and grain-boundary resistance;
(c) wide electrochemical stability window;
(d) chemical stability in the presence of electrodes, especially metallic Li anodes;
(e) matching thermal expansion coefficients with both electrodes;
(f) low cost, high throughput, easy synthesis, and being environment-friendly.

A solid electrolyte has three kinds of interfaces in the solid-state battery, which are as follows:

(1) "surface" exposed a gas or water phase in the surrounding environment,
(2) "interface" faced a solid matrix of another composition,
(3) "grain boundary" within a crystal of the same composition.

Therefore, issues with solid electrolytes mainly arise from the above three aspects and internal structure. In the following, we will discuss every single issue in detail.

Bulk restriction: Crystalline ceramic electrolytes with different lattice parameters can be prevented from achieving high ionic conductivity and transference number because of insufficient sites to receive mobile ions, confined conductive pathways, and high activation energy and migration barriers.

Interface restrictions: The interface between a solid electrolyte and electrode should have a free volume to permit their expansion, particularly at elevated temperature or volume change of the electrode during charge/discharge. Because of the great difference in expansion coefficient and interfacial stress between these two solids, their contact tends to be loose. It gets worse during long-term operation, ultimately creating high interface resistance. As a result, the cycling stability and high-power performance of the cell decrease. Stack pressure is deemed necessary to ensure proper contact of electrodes with the solid electrolyte and also inhibit delamination during cycling due to electrode expansion and contraction. Since the fabrication pressure influences the electrolyte porosity, the fabrication pressure also has a large impact on cycling stability and rate performance of the cell.

The compatibility: Solid-state electrolytes with electrodes are still poor. For example, in a Li–metal battery, inhomogeneity of the solid electrolyte resulted in an uneven distribution of current and variation of surface potential, generating a Li "dendrite." This dendrite penetrates the solid electrolyte, particularly a thin-film electrolyte, and lower battery safety because it could lead to a fire or explosion.

Grain-boundary restrictions: The ionic conductivity of a polycrystalline ceramic is generally much lower than that of a single crystal because of grain boundaries. Grain boundaries block the continuous domains for ion migration, resulting in a sizable grain-boundary resistance. Thus, the charge-transfer region at a grain boundary can influence ion migration and then ionic conductivity. An advantage of sulfide electrolytes is their low-grain-boundary resistance compared with that of some oxide electrolytes, which possess a grain-boundary resistance too high for practical batteries.

Besides the above four aspects, the cost, efficiency, and operation temperature in applying solid electrolytes should also be taken into account.

7.6.2 NASICON-Type Lithium Electrolytes

7.6.2.1 NASICON-Type Lithium Electrolytes for Lithium–Air Batteries

NASICON is an acronym for sodium (**Na**) **S**uper **I**onic**CON**ductor, which usually refers to a family of solids with the chemical formula $Na_{1+x}Zr_2Si_xP_{3-x}O_{12}$, $0 < x < 3$. It is also used for similar compounds where Na, Zr, and/or Si are replaced by isovalent elements, such as Li. NASICON compounds have high electrical conductivities, on the order of 10^{-3} S cm^{-1}, which rival those of liquid electrolytes. They are caused by hopping of Na ions among interstitial sites of the NASICON crystal lattice.

The first Li–air rechargeable battery was introduced in 1996 by Abram and was fabricated through the combination of a Li anode, a gel polymer electrolyte, and a thin carbon composite as a cathode breathing oxygen from air [31]. Lithium–air batteries have the highest known theoretical energy density, which is more than 11.000 Wh kg^{-1}, and comparable to gasoline (13.000 Wh kg^{-1}). The working principle of this type of battery is different from Li-ion batteries and is based on controlled and reversible oxidation of Li to form Li_2O_2. The electrolyte is the basic component of this technology; in fact, its role is to avoid any contact between moisture (eventually absorbed from the air) and the pure Li anode. Therefore, the electrolyte should be a membrane capable to conduct Li but impermeable to water.

Lithium-ion-conducting electrolytes with NASICON type structures are one of the best choices for applications in lithium–air batteries because they are selective lithium-ion conductors and impermeable to water. However, the formation of

Li_2O_2 inside the pores of the positive electrode due to the slow reversible kinetics of Li_2O_2 hinders the application of this technology. In several studies, new types of catalyzers to accelerate the reversible reaction of Li_2O_2 to form lithium have been investigated.

7.6.2.2 NASICON-Type Lithium Electrolytes for Lithium Aqueous Batteries

Aqueous Li–air batteries using an aqueous solution based on LiOH do not have the problem of Li_2O_2 formation because the product of reaction between LiOH and H_2O is stored in the solution. Moreover, LiOH in solution prevents the formation of Li_2CO_3, which poisons the Li–air battery, therefore allowing the battery to operate with untreated air. Also, in lithium aqueous batteries, it is crucial to avoid contact between water and the pure Li anode. The electrolyte membrane needs to be stable and impermeable to the aqueous electrolyte. Again, the NASICON-type Li-conducting membrane seems to be the best separator to date.

Li^+XAlTi_2-$x(PO_4)_3$ (LAPT) has a conductivity of approximately 10^{-4} S cm^{-1}, however, it is not stable in the presence of Li and it should be protected, for example, by using a LiPON film that can be sputtered on it. Unfortunately, LiPON has a very low ion conductivity in the order of 10^{-6} S cm^{-1}. Therefore, laminating LiPON in a thin film may partially solve the problem of its low ionic conductivity, but still, the low-output power of the protected Li LIPON film has to be overcome and it represents a key challenge. A successful approach is to laminate LAPT with PEO-based polymer electrolytes, which results in covering and isolating LAPT from Li while increasing the conductivity with respect to LiPON. The use of additives may further improve the electrochemical cell performance.

7.6.3 Glass Electrolytes

A Li-ion battery with a solid-state electrolyte is very promising not only for the miniaturization of electric appliances but also for larger applications, such as EVs, HEVs, and stationary power. There are several safety problems still present in Li-ion batteries, which use flammable organic liquid electrolytes, gel polymers, and polymer electrolytes. All-solid-state lithium secondary battery systems using nonflammable inorganic solid electrolytes may be the correct answer to safety issues. The ultimate goal is to have high reliability, high-energy density, and high safety. In this regard, inorganic solid electrolytes are very promising because of the wide range of compositions, isotropic Li conduction, no grain boundaries, easy film formation, and nonflammability. Inorganic glassy electrolytes have a higher conductivity than crystal-based due to the so-called "open structure." Single cation conduction is realized because glassy materials belong to the "decoupled systems" in which the mode of ion conduction relaxation is decoupled from the mode of structural relaxation.

Superionic-conducting crystals as a metastable phase are easily formed from inorganic glassy electrolytes. The strategy to improve conductivity of glassy-based electrolytes is to increase Li-ion conductivity and use counter anions with high polarizability. Glassy electrolytes can be processed by ball milling to directly obtain fine powder. The glassy-based powder can be mixed with cathode powder to obtain highly efficient electrodes. All-solid-state Li-ion batteries using electrodes based on $LiCoO_2$, solid-electrolyte $Li_2SP_2S_5$ glass–ceramics, and anode indium or $SnS–P_2S_5$ have been fabricated at the laboratory scale and results are quite impressive. Excellent cycle performances are obtained with no loss of capacity during long cycling, up to 500 cycles. The main characteristic of glass–ceramics with high conductivity is their dense microstructure, which promotes a smooth charge–discharge reaction in the solid/solid interface between the electrolyte and electrode, resulting in exceptional reversibility and cycling performance.

7.6.4 Glass–Ceramics Electrolyte

Ion-conducting ceramics can be employed in several electrochemical devices, including all-solid-state Li-ion batteries. For this kind of application, they also show a very interesting phenomenon: the single-ion-conducting feature, which means transport number equal to 1 for Li. In particular, the Li is not solvated with solvent and therefore undergoes a better intercalation/de-intercalation process while avoiding diffusion at the interface (see Figure 7.6):

Some ion-conducting ceramics can represent the solution to safety issues in Li-ion batteries because they allow removing the combustible organic solvent of the nonaqueous electrolyte commonly used in Li-ion batteries. Among them, glass–ceramics and Li_2S sulfide electrolytes are promising materials. In this regard, cells based on a $LiCoO_2$ cathode, sulfide electrolyte, and graphitic carbon anode have been tested. The rate-determining step was found to be at the $LiCoO_2$/sulfide interface where a highly resistive space charge and lithium deficient layer are formed. An interface structure with an oxide solid-electrolyte buffer seems a viable solution to solve the problem. The oxide solid-electrolyte buffer suppresses the development of the space-charge layer and thus enhances the high-rate capability

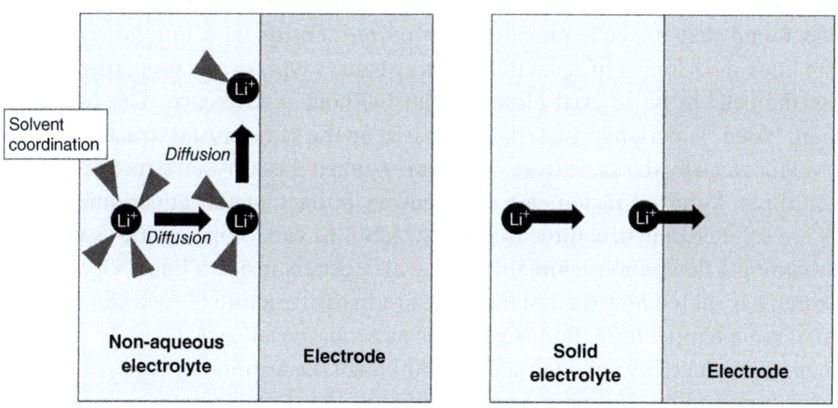

Figure 7.6 Lithium diffusion on other nonaqueous electrolytes and single-lithium diffusion of lithium-ion in glass–ceramic electrolyte.

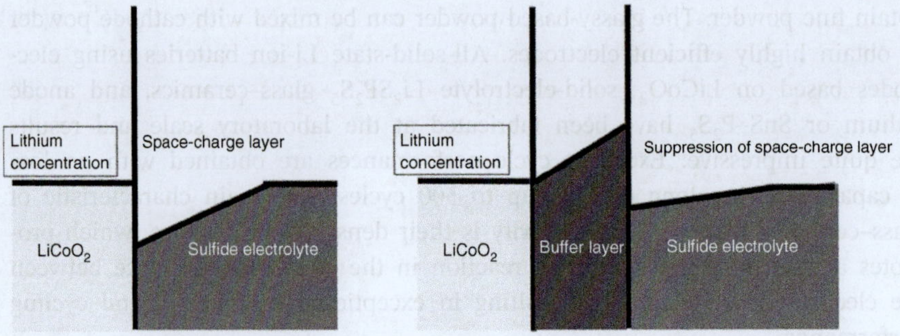

Figure 7.7 Representation of the effect of Li concentration on Sulfide electrolyte-based battery with and without buffer layer.

of the cells. Figure 7.7 represents the conduction model in these solid electrolytes, the line is referred to as Li concentration profile in solid electrolyte, the buffer, and the electrode [32].

7.6.5 LGPS Family

In 2016, Yuki Kato and Ryoji Kanno in collaboration with colleagues from Toyota Motor Corporation, Tokyo Institute of Technology, and High-Energy Accelerator Research Organization Japan (KEK), have successfully designed and tested novel high power and energy batteries based on solid-state electrolyte [33].

The development of the two new lithium-based superionic conductor materials (structures: $Li_{9.54}Si_{1.74}P_{1.44}S_{11.7}Cl_{0.3}$ and $Li_{9.6}P_3S_{12}$), which are part of the LGPS family, represents a leap forward in the creation of useable solid-state batteries. The researchers used their two new solid electrolytes to create two battery cell types; one high-voltage cell and one cell designed to work under large currents. Both all-solid-state cell types exhibited superior performance compared with lithium-ion batteries, operating very well at temperatures between 30 and 100 °C. It was found that the cells provided high-power density, with ultrafast-charging capabilities and a longer lifespan than existing battery types. Their properties would allow the cells to be stacked close together without interference. The two new lithium-based "superionic" materials are based on the same crystal structure previously discovered by the same team. They have studied these crystal structures using Synchrotron X-ray diffractometer and neutron diffractometer. Superionic materials are solid-crystal structures through which ions can "hop" easily, essentially maintaining a flow of ions similar to that which occurs inside a liquid electrolyte. Research has shown how the Li-ions move fast in the structure of such compounds even at room temperature. Both superionic materials developed showed extremely high ionic conductivity and high stability. Although the technology requires further development before it is commercially available, these promising results indicate that all-solid-state batteries may soon provide a much-needed boost to applications requiring stable, long-life energy storage.

7.7 Solid-State Battery Companies

In conclusion, an overview is given of some of the most active companies in the world that are aiming to produce solid-state batteries for the electronics and automotive market by deploying novel solid-state electrolytes. The description is based on the information provided on the company website and available on the internet.

Solid power [34] (USA-based company) is a developer of all-solid-state rechargeable batteries for electric vehicles and mobile power markets. Founded in 2012 as a spin-out from the University of Colorado Boulder, Solid Power replaces the flammable liquid electrolyte in a conventional Li-ion battery with a proprietary sulfide solid electrolyte. The company statement is that it can provide a 50–75% increase in energy density compared to the best available rechargeable batteries, enable cheaper, more energy-dense battery pack designs, and higher compatibility with traditional Li-ion manufacturing processes. Solid Power has extensive partnerships with BMW and Ford to jointly develop all-solid-state batteries. Solid Power is backed by prominent investors, including Samsung, Hyundai, Ford, Volta Energy Technologies, and Solvay.

SES [35] (USA-based company) is developing and manufacturing high-performance Li-Metal batteries for automotive and transportation applications. Founded in 2012 (formerly known as Solid Energy Systems) by Dr. Qichao Hu who is the founder and has developed a novel semisolid ionic liquid electrolyte [36]. SES is headquartered in Singapore with operations in Boston, Shanghai, and Seoul. Its shareholders include General Motors, SK, Temasek, Tianqi Lithium, Applied Materials, Vertex, and Shanghai Auto.

QuantumScape [37] (USA-based company) was founded in 2010 by Jagdeep Singh, Tim Holme, and Professor Fritz Prinz of Stanford University and is developing oxide and sulfuric ceramic-based solid-state electrolytes. QuantumScape design is "anode-free" in that the battery is manufactured anode free in a discharged state, and the anode forms in situ on the first charge. Investors include Bill Gates and Volkswagen.

Ionic materials [38] (USA-based company): The Founder, Dr. Mike Zimmerman, PhD, is a Professor at Tufts University and inventor of a new polymer electrolyte for solid-state battery application. The patents indicate that the polymer is from crystalline resins; for example, polyphenylene sulfide, polyphenylene oxide, polyether ether ketone, and polysulfone. Ionic materials mix the resins with ionic materials and dopants and then extrude them into a film. Graphite is the anode material of preference for solid-state batteries used by Ionic materials.

ProLogium [39] (Taiwan-based company): ProLogium Technology(PLG) was established in Taipei City, Taiwan back in 2006. ProLogium adopts solid-state ceramic electrolytes rather than liquid/gel ones of Li–polymer batteries (LPB). The company claims that no leakage occurs and no flammable materials are used inside and that no short-circuiting will happen under the normal usage. The solid-state ceramic electrolyte also has good thermal stability, including 200–260 °C for 3–10 seconds. There is no thermal runaway at 350 °C or salting-out at low temperatures. Also, this battery does not melt, whereas LPB will melt at about 120–150 °C. Therefore, even if it is damaged by folding, hitting, penetrating, and heating (250 °C, 5 hours), its solid-state batteries do not catch fire or explode. ProLogium uses graphite or a combination of graphite/silicon oxide-based anodes.

Samsung [40] (Korean-based company) presented on March 9, 2020, in London, a study on high-performance, long-lasting all-solid-state batteries using sulfide electrolytes [41]. Samsung's researchers proposed utilizing, for the first time, a silver–carbon (Ag–C) composite layer as the anode. The team found that incorporating an Ag–C layer into a prototype pouch cell enabled the battery to support a larger capacity, a longer cycle life and enhanced its overall safety. Measuring just 5 μm (micrometers) thick, the ultrathin Ag–C nanocomposite anode layer allowed the team to reduce the anode thickness and increase energy density up to 900 Wh l^{-1}.

Murata [42] (Japanese-based company) developed, in 2021, a solid-state battery using an oxide ceramic electrolyte instead of the electrolytic solution utilized in conventional batteries. The new battery delivers safety and durability far superior to conventional Li-ion secondary batteries, making it ideal for hearable devices, such as wireless earphones that require a high degree of safety and are intended to be used for extended periods of time, as well as a good match for the diverse needs of the emerging Internet of Things (IoT) society.

Toyota [43] (Japanese-based company) has released plans to build a prototype electric vehicle in 2021 powered by solid-state batteries. Mitsui Mining and Smelting (Mitsui Kinzoku) has announced that it will build a pilot facility for the production of solid-state electrolytes. Idemitsu Kosan, a Japanese oil company, is preparing to install a solid-state electrolyte production line at one of its Japanese facilities, while Sumitomo Chemical has also announced that it will begin manufacturing solid-state battery materials in the near future. Most solid-state electrolytes are produced by solidifying sulfides on an industrial level. Toyota and Panasonic have decided to establish a joint venture for the development of solid-state batteries [44].

7.8 Conclusions

Much progress has been made in improving the performance of Li-ion batteries since their release into the worldwide market in 1991. A lot of research has been done to improve the safety of nonaqueous electrolytes, increase their stability in a wide range of temperatures (−20 to 70°C), and their resistance to electrochemical decomposition at the cathode and anode. Nonaqueous electrolytes have become the standard electrolyte solution for Li-ion batteries used in the electronic industries. Thirty years of mass production make this technology quite reliable and standardized together with the industrial process of manufacturing. The introduction of gel polymer batteries in 1999 was another important step in the history of Li-ion batteries. The increased energy density and easy shape have been appreciated by the electronic industry. However, Li-ion batteries based on gel polymer electrolytes still suffer from low-specific power. The successful application of Li-ion technology in the automotive market witnessed, in recent years, has given a new driving force to the technology development. Nonaqueous, polymer, and gel polymer electrolytes have been the main choice for transport applications, even if most of the successful EV cars released on the market are using nonaqueous electrolytes. Surely, the automotive market needs to increase both safety and energy density standards, the latter might be accomplished by using new high-voltage cathode materials. The consequence is that new kinds of electrolytes need to be developed to increase the electrochemical stability window from 4 V to 6 V, and the priority while developing these new electrolytes should be safety. The technology beyond lithium-ion batteries also poses new challenges: Li–air batteries and Li aqueous batteries need moisture and water-impermeable electrolytes, such as NASICON-type structures as Li conductors. In recent years, strong interest has been manifested in the battery industry toward all solid electrolytes owing to their safety, stability, and single Li diffusion (transport number of lithium equal to 1). Glass electrolytes, glass–ceramic, ceramic oxide, and polymer-based solid-state electrolytes are under deep investigation as they show very promising results for applications in the automotive industry and eventually for the emerging stationary power industry to increase safety and high-energy density. In the following chapter, another type of promising Li battery chemistry, Li–S, that can employ a pure Li anode is discussed.

Acknowledgment

The author would like to acknowledge the inputs from Dr. Ermanno Miele and Dr. Remo Proietti Zaccaria who independently reviewed this chapter and Dr. Subrahmanyam Goriparti for the figures layout.

References

1 Xu, K. (2004). Nonaqueous liquid electrolytes for lithium-based rechargeable batteries. *Chem. Rev.* 104 (10): 4303–4418.

2 Zhang, S.S. (2006). A review on electrolyte additives for lithium-ion batteries. *J. Power Sources* 162 (2): 1379–1394.

3 Xu, K. (2010). Electrolytes and interphasial chemistry in Li ion devices. *Energies* 3 (1): 135–154.

4 Capiglia, C. et al. (1999). 7Li and 19F diffusion coefficients and thermal properties of non-aqueous electrolyte solutions for rechargeable lithium batteries. *J. Power Sources* 81–82 (0): 859–862.

5 Park, M., Zhang, X., Chung, M. et al. (2010). A review of conduction phenomena in Li-ion batteries. *J. Power Sources* 195 (24): 7904–7929.

6 Aravindan, V., Gnanaraj, J., Madhavi, S., and Liu, H.-K. (2011). Lithium-ion conducting electrolyte salts for lithium batteries. *Chem. Eur. J.* 17 (51): 14326–14346.

7 Xu, K., Zhang, S.S., Lee, U. et al. (2005). LiBOB: Is it an alternative salt for lithium ion chemistry? *J. Power Sources* 146 (1-2): 79–85.

8 Kaneko, H., Sekine, K., and Takamura, T. (2005). Power capability improvement of LiBOB/PC electrolyte for Li-ion batteries. *J. Power Sources* 146 (1-2): 142–145.

9 Gozdz, A.S., Shmutz, C.N., Tarascon, J.M., and Warren, P.C. (1995). Polymeric electrolytic cell separator membrane. US Patent 5,418,091, issued 23 May 1995.

10 Gozdz, A.S., Shmutz, C.N., Tarascon, J.M., and Warren, P.C. (1995). Method of making an electrolyte activatable lithium-ion rechargeable battery cell. US Patent 5,456,000, issued 10 October 1995.

11 Capiglia, C. et al. (2000). Structure and transport properties of polymer gel electrolytes based on PVdF-HFP and LiN $(C_2 F_5SO_2)_2$. *Solid State Ionics* 131 (3): 291–299.

12 Capiglia, C., Saito, Y., Yamamoto, H. et al. (2000). Transport properties and microstructure of gel polymer electrolytes. *Electrochim. Acta* 45 (8): 1341–1345.

13 Diaw, M., Chagnes, A., Carré, B. et al. (2005). Mixed ionic liquid as electrolyte for lithium batteries. *J. Power Sources* 146 (1-2): 682–684.

14 Fedorov, M.V. and Kornyshev, A.A. (2014). Ionic liquids at electrified interfaces. *Chem. Rev.* 114 (5): 2978–3036.

15 Wu, T.-Y., Hao, L., Chen, P.-R. et al. (2013). Ionic conductivity and transporting properties in LiTFSI-doped bis (trifluoromethanesulfonyl) imide-based ionic liquid electrolyte. *Int. J. Electrochem. Sci.* 8: 2606–2624.

16 Kamaya, N., Homma, K., Yamakawa, Y. et al. (2011). A lithium superionic conductor. *Nat. Mater.* 10: 682–686.

17 Fenton, D., Parker, J., and Wright, P. (1973). Complexes of alkali metal ions with poly (ethylene oxide). *Polymer* 14 (11): 589.

18 Armand, M., Chabogno, J.M., and Duclot, M.J. (1979). Fast ion transport in solids: electrodes, and electrolytes. In: *Proceedings of the International Conference on Fast Ion Transport in Solids, Electrodes, and Electrolytes, Lake Geneva, Wisconsin, USA (21–25 May 1979)* (ed. P. Vashishta, J.N. Mundy and G.K. Shenoy), 131. North-Holland, Amsterdam.

19 MacGlashan, G.S., Andreev, Y.G., and Bruce, P.G. (1999). Structure of the polymer electrolyte poly(ethylene oxide)$_6$:$LiAsF_6$. *Nature* 398: 792–794.

20 Chung, S.H. et al. (2001). Enhancement of ion transport in polymer electrolytes by addition of nanoscale inorganic oxides. *J. Power Sources* 97–98 (0): 644–648.

21 Francisco, B.E., Stoldt, C.R., and M'Peko, J.-C. (2014). Lithium-ion trapping from local structural distortions in sodium super ionic conductor (NASICON) electrolytes. *Chem. Mater.* 26 (16): 4741–4749.

22 Nanjundaswamy, K.S. et al. (1996). Synthesis, redox potential evaluation and electrochemical characteristics of NASICON-related-3D framework compounds. *Solid State Ionics* 92 (1–2): 1–10.

23 Savitha, T. et al. (2006). Structural and ionic transport properties of $Li_2AlZr[PO_4]_3$. *J. Power Sources* 157 (1): 533–536.

24 Sengwa, R.J., Dhatarwal, P., and Choudhary, S. (2014). Role of preparation methods on the structural and dielectric properties of plasticized polymer blend

electrolytes: correlation between ionic conductivity and dielectric parameters. *Electrochim. Acta* 142: 359–370.

25 Syzdek, J. et al. (2010). Detailed studies on the fillers modification and their influence on composite, poly(oxyethylene)-based polymeric electrolytes. *Electrochim. Acta* 55 (4): 1314–1322.

26 Thangadurai, V. and Weppner, W. (2002). Solid state lithium ion conductors: design considerations by thermodynamic approach. *Ionics* 8 (3-4): 281–292.

27 Xu, X., Wen, Z., Yang, X. et al. (2006). High lithium ion conductivity glass-ceramics in Li_2O–Al_2O_3–TiO_2–P_2O_5 from nanoscaled glassy powders by mechanical milling. *Solid State Ionics* 177 (26-32): 2611–2615.

28 Capiglia, C., Mustarelli, P., Quartarone, E. et al. (1999). Effects of nanoscale SiO_2 on the thermal and transport properties of solvent-free, poly (ethylene oxide)(PEO)-based polymer electrolytes. *Solid State Ionics* 118 (1): 73–79.

29 Quartarone, E. and Mustarelli, P. (2011). Electrolytes for solid-state lithium rechargeable batteries: recent advances and perspectives. *Chem. Soc. Rev.* 40 (5): 2525.

30 Croce, F., Appetecchi, G.B., Persi, L., and Scrosati, B. (1998). Nanocomposite polymer electrolytes for lithium batteries. *Nature* 394 (6692): 456–458.

31 Abraham, K.M. and Jiang, Z. (1996). A polymer electrolyte-based rechargeable lithium/oxygen battery. *J. Electrochem. Soc.* 143 (1): 1–5.

32 Takada, K. et al. (2008). Interfacial modification for high-power solid-state lithium batteries. *Solid State Ionics* 179 (27-32): 1333–1337.

33 Kato, Y., Hori, S., Saito, T. et al. (2016). High-power all-solid-state batteries using sulfide superionic conductors. *Nat. Energy* 1 (4): 16030.

34 http://www.solidpowerbattery.com/ (accessed 29 June 2021).

35 https://www.ses.ai/ (accessed 29 June 2021).

36 MIT News on Campus and around the word, Doubling battery power of consumer electronics, Rob Matheson | MIT News Office Publication Date: August 16, 2016 https://news.mit.edu/2016/lithium-metal-batteries-double-power-consumer-electronics-0817.

37 https://ir.quantumscape.com/home/default.aspx (accessed 7 July 2021).

38 https://ionicmaterials.com/ (accessed 6 July 2021).

39 http://www.prologium.com (accessed 29 June 2021).

40 https://news.samsung.com/global/samsung-presents-groundbreaking-all-solid-state-battery-technology-to-nature-energy (accessed 7 July 2021).

41 Lee, Y.G., Fujiki, S., Jung, C. et al. (2020). High-energy long-cycling all-solid-state lithium metal batteries enabled by silver–carbon composite anodes. *Nat. Energy* 5: 299–308.

42 https://www.murata.com/en-eu/news/batteries/solid_state/2019/0626 (accessed 7 July 2021).

43 https://www.just-auto.com/analysis/toyota-announces-solid-state-battery-ev-prototype-for-2021 (accessed 7 July 2021).

44 https://global.toyota/en/newsroom/corporate/31477926.html (accessed 7 July 2021).

8

Developments in Lithium–Sulfur Batteries

R. Vasant Kumar and Kai Xi

Department of Materials Science and Metallurgy, University of Cambridge, 27 Charles Babbage Road, Cambridge CB3 0FS, UK

8.1 Introduction to Lithium–Sulfur Batteries

Lithium–sulfur (Li–S) is among the lithium battery chemistry with significant commercial interest since it has a very high theoretical specific energy of 2600 Wh kg^{-1} which is significantly higher compared to 130–280 Wh kg^{-1} nominally provided by well-established Li-ion batteries (LIBs). This difference arises from the fact that commercial LIBs employ a graphitic carbon anode, which is four to six times heavier than pure Li, and a transition-metal oxide (such as lithium cobalt oxide ($LiCoO_2$), lithium nickel manganese cobalt oxides (NMCs), or lithium iron phosphate ($LiFePO_4$)) cathode which can at the best support Li in the weight ratio of 1 : 10, 1 : 12, and 1 : 25, respectively. A Li–S battery is based on a pure Li anode with a 1 : 1 weight ratio and a sulfur cathode with a theoretical weight ratio of 1 : 2.3. In practice, both LIBs and Li–S batteries are capable of only partially utilizing the electrodes [1, 2].

When compared with LIB development, the current phase of Li–S activity has only recently begun. With further research and development, a massive injection of energy density is potentially available. Thus, the starting point for the Li–S battery is already nearly equivalent to the maximal specific energy density of commercial LIBs. Currently available Li–S prototypes are already in the 250–350 Wh kg^{-1} range and are poised to achieve a target range of 400–700 Wh kg^{-1} in the next 3–10 years (Figure 8.1). It is instructive to compare the rate of progress for the Li-ion battery since its definitive commercial inception in 1995 with that of a speculative scenario for Li–S, assuming its practical starting point in the year 2015 (Figure 8.2). Even though energy densities over 400 Wh kg^{-1} have been achieved for Li–S batteries, the capacity fade with cycling is still an outstanding issue to be solved. However, with concerted research investment and trials, the strides can promise to be more rapid than LIBs after take-off.

A Li–S battery can, therefore, surpass the current storage capacity of LIBs and lead to a new revolution in battery technology. The price of simultaneously achieving

Rechargeable Ion Batteries: Materials, Design, and Applications of Li-Ion Cells and Beyond, First Edition.
Edited by Katerina E. Aifantis, R. Vasant Kumar, and Pu Hu.

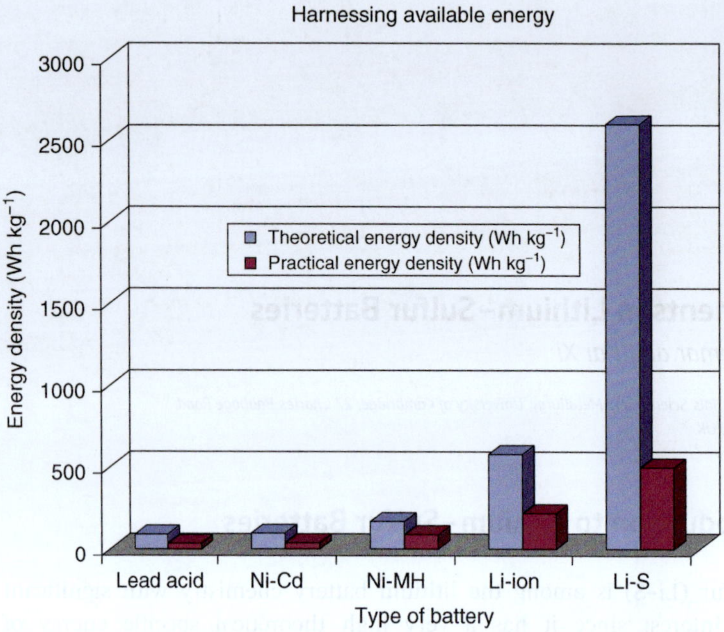

Figure 8.1 Even at a low% of theoretical energy usage, Li–S cells have a high-energy density.

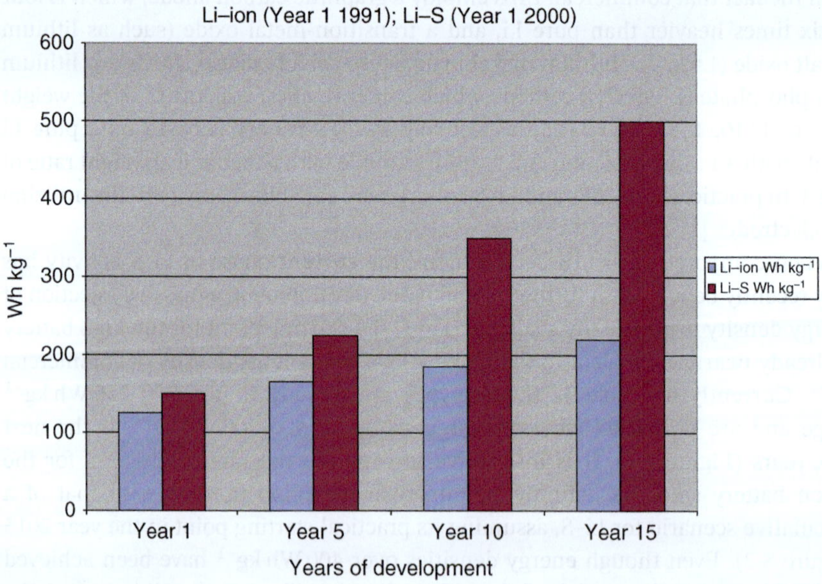

Figure 8.2 Current demonstrations of Li–S batteries have achieved over 400 Wh kg^{-1} of energy densities; however, the capacity fade with cycling is still being solved, but the strides can prove to be more rapid than Li-ion batteries after take-off.

both high gravimetric and volumetric energy densities beyond 500 Wh kg^{-1} and 500 Wh l^{-1}, respectively, is worth the technological, manufacturing, and commercial effort that is currently and increasingly undertaken at rising rates. A Li–S battery can also produce a high-power density comparable to that provided by Ni–Cd batteries, which makes it very attractive for high-energy and high-power applications. But unlike the Ni–Cd battery, a Li–S chemistry is not known to suffer from the memory effect and is very tolerant to overcharging, not to mention the environmental benefit of avoiding toxic chemical elements, such as Cd.

The Li–S battery can become the new standard bearer for another battery revolution in the next 15 years, due to its potential availability of distinctively high-energy density, which puts the Li–S chemistry in a league of its own. For example, the primary criteria for Advanced Battery Technologies for Electric-based transportation stipulate the energy density requirement at >400 Wh kg^{-1} with specific power of 400 W kg^{-1} at the ultimate price of US$0.1 per Wh. The Li–S battery can approach these targets.

The following factors that the Li–S battery can offer provide convincing pointers for potentially exponential growth:

- Increasing demand for considerable weight reduction and/or higher runtime in portable electronic devices, drones, wireless digital devices, power tools, and medical applications.
- Growing interest in electrical vehicles (EVs) and hybrid electrical vehicles (HEVs) requires both high-energy and -power densities.
- A lower voltage trend of consuming electronic products (2.1 V for a Li–S cell vs. 3.8 V for a LIB).
- Li–S batteries can be made by the same or similar manufacturing facilities as LIBs, which can potentially lower the development capital costs.
- Li–S cells are based on intrinsically safer and lower toxic materials than Li-ion, nickel–metal hydride (Ni–MH), or nickel–cadmium (Ni–Cd) batteries.

Although the Li–S battery recently has become a hot spot of technological development, the potential applicability of the Li–S cell as a power source had been considered more than three decades ago. Its development though was impeded by the difficulty in the safe use of pure Li metal as an anode on one hand and even more critically the electrochemical use of S without encountering charge transfer and diffusional problems. Furthermore, the formation of Li dendrites on the Li-anode during recharge, and the loss of active S at the cathode and into the electrolyte, resulted in a rapid fade of the capacity restricting the cycle life to <500 in commercial battery formats (such as pouch cells), under application-relevant crucial conditions, including high S loading, low Li excess, and use of a lean electrolyte. It should of course be noted that for the laboratory coin cell format it has been possible to exceed 1000 cycles.

Commercial developments in other Li-battery technologies, initially primary cells in the 1980s and later rechargeable Li-ion cells in the 1990s, have on one hand led to advantages in initiating research on the Li–S battery, but on the other hand, led to new possible development avenues for realizing the potential of the Li–S battery. In particular, considerable research in the arena of Li sources has (i) directly helped in increasing the knowledge of the safe use of Li; (ii) led to several promising avenues for selecting suitable electrolytes, some of which can be modified for the Li–S cells; in fact the lower voltage of a Li–S cell is more forgiving with respect to electrolyte stability; (iii) resulted in fabrication and processing technologies that can be used to assemble Li–S batteries; (iv) opened up electronics/microprocessors/sensors for managing battery cycling and safety. The early industrially driven work of Li–S batteries in the USA (Sion Power and PolyPlus), Russia (Ufa Academy of Sciences), UK (Oxis Energy), and Korea (Samsung) is notable for taking R&I to the next level.

8.2 Electrochemical Principles

As a lightweight element, sulfur can react with metallic lithium to form Li_2S with a large negative free-energy change such that this reaction can potentially be harnessed in a battery with a two-electron reaction with lithium as the anode and sulfur as the cathode. A Li–S battery has a high theoretical capacity of 1675 mAh g^{-1} of sulfur and a high theoretical energy density (2600 Wh kg^{-1} or 2800 Wh l^{-1}) (at a nominal cell voltage of 2.1 V) of cell weight, which has the potential to offer a practical energy density at least double the value in comparison with the present LIBs, as mentioned in the introduction. A comparison of different battery technologies in terms of energy density is shown in Chapter 3. In addition, elemental sulfur has the advantages of being low cost, nontoxic, and available in natural abundance or as a waste product from the chemical industry. Therefore, it is a very promising candidate as a cathode material for the next generation of high-energy-density rechargeable batteries.

In a typical Li-battery system, the cathode is the main cost component (approximately 40% of the total cell cost), while the remaining approximately 60% costs account for the rest of the components, such as the anode, electrolyte, separator, auxiliary components, plus the fabrication costs (Table 8.1). In addition to superior energy densities, Li–S batteries can provide significant cost savings relative to LIBs,

Table 8.1 Cost breakdown of commercial Li batteries.

Item	Cost (%)
Cathode	35–40
Electrolyte	10–15
Separator	7.5–10
Anode	7.5–15
Auxiliary components; manufacturing	20–30

Source: Adapted from Mauler et al. [3].

Table 8.2 Comparison of different cathode materials.

Cathode	Moles of host per mole of Li	Weight of host per g of Li	Nominal capacity, mAh g^{-1} of the host	Relative Cost relative to $LiCoO_2$ based on 2020 prices
$LiCoO_2$	2	26	146–200	1
$LiMn_2O_4$	1.5	37.5	127	0.17
$LiFePO_4$	2	40	98	0.3
NMCs	2	24–28	120–200	0.7–0.8
$LiMnPO_4$	1	25	150	0.5
Sulfur	2	9.2	>500–800 (target) (theoretical 1675)	0.003

due to the very low cost of S (approximately 0.1 $ kg^{-1} for S at 2020 prices), which lowers by about two orders of magnitude the cathode cost in comparison with the known cathodes used in Li-ion cells (e.g. approximately 30 $ kg^{-1} for $LiCoO_2$ at 2020 prices; Table 8.2). The anode cost of pure Li is similar to or marginally higher than the C-based anode in a LIB. These advantages can be combined with significantly greater opportunities for recycling cell components at the end of the product life in Li–S batteries compared with LIBs.

In Li–S cells, the porous sulfur composite cathode (S powder mixed with binder and carbon) is coated onto a thin metallized substrate, such as aluminum on a polymeric film. The anode can be prepared as a thin Li foil or as a vacuum-coated thin stabilized film on a metallized (copper) polymeric substrate. Both electrodes are separated by a polymeric separator immersed in the solvent/electrolyte system. The cathode is often described as a liquid cathode system, as the sulfur and the intermediate discharge product polysulfides are dissolved in the ether-based electrolyte. In the process of discharge, the S—S bonds are cleaved with the assistance of Li, resulting in a reductive decrease in the oxidation number of S. The S—S bonds are reformed upon charging.

The electrochemical reduction of S in a Li–S cell is a complicated multistep process from $S_8 \rightarrow Li_2S_8 \rightarrow Li_2S_6 \rightarrow Li_2S_4 \rightarrow (Li_2S_3) \rightarrow Li_2S_2 \rightarrow Li_2S$ (Figures 8.3 and 8.4). The most reduced state of sulfur Li_2S cannot be easily accessed in the electrochemical cycle due to its insolubility in liquid electrolytes [5]. Although Li_2S_2 is also relatively insoluble, it can be chemically leached into the electrolyte by reacting with higher-order polysulfides, which in turn are all soluble to varying degrees depending on the electrolyte composition and the relative amount. As a matter of fact, a key in choosing the electrolyte is to ensure that it can serve as a solvent for most of the reaction products. However, this leads to some other severe issues. The plateau of 2.1 V (see Figure 8.3) is due to two-phase equilibria between a solid lithium sulfide phase and the dissolved lithium polysulfide, or possibly between two solid phases. A solid lithium sulfide phase can passivate the carbon particles (added to provide

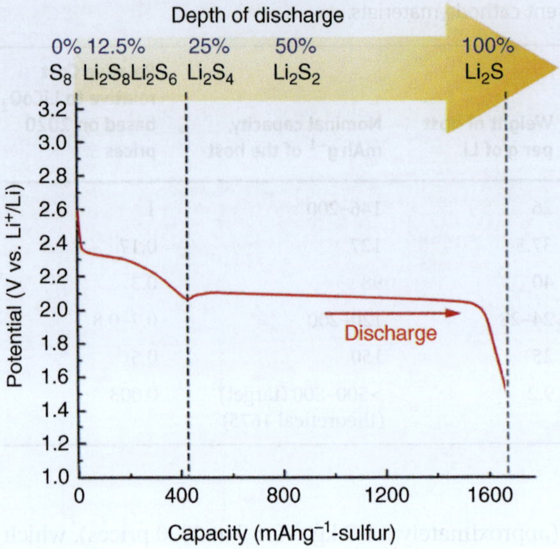

Figure 8.3 Electrochemical reduction of sulfur to various polysulfides in a Li–S cell during discharge.

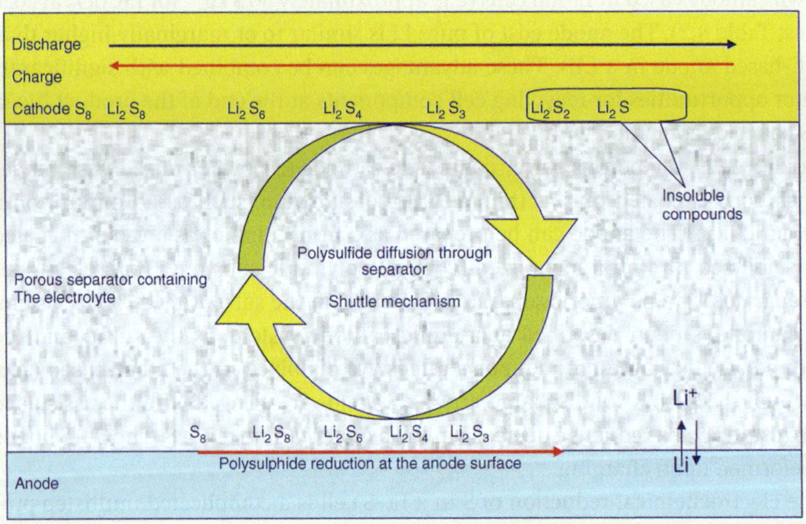

Figure 8.4 Representation of "Shuttle Mechanism." Source: Reproduced with permission from James et al. [4]/Elsevier.

electronic conductivity) in the electrode, hence, preventing charge transfer during the charge and subsequent discharge processes.

The electrolytes normally used in a Li-ion battery are unsuitable for a Li–S battery because the soluble lithium polysulfides are quite unstable in organic carbonate-based electrolytes (EC/PC/DMC). The electrolyte for a Li–S cell must consist of a weak polar solvent capable of dissolving sulfur and a strong polar solvent group capable of dissolving the lithium polysulfides. The weak polar solvents (with a relative permittivity $\varepsilon_r < 15$) are selected from aryl compounds, cyclic or noncyclic ether, and noncyclic carbonate compounds, such as xylene and tetrahydrofuran. The strong polar solvents (with $\varepsilon_r > 15$) are selected from cyclic carbonates, such as sulfoxides, lactones, ketones, esters, and acetonitrile. The electrolyte must also contain a lithium protection solvent group which allows the formation of a good protective layer on the Li electrode surface and one or more Li salt(s) to help achieve adequate Li-ion conductivity. Examples of Li protection solvents are saturated or unsaturated ether compounds, ethylene oxides, and various heterocyclic compounds containing O and/or S, such that Li_2S or Li_2O protection layers are readily formed.

At 100% utilization, the theoretical capacity of sulfur calculated using Faraday's laws is 1675 mAh g^{-1} as a result of complete reduction of S_8 to Li_2S. A decrease in sulfur capacity during cycling also arises due to the irreversible dissolution of polysulfides in the electrolyte which is shuttled away toward the anode followed by parasitic reactions that can take place directly with the Li to form the more oxidized form of the polysulfide, and also due to the formation of electrochemically inactive Li_2S [4]. Parasitic reactions of soluble polysulfides among themselves also lead to significant self-discharge. The challenge is to achieve the required > 1000 cycle life with a self-discharge that is less than 5% per month.

8.3 Sulfur Utilization and Cycle Life

The foremost technical issues to overcome for Li–S batteries are poor sulfur utilization (i.e. ratio of actual to theoretical gravimetric capacity) and low cycle life, which result from parasitic loss of sulfur by the shuttle mechanism, poor microstructure of the cathode for accommodating the discharge and charge products, high solubility of some of the polysulfides in the solvent, poor conductivity of elemental sulfur and solid lithium sulfide products, low conductivity and/or high viscosity of the solvent–electrolyte system, incomplete reduction of sulfur in some solvent systems, corrosion of the Li metal anode in the solvent-electrolyte phase and formation of passivating Li_2S and/or unfavorable solid-electrolyte interface (SEI) on the anode. It is generally thought that the soluble high oxidation state of lithium polysulfides as well as soluble sulfur, which are generated at the cathode, diffuse to the lithium anode where they react directly with the lithium via a parasitic reaction to recreate the low oxidation state lithium polysulfides. These species can diffuse back to the sulfur cathode to generate the high oxidation state lithium polysulfides again, causing loss of sulfur utilization accompanied by lithium corrosion. This mechanism, referred to as the shuttle mechanism, has been directly implicated as a major cause for low sulfur utilization from the first discharge, further exacerbated in subsequent charge–discharge cycles because of the above factors. It is therefore not surprising that any approach that can minimize these problems can produce a higher S utilization and improved cycle life. It is, however, interesting to note that the shuttle mechanism (Figure 8.4) can be beneficial for offering overcharge protection during charging [4].

During the first discharge process in the Li–S cells, the elemental sulfur in the solid phase (S_8(s)) in the cathode is first dissolved in the electrolyte as S_8(l) and then electrochemically reduced by Li^+ ions (which diffuse from the anode through the electrolyte) to sulfide ions with progressively lower states of oxidation on the surface of the electronically conductive material (e.g. carbon), followed and accompanied by coupled chemical reactions between the sulfide ions. The coupled chemical reactions act to disproportionate the more reduced sulfide ions with the sulfur component in a higher oxidation state to form an intermediate oxidation state component. These chemical reactions can take place within the electrolyte solvent in a facile manner for soluble polysulfide chains. Electrochemical and corresponding coupled chemical reactions are dependent greatly on the depth of discharge. They are briefly summarized with some important reactions in Figure 8.5 [6].

Region I in broad terms represents the electrochemical reduction of soluble sulfides at the interface between the liquid phase (solvent–electrolyte system) and the conducting phase (usually carbon) to form a reduced sulfide product that is also soluble. Within Region I, variation in the potential from 2.4 to 2.1 V may exhibit many detailed features depending upon factors, such as solubility of the starting

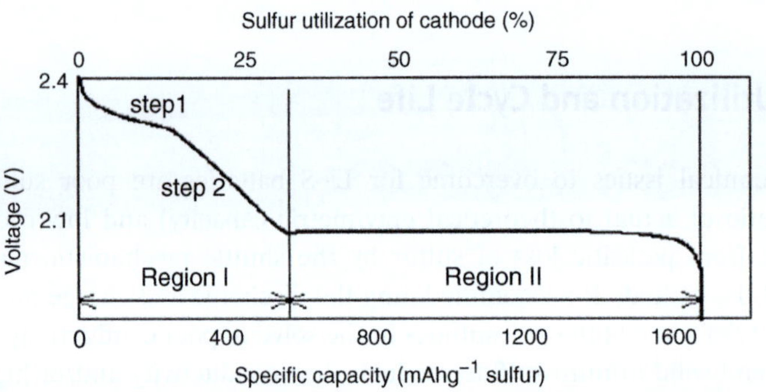

Depth of discharge	Electrode reaction	Coupled chemical reaction
I	$S_{8(l)} + 2e^- \rightleftharpoons S_8^{2-}$ $3S_8^{2-} + 2e^- \rightleftharpoons 4S_6^{2-}$ $3S_6^{2-} + 2e^- \rightleftharpoons 3S_4^{2-}$ $3S_4^{2-} + 2e^- \rightleftharpoons 4S_3^{2-}$	$S_{8(s)} \rightleftharpoons S_{8(l)}$ $4S_8^{2-} \rightleftharpoons 4S_6^{2-} + S_{8(l)}$ $4S_6^{2-} \rightleftharpoons 4S_4^{2-} + S_{8(l)}$
II	$2S_3^{2-} + 2e^- \rightleftharpoons 3S_2^{2-}$ $S_2^{2-} + 2e^- \rightleftharpoons 2S^{2-}(\downarrow)$	$2S_4^{2-} \longrightarrow S_5^{2-} + S_3^{2-}$ $2S_5^{2-} \longrightarrow S_6^{2-} + S_4^{2-}$ $2S_3^{2-} \longrightarrow S_4^{2-} + S_2^{2-}(\downarrow)$ $2S_3^{2-} \longrightarrow S_5^{2-} + S^{2-}(\downarrow)$

Figure 8.5 Electrochemical reduction mechanism of sulfur electrode as a function of depth of discharge in Li–S cells, corresponding to 33.3% S utilization in Region I (also shown are the coupled chemical reactions). Source: Reproduced with permission from Kolosnitsyn et al. [6]/Elsevier.

sulfur and the various high-order lithium polysulfide products from Li_2S_8 to Li_2S_3 (33% S utilization in Region I) or even greater depth of reduction under certain situations from Li_2S_8 to Li_2S_2 (this can account for up to 50% S utilization in Region I) in the liquid electrolyte–solvent phase. Region II represents the presence of two solid products that are either saturated in the electrolyte–solvent phase (e.g. Li_2S_3 or Li_2S_2) or insoluble (e.g. Li_2S), giving rise to the plateau region (two almost overlapping plateaus for 66.7% S utilization and one plateau for 50% S utilization in Region II). The voltage is expected to decrease rapidly at the end of Region II when all the S is consumed (electrochemically at full utilization or chemically in parasitic reactions) or when insoluble sulfides cover the cathode and can cut-off charge transfer reactions.

Region I can be further subdivided into steps 1 and 2 following the transition from the reduction sequence $S_8 \rightarrow Li_2S_8 \rightarrow Li_2S_6 \rightarrow Li_2S_4$ in step 1 to the sequence $Li_2S_4 \rightarrow Li_2S_3$ (and possibly to $\rightarrow Li_2S_2$) in step 2. It must be noted that the sequence is not followed in an orderly manner, as an example Li_2S_4 could be produced before all S_8 has reacted. Depending upon the solvent used, solubility of the sulfide may be lower, and thus Li_2S_3 precipitation can take place readily and the potential plateau can start after 33.3% S utilization (nominally) rather than after 50%, corresponding to Li_2S_2/Li_2S as the main solid products. The coupled chemical reaction can further disrupt the completion of each reaction in the sequence by reforming the high oxidation state lithium polysulfides. Nominally Region I should theoretically account for 33.3–50% S utilization and the remaining 50–66.7% are accounted for by Region II, depending upon the solubility limits of the various lithium polysulfides in the electrolyte–solvent used. A theoretical example is shown in Figure 8.5 assuming a nominal 33.3% S utilization in Region I and 66.7% S utilization in Region II. In this scheme, the plateau in Region II should consist essentially of two plateaus corresponding to Li_2S_2/Li_2S_3 and Li_2S_2/Li_2S. Frequently, because of relatively small differences in the electrode potentials corresponding to the two equilibria and from ongoing chemical reactions enhanced by shuttling from the anode compartment, the plateau in Region II can overlap forming a pseudo-plateau [7, 8]. A summarized version of the above details is described in Figure 8.6 [8].

Therefore, active sulfur can exist in both liquid and solid phases. The solid phase Li_2S_2/Li_2S can form by the irreversible precipitation from the electrolyte in the process of discharge, of which at least some of the particles are difficult to oxidize in the charging process. Available sulfur will continuously decrease if S is lost into unreactive solid sulfides in this manner with progressive cycling. When insulating solid sulfides cover the conducting carbon particles, conducting paths are also lost, which lead to an increase in cell impedance and further loss in energy density.

Many research articles have reported various types of sulfur composite systems as the active materials for Li–S batteries using organic liquid electrolytes [6–10]. Although a considerable enhancement in the performance of Li–S batteries using an organic liquid solvent-based electrolyte has been achieved, there still exists a critical problem in rate capability and cycle life characteristics [11], and therefore several innovative approaches have been investigated [12–15].

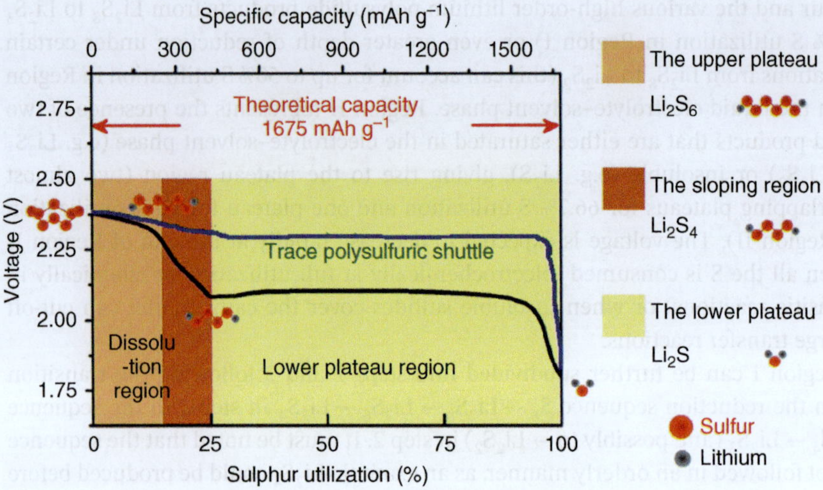

Figure 8.6 Typical discharge–charge profile for a Li–S cell, indicating the type of sulfides during the process. Source: Reproduced with permission from Su et al. [8]/Springer Nature.

8.4 Potential Solutions to Hurdles

Sulfur by itself is an ineffective cathode material due to its highly electronically and ionically insulating nature with a conductivity of $5 \times 10^{-30}\,\text{S cm}^{-1}$ at 25 °C. To increase the surface area for the reaction to occur and to facilitate electron transport, sulfur encapsulation in conductive and nanoscale architectures has been employed. In addition, the insoluble electrical insulators Li_2S and Li_2S_2 might coat the electrodes completely during the reaction and impede further lithiation. The mass densities of S and Li_2S are 2.07 and $1.66\,\text{g cm}^{-3}$, respectively; thus, the volume expansion at the end of full discharge is as large as 80% [12]. During charge, the volume reduction along with the accompanying mechanical stress is adequately large to detrimentally influence the cathode morphology. Another major challenge is that polysulfides diffuse out of the conductive matrix during discharge and do not return during charge, leading to a reduction in active material. Hence the capacity is reduced rapidly with every charge–discharge cycle. Optimization of porous conductive materials to trap S inside the cathode is the major challenge in producing efficient rechargeable Li–S batteries.

As illustrated in Figure 8.7, elemental sulfur arranges itself in a crown-like S_8 structure. In a Li–S cell, the sulfur's electrochemical reduction (Figure 8.8) occurs in two mainly thermodynamic regions expressed as redox reactions

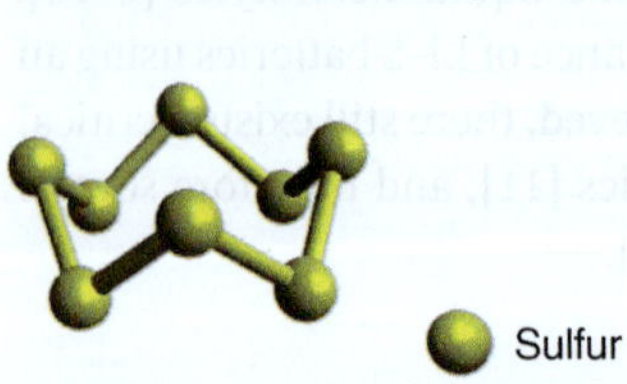

Figure 8.7 Structural representation of a sulfur crown molecule.

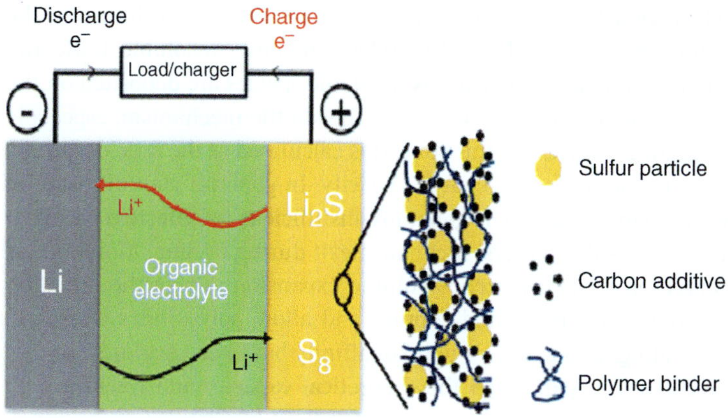

Figure 8.8 Illustration of the charge (red)/discharge (black) process in a lithium–sulfur cell consisting of a lithium metal anode, organic electrolyte, and sulfur composite cathode. Source: Reproduced with permission from Manthiram et al. [16]/John Wiley & Sons.

involving Li–S components. Each electrochemical step combines both chemical and electrochemical reactions in a narrow potential range [16]. The first electrochemical reduction step of sulfur starting at 2.4 V includes the reactions:

$$S_8 + 2e \rightarrow S_8^{2-} \quad \text{(electrochemical)} \tag{8.1}$$

$$S_8^{2-} \leftrightarrow S_6^{2-} + \tfrac{1}{4} S_8 \quad \text{(chemical)} \tag{8.2}$$

$$S_6^{2-} \rightarrow S_4^{2-} + \tfrac{1}{4} S_8 \quad \text{(chemical)} \tag{8.3}$$

The properties of the electrolyte have a very significant influence on the stability of di-anions S_8^{2-} that are formed in reaction 1. The rise of solvent viscosity could increase the S_8^{2-} stability due to the "cage effect" from solvent molecules. During the first step of the electrochemical reduction process, the capacity would increase due to the regeneration of sulfur by dissociation of anion S_8^{2-}.

The second electrochemical region for sulfur at 2.1 V may include some or all the following reactions:

$$2S_6^{2-} \rightarrow S_7^{2-} + S_5^{2-} \quad \text{(chemically slow)} \tag{8.4}$$

$$5S_7^{2-} + 4e \rightarrow 7\,S_5^{2-} \quad \text{(electrochemically fast)} \tag{8.5}$$

$$2S_5^{2-} \rightarrow S_7^{2-} + S_3^{2-} \quad \text{(chemically slow)} \tag{8.6}$$

$$2S_3^{2-} \rightarrow S_5^{2-} + S^{2-} \quad \text{(chemically fast)} \tag{8.7}$$

Further reduction of long-chain polysulfides dissolved in the electrolyte in this region is listed above. The polysulfides formed in the first step are generated from disproportionation reactions, leading to the formation of long- and short-chain polysulfides [12]. Long-chain polysulfides undergo further electrochemical reduction. In the formation of lower polysulfides, poorly soluble components in the electrolyte are produced and formed in solid phases. In this way, Li_2S would deposit and accumulate on the surface of the cathode, which could cause its capacity to fade.

As mentioned previously, the dissolved lithium polysulfides can diffuse to the lithium anode, forming lower-order polysulfides, including insoluble Li_2S_2 and Li_2S, resulting in a rapid fade in capacity, as insoluble species are deposited on the anode, and low Coulombic efficiency (CE) from the shuttle mechanism, especially at low charge/discharge rates. CE in this context is calculated as the ratio of specific discharge capacity to the preceding charge capacity. In general, melt diffusion of S into prefabricated nanostructures is generally insufficient for effective encapsulation of sulfur – where sulfur enters, it can also exit during cycling. Moreover, as the sulfur discharge products experience volume expansion during the lithiation process, surrounding structures would rupture and allow polysulfides to escape. For polysulfides, coatings and additives with an affinity have been explored, so that the electrochemical results can reach the theoretical capacity and retain a high capacity with cycling.

Developing optimized sulfur cathode structures can help overcome many of the challenges mentioned above. Reports in the literature have proposed various innovative approaches, which mainly focus on achieving one or more of the targets given below:

- Introducing a relatively large conductive surface area for electrochemical reactions and allowing for the deposition of insoluble products.
- Trapping intermediate product polysulfides by physical and/or chemical interaction approaches to prevent loss of active materials and to maintain cycling stability.
- Providing sufficient space for buffering the sulfur's volume expansion (approximately 80%) during reactions.
- Using electrocatalysts to decrease overpotentials.

The nanostructured configurations of electrodes have been a crucial factor in development of Li–S batteries. Investigations into nanostructured composite cathodes can be categorized in the following groups based on the chemical composition and structure of the cathode – (i) porous carbon–sulfur composites, (ii) graphene or graphene oxide–sulfur composites, (iii) carbon nanotube (CNT) or carbon nanofiber–sulfur composites, (iv) polymer–sulfur composites, and (v) metal oxide–sulfur composites. In general, the first three groups utilize carbon as a conductive matrix to capture polysulfide species and to promote redox reactions. Each respective geometry has its own pros and cons. Conductive polymers have great potential due to their flexibility, which may potentially reduce the strain of volumetric expansion. Moreover, porous metal oxides are another promising material, which lowers the capacity fade during cycling [17]. In the following sections, examples for each of the above materials systems will be given.

8.5 Carbon Materials

Research in carbon materials can be generally divided into three categories – porous carbon, two-dimensional (2D) nanostructures (e.g. graphene), and one-dimensional (1D) nanostructures (e.g. CNTs).

8.5.1 Porous Carbon

Most research within porous carbon has focused on mesoporous and microporous carbon, in which the pore diameter is 2–50 and <2 nm, respectively. The larger pores in mesoporous carbon facilitate rapid Li^+ transport through good electrolyte penetration but are also large enough to allow polysulfides to escape, leading to active material sulfur loss with cycling. Bimodal pore structures, therefore, with micropores in the carbon framework that houses the mesopores have been developed. During impregnation, the sulfur preferentially fills the smaller pores first, thereby theoretically entrapping the polysulfide species during cycling, while maintaining efficient Li^+ transport in the larger mesopores. However, the pore size seems to be less significant than the carbon structure below the macropore scale. This is largely because the electronically conductive micropores facilitate reduction of polysulfides to insoluble sulfides within the pores during the charge cycle.

Once the pores are filled, their ability to enhance performance is diminished. For both microporous and mesoporous systems, better performance has been achieved by using spheres instead of 3D foams or 2D systems. Hollow carbon capsules with pores of approximately 3 nm have shown to effectively contain the sulfur species, while allowing Li^+ to diffuse through them [18]. The cathode material has shown to maintain a discharge capacity of approximately 975 $mAh\,g^{-1}$ after 100 cycles at a 0.5 C rate and an impressive 450 $mAh\,g^{-1}$ at a high rate of 3 C (Figure 8.9).

In another approach, small molecules of sulfur (S_{2-4}) confined in a microporous carbon matrix (approximately 0.5 nm average pore size) could help avoid the unfavorable transition from S_8 to $S_4{}^{2-}$ during discharge and the reverse transition during charge [19]. As shown in Figure 8.10, a single long plateau at approximately 1.9 V was observed providing 1670 $mAh\,g^{-1}$ initial capacity and good cycle stability of 1149 $mAh\,g^{-1}$ capacity after 200 cycles. It was also found that when the encapsulated sulfur in the carbon-microsphere-based composite was thermally treated, the reversible capacity was improved [20] since the sulfur and the products were constrained inside the microspheres thereby electrochemically stabilizing the cathode components (Figure 8.11).

8.5.2 Graphene

Graphite derivatives have been a major focus in the category of two-dimensional frameworks [21]. Graphene shows a very high conductivity and large surface area, which is easily functionalized into hydrophilic or hydrophobic structure. However, graphene has a much higher affinity for itself than for sulfur, resulting in low gravimetric percentages of sulfur using melt diffusion alone. Compared with graphene, graphene oxide (GO) tends to be more favorable. GO has generally been of great interest for two reasons – more cost-effective production than graphene and greater affinity for polysulfides due to the hydroxyl groups on the GO surface. Melt diffusion of sulfur nanoparticles into GO has shown to result in composites of 66 wt% sulfur with a reversible capacity of approximately 950 $mAh\,g^{-1}$ after 50 cycles at 0.1 C.

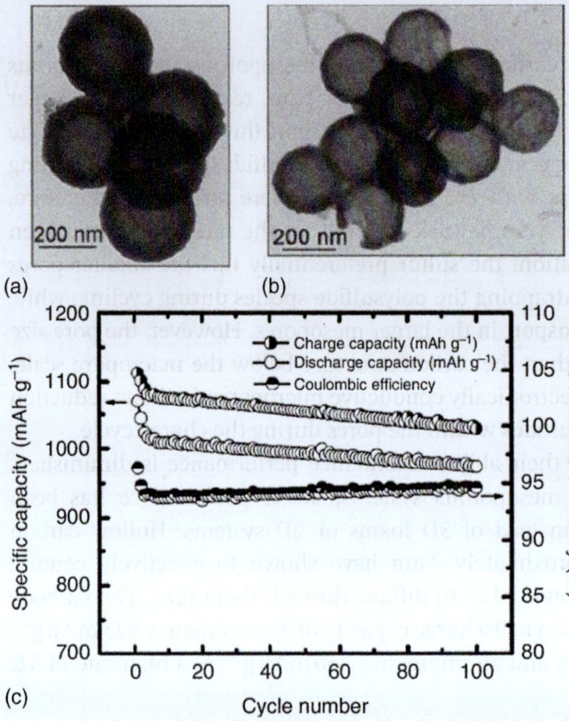

Figure 8.9 Transmission electron microscopes (TEM) images of (a) mesoporous hollow carbon spheres and (b) sulfur-encapsulated carbon spheres. (c) Charge/discharge cycling performance and Coulombic efficiency (CE) over 100 cycles at 0.5 C. Source: Reproduced with permission from Jayaprakash et al. [18]/John Wiley & Sons.

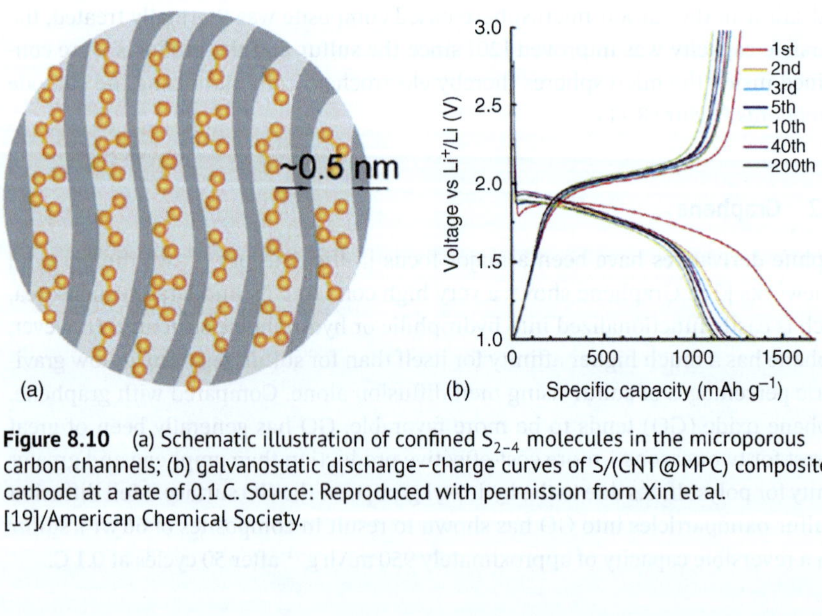

Figure 8.10 (a) Schematic illustration of confined S_{2-4} molecules in the microporous carbon channels; (b) galvanostatic discharge–charge curves of S/(CNT@MPC) composite cathode at a rate of 0.1 C. Source: Reproduced with permission from Xin et al. [19]/American Chemical Society.

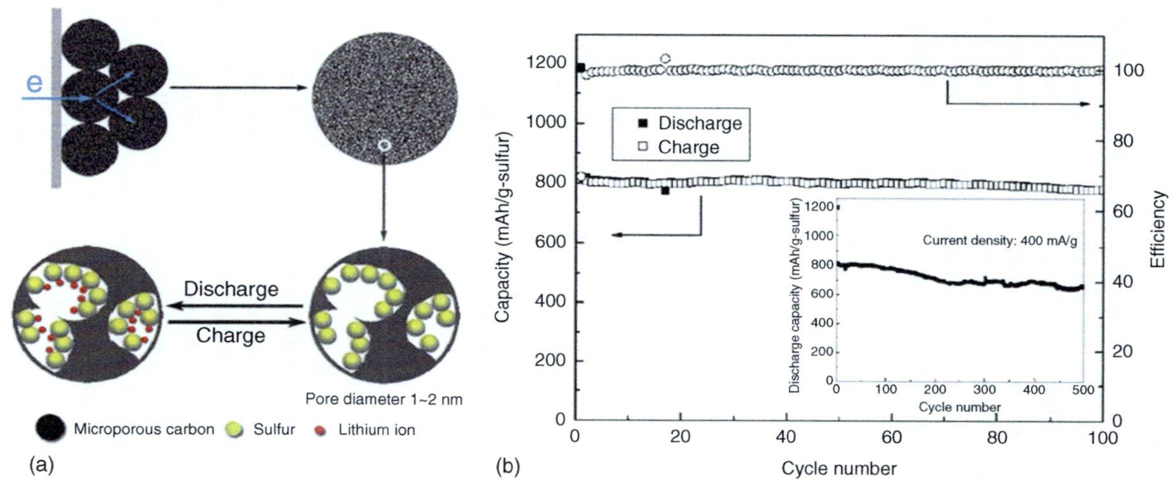

Figure 8.11 (a) Schematic illustration of the constrained electrochemical reaction process inside the micropores of the sulfur–carbon sphere composite cathode; (b) cycling performance of the composite at the current density of 400 mA g^{-1}. The long cycle life (500 cycles) of the composite at 400 mA g^{-1} is inserted in (b). Source: Reproduced with permission from Zhang et al. [20]/Royal Society of Chemistry.

Other work has focused on increasing the sulfur loading by thermal or chemical expansion of the GO. Chemical reduction followed by hydrothermal reaction of GO has been applied to block aggregation of monolayers, which could increase sulfur loading to 75.2 wt%. The expansion also led to polysulfide loss, accompanied by 37% reduction in the discharge capacity after 10 cycles at 0.6 C. Thermal expansion of GO has shown to increase sulfur loading, although the electrochemical performance decreases with a sulfur ratio >60 wt%. Thermal reduction after expansion restores some sp^2 character to GO, facilitating efficient electron transport and increasing the initial discharge capacity to 1210 mAh g^{-1} [22]. Nevertheless, the performance still declines sharply by approximately 0.4% per cycle at 0.17 C. Reduced graphene oxide (RGO) has also been used to coat micron-sized polysulfide in acid for producing a composite of 87 wt% sulfur. However, the relatively large size of the particles limited sulfur utilization, resulting in an initial discharge of 705 mAh g^{-1} at a C/5 discharge rate, stabilizing after 50 cycles to approximately 500 mAh g^{-1} [23].

Recently, the design of an ultralight Li–S battery cathode was reported which is based on loading sulfur onto a free-standing porous and interconnected 3D network of a few-layer graphene (FLG) foam by using a simple melt-solution infiltration with sulfur. The FLG foam was synthesized from a catalyst metal template, by chemical vapor deposition (CVD), which is a promising route for making high-quality graphene. Excellent electron and ion transport are facilitated by the 3D interconnected structure of the FLG foam. In addition, components such as metal current collectors, conducting additives such as C-black, and binders such as polytetrafluoroethylene (PTFE) from the cathode system could be eliminated, thus greatly decreasing the weight by 20–25%. A cell based upon the sulfur/FLG foam 3D network structure exhibited excellent high-rate cycle stability and a high Coulombic efficiency over 400 cycles in a Li–S battery (Figure 8.12) [24].

8.5.3 Carbon Nanotube

One-dimensional carbon nanomaterials have also been of considerable interest in this field. Using a layer of multi-walled carbon nanotubes (MWCNTs) as the separator prevented the polysulfides from diffusing through it and reaching the Li metal anode [25]. The MWCNT paper separator was prepared by ultrasonic dispersion of MWCNTs followed by vacuum filtration without using binders. This paper was readily peeled off the filtration membrane after it was formed. The insertion of the MWCNT interlayer acted like a pseudo-upper current collector. Therefore, the charge transfer resistance dramatically decreased from 277 to 38 Ω cm^2. Scanning electron microscopy (SEM) analysis confirmed that the MWCNT skeleton acted like a sorption matrix to capture and retain the polysulfide species and provided abundant reaction sites [25], significantly improving the battery performance. A Li–S cell with MWCNT interlayer shows a specific capacity of 962 mAh g^{-1} after 50 cycles at a rate of C/5. In contrast, the raw Li–S battery exhibits only about 330 mAh g^{-1} after 50 cycles under the same conditions. In addition, the discharge capacity increased from approximately 0 to 804 mAh g^{-1} after 100 cycles at a high rate of 1 C (Figure 8.13).

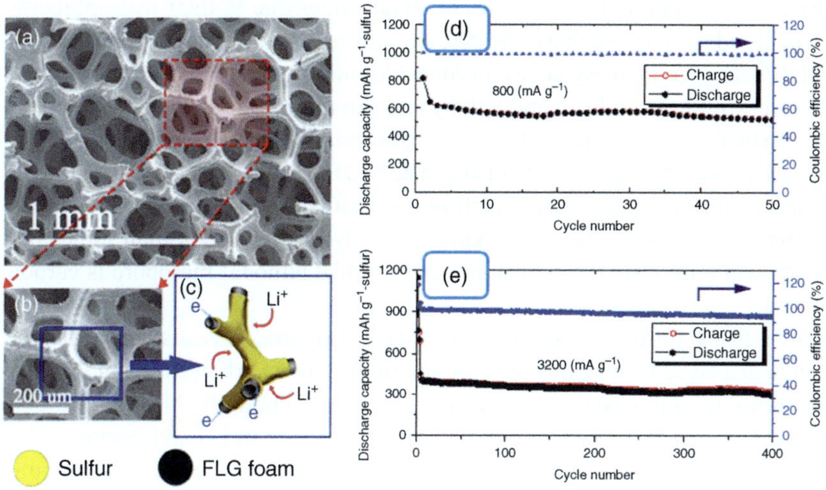

Figure 8.12 (a, b) SEM images of S–FLG foam; (c) scheme of the fast 3D electron/Li^+ transfer pathway inside the S–FLG foam; (d, e) cycle performance and Coulombic efficiency of S–FLG foam cathode different specific currents. Source: Reproduced with permission from Xi et al. [24]/Royal Society of Chemistry.

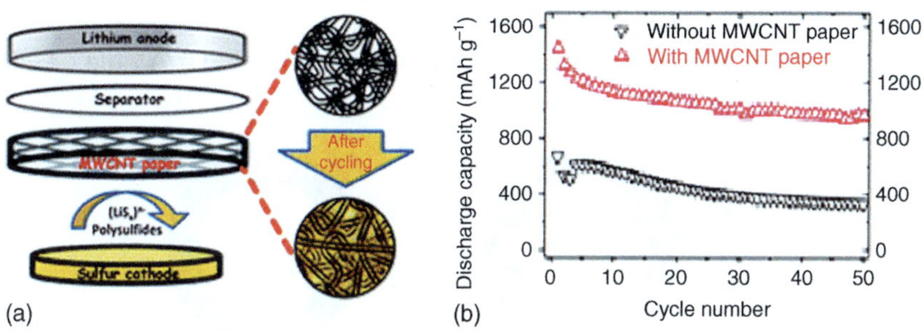

Figure 8.13 (a) Schematic illustration of the Li–S battery with a MWCNT separator. (b) Enhanced discharge capacity and cycling performance over 50 cycles at 0.2 C with the MWCNT paper. Source: Reproduced with permission from Su et al. [25]/Royal Society of Chemistry.

Good sulfur content (up to 70 wt%) was easily achieved resulting in capacities >800 mAh g^{-1} of the total cathode weight when MWCNTs were vertically aligned on a metal substrate using a special CVD process [26]. However, the shuttle mechanism was enhanced by the alignment which could be controlled by selected Li salts (e.g. $LiNO_3$) in the electrolyte allowing formation of SEI layers on the Li-anode but still permitting ionic and electronic transport.

8.6 Metal Oxides

The chemical interactions between metal oxides and sulfur/lithium polysulfides have not yet been fully understood. The electrostatic interactions are enabled between the surfaces and polysulfides of metal oxides due to their hydrophilicity. Metal oxide adsorbents, such as silicates, aluminum, and vanadium oxides, as well as metal chalcogenides, have been considered for inhibiting polysulfide dissolution. Metal oxides suffer from low conductivity and can thereby limit electron transport for the reduction of sulfur. Studies using mesoporous metal–organic framework (MOF) as a host material are promising. The MOFs can have large pore volumes with meso- and micro-pore sizes (such as approximately 25–29 and 5–9 Å) which allow the structure to house some liquid electrolyte, providing a high ionic conductivity, while limiting the diffusion process – like bimodal mesoporous carbon structures.

Sulfur-infused SBA-15, a mesoporous silica structure, was also investigated. The MOF displayed better capacity retention than similar carbon confinements but did not reach the same peak. The SBA-15 outperformed both the MOF and carbon structure, but only in a composite containing 50 wt% carbon [27–29]. Metal oxides generally perform quite poorly without the incorporation of a substantial amount of carbon or other conductive material. But one exception was a sulfur–TiO_2 yolk–shell nanoarchitecture with an internal void space which only incorporated 15 wt% carbon black into the electrodes [27]. A TiO_2 coating was employed on sulfur nanoparticles through hydrolysis of a sol–gel precursor, followed by partial dissolution of the sulfur by toluene [27], allowing the yolk sufficient internal void space to accommodate volume expansion. As shown in Figure 8.14, this structure illustrated an initial specific capacity of 1030 mAh g^{-1} at 0.5 C and CE of

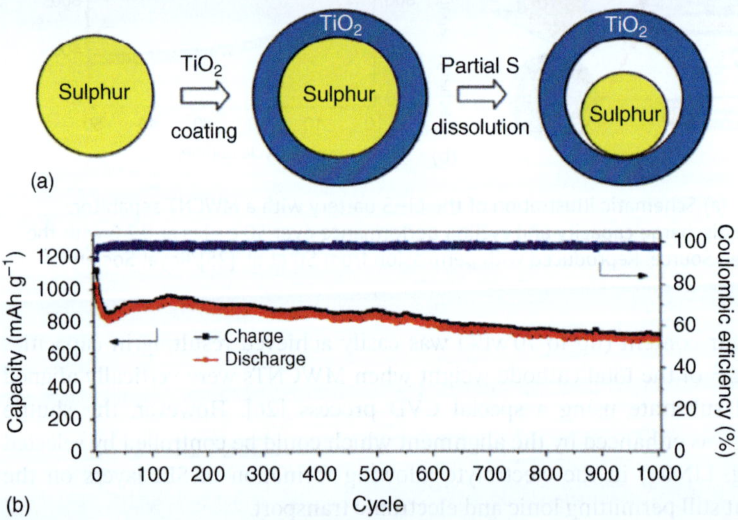

Figure 8.14 (a) Schematic illustration of the construction of the sulfur–TiO_2 yolk–shell architecture. (b) Cycling performance and Coulombic efficiency (CE) of the material over 1000 cycles at a rate of 0.5 C. Source: Reproduced with permission from She et al. [27]/Springer Nature.

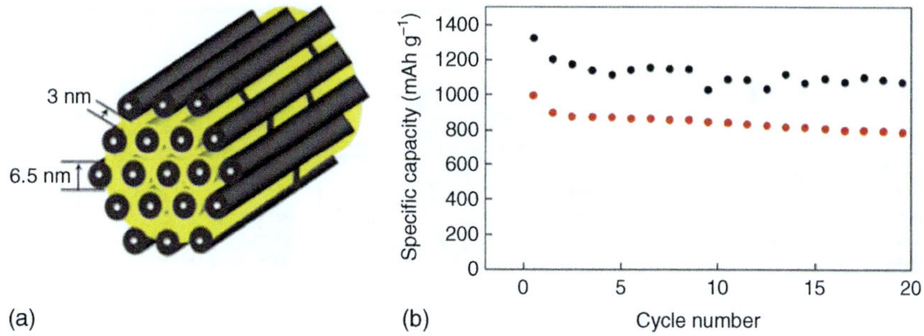

Figure 8.15 (a) Schematic illustration of the sulfur (yellow) confined in CMK-3, a highly ordered mesoporous carbon; (b) cycling performance comparison of CMK-3/S-PEG (upper points, in black) vs. CMK-3/S (lower points, in red) at 168 mA g^{-1} at room temperature. Source: Reproduced with permission from Ji et al. [28]/Springer Nature.

98.4% over 1000 cycles. In addition, it is impressive that the capacity decay was only 0.033% per cycle over 1000 cycles.

Metal oxides have sometimes been used to produce a highly ordered mesoporous carbon structure, such as carbon mesostructured (Korea) (CMK)-3 [28], exhibiting a uniform pore diameter, very high pore volume, and a well-interconnected porous structure with good electrical conductivity. Siliceous SBA-15 was used as a hard template producing a replica consisting of 6.5 nm-thick carbon nanorods containing 3–4 nm channels that separated the rods, resulting in a 2D hexagonal-type structure (Figure 8.15a). To stabilize the nanoarchitecture and trap polysulfides, the sulfur/CMK external surface was modified by polyethylene glycol (PEG). The capacity displayed was reversible and stable over 20 cycles showing much promise for a well-defined mesoporous morphology within a composite structure (Figure 8.15b).

On combining carbon derived from a SBA template with a silica colloid monolith it was reported that the mesoporous structure was able to provide an effective internal reservoir for the sulfur-infiltrated cathode in a Li–S battery [29]. As a result, the cycling stability was significantly improved. The authors demonstrated that improvement arose from an increased usage of the larger mesopores (>10 nm) in the carbon and a decreased shuttle effect (Figure 8.16).

Variable hierarchically porous carbons were synthesized from direct carbonization of four different selected MOF precursors [30]. The authors reported the first investigation of this kind of hierarchically porous carbon from MOFs for the fabrication of cathodes in Li–S batteries. The impact of the pore volume and pore size distribution of the carbonized MOFs loaded with sulfur on the capacity and the cycle life of Li–S batteries was demonstrated (Figure 8.17). In the future, this work will be able to provide a roadmap for developing carbon materials using templates and precursors with an optimum hierarchically high-volume porous structure that has both mesopores and micropores for good sulfur loading and electrochemical utilization.

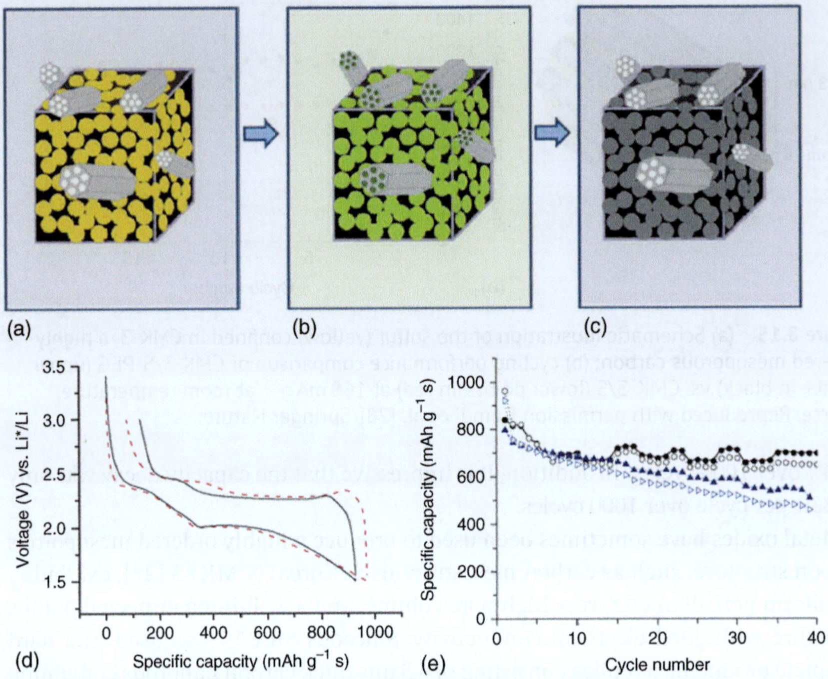

Figure 8.16 Schematic illustration of the concept of the "polysulfide reservoir" afforded by the SBA-15 in the supply chain management (SCM)/S electrode. (a) Sulfur electrode embedded with SBA-15 before discharge; (b) absorption of polysulfides by SBA-15 during discharge process; (c) release of polysulfides by SBA-15 at the end of discharge; (d) comparison of the discharge curves of the first cycle of SCM/S (black solid line) and SCM/S with the SBA-15 additive (red dotted line); (e) comparison of the cycling performance of SCM/S (blue) and SCM/S with SBA-15 additive (black) showing the capacity and cycle stabilization effect of SBA-15. Source: Reproduced with permission from Ji et al. [29]/Springer Nature.

8.7 Polymers

Polymers have been used as carbon substitutes in composite cathodes for Li–S batteries [31]. The merits of polymers can be that they do not require a high-temperature processing as carbon nanostructures do (>600 °C for carbonization), enabling the synthesis of structures that are tricky to achieve with carbon–sulfur composites, such as conformal coatings to trap lithium polysulfides. In addition, functional groups along the polymer chain can aid in weak binding of polysulfides, adding chemical confinement to a purely physical containment. Moreover, soft polymer matrices can buffer the volumetric change associated with the conversion of sulfur to lithium sulfides, and vice versa. Polymer coatings of sulfur particles via *in situ* chemical oxidative polymerization can show relatively good performance in terms of Coulombic efficiency.

An interesting polymeric architecture incorporated with sulfur can also be achieved by using vulcanization to harden rubber-like materials by treating

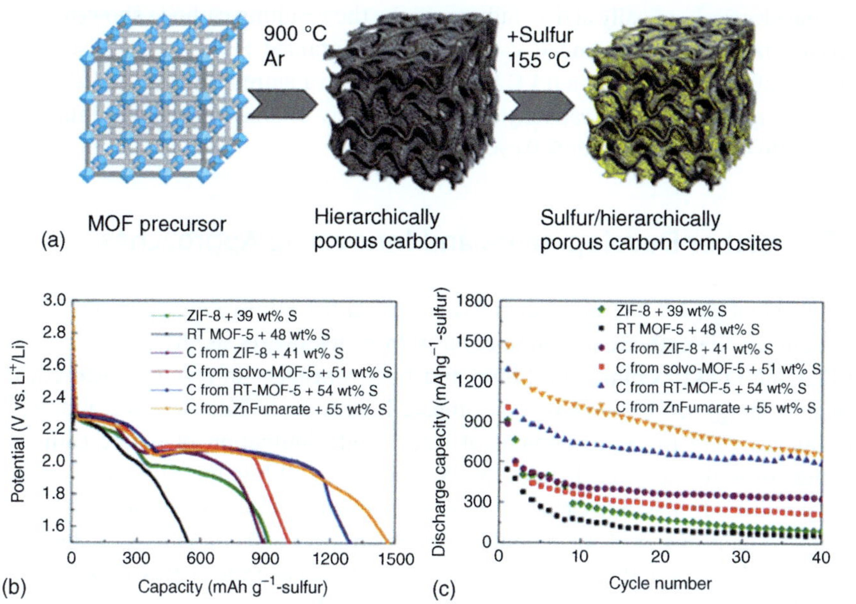

Figure 8.17 (a) Scheme illustration of sulfur/hierarchical porous carbon composite preparation. The first discharge curves (b) and cycling performance (c) of samples at a specific current of 400 mA g^{-1}. Source: Reproduced with permission from Xi et al. [30]/Royal Society of Chemistry.

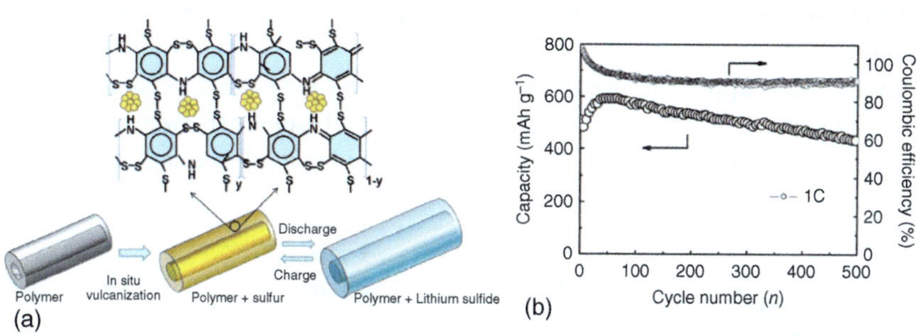

Figure 8.18 (a) Schematic illustration of the construction and charge/discharge process of the PANI-NT/sulfur composite. (b) Cycling performance and Coulombic efficiency (CE) over 500 cycles at a rate of 1 C. Source: Reproduced with permission from Xiao et al. [31]/John Wiley & Sons.

them with sulfur at high temperatures employed with self-assembled polyaniline nanotubes (PANI-NTs) at 280 °C (Figure 8.18a). As shown in Figure 8.18b, the discharge capacity of such an electrode had a gradual capacity increase in the initial several tens of cycles, indicating that the PANI-NT/sulfur composite needs an activation step [31]. The possible reason is the surface area of the polymer host is lower than that of a carbon matrix, and therefore, the electrolyte takes more time to diffuse into the internal surfaces of the polymers. Only when they are in contact with the electrolyte can the deeply buried sulfur and disulfide bonds gradually become electrochemically active. Subsequently, the resulting cathode showed little capacity fading upon extended cycling and maintained 837 mAh g^{-1} discharge capacity after 100 cycles at a 0.1 C [31]; however, in Figure 8.18 it is seen that, at 1 C, after reaching a peak capacity of approximately 600 mAh g^{-1}, it continuously decayed to 400 mAh g^{-1} after 500 cycles.

8.8 Further Developments and Innovative Approaches

In a novel method, Li_2S, rather than S, was embedded within a mesoporous carbon structure as the cathode in conjunction with Si nanowires as the anode instead of metallic Li [32]. Combining Li-ion and Li–S concepts can result in theoretical specific energy of 1550 Wh kg^{-1}. Compared with Li–S batteries, the reported advantages were the improvement of safety and elimination of severe cathode damage from the volume changes that S experiences. While the cathodic damage was minimized, it is not clear yet how the anodic damage associated with the volume expansions of Si was handled, which are over 300%. Figure 8.19 shows a schematic diagram of the cell along with electrochemical performance data.

As shown in Figure 8.20, to alleviate the issues of low-sulfur utilization and poor cycle life, a bifunctional microporous carbon paper was inserted between the sulfur composite cathode and the polypropylene (PP)-based separator (Celgard) [33], allowing for a decrease in the charge transfer resistance and an improvement in the cycling performance, possibly arising from the enhanced retention of soluble polysulfide species. Stable capacities over 700 mAh g^{-1} at 2 C were obtained for 150 cycles.

In another new approach, the Li metallic anode was hybridized by an electrical connection with a graphite layer placed in front of the Li anode (between the separator and the Li anode) (Figure 8.21a) [34]. The aim of using this integrated anode

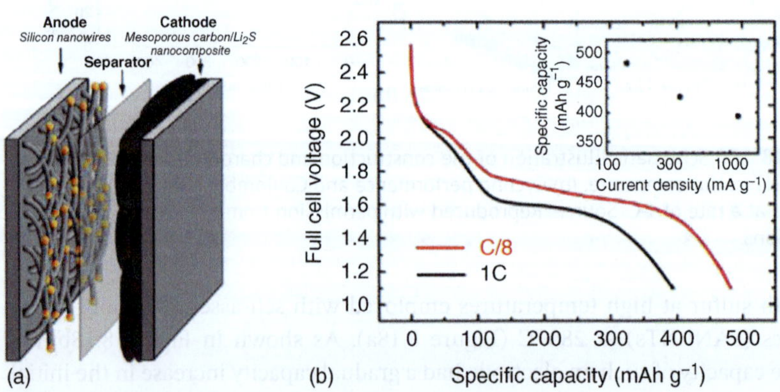

Figure 8.19 (a) Schematic illustration of a CMK-3 mesoporous carbon-embedded Li_2S/silicon nanowire battery; (b) first galvanostatic discharge curves of the full cell at various rates. The inset in (b) is a plot of the first discharge specific capacity of full batteries operating at different current densities. Source: Reproduced with permission from Yang et al. [32]/American Chemical Society.

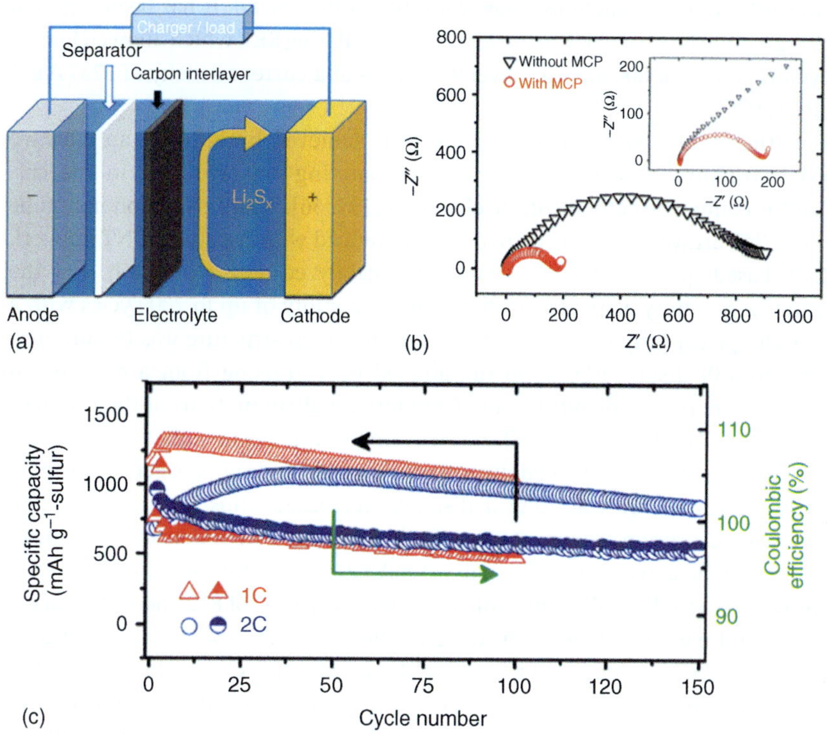

Figure 8.20 (a) Schematic illustration of a Li–S cell with a bifunctional microporous carbon interlayer inserted between the sulfur cathode and the separator; (b) electrochemical impedance spectroscopy plots (EIS) of Li–S batteries with/without microporous carbon paper (MCP); (c) cycling performance and Coulombic efficiency (CE) of the batteries with MCP at 1 and 2 C. Source: Reproduced with permission from Su and Manthiram [33]/Springer Nature.

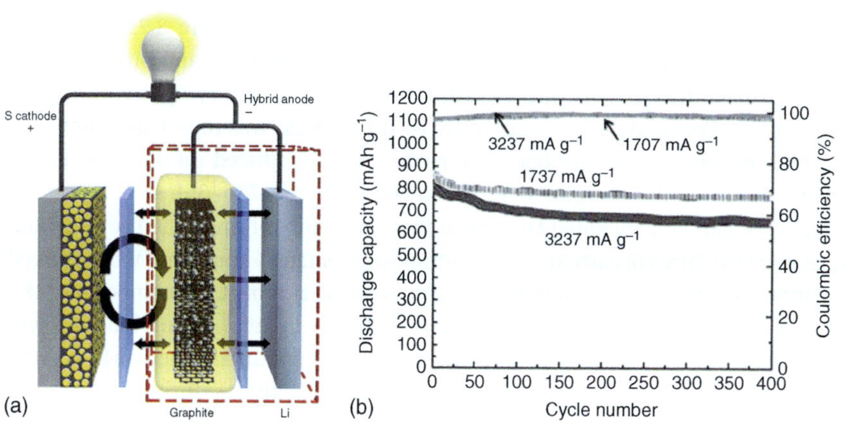

Figure 8.21 (a) Schematic illustration of the hybrid anode designed to manipulate the surface reactions in Li–S cells; (b) cycling performance and Coulombic efficiencies of hybrid Li–S batteries. Source: Reproduced with permission from Huang et al. [34]/Springer Nature.

was to shift the redox reactions away from the lithium metal. It has been reported that the graphite layer acted as an artificial self-regulating SEI, which allowed for stable cycling results of 800 mAh g^{-1} over 400 cycles at a current density of 1737 A g^{-1} (Figure 8.21b).

In another study, good cycling stability was achieved by utilizing an effective material structural design rather than incorporating any additives in the carbon/sulfur composite cathode or the electrolyte [35, 36]. A sulfur/carbon nanotube array (S/CNT array) composite cathode was produced within a stable CNT network by CVD-based growth of CNT arrays on a catalyzed substrate, such that the resulting Li–S battery achieved electrochemical stability of up to 200 cycles with a high discharge capacity retention. The aim of this microstructure was to minimize the wall-to-wall distance between the aligned CNTs arising from a decrease in the reaction energy of the adsorption. Computational simulations at the interface between the polysulfides and CNT surface (single-wall CNTs suffice to prove the principle) were carried out to confirm the improved binding between C and S in the polysulfides as the wall-to-wall distance decreased, thus favoring retention and surface adsorption (Figure 8.22a–e)[36]. By using a high-density CNT array, an initial specific capacity of 1340 and 831 mAh g^{-1} (of sulfur and electrode mass, respectively) was achieved, along with a discharge capacity of 812 and 503 mAh g^{-1} (of sulfur and electrode mass, respectively) after the 200th cycle, with a capacity decay of only 0.054% per cycle (Figure 8.22f)[36]. This good electrochemical performance could be attributed to the presence of an array of relatively high-density CNT scaffolds with an orderly structure (Figure 8.22g) that can adsorb the polysulfides and accommodate the volume expansion of sulfur even within the basic organic ether electrolyte (without any of the normal $LiNO_3$ additive), which is known for its high sulfur solubility. Thus, effective sulfur confinement is a potentially rewarding approach for improving cycle stability.

Further improvements were obtained by improving the wall quality of the CNTs grown as <7 nm CNT layers on a thin layer (<3 nm) of a transition-metal catalyst coated within a 3D porous scaffold (see SEM images in Figure 8.23a) made of metallic aluminum foam [36]. The cathode does not require any binders or an additional metallic current collector; thus, the electrode is very light with a good conducting network. A Li–S battery made by this process was reported to provide a threefold increase in the number of cycles to >600, a reduced internal resistance, and an increase in the charging rate, as shown in Figure 8.23b [36].

Overall, it can be concluded that given the availability of diversified sources and adjustable structures of carbon, carbon composites (such as graphene/MOF-derived microporous carbon composite [37]), and amorphous carbon/CNT composites [38], the opportunities for developing composites that can host sulfur have grown. Such carbon composite materials provide an efficient conductive network, prevent or minimize the dissolution of sulfur and lithium polysulfides during cycling, and accommodate the large volume changes that can help to ensure structural stability. The development of facile and large-scale synthesis routes to prepare homogeneous cathodes based on sulfur and carbon composites is of immense importance for the future commercial production of lithium–sulfur batteries.

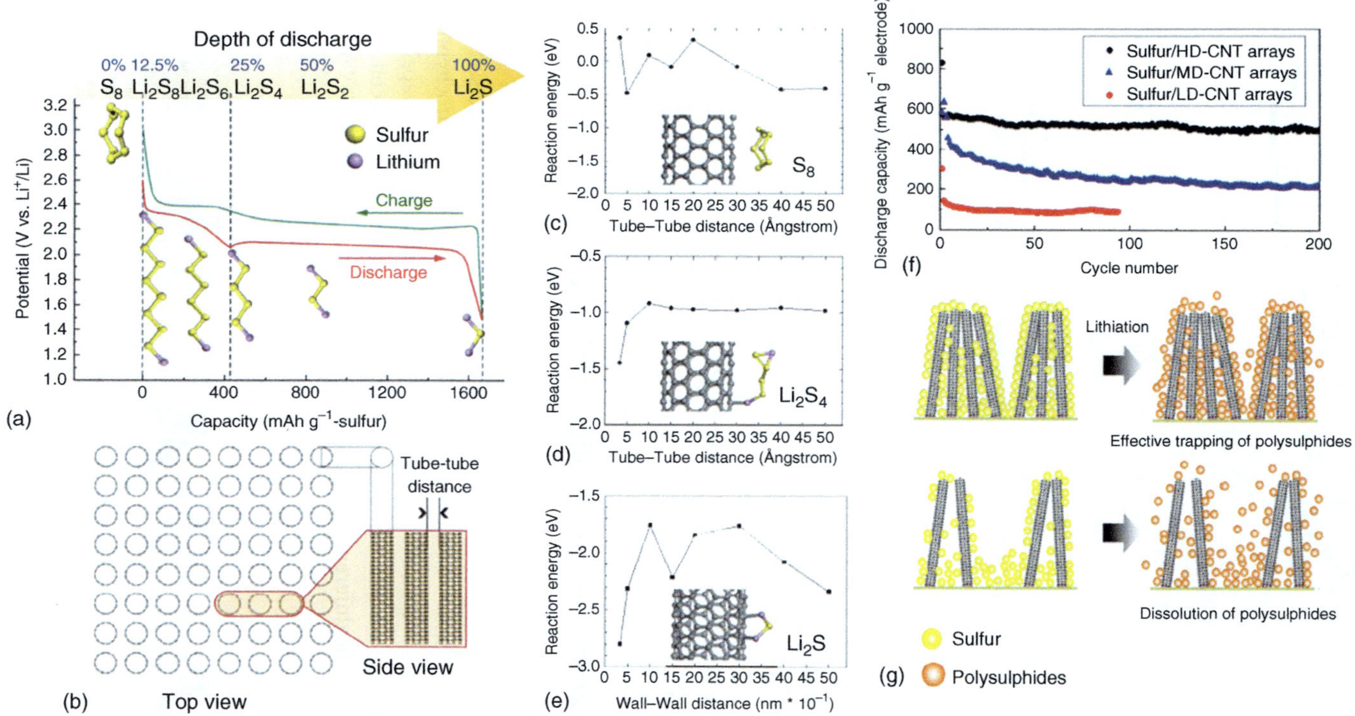

Figure 8.22 Schematic of the lithiation process in sulfur/CNT array composites. (a) The discharge–charge voltage profile and chemistry of sulfur cathode in the organic electrolyte. (b) Carbon nanotube arrays used in the simulation. S_8(c), Li_2S_4 (d), Li_2S (e) adsorptions on (6, 6) single-wall nanotube array. (f) The cycle performance of the electrode at a specific current of 400 mA g^{-1} in the electrolyte ($LiN(CF_3SO_2)_2$/DOL:DME = 1 : 1(v/v)). (g) Schematic of the lithiation process in sulfur/CNT array composites. Source: Reproduced with permission from Xi et al. [35]/Elsevier.

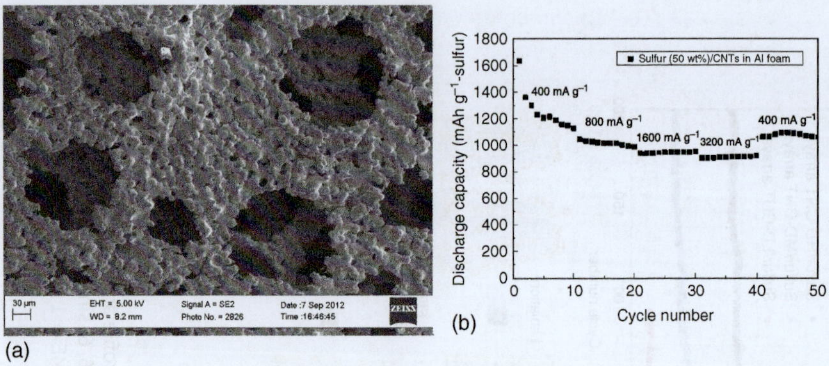

Figure 8.23 SEM images of Al foam (a) which can directly support Fe catalyst for growing CNT and infiltrating S; (b) Good rate capability demonstrated. Source: Reproduced with permission from Zhang et al. [36].

Ball-mill technology, which is widely used for grinding materials in industrial applications, has advantages for large-scale synthesis of sulfur/carbon composite with high consistency [38, 39]. In addition, chemical adsorption, which offers a stronger interaction than physical adsorption, can be used to restrict the dissolution of polysulfides from sulfur cathodes. Therefore, nitrogen, sulfur co-doped (N, S-*co*-doped) sponge-like graphene [40] and B_2O_3/carbon microtube composite [41] have been put forward as electroactive interlayers for Li–S batteries to control the diffusion of polysulfides.

8.9 Key Parameters for Application Prospects

Great endeavors have been made to develop high-energy Li–S batteries with a prime focus on high capacity, high sulfur utilization, and high cathode stability. However, most lab-scale research use coin-type cells to test the electrochemical performance. The comparable performances, such as cycle life, rate capability, and specific energy, are difficult to be replicated at the prototype pouch cell level. Figure 8.24 shows the differences in mass distribution of components in Li–S pouch

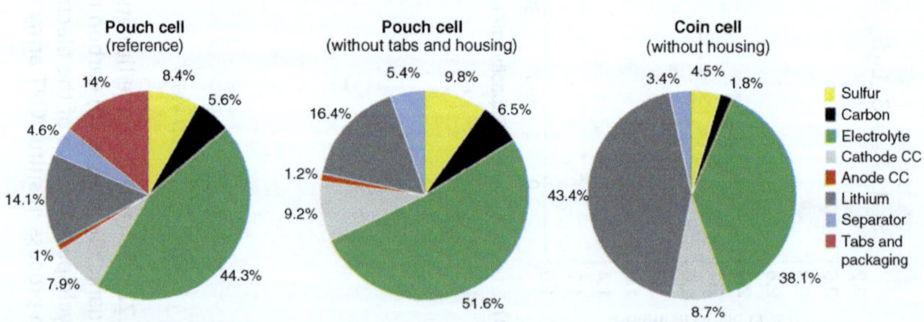

Figure 8.24 Mass distribution of Li–S pouch and coin cell. Source: Reproduced with permission from Dörfler et al. [42]/Elsevier.

and coin cells [42]. Therefore, commercial battery formats, such as a multilayer pouch, cylindrical, and prismatic-hardcase cells, should be used to help more realistic prospects for translating academic research results into future commercial viability [42, 43]. To commercialize Li–S batteries, S loading (S areal loading in $mg\,cm^{-2}$), content, utilization, N/P (the areal capacity ratio of the negative and positive electrodes), and the electrolyte/sulfur (E/S) ratio are key parameters defining the energy densities of pouch cells (Figure 8.25) [42–44]. Accordingly, future research and commercial efforts should aim at understanding the degradation phenomena based on using application-relevant parameters, including high sulfur loading, lowering of excess Li, and lean electrolyte volume. Some crucial metrics for developing and commercializing high-energy Li–S batteries are presented in the following sections.

8.9.1 Functionalized Cathode Materials

The employment of various carbon-based hosts helps in providing impressive gravimetric energy density with enhanced capacity retention and cyclability. But these composite cathodes show a relatively limited volumetric energy density due to the low tap density of nanocarbon (typically $<0.6\,g\,cm^{-3}$), while the mass density of bulk sulfur (approximately $2.07\,g\,cm^{-3}$) is less than half of typical metal oxide cathodes in LIBs (e.g. $5.1\,g\,cm^{-3}$ for $LiCoO_2$). Therefore, building a compact cathode with high sulfur loading is preferable for better volumetric performance by minimizing the nanocarbon content and/or employing a conductive, high-mass-density host in the cathode (Figure 8.25 and 8.26).

Various metal-based compounds, such as oxides, sulfides, nitrides, and carbides, have been developed as high-mass-density hosts for Li–S cells [45–48]. In addition, some metal chalcogenide hosts could contribute extra capacity because they can react with Li at sulfur's electrochemical window (1.7–2.8 V, vs. Li/Li^+) (Figure 8.26b) [46–48]. For example, an FeS_2/FeS/S composite [47] with a high tap density of $2.04\,g\,cm^{-3}$ was developed by a scalable ball-milling route (Figure 8.26c). As a cathode material in Li–S batteries, this composite electrode demonstrated both a high reversible capacity of $1044.7\,mAh\,g^{-1}$ and a volumetric capacity of $2131.1\,Ah\,l^{-1}$ after 30 cycles (Figure 8.26d). The *in situ* formed FeS/FeS_2 contributed to the extra capacity (Figure 8.26e). FeS/FeS_2 also provides binding sites to anchor polysulfides and promote electrocatalytic activity for sulfur redox reactions. This work offers new insights into the development of high volumetric energy density Li–S batteries based on multifunctional sulfur cathodes.

A combined intercalation–conversion hybrid cathode has been synthesized by incorporating the intercalating Mo_6S_8 with conversion sulfur [48]. The Mo_6S_8/sulfur composite cathode is composed of 42.25 wt% Mo_6S_8 and 42.75 wt% sulfur, lowering the ratio of carbon to 10 wt% and binder to 5 wt%. The pouch cell, based on the hybrid cathode with approximately 200% excess of Li metal anode, delivered a gravimetric energy density of $366\,Wh\,kg^{-1}$ and a volumetric energy density of $581\,Wh\,l^{-1}$. The Chevrel-phase Mo_6S_8 exhibited extra capacity due to lithium intercalation reactions. These reports provide effective strategies to solve the

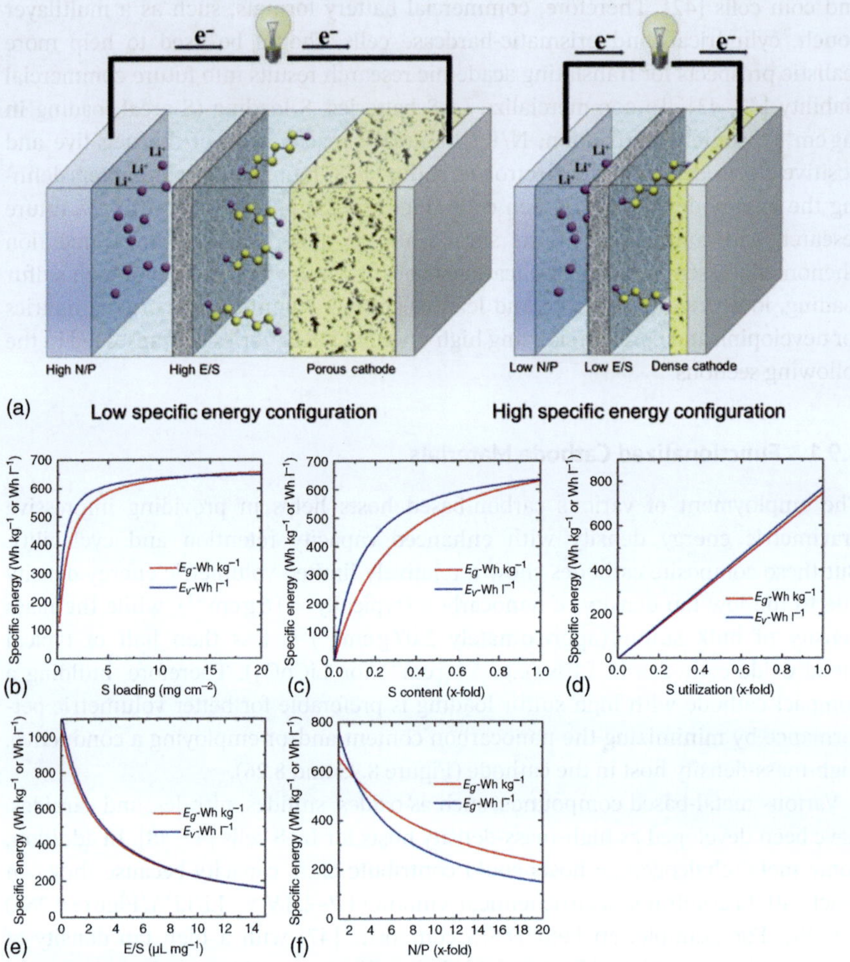

Figure 8.25 (a) Schematic showing the main issues leading to a low-energy-density configuration and the key changes to overcome these limitations toward a high-energy-density battery. Source: Reproduced with permission from Bhargav et al. [44]/Elsevier. Predicted gravimetric and volumetric energy. (b) S loading, (c) S content, (d) S utilization, (e) E/S, (f) N/P. Each graph is plotted as a function of one parameter by fixing the other four parameters. S loading: 6 mg cm^{-2}, S content: 0.8, S utilization: 0.8, E/S ratio: 2 μl mg^{-1}, N/P ratio: 2. Source: (b, c, d, e) Reproduced with permission from Robinson et al. [43]/IOP Publishing.

major problems, such as the use of very high carbon content, cathode porosity, E/S, and N/P. Other dense metal-based compounds as sulfur hosts are worthy of further examination to develop systems that can provide both high gravimetric and volumetric energy densities in Li–S batteries.

8.9.2 Redox Conversion Catalysts

To propel the interfacial reaction kinetics and control polysulfide shuttling, sulfur host materials need to have high electronic conductivity and sulfophilic surface

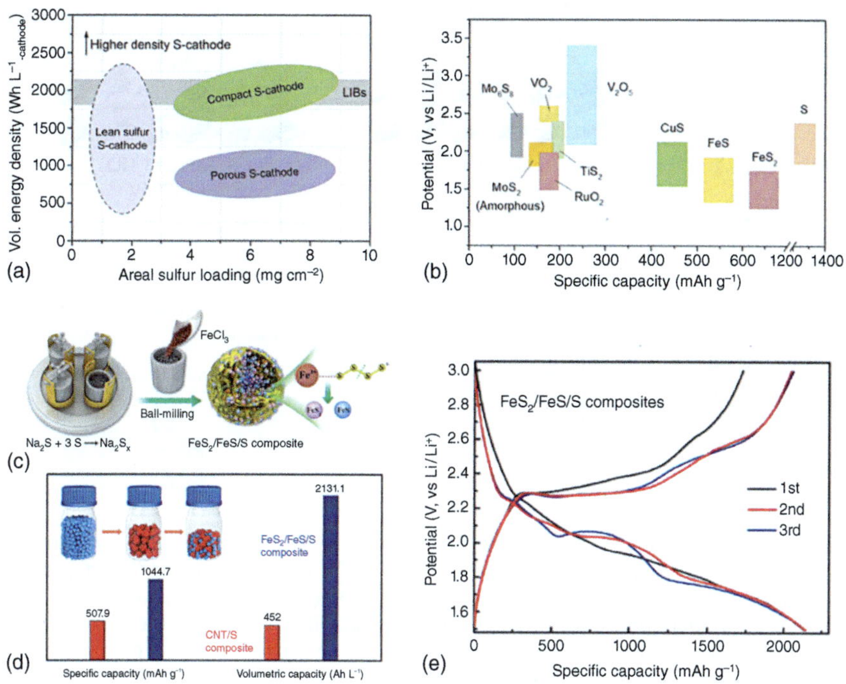

Figure 8.26 (a) Projected volumetric energy density of various S-cathodes as a function of areal sulfur loading. Source: Reproduced with permission from Lin et al. [45]/John Wiley & Sons. (b) Potential (vs. Li/Li^+) vs. specific capacity of several active metal oxides and sulfides. Source: Reproduced with permission from Liu et al. [46]/John Wiley & Sons. (c) Synthesis schematic of FeS_2/FeS/S composites; (d) Comparisons of specific and volumetric capacity between FeS_2/FeS/S and CNT/S composites; (e) Discharge and charge curves of FeS_2/FeS/S composites. Source: (c, d and e) Reproduced with permission from Xi et al. [47]/John Wiley & Sons.

properties. Current research has emerged with focus on a lean electrolyte battery, as the electrolyte/sulfur (E/S) ratio strongly affects the practical energy density. Under lean electrolyte conditions (E/S < 3 μl mg^{-1}), the Li–S battery requires highly efficient electrocatalysts to promote the relatively sluggish redox kinetics.

As shown in Figure 8.27a–e, the improved electrocatalytic activity can be achieved by introducing oxygen vacancies or forming the heterojunction in an electrocatalyst for fast polysulfide redox kinetics. For example, through a defect-rich CNT@TiO_{2-x} heterojunction electrocatalyst [49], the inherent electron band structure of TiO_2 can be altered, and thus enable enhanced conductivity, stronger binding with polysulfides, and improved electrocatalytic activity for fast redox reactions. The surface vacancies present in TiO_{2-x} nanosheets improve carrier densities and create higher activity-trapping sites for polysulfide adsorption. The special heterostructure of CNT@TiO_{2-x} also helps to achieve favorable catalytic kinetics and low reaction energy barriers for efficient surface electrocatalytic reactions. Adopting data from single-atom metal electrocatalysis is also an effective strategy for optimizing the catalytic behaviors in Li-S chemistry. The electronic structure calculations have proven that the d-p orbital hybridization between the single-atom metal

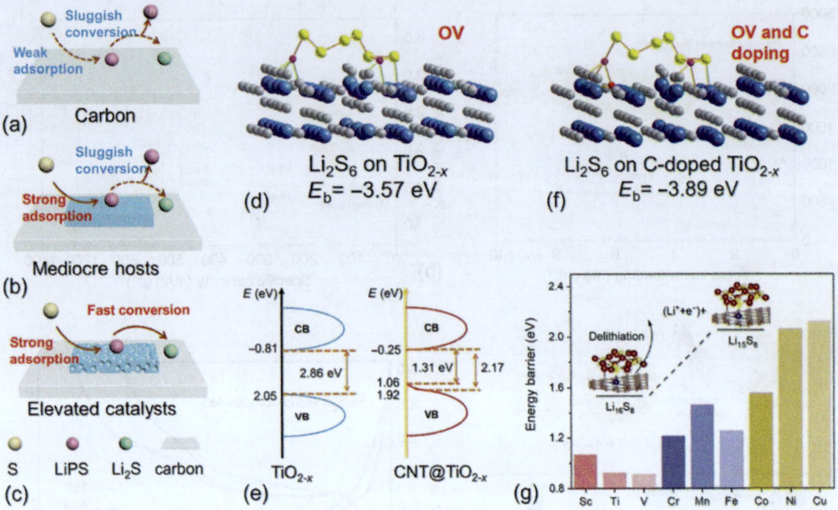

Figure 8.27 (a–c) Illustration of the reaction process of lithium polysulfides on various host surfaces; Optimized geometries of (d) Li_2S_6 on TiO_{2-x} (110) surface and (f) C-doped TiO_{2-x} (110) surface; (e) band diagram for TiO_{2-x} and CNT@TiO_{2-x}. Source: (a–e) Reproduced with permission from Wang et al. [49]/John Wiley & Sons. (g) The delithiation energy barrier during the $(Li_2S)_8$ oxidation process. Source: Reproduced with permission from Han et al. [50]/John Wiley & Sons.

and the sulfur species can be adjusted to accelerate the polysulfide redox kinetics (Figure 8.27f) [50]. Among various SAC (single-atom Mn, Cu, Cr, and Ti)-embedded 3D carbon foams (SACs@CFs), the SATi@CF has proven to possess the lowest electrochemical barrier to polysulfide reduction in discharge and Li_2S oxidation during charge, resulting in the highest electrocatalytic activity to promote battery performance. These results are mainly attributed to transition metals with lower atomic numbers that have fewer filled antibonding states and more effective d-p orbital hybridization, hence enabling effective binding to polysulfides or Li_2S and thus lowering the energy barrier for the redox reaction.

Constructing smart cathode architectures is adopted to accelerate redox kinetics based on approaches, such as materials engineering [51], designing the bidirectional electrocatalyst, and oxide–sulfide heterostructure [52], and lowering the activation energy in the multistep sulfur reduction reaction by use of nitrogen and sulfur dual-doped graphene [53].

8.9.3 Lithium Metal Anode

The lithium metal anode is one of the bottlenecks for practical applications of Li–S cells due to the unfavorable side reactions with the electrolyte and the uncontrollable formation of Li dendrites [54–58]. The unstable SEI layers on the lithium metal surface cause severe electrolyte decomposition. As shown in Figure 8.28a, the uncontrollable growth of dendrites stems from the uneven electrodeposition of Li, which decreases the Coulombic/energy efficiency and induces safety concerns

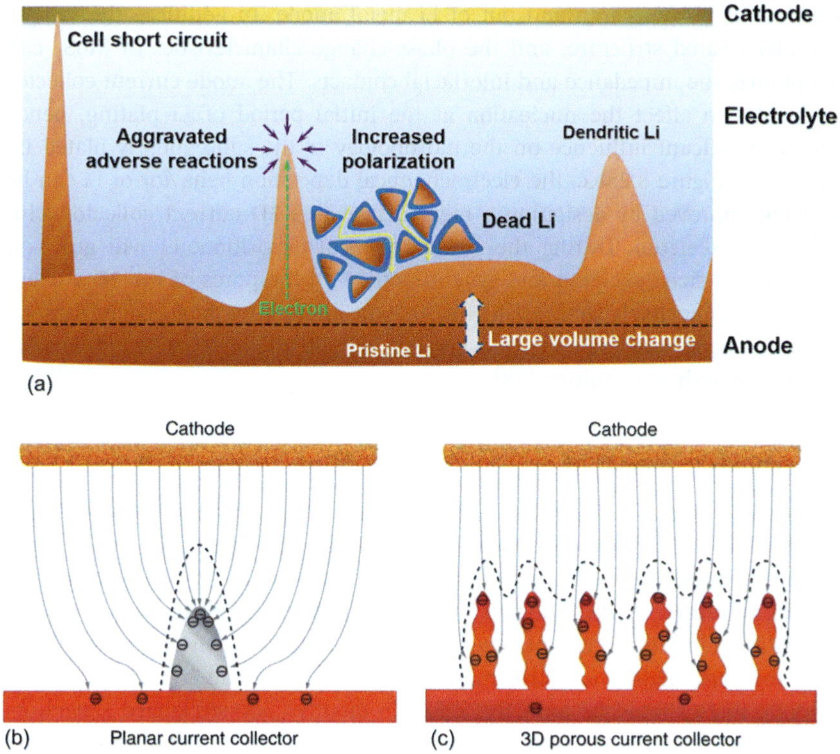

Figure 8.28 (a) Scheme of dilemma for Li metal anode in Li-ion batteries. Source: Reproduced with permission from Cheng et al. [54]/American Chemical Society. Illustration of the proposed electrochemical deposition processes of Li on (b) planar current collector and (c) 3D current collector. Source: (b and c) Reproduced with permission from Yang et al. [58]/Springer Nature.

due to short-circuiting between the two electrodes. In addition, the continuous consumption of both the metallic Li and electrolyte leads to the rapid degradation of the electrochemical performance. Approaches to stabilize the electrolyte/Li interface and control electrochemical deposition of Li should be particularly focused on high current densities and especially in a polysulfide-containing environment. Recent research shows titration gas chromatography can be adopted to quantify the contribution of unreacted Li metal to the total amount of inactive lithium, making lithium plating and stripping more efficient [56]. With the development of stable lithium metal anodes, a lowering of Li excess (N/P <2) can be achieved in the future for the practical application of high-energy Li–S cells.

Creating an artificial protective layer between the Li metal and electrolyte is an effective strategy for stabilizing the Li anode, for example, by adding an atomic 2D layer of MoS_2 onto the surface of Li metal as a protective layer to stabilize Li electrodeposition [54, 57]. Therefore, homogeneous nucleation sites can be achieved (Figure 8.28a). The layered MoS_2 structure with intercalated Li atoms during cycling can reduce the interfacial resistance and facilitate the stable and rapid flow of Li^+ ions into and out of Li metal anode. In addition, the unique atomically layered structure, and the phase-change characteristics of MoS_2 can also optimize the impedance and interfacial contacts. The anode current collector also enables to affect the nucleation at the initial period of Li plating, hence having a significant influence on the morphology of the subsequently plated Li. As shown in Figure 8.28b,c, the electrochemical deposition behavior of Li can be effectively improved by designing a high-surface-area 3D current collector with a submicron skeleton. During the electrochemical deposition, Li will grow on the submicron-sized skeleton and synchronously fill the pores of the 3D current collector, thus effectively suppressing the uncontrollable growth of Li dendrites. As a result, Li–metal batteries can achieve a long cell life and high safety without the consequence of short circuiting [58].

8.9.4 Modified Separators

The separator is an indispensable part of a battery design. It directly contacts and interacts with the cathode and the anode, while facilitating electrolyte and ionic transport between the two. LIBs still use polymer separators, based on polypropylene (PP), polyethylene (PE), and polytetrafluoroethylene (PTFE) alone or in various combinations. But these commercialized separators allow polysulfides to easily diffuse through, leading to active material loss and Li anode corrosion [59–62]. Therefore, there should be an active effort to modify commercialized separators or develop new materials as novel separators to overcome the challenges of the Li–S chemistry flow between the electrodes. As reported by Zhang et al [59], an ideal separator material should fulfill the following conditions – (i) excellent Li-ion transport capacity, especially at high sulfur loading in a cathode; (ii) effectively anchor polysulfides; and (iii) accelerate the catalytic conversion of polysulfides.

A composite material of MOF@GO is prepared as a separator for Li–S batteries [60], which can act as an ion sieve, effectively limiting the polysulfide shuttle to improve sulfur utilization, and MOF groups can improve the electrochemical stability of the battery. The assembled Li–S cell has a capacity decay rate of 0.019% per cycle in 1500 cycles. In another innovative approach, bimetallic carbides Co_3Mo_3C made via simple pyrolysis of metallic precursors and dicyandiamide mixture were used [59]. The results have shown that the bimetallic carbide could greatly promote the kinetics of the polysulfide reaction, which is beneficial for using a high-loading sulfur cathode; thus, the battery could reach an areal capacity of 6.8 mAh cm^{-2} at a sulfur loading of 8.0 mg cm^{-2}. Thus, the separator modification not only effectively limits the shuttle of polysulfides, but also improves the catalytic reaction kinetics of polysulfides, enhancing the utilization of the active material and the energy density of the battery [60].

In a recent study, a single-step hydrothermal method was explored to prepare 2D hexagonal VS_2 flakes, which were used to modify both sides of the PP separator (D-HVS@PP) without blocking the membrane pores [61]. The “sulfophilic” and “lithiophilic” properties of HVS effectively suppressed the shuttle effect and promoted the uniform nucleation and growth of lithium. At a high sulfur loading of

9.24 mg cm^{-2}, a low E/S of 4 µl mg^{-1} and a low N/P of 1.7, an initial specific capacity of up to 8.3 mAh cm^{-2} was achieved, which is over 16 times higher than using a PP separator under the same conditions. This remarkable performance was achieved just at the expense of an 11% increase in the separator weight (0.14 mg cm^{-2}), accounting for only 0.48% of the basic units of the Li–S battery [62].

The design of integrated flexible electrodes has attracted ever-increasing attention. Compared with the traditional discrete stacked batteries, the integrated design favors combining the advantages of various functional units and materials. In addition, it can also reduce the overall volume and mass of cells, which is conducive to the realization of high-energy-density batteries. As shown in Figure 8.29a,b, an integrated three-in-one flexible fibrous membrane [62] was prepared by electrospinning and base-coating techniques. Three functional units, including the cathode, interlayer, and separator, were integrated into one efficient composite. This structure limits the polysulfide shuttle and improves the sulfur utilization,

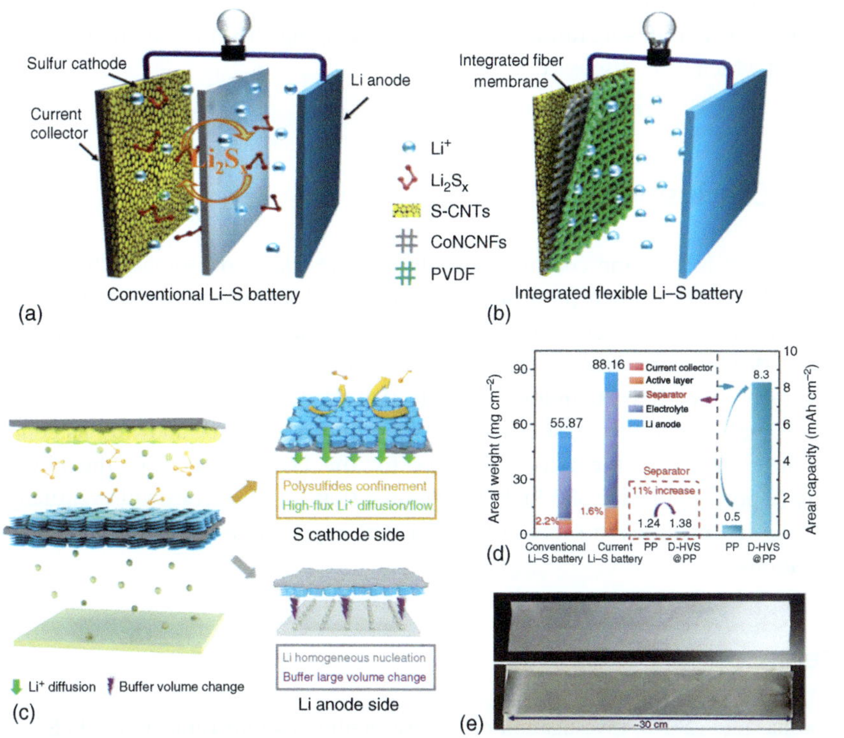

Figure 8.29 (a, b) Schematic illustrations of the working principle of the conventional Li–S battery system and integrated flexible Li–S cell assembling by S-CNTs/CoNCNFs/PVDF fibrous membrane. Source: (a and b) Reproduced with permission from Wang et al. [62]/John Wiley & Sons. (c) Schematic illustrations of the working principle of the Li–S battery with 2D hexagonal VS_2-nanotower-modified separator (D-HVS@PP separators); (d) Comparison of both the areal weight and areal capacity of PP and D-HVS@PP separators; (e) Large-scale fabrication of the modified separator. Source: (c–e) Reproduced with permission from Wang et al. [61]/American Chemical Society.

Li-ion diffusion, and electron transfer, as described schematically in Figure 8.29c,d. A Li–S battery assembled with the integrated membrane achieved a high initial capacity of 1501 mAh g^{-1} at 0.2 C. In addition, a high area capacity of 11.4 mAh cm^{-2} was obtained under the condition of high sulfur loading of 13.2 mg cm^{-2} and low E/S of 6 μl mg^{-1}. Figure 8.29e depicts a fabricated pouch cell [61].

According to calculations, the excess electrolyte and lithium anode significantly damped the energy density of Li–S pouch cells, while the sulfur loading over 10 mg cm^{-2} shows only a minor impact. Considering the poor charge transport inside a thick electrode, a moderate sulfur loading value of 5–10 mg cm^{-2} is more reasonable than maximizing sulfur loading exceeding 10 mg cm^{-2} in a practical Li–S cell. The British Faraday Institute's LiSTAR project proposed that a comparable volumetric energy density of about 600 Wh l^{-1} to state-of-the-art LIBs could be achieved by employing the sulfur loading of 6 mg cm^{-2}, at a sulfur content of 80%, S utilization of 80%, E/S of 2 μl mg^{-1}, and an N/P of 1.5 (50% excess of Li). Specific energy of 589 Wh kg^{-1} can be achieved, which is about two to three times that of commercial LIBs [43].

In addition, a promising future direction is developing all-solid-state Li batteries, aspects of which were discussed in Chapter 7.

8.10 Conclusions

The Li–S cell is awaiting commercialization, while trials and some niche applications have shown much promise. In conclusion, it should be noted that many companies and researchers rejected the idea of cobalt oxide as an electrode in the late 1980s before Asahi Chemicals and Sony, both in Japan, pioneered the Li-ion revolution. It is, therefore, possible that pioneering groups in Li-battery research may devote their research toward the Li–S cell, hence allowing its commercialization. Given the price advantage for the Li–S battery over the Li-ion battery (a potential factor of 0.5), it is suggested that Li–S batteries will eventually replace Li-ion batteries in some of their current applications soon and open new applications.

References

1 Iwona, L.R. (2010). Investigation of microstructure of sulfur cathode of a Li-S battery. PhD Thesis. University of Cambridge.

2 Kumar, R.V. (2006). Future prospects of lithium batteries, Proc. of IX International Conference on Fundamental problems of energy conversion in lithium electrochemical systems, *Ufa, Russian Academy of Sciences*, 222.

3 Mauler, L., Duffner, F., Zeier, W.G., and Leker, J. (2021). Battery cost forecasting: a review of methods and results with an outlook to 2050. *Energy Environ. Sci.* 14: 4712–4739.

4 James, R.A., Yuriy, V.M., and Neal, W. (2004). Li/S fundamental chemistry and application to high performance rechargeable batteries. *Solid State Ionics* 175: 243–245.

5 Cheon, S.E., Choi, S.S., Han, J.S. et al. (2004). Capacity fading mechanisms on cycling a high-capacity secondary sulfur cathode. *J. Electrochem. Soc.* 151 (12): A2067–A2073.

6 Kolosnitsyn, V.S., Karaseva, E.V., Seung, D.Y., and Cho, M.D. (2002). Cycling a sulfur electrode in mixed electrolytes based on sulfolane: effect of ethers. *Russ. J. Electrochem.* 38 (5): 1314–1318.

7 Kolosnitsyn, V.S., Karaseva, E.V., Seung, D.Y., and Cho, M.D. (2003). Cycling a sulfur electrode: effect of physicochemical properties of electrolyte system. *Russ. J. Electrochem.* 39: 1089–1093.

8 Su, Y.S., Fu, Y., Cochell, T., and Manthiram, A. (2013). A strategic approach to recharging Li-S batteries for long cycle life. *Nat. Commun.* 4: 2985.

9 Yu, X.G., Xie, J.Y., Yang, J., and Wang, K. (2004). All solid-state rechargeable lithium cells based on nano-sulfur composite cathodes. *J. Power Sources* 132: 181–186.

10 Wang, J., Chew, S.Y., Zhao, Z.W. et al. (2008). Sulfur-mesoporous carbon composite in conjunction with a novel ionic liquid electrolyte for lithium rechargeable batteries. *Carbon* 46: 229–235.

11 Bruce, P.G., Scrosati, B., and Tarascon, J.M. (2008). Nanomaterials for rechargeable lithium batteries. *Angew. Chem. Int. Ed.* 47: 2930–2946.

12 Ji, X. and Nazar, L.F. (2010). Advances in Li-S batteries. *J. Mater. Chem.* 20: 9821–9826.

13 Bruce, P.G., Freunberger, S.A., Hardwick, L.J., and Tarascon, J.M. (2012). Li-O_2 and Li-S batteries with high energy storage. *Nat. Mater.* 11: 19–29.

14 Ji, X., Lee, K.T., and Nazar, L.F. (2009). A highly ordered nanostructured carbon-sulfur cathode for lithium-sulfur batteries. *Nat. Mater.* 8: 500–506.

15 Wang, H., Yang, Y., Liang, Y. et al. (2011). Graphene-wrapped sulfur particles as a rechargeable lithium-sulfur battery cathode material with high capacity and cycling stability. *Nano Lett.* 11 (7): 2644–2647.

16 Manthiram, A., Fu, Y., and Su, Y. (2013). Challenges and prospects of lithium–sulfur batteries. *Acc. Chem. Res.* 46 (5): 1125–1134.

17 Hinckley, A. (2012). Improvements in Li-S batteries. MPhil thesis. University of Cambridge.

18 Jayaprakash, N., Shen, J., Moganty, S.S. et al. (2011). Porous hollow carbon@sulfur composites for high-power lithium–sulfur batteries. *Angew. Chem. Int. Ed.* 50: 5904–5908.

19 Xin, S., Gu, L., Zhao, N.H. et al. (2012). Smaller sulfur molecules promise better lithium–sulfur batteries. *J. Am. Chem. Soc.* 134: 18510–18513.

20 Zhang, B., Qin, X., Li, G.R., and Gao, X.P. (2010). Enhancement of long stability of sulfur cathode by encapsulating sulfur into micropores of carbon spheres. *Energy Environ. Sci.* 3: 1531–1537.

21 Ji, L., Rao, M., Zheng, H. et al. (2011). Graphene oxide as a sulfur immobilizer in high performance lithium/sulfur cells. *J. Am. Chem. Soc.* 133: 18522–18525.

22 Wang, Y.X., Huang, L., Sun, L.C. et al. (2012). Facile synthesis of a interleaved expanded graphite-embedded sulphur nanocomposite as cathode of Li–S batteries with excellent lithium storage performance. *J. Mater. Chem.* 22: 4744–4750.

23 Evers, S. and Nazar, L.F. (2012). Graphene-enveloped sulfur in a one pot reaction: a cathode with good coulombic efficiency and high practical sulfur content. *Chem. Commun.* 48 (9): 1233–1235.

24 Xi, K., Kidambi, P., Hoffman, S. et al. (2014). Binder free three-dimensional sulfur/few-layer graphene foam cathode with enhanced high-rate capability for rechargeable lithium sulfur batteries. *Nanoscale* 6 (11): 5746–5781.

25 Su, Y. and Manthiram, A. (2012). A new approach to improve cycle performance of rechargeable lithium–sulfur batteries by inserting a free-standing MWCNT interlayer. *Chem. Commun.* 48: 8817.

26 Dorfler, S., Hagen, M., Althues, H. et al. (2012). High capacity vertical aligned carbon nanotube/sulfur composite cathodes for lithium–sulfur batteries. *Chem. Commun.* 48: 4097–4099.

27 She, Z.W., Li, W., Cha, J.J. et al. (2013). Sulphur–TiO_2 yolk–shell nanoarchitecture with internal void space for long-cycle lithium–sulphur batteries. *Nat. Commun.* 4: 1331.

28 Ji, X., Lee, K.T., and Nazar, L.F. (2009). A highly ordered nanostructured carbon-sulphur cathode for lithium-sulphur batteries. *Nat. Mater.* 8: 500–506.

29 Ji, X., Evers, S., Black, R., and Nazar, L.F. (2011). Stabilizing lithium-sulphur cathodes using polysulphide reservoirs. *Nat. Commun.* 2: 235–241.

30 Xi, K., Cao, S., Peng, X. et al. (2013). Carbon with hierarchical pores from carbonized metal organic frameworks for lithium-sulfur batteries. *Chem. Commun.* 49 (22): 2192–2194.

31 Xiao, L., Cao, Y., Xiao, J. et al. (2012). A soft approach to encapsulate sulfur: polyaniline nanotubes for lithium-sulfur batteries with long cycle life. *Adv. Mater.* 24: 1176–1181.

32 Yang, Y., McDowell, M.T., Jackson, A. et al. (2010). New nanostructured Li_2S/Silicon rechargeable battery with high specific energy. *Nano Lett.* 10: 1486–1491.

33 Su, Y.S. and Manthiram, A. (2012). Lithium-sulfur batteries with a microporous carbon paper as a bifunctional interlayer. *Nat. Commun.* 3: 1166.

34 Huang, C., Xiao, J., Shao, Y. et al. (2014). Manipulating surface reactions in lithium–sulphur batteries using hybrid anode structures. *Nat. Commun.* 5: 3015.

35 Xi, K., Chen, B., Li, L. et al. (2015). Soluble polysulphide sorption using carbon nanotube forest for enhancing cycle performance in a lithium–sulphur battery. *Nano Energy* 12: 538–546.

36 Zhang, C., Xi, K., Chen, B. et al. (2016). Product and process in energy storage structures. UK Patent Application: GB 1301639.9, Filed on 31 Jan 2013; PCT/GB2014/050253, filed on 31 Jan 2014.

37 Chen, R., Zhao, T., Tian, T. et al. (2014). Graphene-wrapped sulfur/metal organic framework (MOF)-derived microporous carbon composite for lithium sulfur batteries. *APL Mater.* 2 (12): 124109.

38 Lai, C., Wu, Z., Gu, X. et al. (2016). Reinforced conductive confinement of sulfur for robust and high performance lithium-sulfur batteries. *ACS Appl. Mater. Interfaces* 7 (43): 23885–23892.

39 Su, Z., Gao, C., Li, H. et al. (2014). Graphene wrapped sulfur-based composite cathodes: ball-milling synthesis and high discharge capacity. *RSC Adv.* 4: 48438–48442.

40 Xing, L., Xi, K., Li, Q. et al. (2016). Nitrogen, sulfur-co-doped graphene sponge as electroactive carbon interlayer for high-energy and -power lithium-sulfur batteries. *J. Power Sources* 303: 22–28.

41 Su, Z., Tong, C., He, D. et al. (2016). Ultra-small B_2O_3 nanocrystals grown in situ on highly porous carbon microtubes for lithium-iodine and lithium-sulfur batteries. *J. Mater. Chem. A* 4 (22): 8541–8547.

42 Dörfler, S., Althues, H., Härtel, P. et al. (2020). Challenges and key parameters of lithium-sulfur batteries on pouch cell level. *Joule* 4 (3): 539–554.

43 Robinson, J.B., Xi, K., Kumar, R.V. et al. (2021). 2021 roadmap on lithium sulfur batteries. *J. Phys. Energy* 3 (3): 031501.

44 Bhargav, A., He, J., Gupta, A., and Manthiram, A. (2020). Lithium-sulfur batteries: attaining the critical metrics. *Joule* 4 (2): 285–291.

45 Liu, Y.T., Liu, S., Li, G.R. et al. (2020). High volumetric energy density sulfur cathode with heavy and catalytic metal oxide host for lithium-sulfur battery. *Adv. Sci.* 7 (12): 1903693.

46 Liu, Y.T., Liu, S., Li, G.R., and Gao, X.P. (2021). Strategy of enhancing the volumetric energy density for lithium-sulfur batteries. *Adv. Mater.* 33 (8): 2003955.

47 Xi, K., He, D., Harris, C. et al. (2019). Enhanced sulfur transformation by multifunctional FeS_2/FeS/S composites for high-volumetric capacity cathodes in lithium-sulfur batteries. *Adv. Sci.* 6 (6): 1800815.

48 Xue, W., Shi, Z., Suo, L. et al. (2019). Intercalation-conversion hybrid cathodes enabling Li–S full-cell architectures with jointly superior gravimetric and volumetric energy densities. *Nat. Energy* 4 (5): 374–382.

49 Wang, Y., Zhang, R., Chen, J. et al. (2019). Enhancing catalytic activity of titanium oxide in lithium–sulfur batteries by band engineering. *Adv. Energy Mater.* 9 (24): 1900953.

50 Han, Z., Zhao, S., Xiao, J. et al. (2021). Engineering d-p orbital hybridization in single-atom metal-embedded three-dimensional electrodes for Li-S batteries. *Adv. Mater.* 2105947.

51 Pang, Q., Liang, X., Kwok, C.Y., and Nazar, L.F. (2016). Advances in lithium-sulfur batteries based on multifunctional cathodes and electrolytes. *Nat. Energy* 1 (9): 1–11.

52 Wang, R., Luo, C., Wang, T. et al. (2020). Bidirectional catalysts for liquid-solid redox conversion in lithium-sulfur batteries. *Adv. Mater.* 32 (32): 2000315.

53 Peng, L., Wei, Z., Wan, C. et al. (2020). A fundamental look at electrocatalytic sulfur reduction reaction. *Nat. Catal.* 3 (9): 762–770.

54 Cheng, X.B., Zhang, R., Zhao, C.Z., and Zhang, Q. (2017). Toward safe lithium metal anode in rechargeable batteries: a review. *Chem. Rev.* 117 (15): 10403–10473.

55 Xu, W., Wang, J., Ding, F. et al. (2014). Lithium metal anodes for rechargeable batteries. *Energy Environ. Sci.* 7 (2): 513–537.

56 Fang, C., Li, J., Zhang, M. et al. (2019). Quantifying inactive lithium in lithium metal batteries. *Nature* 572 (7770): 511–515.

57 Cha, E., Patel, M.D., Park, J. et al. (2018). 2D MoS_2 as an efficient protective layer for lithium metal anodes in high-performance Li-S batteries. *Nat. Nanotechnol.* 13 (4): 337–344.

58 Yang, C.P., Yin, Y.X., Zhang, S.F. et al. (2015). Accommodating lithium into 3D current collectors with a submicron skeleton towards long-life lithium metal anodes. *Nat. Commun.* 6: 8058.

59 Zhang, Z., Wang, J.N., Shao, A.H. et al. (2020). Recyclable cobalt-molybdenum bimetallic carbide modified separator boosts the polysulfide adsorption-catalysis of lithium sulfur battery. *Sci. China Mater.* 63 (12): 2443–2455.

60 Bai, S., Liu, X., Zhu, K. et al. (2016). Metal–organic framework-based separator for lithium–sulfur batteries. *Nat. Energy* 1 (7): 16094.

61 Wang, J., Yi, S., Liu, J. et al. (2020). Suppressing the shuttle effect and dendrite growth in lithium-sulfur batteries. *ACS Nano* 14 (8): 9819–9831.

62 Wang, J., Yang, G., Chen, J. et al. (2019). Flexible and high-loading lithium–sulfur batteries enabled by integrated three-in-one fibrous membranes. *Adv. Energy Mater.* 9 (38): 1902001.

9

Sodium-Ion Batteries

Pu Hu[1] *and Katerina E. Aifantis*[2]

[1] *Wuhan Institute of Technology, School of Materials Science and Engineering, LiuFang Campus, No. 206, Guanggu 1st road, Donghu New & High Technology Development Zone, Wuhan 430205, China*
[2] *University of Florida, Mechanical and Aerospace Engineering, 1064 Center Dr., Gainesville, FL 32611, USA*

9.1 Introduction

Lithium-ion batteries have been successfully employed in energy storage devices, such as portable electronics, hybrid electric vehicles, and electric vehicles, due to their high working voltage, high energy density, high energy storage and conversion efficiency, and long cycle life. The growing electric vehicle and the emerging electric equipment markets are constantly consuming lithium resources and exacerbating further increases in costs. The global consumption of lithium resources is expected to increase from 150 000 tons in 2012 to about 300 000 tons in 2022. The content of lithium used in lithium-ion batteries is increasing at an annual rate of about 25%. Limited resource reserves and increased costs have severely restricted the wide application of lithium-ion batteries in large-scale systems. Particularly, the reserves of lithium resources in the earth's crust are finite and unevenly distributed, which raises further concerns regarding its cost.

Sodium (Na) belongs to the same family as lithium and has similar physical and chemical properties. Therefore, the principles of operation for Na-ion batteries are similar to the well-known Li-ion. Unlike Li, though, Na, resources are widely distributed and rich in reserves. The use of sodium instead of lithium and the commercialization of sodium-ion batteries will greatly reduce costs and meet the growing market demand. Sodium-ion batteries can alleviate the shortage of lithium resources to a certain extent, and supplement the market applications of lithium-ion batteries, especially in large-scale energy storage grids. At the same time, sodium-ion batteries have a high energy storage conversion efficiency and long cycle life, which provide an opportunity to gradually replace the use of lead–acid batteries.

As mentioned, the working principle and components of a Na-ion battery are similar to those of a Li-ion battery. Na-ion batteries are mainly composed of the cathode, anode, electrolyte, separator, and current collectors. As shown in Figure 9.1, when charging, electrons flow through an external circuit from the cathode to the anode. As a result, sodium-ions are extracted from the cathode, which has a higher

Rechargeable Ion Batteries: Materials, Design, and Applications of Li-Ion Cells and Beyond, First Edition.
Edited by Katerina E. Aifantis, R. Vasant Kumar, and Pu Hu.

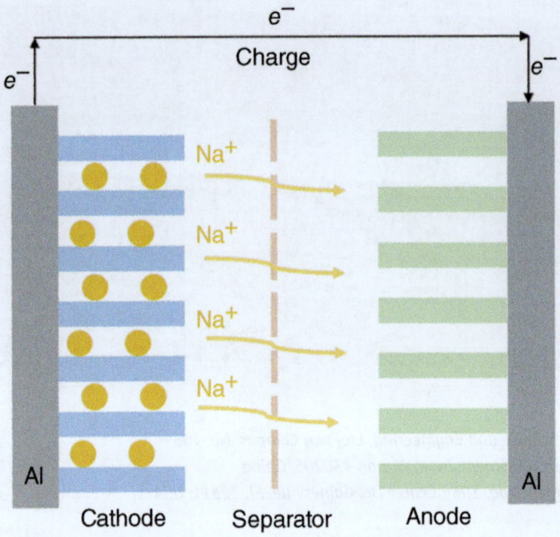

Figure 9.1 The working principle and components of Na-ion battery are similar to the rocking-chair principle of Li-ion batteries.

potential range, and are inserted into the negative electrode material through the electrolyte. At this time, the positive electrode is in a sodium-poor state and the negative electrode is in a sodium-rich state. The discharge process is just the opposite: electrons flow externally to the cathode, and Na-ions are attracted back to their original state. The positive electrode then is in a sodium-rich state and the negative electrode is in a sodium-depleted state. The potential difference between the positive electrode and the negative electrode determines the electromotive force of the battery. Therefore, during the charge and discharge process, Na-ions complete the "intercalation-extraction" reaction from the positive and negative materials, and the Na-ions move between the electrodes. This type of battery, similarly to the case of Li-ion batteries and is called a "rocking-chair" battery (Figure 9.1).

With $NaMO_2$ as the positive electrode and hard carbon as the negative electrode, the electrode and battery reactions are, respectively, expressed as:

Cathode:

$$NaMO_2 \leftrightarrow Na_{1-x}MO_2 + xNa^+ + xe^-$$

Anode:

$$nC + xNa^+ \leftrightarrow Na_xC_n$$

Cell:

$$NaMO_2 + nC \leftrightarrow Na_{1-x}MO_2 + Na_xC_n$$

The historical development of sodium-ion batteries can be traced back to the 1980s, at the same time as lithium-ion batteries. Early research on lithium-ion batteries and sodium-ion batteries was almost at the same level. In 1970, the first report on the electrochemical behavior of Li-ion as a charge carrier embedded in TiS_2 materials at room temperature was reported [1]. On this basis, in 1980, the electrochemical study of the deintercalation of Na-ion as a charge carrier in TiS_2 materials was reported for the first time [2]. In the same year, $LiCoO_2$, a cathode material for Li-ion batteries, was developed and exhibited excellent electrochemical

Figure 9.2 The charge–discharge curves of $LiCoO_2$/Li and $NaCoO_2$/Na half-cell. Source: Reproduced with permission from Yabuuchi et al. [5]/American Chemical Society.

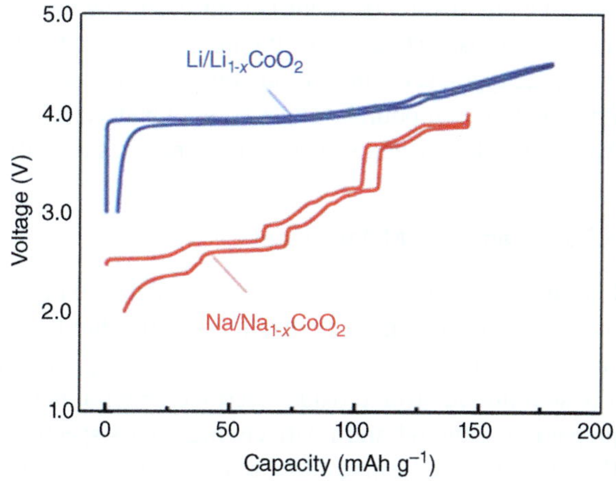

performance and energy density [3]. It is still the main cathode material for many commercial Li-ion batteries. At the same time, Na_xCoO_2 was also developed and explored as a cathode material for Na-ion cells [4]. In the next few decades, a large number of excellent cathodes in combination with graphitic anodes and electrolytes were continuously developed for Li-ion batteries reaching a wide range of applications, as mentioned in Chapter 4. Li-ion batteries, therefore, experienced a period of prosperity, while the development of Na-ion batteries is at a standstill at this time (Figure 9.2).

Compared with Li-ion batteries, Na-ion batteries have the following advantages:

(1) Abundant sodium resources (the global sodium resources are 23 billion tons, while the lithium resources are only 130 million tons), and the cost of electrode materials and electrolytes is low (sodium: US$ 150/ton and lithium: US$ 5000/ton).
(2) Both the anode and cathode materials can use aluminum foil as the current collector, avoiding the use of copper foil (Figure 9.1), which further reduces the production cost of Na-ion batteries (Cu/Li: US$ 1022 vs. Al/Na: US$ 894).
(3) The electrode materials of Na-ion batteries have low toxicity and are environmentally friendly.
(4) In the electrolyte of the same concentration, Na-ions have a relatively higher charge transfer rate, which is conducive to obtaining a superior rate performance.

A main disadvantage of Na-ion batteries is that the ionic radius of Na (102 pm) is larger than that of Li (76 pm), and therefore structural changes during the insertion and de-insertion process are more severe.

9.2 Cathode Materials for Na-Ion Batteries

The cathode material is a key factor affecting the voltage, energy density, power density, safety, and cycle life of Na-ion batteries (SIBs). Therefore, the research and development of cathode materials have become key to SIB research. According to the characteristic structure of the material, the cathode material is mainly divided into three types: sodium-ion transition-metal oxides, polyanionic compounds, and Prussian blue compounds. These types of materials have different chemical compositions and electrochemical behaviors and are summarized below.

9.2.1 Transition-Metal Oxides

Layered transition-metal oxides are considered to be one of the most likely cathode materials for the commercialization of SIBs due to their high theoretical specific capacity, controllable component structure, simple synthesis method, and low price. The sodium-ion layered oxide electrode material has been remarked for its simple preparation method and relatively high-mass-specific capacity. This coupled with the diversity of its composition, that is, different types and different proportions of elements can be substituted at the location of transition metals, allows for a wide range of industrial applications and basic research.

The most common layered compound is composed of a layer of co-edged TMO6 octahedrons, where TM is the transition metal. With its different stacking along the *c*-direction, compounds of various shapes and structures are produced. In 1981, the electrochemical properties of Na_xCoO_2 layered oxide cathode materials were studied and a classification method was proposed based on the sodium-ion polyhedral coordination environment and the stacking mode of oxygen ions [6]. According to the different sodium-ion polyhedral coordination environment, layered oxides are divided into O and P phases, meaning that the coordination environment of sodium-ions is structured in an octahedral (O) and prismatic (P) triangular prism, respectively. The number 2 or 3 represents the minimum number of repetitive layers of oxygen atoms in the transition-metal layer in the smallest oxide cell. For example, the oxygen atoms in the P2 phase material are arranged as ABBA, the oxygen atoms in the P3 phase material are arranged as ABBCCA; while in the O2 phase material, the oxygen atoms are arranged as ABAC, and the oxygen atoms in the O3 phase material are arranged as ABCABC.

Among the sodium-ion layered oxide materials, the most used crystal structures are the O3 (Figure 9.3a) and P2 (Figure 9.3b) phases. The general structural formula is Na_xTMO_2. TM is the transition metal, which can be occupied by one or several different transition-metal elements (such as Fe, Co, Ni, Cu, and Ti) and/or other inactive elements (such as Li, Mg, Sn, and Sb) at the same time. Active transition-metal elements compensate for the charge balance during the de-intercalation of Na-ions by gaining and losing electrons, while inactive elements assist in forming the structural framework and improving the electrochemical properties of the material. Generally, the Na-ion content x in Na_xTMO_2 ranges from 0.5 to 1 mol, and as x changes, Na_xTMO_2 compounds undergo conversions between O and P phases. In general, P2 phase layered oxides have a low-sodium content (~2/3 mol), and O3 phase layered oxides have a high-sodium content (~1 mol). The ability to vary the sodium content, and flexibility in choosing different combinations of transition metals, allow a rich composition diversity for sodium-ion layered oxides. Even very small changes in composition can cause very different electrochemical properties.

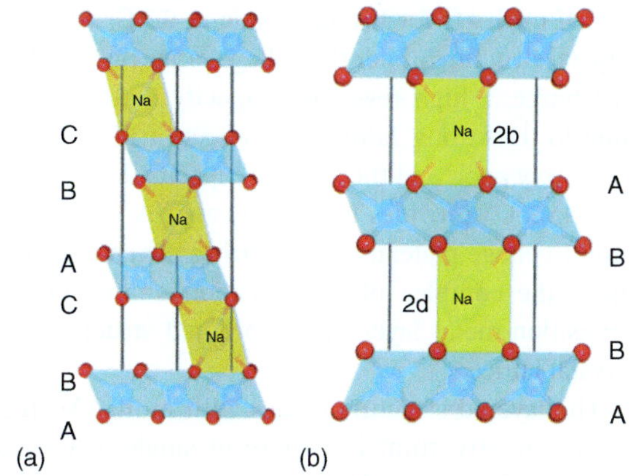

Figure 9.3 The stacking types of (a) O3 and (b) P2 phases in Na_xMO_2. Source: Reproduced with permission from Pan et al. [7]/Royal Society of Chemistry.

9.2.1.1 O3 Phase Cathode Materials

The O3 type configuration belongs to the typical α-$NaFeO_2$ structure (R-$3m$ space group), which has been widely used in Li-ion battery cathodes, such as $LiCoO_2$, ternary materials, and Li-rich materials. Due to the low price of iron and the abundant iron resources in the earth's crust, α-$NaFeO_2$ has also received extensive attention as a cathode material for SIBs. As shown in Figure 9.4a, when the voltage window of α-$NaFeO_2$ is 2.5–4.3 V, the material delivers a reversible specific capacity of about 80 mAh g^{-1} with a large voltage plateau of ~3.3 V, corresponding to the redox reaction of Fe^{3+}/Fe^{4+} during charge/discharge. However, when it is charged to a higher potential of 4.5 V, where $x > 0.5$ in $Na_{1-x}FeO_2$, Fe^{3+} ions in O3 $NaFeO_2$ migrate to the sodium layer, causing irreversible damage to the material structure and attenuation of performance. Acceptable cyclability is, therefore, achieved for $0 < x < 0.45$ in O3-type $Na_{1-x}FeO_2$. A similar phenomenon of irreversible structural changes and corresponding capacity decay, depending on the Na content, are also observed in O3-$NaCrO_2$.

In addition to O3-$NaFeO_2$, O3-$NaMnO_2$ and O3-$NaCoO_2$ are also important cathode materials belonging to the O3 family. O3-$NaMnO_2$ gives a high capacity

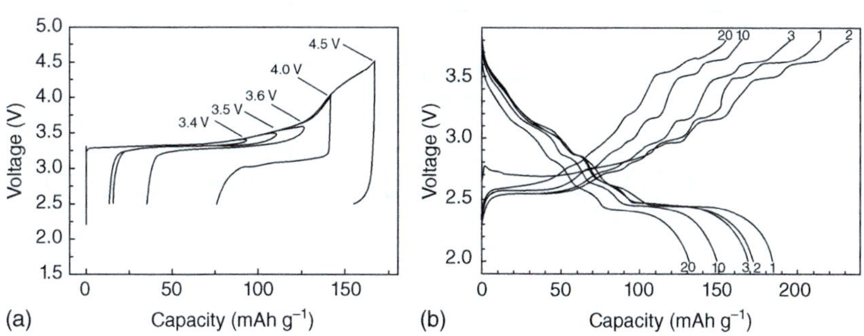

Figure 9.4 (a) Initial charge/discharge curves of α-$NaFeO_2$ cells with different cut-off voltage at a rate of 12 mA g^{-1}. Source: Reproduced with permission from Yabuuchi et al. [8]/Electrochemical Society of Japan. (b) Voltage profile of α-$NaMnO_2$ after multiple cycles at C/10 with potential range of 2.0–3.8 V. Source: Reproduced with permission from Ma et al. [9]/IOP Publishing.

of 195 mAh g^{-1} for the first cycle at 0.1 C, with 0.85 Na being extracted from the crystal structure under the voltage range of 2–3.8 V. As shown in Figure 9.4b, during discharge, a high reversible capacity of 185 mAh g^{-1} can be achieved. However, due to the strong Jahn–Teller effect of Mn^{3+}, the cycle stability of α-$NaMnO_2$ is relatively poor, and the specific capacity decays to 132 mAh g^{-1} after only 20 cycles. O3-$NaCoO_2$ provides a specific capacity of ~140 mAh g^{-1} between 2.5–4.0 V and multi-voltage plateaus occur during the charge–discharge curve [10], which result from the complex phase transition among O3–O′3–P′3 due to the slip of the transition-metal layer or the ordered structure of Na-ions and vacancies in the crystal lattice.

The layered structure oxides with multi-TM in Na_xTMO_2 are explored to overcome the structural instability of single TM oxides. For example, as shown in Figure 9.5, O3-type $NaNi_{0.5}Mn_{0.5}O_2$ can deliver a higher reversible capacity of 185 mAh g^{-1} at the first cycle when charged to 4.5 V, as opposed to 2–3.8 V when only Mn is present ($NaMnO_2$) [11]. Many voltage plateaus can be observed during charge–discharge, implying that the electrode material undergoes multi-phase transitions. However, the electrochemical reaction shows good reversibility except at the plateaus at ~4.0 V. Therefore, the reversibility can be improved by controlling the cutoff voltage. A high reversible capacity of 125 mAh g^{-1} is achieved at a cutoff voltage of 3.8 V.

9.2.1.2 P2 Phase Cathode Materials

Generally, the sodium content in the P2 phase materials ranges between 0.5–0.8, while the valence of the transition is trivalent and tetravalent. The P2 phase layered metal oxides of the monobasic transition metal mainly include P2-$Na_{2/3}VO_2$, P2-$Na_{2/3}MnO_2$, and P2-$Na_{2/3}CoO_2$.

In 1981, Na-storage performance of Na_xCoO_2 layered oxide was reported for the first time [4]. P2-Na_xCoO_2 is stable when the sodium content x is between

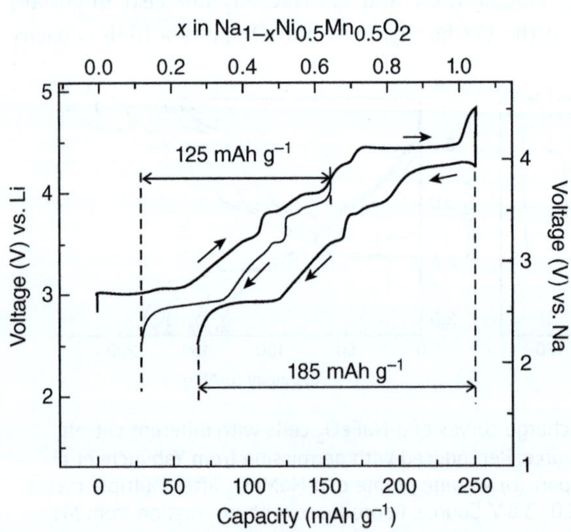

Figure 9.5 Initial charge and discharge curves of the $Na/NaNi_{0.5}Mn_{0.5}O_2$ cell at a rate of 1/50 C (4.8 mA g^{-1}) in the voltage ranges of 2.2–3.8 and 2.2–4.5 V. Source: Reproduced with permission from Lu and Dahn [11]/IOP Publishing.

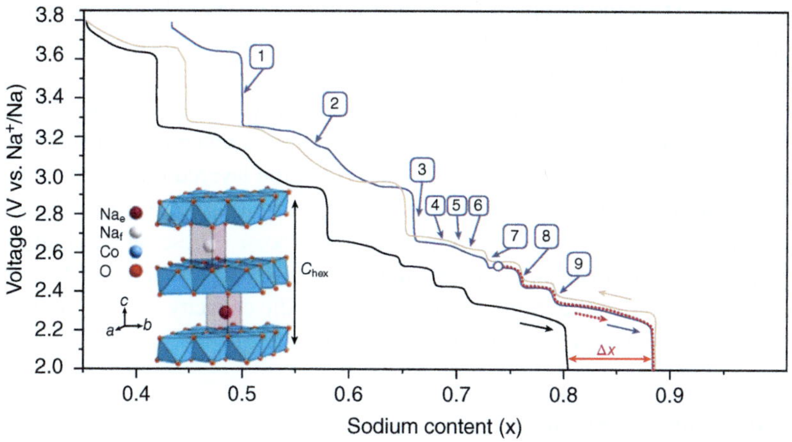

Figure 9.6 Initial charge/discharge curves of P2-$NaCoO_2$ cells. Source: Reproduced from with permission from Berthelot et al. [12]/Springer Nature.

0.48 and 0.83. In addition, the charge–discharge curve of P2-Na_xCoO_2 exhibits multiple plateaus, since the material has an ordered structure of Na-ions and vacancies (Figure 9.6), which cause the material to undergo multiple phase transitions during de-sodiation.

P2-$Na_{2/3}Ni_{1/3}Mn_{2/3}O_2$ has always been a research hotspot in view of its high theoretical specific capacity (173 mAh g^{-1}) and high stability against moist air. As shown in Figure 9.7, almost all the Na^+ can be extracted from P2-$Na_{2/3}Ni_{1/3}Mn_{2/3}O_2$. It delivers 160 mAh g^{-1} for the first cycle, which is close to its theoretical capacity, demonstrating that the material shows high reversibility during Na^+ intercalation/de-intercalation. However, rapid capacity decay caused by the undesirable P2–O2 phase transition (with a large volume change of ≈23%) when charged to above 4.2 V is the main challenge hindering the wide application of P2-$Na_{2/3}Ni_{1/3}Mn_{2/3}O_2$.

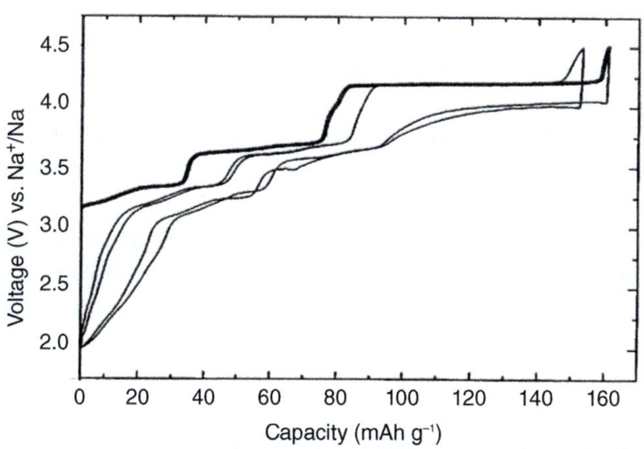

Figure 9.7 Charge/discharge curves of P2-$Na_{2/3}Ni_{1/3}Mn_{2/3}O_2$ cell. Source: Reproduced with permission from Lu and Dahn [11]/IOP Publishing.

It has been shown that the hierarchical engineering of porous nanofibers assembled by nanoparticles can effectively boost the reaction kinetics and stabilize its structure, allowing for ~81% capacity retention after 500 cycles [13].

Although layered transition-metal oxides have a variety of structural combinations, they undergo more complicated structural rearrangements during the de-intercalation of Na-ions than Li-ions. In particular, the layered transition-metal oxide of the P- and the O-phase structures will undergo a mutual conversion through the slip of the laminate during the electrochemical process. These phase-changes are one of the most critical steps in the charge and discharge process of Na-ion batteries. They not only determine the reversible specific capacity and Coulombic efficiency, but also affect the rate and cycle life of the battery.

9.2.2 Polyanionic Compounds

Polyanionic compounds are a class of anionic groups composed of MO_x (M = Fe, Mn, Co, V, Ti, Cr, etc.) polyhedrons, and (XO_4) (X = B, P, S, Si, etc.) tetrahedrons. In these compounds, the polyhedron and the anionic group are connected by a common angle or a common edge to form a framework structure, which is beneficial for Na^+ diffusion in solids. Generally speaking, the stronger the electronegativity of the polyanion, the higher the voltage of the material. The electronegativity sequence of common polyanions is $SiO_4^{4-} < CO_3^{2-} < PO_4^{3-} < F^- < P_2O_7^{4-} < SO_4^{2-}$, therefore, the working voltage of materials from $(SiO_4)^{4-}$ to $(SO_4)^{2-}$ continues to increase. Phosphate materials have attracted great attention in the research of Li-ion batteries due to their high structural stability, thermal stability, and high redox potential generated by the PO_4^{3-} strong induction effect. Among them, $LiFePO_4$ with an olivine structure is considered as one of the most promising cathode materials.

As a typical polyanionic compound, the Na-ion superconductor structure (NASICON) has a large three-dimensional channel structure that can be used for rapid deintercalation of Na-ions. NASICON compounds can be expressed as $AMM'X_3O_{12}$ or $AMM'(XO_4)_3$, where X can be any of the elements P, Si, S, and As, A can be any alkali metal ion (Li^+, Na^+, K^+, etc.) or vacancy, M and M′ positions can be occupied by transition-metal ions from divalent (Zn^{2+}, Cd^{2+}, Ni^{2+}, Mn^{2+}, Co^{2+}, etc.), trivalent (Fe^{3+}, V^{3+}, Cr^{3+}, Al^{3+}, In^{3+}, Ga^{3+}, Y^{3+}, Lu^{3+}, etc.), tetravalent (Ti^{4+}, Zr^{4+}, Hf^{4+}, Sn^{4+}, Ge^{4+}, etc.), and pentavalent (V^{5+}, Nb^{5+}, Ta^{5+}, Sb^{5+}, As^{5+}, etc.) transition-metal ions to achieve a proper charge balance. From the composition point of view, the structure has a high diversity and can provide a tunable electrochemical performance [14]. Therefore, NASICON structure materials are widely designed as cathodes material, anodes material, and solid electrolytes for both Li-ion and Na-ion batteries. For example, in 1976, Hong [15] reported the $Na_{1+x}Zr_2P_{3-x}O_{12}$ ($0 \leq x \leq 3$) sodium fast-ion conductor, which has a conductivity of 6.7×10^{-4} S cm^{-1} and 0.2 S cm^{-1}, at room temperature and 300 °C, respectively. It is expected to be used as a solid electrolyte for solid sodium-ion batteries [15]. For the electrode materials of Na-ion batteries, NASICON structural materials are commonly designed as Na $MM'(XO_4)_3$, where M and M′ positions are active redox ions.

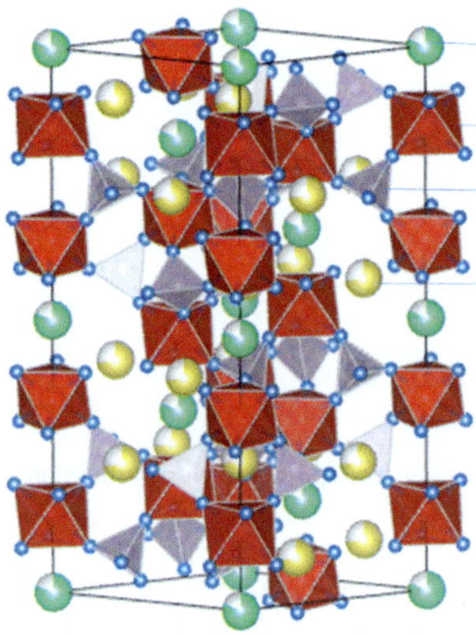

Figure 9.8 Crystal structure of NASICON-typed $Na_3V_2(PO_4)_3$.

A promising NASICON compound for SIB cathodes is $Na_3V_2(PO_4)_3$ due to its stable open 3D framework structure. Its crystal structure at room temperature was reported for the first time in 1978 [16] and is shown in Figure 9.8. Two VO_6 octahedrons and three PO_4 tetrahedrons share a vertex along the c-axis direction to form a $[V_2(PO_4)_3]$ structural unit; each $[V_2(PO_4)_3]$ structural unit is connected to other structural units through a PO_4 tetrahedron to form a NASICON cell with a three-dimensional framework and a large number of suitable Na channels. The structure has a large amount of ion interstitial spaces that facilitate ion diffusion.

Due to its high reversible specific capacity (117 mAh g^{-1}) and stable voltage plateau (average discharge at 3.4 V), $Na_3V_2(PO_4)_3$ has recently attracted wide attention from researchers and is currently the most studied Na-ion cathode among phosphate electrode materials. Its operating voltage is similar to that of $LiFePO_4$ for Li-ion cells and it belongs to the rhombohedral crystal system, and the space group is R-3c. The three Na^+ occupy two different crystallographic positions: the Na1 position at 6b (Na-O six-coordination environment) and the Na2 position at 18e (Na-O eight-coordination environment). This material can have 2 Na^+ react in the voltage range of 2–3.8 V and has good structural stability (Figure 9.9).

In addition, dual-metallic NASICON-based compounds have received extensive attention and are widely investigated. Because V is highly toxic and expensive, it is a good strategy to replace it with other metals. As mentioned earlier, the NASICON structure can be written as $NaMM'(PO_4)_3$, where M and M′ represent different transition metals. Therefore, based on the physical and chemical properties of different metals, multifunctional electrode materials can be flexibly designed. It has been possible to replace half of the V in $Na_3V_2(PO_4)_3$ with Ti, to synthesize $Na_2VTi(PO_4)_3$ [18]. Compared with other V-based compounds, $Na_2VTi(PO4)_3$ shows an excellent cycle stability in water-based SIBs, with the capacity retention being 70% at 10 C for 1000 cycles [19]. Studies have shown that the substitution

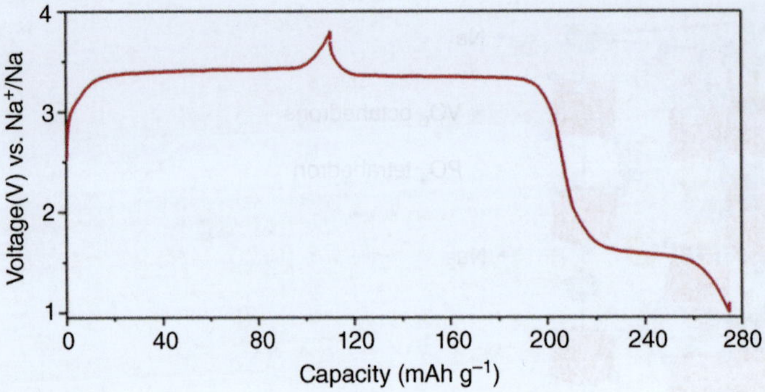

Figure 9.9 Voltage-capacity profile for $Na_3V_2(PO_4)_3$. Source: Reproduced with permission from [17]/American Chemical Society.

of a part of V by Ti can inhibit the dissolution of V, mainly due to the polarization effect of Ti atoms and lower the redox potential, which reduces the rate of side reactions. The capacity of $Na_2VTi(PO_4)_3$ in organic electrolytes can reach a value of 147 mAh g^{-1} under a voltage window of 1.5–4.5 V with three potential plateaus at ~3.4, ~2.1, and ~1.6 V, which correspond to the redox pairs of V^{4+}/V^{3+}, Ti^{4+}/Ti^{3+}, and V^{3+}/V^{2+} [20].

Goodenough and his colleagues replaced V with Mn, which is a low-cost and abundant earth resource, and synthesized another bimetallic NASICON structure, $Na_4MnV(PO_4)_3$, for the first time in 2016 [21], where Mn/VO_6 octahedron and PO_4 tetrahedrons share corners to form a $[MnV(PO_4)_3]^{4-}$ anion framework. The introduction of Mn reduces the material cost and increases the working voltage of $Na_3V_2(PO_4)_3$. At the same time, the presence of V in the material can suppress the structural distortion caused by the Jahn-Teller effect of Mn^{3+} during the charging process, so that it can produce a stable structure, and allow for an excellent cycle performance. This is a very promising cathode, as it has a high voltage, high energy density (Figure 9.10a), and two pairs of redox voltages, 3.3 and 3.6 V, corresponding to V^{3+}/V^{4+} and Mn^{2+}/Mn^{3+}, respectively. When the charging voltage is increased to 4.3 V, the V^{4+} is further oxidized to V^{5+}, and the corresponding voltage plateau occurs at ~4.1 V. 2.8 Na^+ are extracted from the crystal structure of $Na_4MnV(PO_4)_3$, providing the reversible specific capacity of about 150 mAh g^{-1}.

$Na_3MnTi(PO_4)_3$ has the same structure as $Na_3V_2(PO_4)_3$, and both belong to the *R-3C* space group of the hexagonal crystal system. The octahedron composed of MnO_6 and TiO_6 is connected with the PO_4 tetrahedron through the corners to form a three-dimensional frame structure. Again Goodenough applied this material for the first time as a cathode in aqueous SIBs [22]. During the oxidation process from $Na_3MnTi(PO_4)_3$ to $Na_2MnTi(PO_4)_3$, the redox potential of Mn^{3+}/Mn^{2+} is lower than the precipitation potential of oxygen. During the reduction process from $Na_3MnTi(PO_4)_3$ to $Na_4MnTi(PO_4)_3$, the redox potential of Ti^{4+}/Ti^{3+} is higher than the precipitation potential of hydrogen, so $Na_3MnTi(PO_4)_3$ can be used in symmetrical batteries in aqueous electrolytes. $Na_3MnTi(PO_4)_3$ has also been shown to be

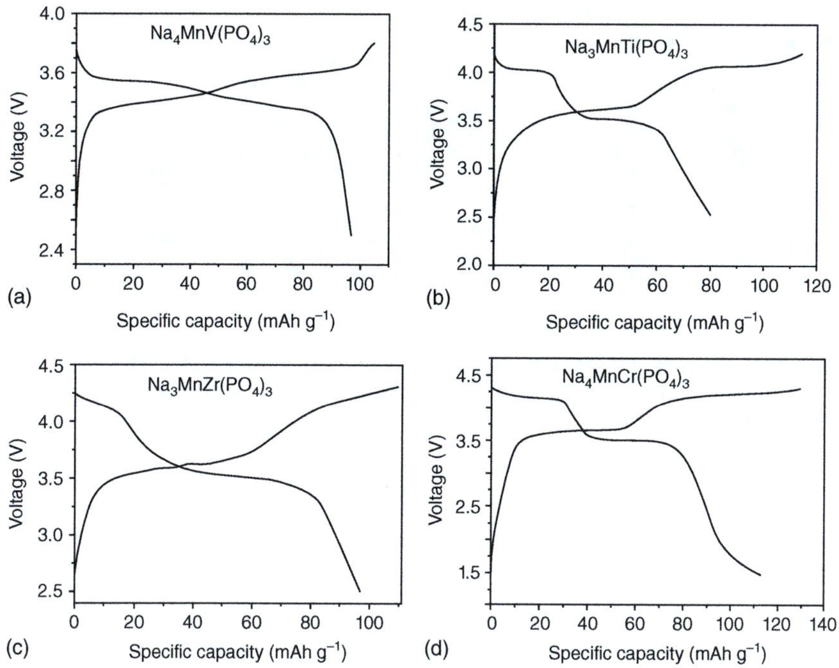

Figure 9.10 Voltage-capacity profile for $Na_4MnV(PO_4)_3$ [21], $Na_3MnTi(PO_4)_3$ [22], $Na_3MnZr(PO_4)_3$ [23], $Na_4MnCr(PO_4)_3$ [24]. Source: Adapted with premission from Zhou et al. [21]/American Chemical Society, Gao et al. [22]/American Chemical Society, Ma et al. [23]/American Chemical Society, Zhang et al. [24]/John Wiley & Sons.

a promising cathode material in organic electrolytes, delivering voltage plateaus of 3.6 and 4.2 V for the redox pairs of Mn^{2+}/Mn^{3+} and Mn^{3+}/Mn^{4+}, respectively (Figure 9.10b). In a wider potential range, it exhibits the three-electron reaction and delivers a high capacity of 160 mAh g^{-1} at 0.2 C under the window of 1.5–4.2 V.

The electrochemical behavior of another bimetallic NASICON material, $Na_3MnZr(PO_4)_3$, is also shown in Figure 9.10c [23]. Again two reduction plateaus are noted at 4.0 and 3.5 V, corresponding to Mn^{4+}/Mn^{3+} and Mn^{3+}/Mn^{2+}. The specific capacity is 105 mAh g^{-1} at 0.1 C.

Theoretically, $Na_4CrMn(PO_4)_3$ is the cathode material with the highest average redox platform when realizing the three-electron reaction, and the theoretical capacity can reach 165 mAh g^{-1}. As shown in Figure 9.10d, the material has an initial charge specific capacity of 141 mAh g^{-1} in the voltage range of 1.5–4.5 V, a discharge specific capacity of 130 mAh g^{-1}, and an energy density that can reach 470 W h kg^{-1} [24].

9.2.3 Prussian Blue Compounds

Prussian Blue Analogs (PBAs) or hexacyanoferrate (FHC) framework compounds are transition-metal hexacyano complexes. The general formula can be written as

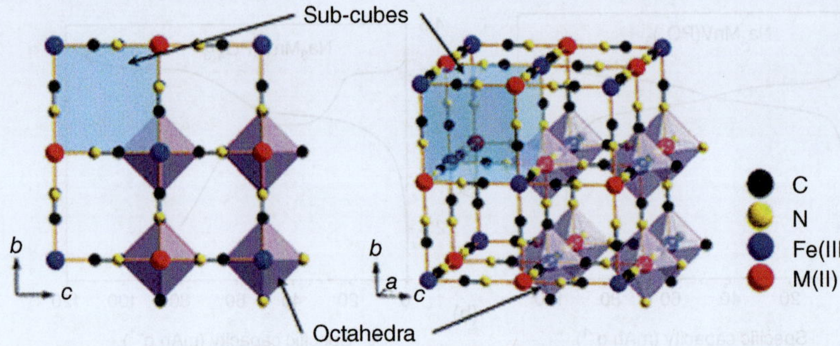

Figure 9.11 Structure of Prussian blue and its analogs: a new framework of electrode materials for sodium batteries. Source: Reproduced with permission from Lu et al. [25]/Royal Society of Chemistry.

$A_xMM'(CN)_6{\cdot}nH_2O$, where A is usually an alkali metal ion, such as Na and K, while M and M′ are generally ions of different valence states, such as Co, Mn, Ni, Fe, Zn, and Cu. Metallic M ions are coordinated with the N-terminal of cyanide CN^-, and metallic M′ ions are combined with the C of CN^-. For PBAs, M′ is usually Fe ion and $A_xMM'(CN)_6{\cdot}nH_2O$ is denoted as MFHC (M = Co, Mn, Ni, Fe, Zn, Cu, etc.). The CN bond opens the faces of the elementary cubes for Na^+ to move between half-filled body-centered positions.

Figure 9.11 shows the structure of PBAs. In this structure, the rigid open framework with large interstitial sites ensures high Na^+ mobility and accommodates the volume variation during Na^+ insertion/extraction. PBA compounds contain two different redox-active centers: $M^{2+/3+}$ and $Fe^{2+/3+}$ couples, both of which can undergo a complete electrochemical redox reaction (when M = Fe, Co, Mn, etc.). The theoretical specific capacity of PBAs is as high as 170 mAh g^{-1} [calculation based on $Na_2FeFe(CN)_6$] because of a theoretical two-electron redox reaction during charge–discharge.

MnHCF is a common and simple example of mono-metal hexacyanoferrate. It exhibits both high specific capacity and redox potentials. Figure 9.12 shows the charge/discharge profiles of MnHCF. Owing to two active redox sites of $Fe^{2+}/^{3+}$ and $Mn^{2+}/^{3+}$ in Fe—C≡N—Mn, MnHCF exhibits the highest reversible capacity among PBAs; however, its capacity retention is poor. As shown in Figure 9.12, the discharge capacity of MnHCF decreases rapidly from 122 to 64 mAh g^{-1}, corresponding to a capacity retention of a meager 52.5% after 100 cycles. The voltage plateaus become shorter, and a larger overpotential is observed upon cycling. A better performance was observed in Ref. [27], where 209 mAh g^{-1} was achieved at a current rate of 40 mA g^{-1} and a 75% capacity retention was sustained over 100 cycles (Figure 9.12b). Although considerable progress has been made in improving the Na^+ insertion performance of the Mn-based PBA frameworks, there still exist a few problems that need to be resolved, such as the Jahn–Teller effect of Mn^{3+}, which often causes the distortion of the lattice and the dissolution of Mn into electrolyte, thus leading to a limited cycling stability.

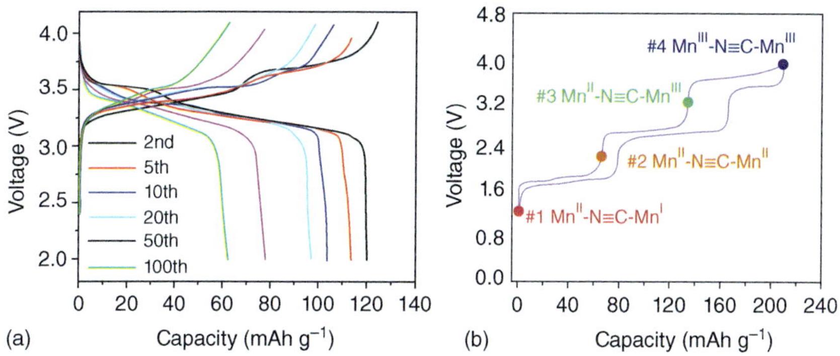

Figure 9.12 The charge/discharge profiles of MnHCF [26] and $Na_2MnMn(CN)_6$ [27]. Source: Adapted with permission from Hu et al. [26]/American Chemical Society, and Lee et al. [28]/Springer Nature.

The capacity decay for MnHCF can also be attributed to the structural changes, which lead to mechanical degradation and crack formation during charge/discharge. Figure 9.13 shows the morphological evolution of MnHCF electrodes during cycling. The MnHCF cathode retained its cubic shape and size upon cycling, however, the morphology of its surface changed and became rougher/coarser. Such morphology changes can be attributed to the dissolution of Mn^{2+} into the organic electrolyte, leading to corrosion of the surface. MnHCF active particles exhibited no obvious signs of fracture after 10 cycles (Figure 9.13a,d) when scanning

Figure 9.13 (a–c) Low-magnification SEM images of MnHCF materials after (a) 10, (b) 20, and (c) 50 cycles. (d–f) High-magnification SEM images of MnHCF materials after (d) 10, (e) 20, and (f) 50 cycles. Source: Reproduced with permission from Hu et al. [26]/American Chemical Society.

electron microscopy imaging was performed. However, small-scale crack initiation and propagation were observed from 20 cycles (Figure 9.13b,e), and after 50 cycles (Figure 9.13c,f), fracture was observed in almost every particle. It is noted though that if transmission electron microscopy was employed, fine cracks may be observed before 20 cycles. This fracture can occur from the volume changes of ~8.5% that MnHCF undergoes upon de-insertion and re-insertion of Na^+.

There are other Prussian blue materials that can also host Na. One example is $Na_2NiFe(CN)_6$ (NiHCF) which shows a high stability due to its zero lattice strain (volume change ~ 0.29%) during sodiation/desodiation. In NiHCF, Ni^{2+} is electrochemically inert and Fe is the only redox-active site, making the reversible capacity of NiHCF low (~60 mAh g^{-1}). It was proposed that an effective approach to solve the mechanical degradation and improve the electrochemical cyclability of MnHCF is by constructing a gradient structure that combines the advantages of both MnHCF (high capacity) and NiHCF (low volume change), as shown in Figure 9.14 [26].

Figure 9.15 shows that even low substitution of Ni resulted in a significant increase in the capacity retention. For the composition of $Na_{1.17}Ni_{0.12}Mn_{0.88}Fe[(CN)_6]_{0.79}\cdot 1.5H_2O$ (g-NiMnHCF), in which Ni gradually increases while Mn decreases from the interior to the particle surface, a high reversible specific capacity

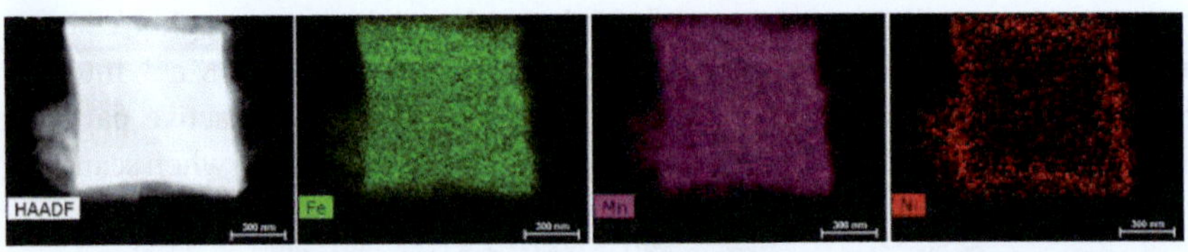

Figure 9.14 STEM images and element mapping for g-($Ni_{0.3}Mn_{0.7}$)HCF before cycling. Source: Reproduced with permission from Hu et al. [26]/American Chemical Society.

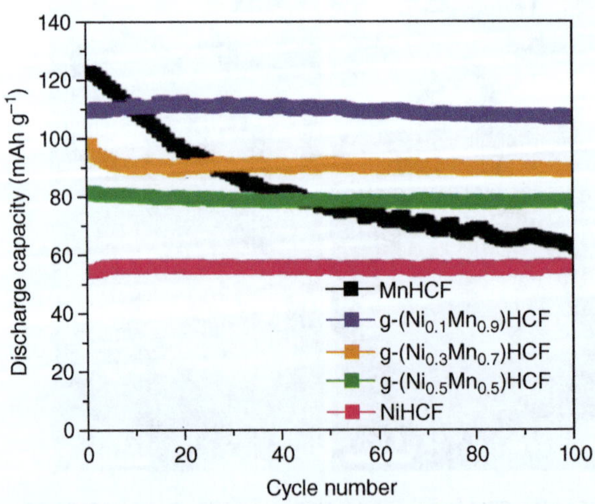

Figure 9.15 Comparison of performance between concentration gradient and homogeneous Prussian blue materials. Source: Reproduced with permission from Hu et al. [26]/American Chemical Society.

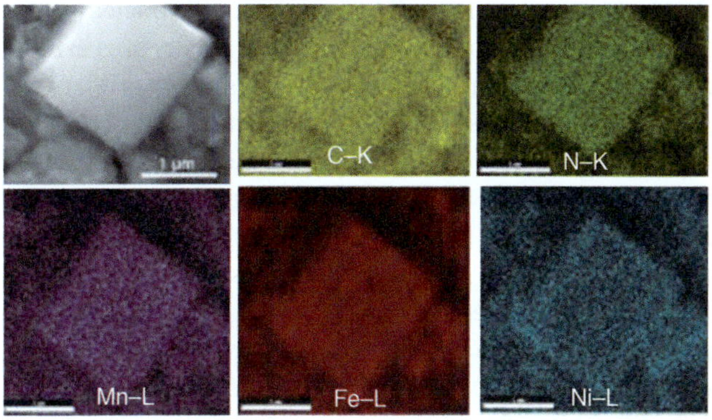

Figure 9.16 SEM images and elemental mapping for g-($Ni_{0.1}Mn_{0.9}$)HCF after 100 cycles. Source: Reproduced with permission from Hu et al. [26]/American Chemical Society.

of 110 mAh g^{-1} at 0.2 C and an outstanding cycling stability (93% retention after 100 cycles) were obtained. This excellent electrochemical response was attributed to the ability of the concentration gradient to minimize the damage that the active material experienced.

Figure 9.16 it is seen that even for the lowest Ni content fracture did not occur after 100 cycles. However, it is interesting to note that after cycling the Ni appeared to be more uniformly distributed.

9.3 Anode Materials for Na-Ion Batteries

A promising anode material should have a lower oxidation–reduction potential than the cathode, small volume expansions upon Na-ion insertion, high electronic conductivity and ion mobility, and good compatibility with the electrolyte. Similarly to the storage mechanisms of Li-ion battery anodes, Na-ion anode materials also involve three types of storage mechanisms: the insertion/extraction reaction mechanism, the alloying/dealloying reaction mechanism, and the conversion reaction mechanism. Since the latter is not commercially preferable it will not be discussed herein.

9.3.1 Carbon-Based Anode Materials

Carbon-based materials are considered to be promising anode materials for Na-ion batteries due to their rich resources, low price, and environmental friendliness. At present, carbon-based anodes mainly include graphite, hard carbon, and soft carbon, and they are the most well-studied intercalation materials. Among them, graphite can form LiC_6 intercalation compounds when used as the anode material of Li-ion batteries, as seen in Chapter 6, showing a high theoretical specific capacity (372 mAh g^{-1}), and is the commercial anode for LIBs. However, due to the thermodynamic instability of the compounds formed between sodium and graphite (such as NaC_6 and NaC_8), Na^+ cannot be reversibly inserted and extracted in the graphite layer like Li^+. Therefore, when graphite is used as the anode, SIBs exhibit a relatively low-specific capacity. In addition, the graphite layer spacing is relatively small, only about 0.34 nm, which also limits the de-intercalation process of Na-ions in the graphite layer. Theoretical calculations have found that the interlayer spacing for the reversible de-intercalation of Na^+ in layered carbon materials requires a spacing greater than 0.37 nm.

Figure 9.17 shows the difference in the intercalation process between Li^+ and Na^+ in graphite. In a standard electrolyte of LIBs (1 M $LiPF_6$ in EC/DMC), the LiC_6 is reversibly formed at potentials below 0.25 V with a theoretical capacity of 372 mAh g^{-1}. The achievable capacities are virtually zero in the case of the analog sodium cell (1 M $NaPF_6$ in EC/DMC). By using a single-solvent electrolyte based on diglyme and lithium (LiOTf) or sodium triflate (NaOTf) as the conductive salt, capacities around 90–100 mAh g^{-1} within the first ten cycles are found. Several steps can be observed in the voltage profile, which are indicative of more complex

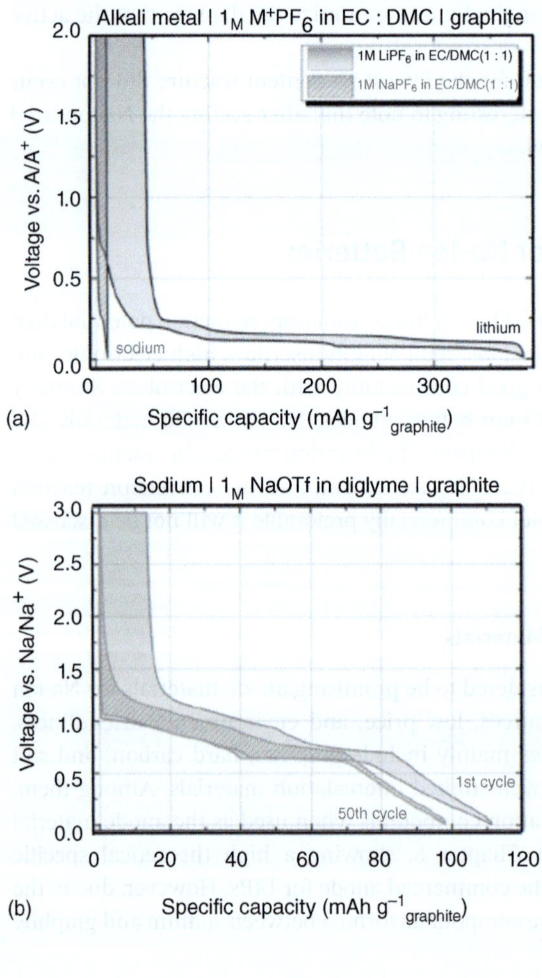

Figure 9.17 Charge–discharge curves of graphite in carbonate-based (a) and diglyme electrolyte (b). Source: Reproduced with permission from Jache and Adelhelm [28]/John Wiley & Sons.

ordering phenomena with different thermodynamic stabilities. When diglyme is used as the solvent of the electrolyte, the sodium-storage capacity of graphite is 100 mAh g^{-1}, at a current density of 37 mA g^{-1}, the cycle life is up to 1000 cycles, and the Coulombic efficiency is greater than 99.87% [28].

When using ether-type electrolytes, the sodium storage mechanism in natural graphite is a combination of a sodium ion-solvent co-intercalation and pseudocapacitance behavior. At a current density of 100 mA g^{-1}, its reversible specific capacity in the first cycle is 150 mAh g^{-1}, and the capacity retention after cycles at 500 mA g^{-1} is 80%. Although the sodium ion-solvent co-intercalation mechanism makes natural graphite exhibit a certain sodium storage performance, its specific capacity is relatively low and the intercalation potential is relatively high, which seriously reduces the energy density of the whole battery and limits its practical application prospects.

At present, hard carbon is the best choice for carbon-based SIB anodes. Hard carbon refers to non-graphitized carbon materials generated from precursors at high temperatures. Compared with layered graphite, hard carbon is composed of graphitized regions with different orientations, showing the structural characteristics of short-range order and long-range disorder. The anisotropic micro-regions are combined together so that the hard carbon material is not easy to peel off during the de-insertion process, and the gaps in the boundaries of different micro-regions can alleviate the stress caused by volume changes and keep the material structure intact. At the same time, these voids also provide additional sites for the storage of Na. Therefore, compared with layered graphite materials, hard carbon allows for a higher specific capacity and cycle stability. Figure 9.18 depicts the charge–discharge curve of a typical hard carbon material. It can be seen that the initial discharge

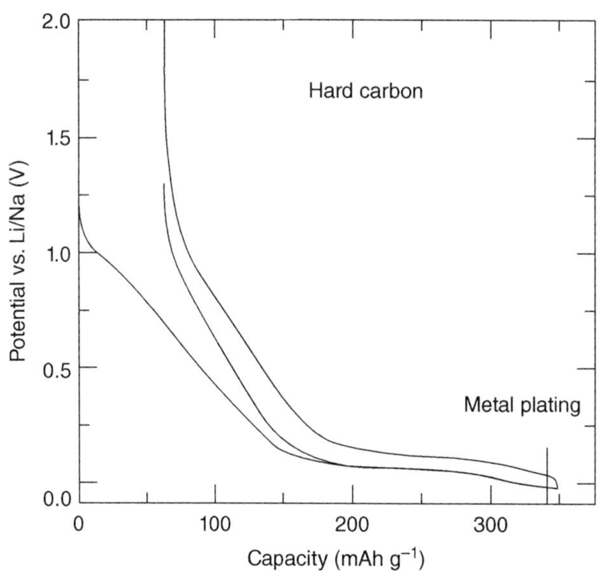

Figure 9.18 Charge–discharge curves of hard carbon. Source: Reproduced with permission from Stevens and Dahn [29]/IOP Publishing.

specific capacity is ~300 mAh g^{-1}, and the discharge curve is mainly composed of a slope below 1 V and a low discharge platform approaching 0 V [29].

9.3.2 Alloying Anodes

The alloying elements for Na are mainly those in groups IV and V, such as Sn, Bi, Si, Ge, Sb, and P. They react with Na to form a variety of sodium–metal alloy phases under multiple steps, accompanying multi-electron transfer. Therefore, they present a higher theoretical specific capacity than intercalation materials. However, the main drawback of these materials is the large volume expansion that take place during the formation of Na-alloys. These repeated volume changes during each electrochemical cycle eventually cause the electrode material to pulverize, which leads to capacity over cycling and battery failure. Since the Na^+ radius is larger than that of Li^+, the volume changes are even higher than those for the case of LIBs (Chapter 6). Figure 9.19 presents the theoretical specific capacity and volume expansion of alloying materials. In order to commercialize any of these candidates, the large volume expansions need to be alleviated, and this can be achieved by using appropriate coatings, as has been done in the case of LIBs. Although Si, P, and As may seem like the best candidates as they provide the highest capacity, they have not received considerable attention, since in addition to the fact that they undergo the larger volume expansions, they also have the lowest conductivities, and good capacities cannot be obtained experimentally.

Sn-based anode materials are the most promising electrode materials in SIBs. The full sodiation of Sn results in the formation of $Na_{15}Sn_4$, which gives a high theoretical specific capacity of up to 847 mAh g^{-1}. In the Na–Sn binary system at room temperature, only the final state of $Na_{15}Sn_4$ is a stable crystalline phase, while other intermediate phases are difficult to be observed due to extreme reactivity or instability, so the intermediate state is uncertain (Figure 9.20); however, it has been shown that Sn gradually transforms into $NaSn_5$, NaSn, Na_9Sn_4, and $Na_{15}Sn_4$ during the sodiation process [31]. The drawbacks that inhibit the commercialization of Sn-based anodes are listed separately below.

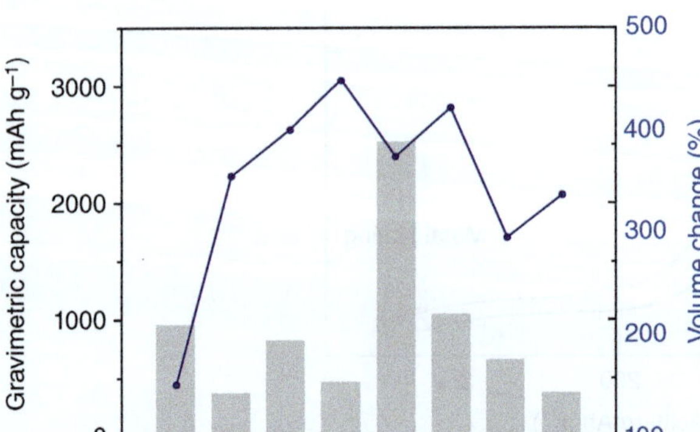

Figure 9.19 Theoretical gravimetric capacity and volume change of alloying anode.

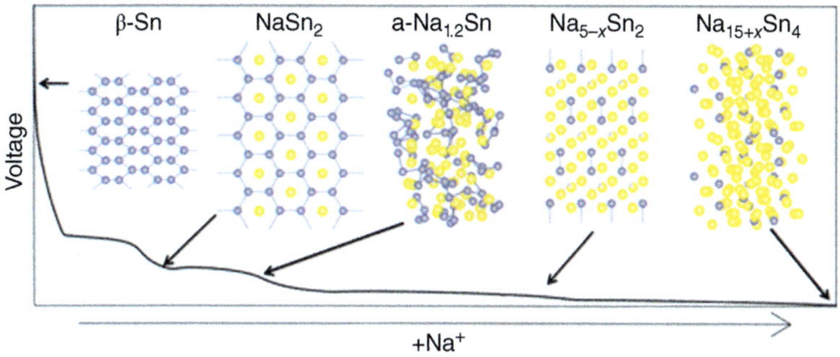

Figure 9.20 Discharge curve for Sn anode and the phase composition of Na_xSn at different discharge of state. Source: Reproduced with permission from Stratford et al. [30]/American Chemical Society.

9.3.2.1 Huge Volume Expansion Causes Active Material Fracture

Sn undergoes a volume expansion of up to 420% when it is completely sodiated to form $Na_{15}Sn_4$, which results in fracture that, in turn, seriously affects the cycle performance. This fracture has been well-documented with *in situ* X-ray probe imaging [32], showing that there is a correlation between the fracture of Sn particles during the sodiation process and the size of the particles (Figure 9.21). It was concluded that if the Sn particle size was less than 0.5 μm, the Sn particles remained intact during the sodiation process. When the Sn particle size was greater than the critical point of 1.6 μm, severe fracture occured during sodiation. These observations were done with X-ray nanotomography and therefore a resolution down to the nanoscale was not possible. Hence, it may be possible that fracture took place for particles less than 0.5 μm.

To better understand the mechanisms by which damage takes place, Sn thin films (0.06 mm thickness) [33] and thick Sn sheets (0.5 mm thickness) [34] have been cycled. The films were directly in contact with the current collector; therefore binder and conductive carbon were not employed and did not affect the expansion of the Sn. Such configurations allow to better analyze the formation of damage during cycling. In Figure 9.22 it is seen that for the case of the Sn thin films, sodiation initially resulted in the formation of pores throughout the film surface and cracks formed upon de-sodiation. Using impedance spectroscopy, the diffusion coefficient was calculated as $3.9 \times 10^{-14}\ cm^2\ s^{-1}$.

For the case of Sn thick sheets, Figure 9.23, in addition to pores and cracks, Sn whiskers also formed during electrochemical cycling. Focused ion beam was used to cut a cross section of the Sn thick sheet after cycling, and SEM illustrated that pores formed within the Sn sheet indicating the distance over which sodiation occurred, since pore formation is a direct result of the damage induced from volume expansions.

9.3.2.2 Volume Expansion Causes Increase the Impedance

There are two main reasons for the sharp increase in the impedance of Sn upon cycling: (i) During sodiation of Sn, the covalently bonded Sn clusters in α-NaSn, namely, Zintl ions (Sn^{4-}), result in a higher resistivity, which is about 106 times that

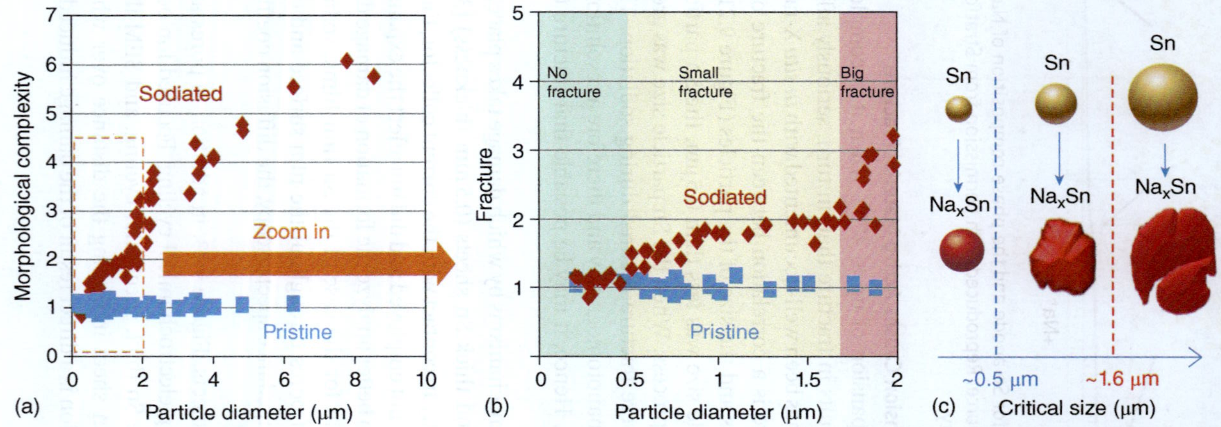

Figure 9.21 (a) Size-dependent morphological complexity for the pristine and sodiated samples. (b) An enlarged plot showing three different fracture zone. (c) A schematic illustration gives the two critical sizes for Sn fracture in sodium-ion battery. Source: Reproduced with permission from Wang et al. [32]/Springer Nature.

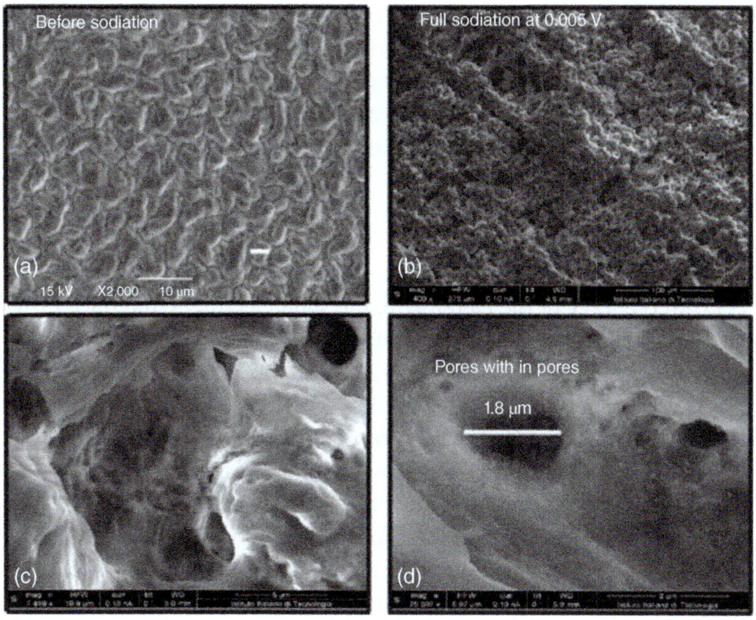

Figure 9.22 SEM images of Sn thin film (a) before cycling and (b) after full sodiation at 0.005 V, (c) and (d) are both magnified images of (b) showing the porous structure of Sn film (c) with nanoparticles on the surface (d). Source: Reproduced with permission from Li et al. [33]/American Chemical Society.

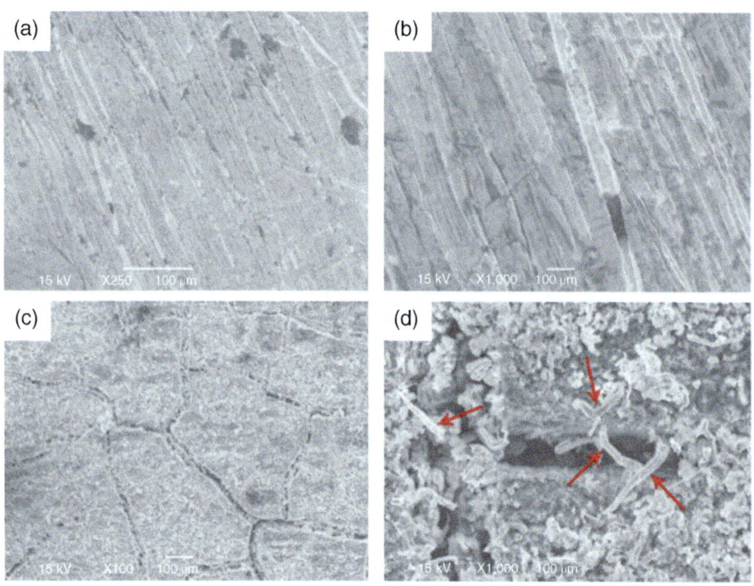

Figure 9.23 SEM images of surface morphology of the thick Sn foil electrode. (a, b) Sn surface before cycling. (c, d) Sn surface in contact with the electrolyte after five cycles, indicating crack and whisker formation. The red arrows in (d) indicate whisker formation on the fractured particles. Source: Reproduced with permission from Li et al. [34]/American Chemical Society.

of Sn. (ii) The stress accumulation in Sn, which results from its volume expansion during sodiation, causes the ionic configuration of Zintl ions to be distorted, which also changes their electronic structure, resulting in a 100-fold increase in the resistance of the sodiated phase. The large resistance increase makes the Sn utilization rate extremely low during high current charging and discharging, resulting in the poor rate performance of Sn metal.

9.3.2.3 Unstable SEI

When the Fermi energy level of the anode material is higher than the lowest unoccupied orbit (LUMO) energy of the electrolyte, the solid electrolyte interface (SEI) film will form. Generally, the voltage plateau of alloy anode materials is lower than 1 V (vs. Na+/Na), making the formation of SEI inevitable. The ability to form a stable SEI film is very important in batteries; however, the SEI for alloy-type negative electrodes is not stable, as has been also seen for LIBs. This is because the stress caused by the huge volume expansion causes the SEI film to rupture once it forms, and the pulverization of the electrode material exposes new surface, leading to the reformation of the SEI film. The unstable SEI membrane not only continuously consumes the electrolyte, resulting in low Coulombic efficiency and even battery failure, but also continues to accumulate and produces an excessively thick SEI membrane, which reduces the ion-transfer efficiency at the electrode–electrolyte interface. Furthermore, Na is more reactive than Li, making the SEI film in SIBs more likely to re-dissolve than the SEI film in LIBs.

9.4 Electrolytes for Na-Ion Batteries

9.4.1 Electrolyte Components

As for LIBs, there are five common types of electrolytes that have been investigated for SIBs, namely, organic, ionic liquids (ILs), aqueous, inorganic solid electrolytes (ISEs), and solid polymer electrolytes (SPEs). Regarding liquid electrolytes, the solvation effect and electric field-driven transport can be regarded as the transport power for Na^+. Since sodium is more reactive than lithium (although the potential of lithium is −3.05 V, which is lower than that of sodium at −2.71 V, the kinetic rate of side reactions in sodium is faster), it is easier to react with organic solvents, so selecting a suitable electrolyte is particularly important. The cost of the electrolyte accounts for about 5% to 10% of the total battery cost.

Organic liquid electrolytes are used in most state-of-the-art SIBs and are mainly composed of sodium salt, solvent, and additives, and based on the composition of the solvent they can be divided into two categories: ester organic electrolytes and ether organic electrolytes.

The ideal sodium salt should have the ability to completely dissolve and dissociate in the electrolyte. In addition, the sodium salt should remain electrochemically inert to other battery components, such as electrodes, separators, and current collectors. At the same time, the SEI formed by the degradation of sodium salt on the surface of the electrode is also extremely important for its electrochemical stability.

$NaClO_4$ and $NaPF_6$ salts are the two most commonly studied sodium salts. Since many properties of the salts depend more on anions rather than cations, the advantages and disadvantages of sodium and lithium salts are mostly the same. ClO_4^- is a strong oxidant, which will have great limitations in the applications of SIBs; PF_6^- is the preferred negative ion for SIBs, but there are serious safety issues, especially since in the presence of high temperature and water, it will be hydrolyzed to produce PF_5, POF_3, and HF. FSI^- and $TFSI^-$ have a high solubility, but they will also corrode the aluminum current collector.

9.4.2 Ester-Based Organic Electrolyte

Ester organic solvents for SIBs are similar to those for LIBs, mainly: dimethyl carbonate (DMC), diethyl carbonate (DEC), ethylene carbonate (EC), propylene carbonate (PC), etc. In order to identify suitable electrolyte formulations for the Na-ion chemistry, various electrolytes have been tested, which contain diverse solvent mixtures (cyclic, acyclic carbonates, glymes) and Na-based salts having either F-based or perchlorate anions. Their viscosity, ionic conductivity, thermal and electrochemical stability were measured, as shown in Figure 9.24. The conductivity values for PC-based electrolytes with various sodium salts are relatively similar, ranging from 7.98 for $NaPF_6$ to $6.2\,mS\,cm^{-1}$ for NaTFSI and being intermediate for $NaClO_4$. This dependence upon the nature of the anion is consistent with what was previously reported for Li-based electrolytes, which was ascribed to the lower polarizing character of the PF_6^- anions, which would improve salt dissociation and enhance the ionic mobility. In contrast, larger ionic conductivity variations were found by fixing the salt ($NaClO_4$) and changing the nature of the solvents by varying the viscosity (σ) values following the trend EC : DME > EC : DMC > EC : PC > EC : Triglyme > EC : DEC > PC > Triglyme ≫ DME, DMC, DEC. The electrochemical stability of binary solvent-based electrolytes decreases following the trend EC : PC > EC : DMC > EC : DME > EC : DEC > EC : Triglyme, and hence the trend for EC-based binary solvents differs from that observed for the co-solvent alone. This indicates that the electrochemical stability of a solvent mixture does not strictly depend on that of each co-solvent since synergetic effects do occur when solvents are mixed [35].

The electrochemical performance of SIBs was found to be remarkably dependent on the electrolyte solvent used, EC : PC emerging as clearly the optimum (Figure 9.25). When PC and EC/DEC mixed solution is used, the hard-carbon electrodes exhibit a high reversible capacity of more than $200\,mAh\,g^{-1}$ with an excellent capacity retention over 100 cycles. On the other hand, in the case of EC/DMC and EC/ethyl methyl carbonate (EMC), the cycle performance was poor [35].

9.4.3 Ether-Based Organic Electrolyte

Ether electrolytes are widely used in SIBs since ether solvents have a better oxidation and reduction resistance in the SIB system. Compared with ester electrolytes, ethers can produce thinner, more stable, and denser SEI films on the

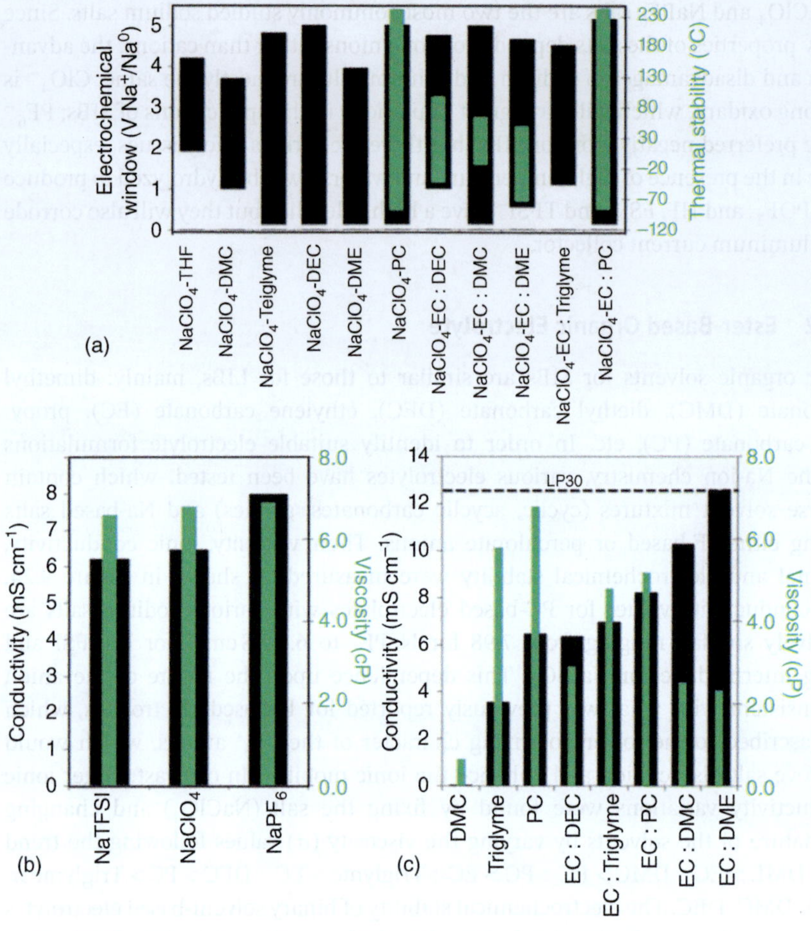

Figure 9.24 Comparison of electrochemical window, thermal stability, and conductivity values in various electrolytes. Source: Reproduced with permission from Ponrouch et al. [35]/Royal Society of Chemistry.

anode electrode surface. However, ether electrolytes are rarely used in LIBs, since they cannot form a sufficiently dense SEI on the electrodes and become unstable above 4 V.

Particularly, graphitic and alloying anodes show a much better cycling stability in ether-based electrolytes than in ester-based electrolytes. Due to the larger Na-ion radius compared with the layer spacing between graphite layers, the graphite anode exhibits limited electrochemical behavior for Na-ion storage. However, when the Na-ion solvated substance is formed in the electrolyte with ether solvent molecules, it can be embedded in the graphite layer, thereby obtaining a relatively ideal capacity and stability. The graphite anode can deliver a reversible capacity of ≈150 mAh g^{-1} with a cyclic stability for 2500 cycles [37].

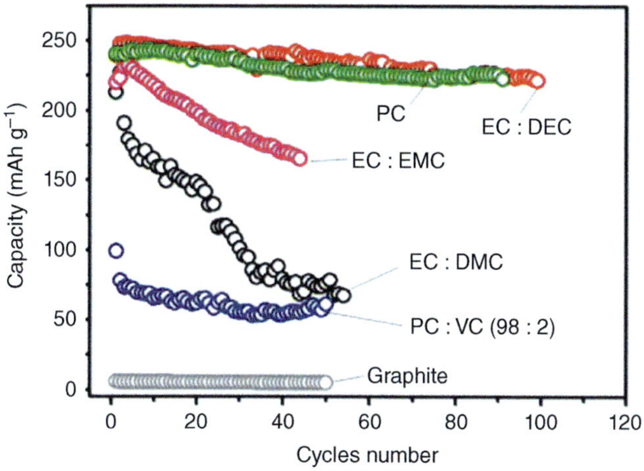

Figure 9.25 Capacity retention for hard-carbon electrodes with 1 mol dm^{-3} $NaClO_4$ dissolved in different solvent mixtures. Source: Reproduced with permission from Dahbi et al. [36]/Royal Society of Chemistry.

It has been shown that Sn also has a higher stability in ether-based electrolytes during sodiation [38] since the SEI that forms on the Sn limits its damage. At the same time, the intermediate product (Na–Sn alloy) has a better stability in the ether-based electrolyte, so a smaller polarization voltage is achieved, and the electrochemical performance is greatly improved.

Figure 9.26a–d compares the voltage profiles when Na/TiS_2 cells were cycled using different electrolytes. A similar initial was obtained capacity for the different electrolytes. However, the cells using carbonate-based electrolytes (1 M $NaClO_4$-EC/DMC [Figure 9.26a], 1 M $NaPF_6$-EC/DMC [Figure 9.26b]) displayed a better capacity retention than those using ether-based electrolytes (1 M NaTFSI-TEGDME [Figure 9.27c] and 1 M $NaPF_6$–DGME [Figure 9.26d]). The voltage curves for the $NaPF_6$–DGME electrolyte exhibited more plateaus, which implies multi-phase transformations of the electrode material during the charge/discharge process. Moreover, the plateau at ~2.1 V decreased upon cycling and almost disappeared after 20 cycles in the TEGDME and DGME-based electrolytes. The cells with carbonate-based electrolyte exhibited a stable voltage profile during cycling, and the potential curves maintained the same shape and two stable flat plateaus at the initial cycle. Specifically, employing $NaFP_6$ in EC/EMC as the electrolyte allowed for the most promising electrochemical properties reported in the literature for TiS_2 up to date: a reversible capacity of 203 mAh g^{-1} at 0.2 C and 88 mAh g^{-1} at 10 C with a capacity retention of 92% over 50 cycles. A significant capacity fade, however, was still observed with continuous cycling, and this was due in part to the strains that the material underwent during sodiation, which resulted in fracture, as seen in Figure 9.27. It is interesting to note that TiS_2 was the initial cathode proposed to be used in secondary Li cells. More information on the evolution of damage in TiS_2 during ion insertion is given in Chapter 10.

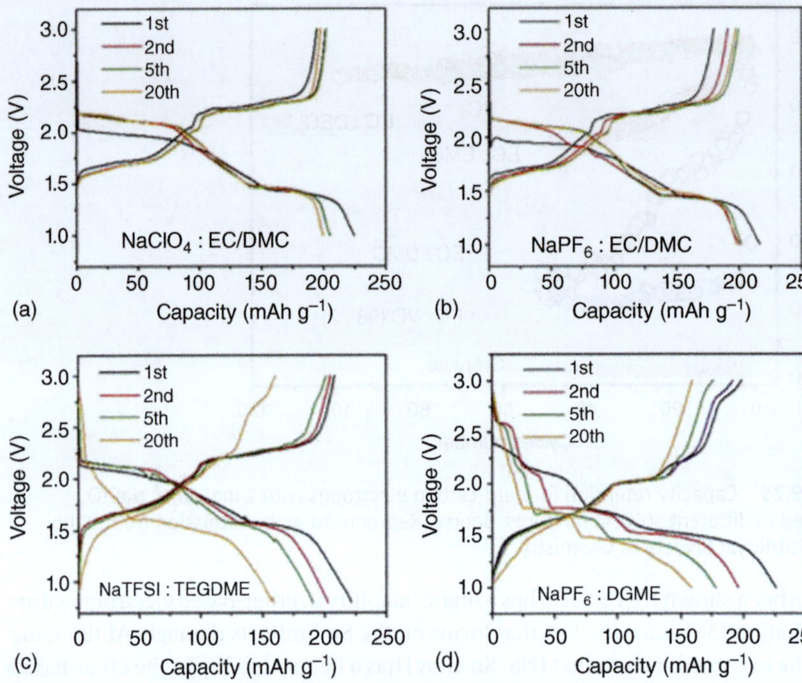

Figure 9.26 Comparison of the discharge–charge curves of TiS_2 electrodes in different electrolytes. (a) $NaClO_4$: EC/DMC, (b) $NaPF_6$: EC/DMC, (c) NaTFSI: TEGDME, and (d) NaPF6: DGME. Source: Reproduced with permission from Hu et al. [39]/Elsevier.

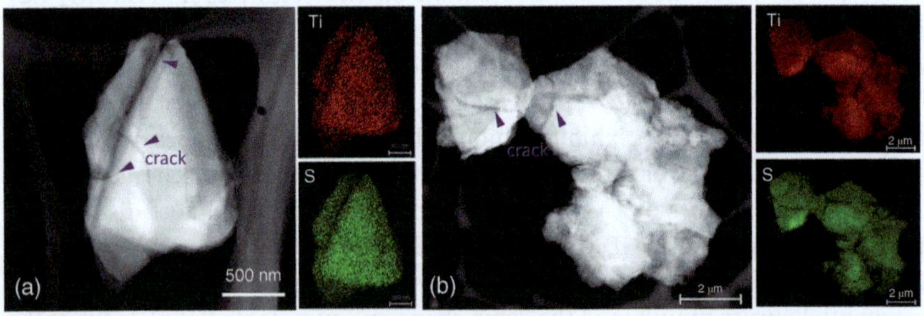

Figure 9.27 STEM images and the corresponding elements map for TiS_2 electrode after (a) 100 cycles and (b) 500 cycles. Source: Reproduced with permission from Hu et al. [39]/Elsevier.

9.5 Industrialization of SIBs

9.5.1 Status of Industrialization of SIBs

Since 2010, SIBs began receiving extensive attention from academia and industry. In 2011, after the establishment of the world's first company (British FARADION) focusing on the industrialization of SIBs, sodium-ion related research entered a

Table 9.1 The world's major sodium-ion battery manufacturers.

Nation	Company/ Organization	Sodium-ion battery products	Core indicators
China	CATL	Sodium-ion battery square shell; Cathode Material-Prussian White; Anode material-hard carbon.	160 Wh kg^{-1}, charging at room temperature for 15 minutes, the power can reach more than 80%, and the discharge retention rate at −20 °C is more than 90%.
US	Natron energy	Sodium-ion battery–symmetrical water-based battery; cathode material-high-rate Prussian blue.	50 Wh l^{-1}, cycled 10 000 times at 2 C.
UK	FARADION	Sodium-ion battery soft pack; cathode material-Ni-based layered oxide; anode material–hard carbon.	140 Wh kg^{-1}, 80% DOD cycle life exceeds 1000 times.
France	NAIADES	Sodium-ion battery-cylindrical; cathode material-sodium fluorophosphate; anode material-hard carbon.	90 Wh kg^{-1}, after 4000 cycles at 1 C, capacity retention 80%.
Sweden	ALTRIS	Cathode material-Prussian white.	—
Japan	Toyota Motor Corporation	Cathode material	—
Japan	Kishida Chemistry	Cathode material-transition metal oxide; electrolyte.	—
Australia	University of Wollongong	Cathode material-Prussian blue; anode material-hard carbon.	—

new era. At present, SIBs have gradually begun to move from the laboratory to the practical stage. More than 20 companies in the world have been commercializing sodium-ion batteries and have made important progress, even though they have yet to penetrate the market.

Table 9.1 compares the Na-ion battery material systems used by different companies. The cathode mainly consists of layered oxides (such as copper–iron–manganese and nickel–iron–manganese ternary materials), polyanionic compounds (such as sodium vanadium fluorophosphate), and Prussian white. The anode material system mainly includes soft carbon, hard carbon, and composite amorphous carbon materials.

As mentioned, the British FARADION company was the first that began the development and industrialization of SIBs. Its cathode material is Ni-based-layered oxide, and the anode material is hard carbon. The specific energy of the battery reaches 140 W h kg^{-1}, the average working voltage is 3.2 V, and the cycle life under 80% depth of discharge (DOD) can be predicted to exceed 1000 cycles. The high-rate water-based Na-ion battery developed by Natron Energy of the United States using

Prussian white materials has a life of 10 000 cycles at a rate of the 2 C, but the compaction density of Prussian white cathode materials is low, and its volume specific energy only 50 W h l^{-1}. NAIADES uses sodium vanadium fluorophosphate/hard carbon as the cathode and anode materials, and the battery has a working voltage of 3.7 V, specific energy of 90 W h kg^{-1}, and a life of 4000 cycles at 1 C. Since the electronic conductivity of the material (cathode or anode) is low, a carbon coating is used and the dimensions are reduced to the nanoscale and the compaction density is low. The SIBs developed by CATL use Prussian white and hard carbon materials as the cathode and anode electrodes, respectively, and the specific energy exceeds 160 W h kg^{-1}.

At present, the SIBs of several world manufacturers have achieved energy densities of 90–160 Wh kg^{-1}, which is much higher than the 30–50 Wh kg^{-1} of lead–acid batteries, and is equivalent to the 120–200 Wh kg^{-1} of LIBs using a $LeFePO_4$ cathode. In terms of cycle life, sodium batteries have achieved more than 1000 cycles, which is far beyond the 300 cycles of lead–acid batteries. Their cost, however, is comparable to that of LIBs.

9.5.2 Challenges of the Industrialization of Sodium-Ion Batteries

To a certain extent, the SIB industry can learn from that of LIBs. It can, therefore, be said that SIBs "stand on the shoulders of giants." However, we should also be aware that there are still some challenges in the process of developing Na-ion battery products and realizing their industrialization.

(1) At present, SIBs are in a state of parallel development of multiple material systems, and the processing performance of some of the cathode and anode material systems needs to be further improved. Among them, the hard carbon anode material has problems, such as low Coulombic efficiency in the initial cycle and unclear sodium-storage mechanisms. In addition, the research and development of electrolyte systems that match the cathode and anode materials are insufficient.
(2) Although most of the inactive materials (current collectors, binders, conductive agents, separators, casings, etc.) of SIBs can learn from the mature industry chain of Li-ion batteries, large-scale supply channels for active materials are still missing. Therefore, the source stability cannot be guaranteed, which in turn affects the stability of the production process and product quality.
(3) Compared with LIBs, the existing SIB system has a lower energy density; the amount of inactive substances and the cost ratio per unit energy density do not allow the cost advantage of the active materials to be fully utilized.
(4) Currently, an official release of standards and specifications related to SIBs does not exist, which affects the standardization of the SIB manufacturing processes and the consistency of product quality. This also leads to a difficulty in producing uniform and standardized products between different companies, which is not conducive to a product market promotion and cost reduction.

9.6 Conclusions

At present, Na-ion batteries are still in the initial stage of industrialization. The properties of both the cathode and anode materials need to be optimized, in order to further improve the energy density and safety of these batteries. The continuous advancements in Li-ion, Li-S, and Li solid-state systems will significantly aid in commercializing economically viable and environmentally friendly Na systems that will be particularly used in grid energy storage.

Acknowledgments

The authors are grateful to the National Science Foundation for supporting this work through the CMMI grant [CMMI-1762602].

References

1 Whittingham, M.S. (1978). *Prog. Solid State Chem.* 12 (1): 41–99.
2 Newman, G.H. and Klemann, L.P. (1980). *J. Electrochem. Soc.* 127 (10): 2097.
3 Mizushima, K., Jones, P.C., Wiseman, P.J., and Goodenough, J.B. (1980). *Mater. Res. Bull.* 15 (6): 783–789.
4 Delmas, C., Braconnier, J.-J., Fouassier, C., and Hagenmuller, P. (1981). *Solid State Ionics* 3: 165–169.
5 Yabuuchi, N., Kubota, K., Dahbi, M., and Komaba, S. (2014). *Chem. Rev.* 114 (23): 11636–11682.
6 Delmas, C., Fouassier, C., and Hagenmuller, P. (1980). *Physica B+C* 99 (1-4): 81–85.
7 Pan, H., Hu, Y.-S., and Chen, L. (2013). *Energy Environ. Sci.* 6 (8): 2338.
8 Yabuuchi, N., Yoshida, H., and Komaba, S. (2012). *Electrochemistry* 80 (10): 716–719.
9 Ma, X., Chen, H., and Ceder, G. (2011). *J. Electrochem. Soc.* 158 (12): A1307.
10 Niwa, H., Shibata, T., Imai, Y. et al. (2017). *Batteries* 3 (4): 5.
11 Lu, Z. and Dahn, J. (2001). *J. Electrochem. Soc.* 148 (11): A1225.
12 Berthelot, R., Carlier, D., and Delmas, C. (2011). *Nat. Mater.* 10 (1): 74–80.
13 Liu, Y., Shen, Q., Zhao, X. et al. (2019). *Adv. Funct. Mater.* 30 (6): 1907837.
14 Rajagopalan, R., Zhang, Z., Tang, Y. et al. (2021). *Energy Storage Mater.* 34: 171–193.
15 Hong, H.-P. (1976). *Mater. Res. Bull.* 11 (2): 173–182.
16 Delmas, C., Olazcuaga, R., Cherkaoui, F. et al. (1978). *Comptes Rendus Hebdomadaires Des Seances De L Academie Des Sciences Serie C* 287 (5): 169–171.
17 Qiu, Q., Li, C., Liu, H. et al. (2021). *J. Phys. Chem. C* 125 (28): 15200–15209.
18 Mason, C.W. and Lange, F. (2015). *ECS Electrochem. Lett.* 4 (8): A79–A82.
19 Wang, H., Zhang, T., Chen, C. et al. (2018). *Nano Res.* 11 (1): 490–498.

20 Wang, D., Bie, X., Fu, Q. et al. (2017). *Nat. Commun.* 8 (1): 1–7.
21 Zhou, W., Xue, L., Lu, X. et al. (2016). *Nano Lett.* 16 (12): 7836–7841.
22 Gao, H., Li, Y., Park, K., and Goodenough, J.B. (2016). *Chem. Mater.* 28 (18): 6553–6559.
23 Ma, X., Wu, X., Liu, Y. et al. (2021). *ACS Appl. Mater. Interfaces* 13 (18): 21390–21400.
24 Zhang, J., Liu, Y., Zhao, X. et al. (2020). *Adv. Mater.* 32 (11): e1906348.
25 Lu, Y., Wang, L., Cheng, J., and Goodenough, J.B. (2012). *Chem. Commun.* 48 (52): 6544–6546.
26 Hu, P., Peng, W., Wang, B. et al. (2019). *ACS Energy Lett.* 5 (1): 100–108.
27 Lee, H.W., Wang, R.Y., Pasta, M. et al. (2014). *Nat. Commun.* 5: 5280.
28 Jache, B. and Adelhelm, P. (2014). *Angew. Chem.* 126 (38): 10333–10337.
29 Stevens, D. and Dahn, J. (2001). *J. Electrochem. Soc.* 148 (8): A803.
30 Stratford, J.M., Mayo, M., Allan, P.K. et al. (2017). *J. Am. Chem. Soc.* 139 (21): 7273–7286.
31 Chevrier, V.L. and Ceder, G. (2011). *J. Electrochem. Soc.* 158 (9): A1011.
32 Wang, J., Eng, C., Chen-Wiegart, Y.-c.K., and Wang, J. (2015). *Nat. Commun.* 6 (1): 1–9.
33 Li, T., Gulzar, U., Bai, X. et al. (2019). *ACS Appl. Energy Mater.* 2 (1): 860–866.
34 Li, T., Gulzar, U., Proietti Zaccaria, R. et al. (2019). *J. Phys. Chem. C* 123 (24): 15244–15250.
35 Ponrouch, A., Marchante, E., Courty, M. et al. (2012). *Energy Environ. Sci.* 5 (9): 8572–8583.
36 Dahbi, M., Yabuuchi, N., Kubota, K. et al. (2014). *Phys. Chem. Chem. Phys.* 16 (29): 15007–15028.
37 Kim, H., Hong, J., Park, Y.U. et al. (2015). *Adv. Funct. Mater.* 25 (4): 534–541.
38 Zhang, B., Rousse, G., Foix, D. et al. (2016). *Adv. Mater.* 28 (44): 9824–9830.
39 Hu, P., Wang, B., Xiao, D., and Aifantis, K. (2019). *Nano Energy* 63: 103820.

10

Modeling Ion Insertion for Predicting Next-Generation Electrodes

Bo Wang[1], *Fei Shuang*[2], *and Katerina E. Aifantis*[1]

[1] *University of Florida, Mechanical and Aerospace Engineering, 1064 Center, Gainesville, FL 32611, USA*
[2] *The University of Texas at San Antonio, Department of Mechanical Engineering, 1 UTSA Circle, San Antonio, Texas 78249, USA*

10.1 Introduction

As described in Chapter 6, graphite is the conventional and commercially used anode in Li-ion batteries (LIBs), due to its stable mechanical and electrochemical performance, despite its limited theoretical capacity of 372 mAh g^{-1}. Even though Si provides a capacity that is approximately 10 times higher (3579 mAh g^{-1}), it experiences volume expansions that are nearly 400% during the formation of Li–Si alloys (Li_xSi, $0 < x < 3.75$ at room temperature; at higher temperatures x can reach 4.4), and respective contraction takes place during the delithiation process. Naturally, these volume changes are accompanied by changes in the lattice spacing, and diffusion-induced stresses can occur. Especially for active materials with a high theoretical capacity, phase transformations are often involved. As Li-ions are inserted into the active material (Si), the outer surface of the particle becomes Li–Si, while its core remains pure Si. The high stress arising from the mismatch between the expanded (Li–Si) and non-expanded (Si) regions can lead to failure and fracture of the active particles and strongly affects the capacity and cycle life, as was documented in Chapter 6.

To commercialize these promising electrode materials, it is necessary to understand the mechanical stresses that develop throughout the lithiation/delithiation process. An overview of existing theoretical and computational approaches for doing so is summarized in this chapter. Since similar volume changes that take place during lithiation are also present for sodiation in Na-ion batteries (SIBs), the frameworks below can also be applied for such systems as well, and in fact for any metal-ion battery system. However, as most of these models are targeted toward LIBs, the term Li-ion will be commonly used in the sequel.

Rechargeable Ion Batteries: Materials, Design, and Applications of Li-Ion Cells and Beyond, First Edition.
Edited by Katerina E. Aifantis, R. Vasant Kumar, and Pu Hu.

10.2 The Role of Mechanics in Batteries

As described in Chapter 1, the energy density, power density, and cycle life of batteries are closely related to the continuous ion-induced stresses that take place during the charge/discharge cycle. Excessive stresses result in dislocations, and as they accumulate, plastic deformation occurs causing the electrodes to fail by fatigue or fracture under repeated cycling. Understanding the stress evolution process is crucial for revealing the damage mechanisms.

If the material undergoes phase transformations during ion insertion, it can lead to a high tensile hoop stress as the outer shell shrinks while the interior core (unlithiated/unsodiated) preserves its volume. This phenomenon can cause cracks in the particle to grow since the tensile hoop stress is the crack-driving force. Figure 10.1 shows the morphological changes of a Si electrode particle during the lithiation stage. It can be seen clearly that cracks initiate and propagate during the insertion process due to the higher than 300% volume expansion. It is reminded that cracks do not only occur in high-volume expanding active materials, but can also take place in electrode materials (cathodes) with relatively smaller volume changes.

The volumetric changes can also result in the detachment of the active materials from the current collector, after which they can no longer contribute to the electrode capacity. The three failure mechanisms that can lead to cell fracture and capacity fade are shown in Figure 10.1.

In addition to the mechanical instabilities that occur in the active materials, the formation of a solid electrolyte interface (SEI) layer on the electrode surfaces is also responsible for cell fracture and capacity decay. The SEI (composition of Li_2CO_3, LiF, $ROCO_2Li$, Li_2O, etc.) normally acts as a barrier between the electrolyte solution

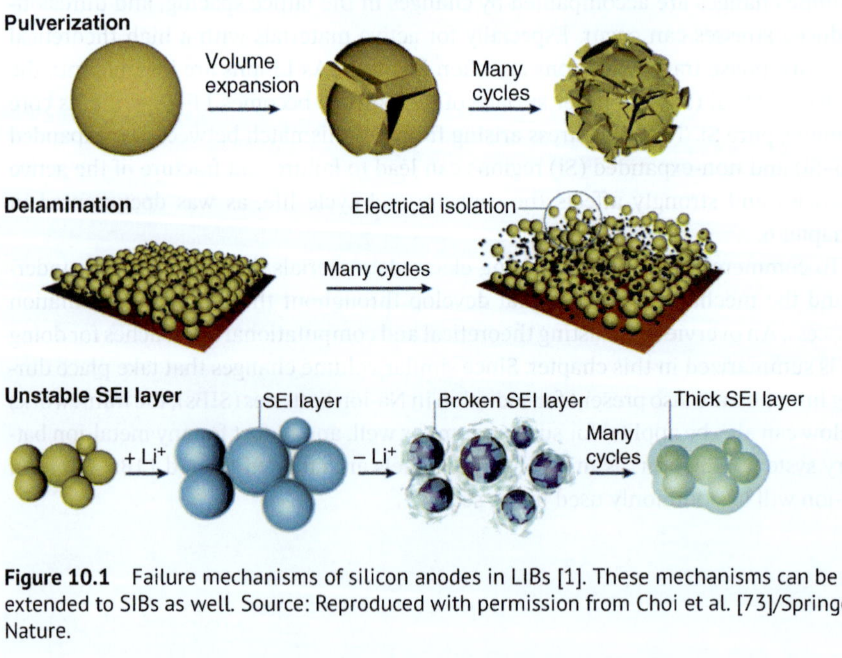

Figure 10.1 Failure mechanisms of silicon anodes in LIBs [1]. These mechanisms can be extended to SIBs as well. Source: Reproduced with permission from Choi et al. [73]/Springer Nature.

and the electrode. It is thermodynamically favorable since it is ionically conductive and electronically insulating, thus preventing further chemical side reactions. The SEI layer formed on some active material surfaces, such as Si, is not flexible enough to withstand the dramatic volume expansions, and thus fracture occurs, exposing more Si surface to chemical attack by the electrolyte [2]. Therefore, the SEI layer continuously forms and regrows accompanied by the continuous consumption of active lithium and electrolyte.

There are multiple interfaces inside a battery system where reactions take place: phase interfaces (e.g. $LiFePO_4/FePO_4$, and $Li_{4.4}Si/Si$), gain boundaries (the boundaries between different grains in a polycrystalline electrode), solid–solid interfaces (active materials, conductive fillers, polymer binders, and current collectors), and electrode–electrolyte interfaces (SEI layer). For electrodes that undergo phase transformations, there is a sharp phase boundary between the lithiated phase and the pristine phase. Steep Li-concentration gradients across the boundary between the lithiated and unlithiated phases lead to mismatch strains and severe stresses. If there are massive multiple grain boundaries in the electrodes, some lithium ions can be stored at the grain boundaries. The chemo-mechanical (CM) properties, such as the ion transport and the stability of the electrodes, may also be affected by such interfaces. These solid–solid interfaces include spatial constraints by adjacent particles, polymer binders, conductive additives, current collectors, and matrix coatings. These natural boundaries are the major sources of stress generations, and therefore, in order to have a better understanding of the stress generation inside a battery system, it is necessary to take these factors into consideration. In this sense, the mechanics in the full electrode are also important.

In summary, the stresses during Li-ion diffusion can arise from a variety of reasons, which are as follows:

(i) Physical constraints (Figure 10.2a,b) on the expansion of the active materials. These constraints include spatial constraints by adjacent particles, polymer binders, conductive additives, current collectors, and matrix coatings. These natural boundaries are the major sources of stress generation [3].
(ii) Volume mismatch (Figure 10.2c,d) between the lithiated and pristine phases. This can be thought of as a concentration-gradient-induced stress. A steep Li-concentration gradient across the interface between the lithiated and unlithiated phases leads to mismatch strains and severe stresses since a gradient in concentration results in heterogeneous volume expansions and contractions. Heterogeneous swelling and shrinkage of the anode and the cathode can lead to severe stresses. This is because the battery has a different volume when it is fully charged compared to when it is completely discharged. The battery pack needs to accommodate such deformation [3].

10.2.1 Initial Modeling of Damage Using Fracture Mechanics

Mechanicians initially tried to model the volume expansions of active materials during lithiation using purely mechanics frameworks, which did not account

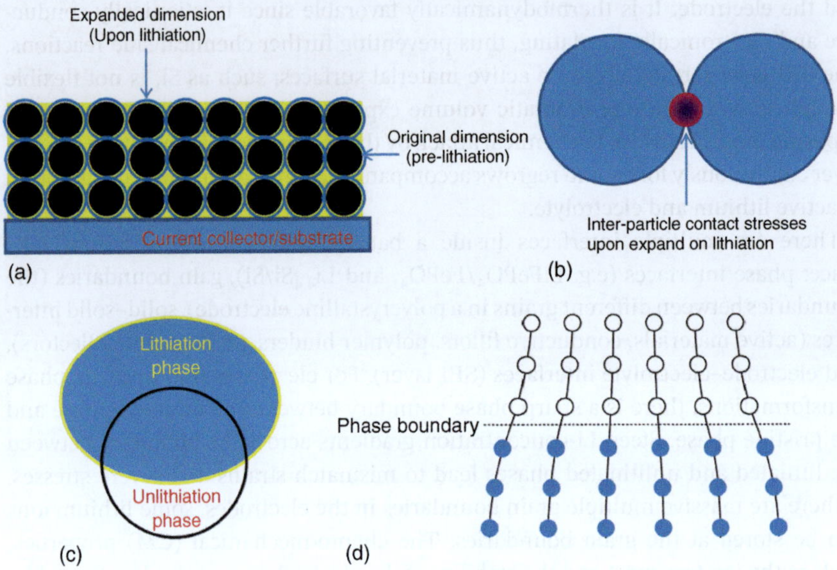

Figure 10.2 Schematic illustrations of (a) constraining effects of neighboring active particles, inactive matrix, and current collector/substrate on the expanding active particles upon lithiation of "porous composite" electrode; (b) stress generated at the inter-particle contact between the expanding electrode particles upon lithiation; (c) Li-concentration gradient between lithiated and unlithiated portions of a particle, resulting in the development of stress discontinuities. (d) Mismatch between crystalline phases. Source: Adapted with permission from Mukhopadhyay and Sheldon [3]/Elsevier.

for Li-ion diffusion. Initial theoretical consideration of Li-battery anodes done by Wolfenstine [4] suggested that decreasing the particle size is not a practical solution to solve the mechanical instability problem of Li alloys. Further theoretical considerations were done by Huggins and Nix in 2001 [5]. The configuration they considered was that of a bilayer system, by which a Sn anode was deposited on a Cu substrate. They used a simple one-dimensional decrepitation model, predicting a terminal particle size of the order of a fraction of a micron. This model was concerned with internal stress developments and fracture [6] of an epitaxial thin film on a compliant substrate. Therefore, it was not applicable for the active nanoparticles (NPs) with radial geometries. Also, it was not possible to develop precise design criteria.

Aifantis and coworkers were the first who modeled the mechanical behavior of spherical active anode particles undergoing severe volume expansions beginning in 2003 [7] and further motivated the mechanics community to try and understand the mechanical behavior of such materials. The configuration they considered was an active particle surrounded by a matrix material that could constrain the volume expansion. This could be the binder, a coating, or the solid electrolyte in new solid-state systems. As a first step, they used either plane stress or plain strain conditions to look at the purely elastic response of the unit cell upon maximum Li insertion. This plane stress condition essentially corresponds to a thin film (active/less active nanostructured) anode, which, however, has no effect from the substrate. In continuing, to capture the effect of crack formation, they inserted a damage zone at the active site – matrix interface, which accounted for the fracture that takes place, and by applying linear elastic fracture mechanics, they were able to predict the energy-release rate and stability index during crack growth, indicating the distance over which stable crack growth would occur [8]. This work allowed the development of design criteria [9], for selecting material properties and configurations (thin-film/cylindrical/spherical particles) that would limit damage. Although the model did not explicitly account for ion insertion and de-insertion, or the interface between the particle and matrix, it was in agreement with experimental data. Specifically, the stability diagrams indicated that cracking in thin films was highly unstable, while for spherical and cylindrical particles it was very stable. This was in agreement with experimental data, which have shown that fracture in Si and Sn thin films occurs after the first cycle, as seen in Chapter 6. Using further linear elastic mechanics considerations [10], it was shown that increasing the aspect ratio of the active particles increases their resistance to delamination from the matrix during volume expansions upon maximum Li-ion insertion.

Understanding the configuration/geometry of the active particles is important, but as seen in Chapter 6, particle size also plays a key role in the mechanical degradation that will take place during lithiation. The most efficient theories that can capture size effects are gradient theories, which were initially introduced for plasticity [11–13] and elasticity [14], hence a gradient damage model was employed to account for the effect of particle size and interparticle spacing on damage in Sn-based anodes. The system considered was Sn spherical particles embedded in a matrix material to constrain their volume expansion, and by comparing the damage ratio it was possible to select the material that would most efficiently limit damage. As seen in Figure 10.3, graphene was predicted to be the most promising matrix.

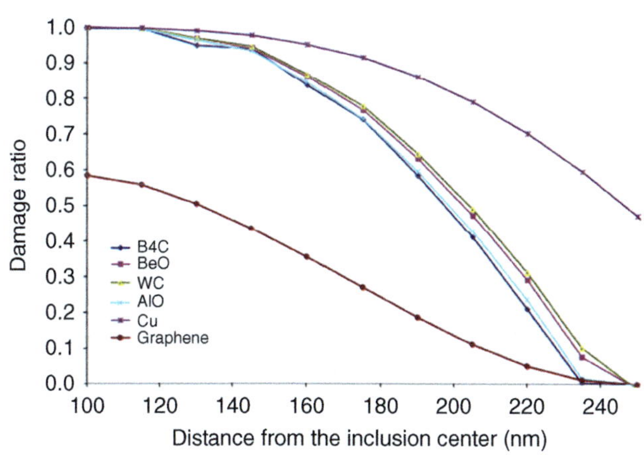

Figure 10.3 Damage distribution along the radial distance from the active particle center. Different materials were considered as the surrounding matrix. The radius of the spherical inclusion is 100 nm and the interparticle spacing (distance between particle centers) is 1000 nm. Source: Reproduced with permission from Dimitrijevic et al. [15]/Elsevier.

Comparing the response for different particle sizes indicated that when graphene was used as the matrix, maintaining a particle size of 20 nm and an interparticle spacing that is greater than 30 nm resulted in minimal damage and therefore would be a most promising anode configuration, which is in agreement with experimental studies on Sn/graphene composite anodes [15]. It is interesting to note that in the initial models by Aifantis et al., some ~20 years ago [9], glass materials were examined as a matrix, into which active particles were embedded. At that time, such a configuration was not employed experimentally, however, in current solid-state batteries, it is indeed proposed to directly embed the active material in the solid electrolyte, which can be a glass material.

10.3 Accounting for Li-Ion Diffusion

10.3.1 Modeling the Diffusion-Induced Stress in Single Particles

Multiple models have been developed to study diffusion-induced stresses (DISs) in materials, well before this was of interest in LIBs. The phenomenon of DIS was first studied by employing an analogy between thermal stress and DIS to analyze the transverse stresses introduced by solute lattice contraction of boron and phosphorus in a thin silicon plate during mass transport [16]. The stress generation due to diffusion was captured through lattice deformation analysis and it was applied to observe the effect of phase transformations on DISs. Li provided a number of analytical solutions to DIS problems in elastic media of simple spherical and cylindrical geometries [17]. Larché and Cahn investigated the DISs arising from the inhomogeneous concentration in materials [18, 19]. Lee and coworkers studied the effect of DISs on diffusion in hollow cylinders [20, 21] and square sandwich composites [22]. Later on, numerous researchers followed this analogy to investigate the DISs in cylindrical electrodes [23, 24], nanowire electrodes [25], spherical electrode particles [26–28], and layered electrode plates [29].

Most recently, based on Gibbs free-energy and mass-transport models, a coupled model was proposed for the lithium diffusion and stress generation in a particle of active material undergoing a phase change [30]. The model can be used to compute the lithium concentration and stress in storage particles for an anode at fixed Li-extraction and insertion rates. To reveal the intrinsic relation between large deformations and Li-ion insertion/extraction, fully coupled diffusion–deformation models have been developed [31–33]. These frameworks relate continuum mechanics with Cahn–Hilliard diffusion models, and account for transient diffusion of lithium and accompanying large elastic–plastic deformations; however, they are quite complex and cannot be easily related to experiments.

10.3.1.1 Analytical Modeling of DISs Under Elastic Deformation

TiS_2 is one of the few materials that can be used as an electrode in both LIBs and SIBs; its examination can, therefore, illustrate the differences that arise based on the ion being intercalated. In a study by Hu et al. [34], a comprehensive comparison of Li-ion and Na-ion insertion in TiS_2 was performed, both experimentally and theoretically. TiS_2 undergoes a phase transformation after sodiation, leading to a sharp concentration profile near the phase boundary between the sodiated shell and pristine core. However, there is no phase change during lithiation, which means that the concentration is relatively smooth and continuous across the whole TiS_2 particle. In this section, we summarize the three-dimensional spherical and analytical model that was used to capture the concentration profile and stress evolution during the first lithiation/sodiation process [34]. The diffusion of Li^+ or Na^+ is assumed to be isotropic and is, therefore, a function of the radius only, so the concentration is also taken to be a function of:

$$c = c(r)$$

During ion insertion, a phase transformation takes place, and therefore the Fickian diffusion equation with a nonlinear concentration-dependent diffusivity [35] is used to capture the sharp phase boundary. If the chemo-mechanical coupling effects are taken into consideration, the chemical potential coupled with the stress [32] needs to be adopted. Alternatively, as indicated in Ref. [36], the concentration growth can be modeled by a generalized logistic function:

$$c(r) = \frac{1}{\left[1 + Qe^{-B(r-r_c)}\right]^{1/\alpha}} \tag{10.1}$$

where c is the normalized concentration of Li^+ or Na^+ and ranges from 0 (no ions) to 1 (full insertion); the two limits $c = 0$ and $c = 1$, correspond to the pristine TiS_2 and lithiated $LiTiS_2$ (or sodiated $NaTiS_2$), respectively. The physical meaning of the parameters Q, B, and α is given in Table 10.1, and r_c represents the position where $c = 0.5$. The aim was to study the stress evolution of sodiation vs. lithiation, rather than the whole dynamic ion-insertion process. Therefore, the concentration was approximated by the generalized logistic equation, which simplifies the computation, while still capturing the phase transformation that takes place in the case of sodiation.

Due to spherical symmetry, there are three nonzero stress components in the particle, i.e. the hoop (tangential) stresses ($\sigma_\theta = \sigma_\varphi$) and the radial stress (σ_r). By using

Table 10.1 Parameters for simulating the stress evolution during lithiation and sodiation.

Symbols	Physical meaning	Numeric values	
		Li	Na
Q	A parameter that in part determines the point of inflection	1	1
B	Growth rate	5	100
α	A variable that fixes the point of inflection	1	1
E	Young's modulus (GPa)	90	75
υ	Poisson's ratio	0.2	0.2
β	Chemical expansion coefficient	0.088/3	0.177/3

an analogy to thermal expansion, the constitutive relations for the stresses (σ_r and σ_θ) due to ion diffusion can be written as:

$$\left.\begin{aligned}\sigma_r &= \frac{E}{(1+\upsilon)(1-2\upsilon)}\left[(1-\upsilon)\frac{du}{dr}+2\upsilon\frac{u}{r}-(1+\upsilon)\beta c\right]\\ \sigma_\theta &= \frac{E}{(1+\upsilon)(1-2\upsilon)}\left[\frac{u}{r}+\upsilon\frac{du}{dr}-(1+\upsilon)\beta c\right]\end{aligned}\right\} \tag{10.2}$$

where E is Young's modulus, v is Poisson's ratio, u is the radial displacement, β is the expansion coefficient of the particle, and c is the ion concentration.

The stress-equilibrium equation for a spherically symmetric particle is:

$$\frac{d\sigma_r}{dr}+\frac{2(\sigma_r-\sigma_\theta)}{r}=0 \tag{10.3}$$

The boundary conditions (BCs) are: $u(r=0)=0$ and $\sigma_r(r=R)=0$. Therefore, the stress field in the particle is:

$$\left.\begin{aligned}\sigma_r &= \frac{2\beta E}{1-\upsilon}\left(\frac{1}{R^3}\int_0^R cr^2\mathrm{d}r-\frac{1}{r^3}\int_0^r cr^2\mathrm{d}r\right)\\ \sigma_\theta &= \frac{2\beta E}{1-\upsilon}\left(\frac{1}{R^3}\int_0^R cr^2\mathrm{d}r+\frac{1}{2r^3}\int_0^r cr^2\mathrm{d}r-\frac{c}{2}\right)\end{aligned}\right\} \tag{10.4}$$

The concentration distribution is approximated by the logistic function as given in Eq. (10.1).

Based on the above equations, the stress field can be obtained at various locations in the particle and at different stages of ion insertion. For illustration purposes, we plot the solutions when r_c = 25%, 50%, and 75%, respectively. The parameters of the simulation are given in Table 10.1, and it is noted that the values of the expansion coefficient β correspond to the volume changes due to lithiation and sodiation, which are 8.8% and 17.7%, respectively, according to experimental measurements.

Figure 10.4a,b shows the schematic diagrams of the stress components in a spherical free-standing particle due to sodiation (Figure 10.4a)/lithiation (Figure 10.4b). Sodiation results in a large volume expansion of 17.7%, so there is a high mismatch between the sodiated (shell) and pristine phases (core). Also, there is a phase transformation that takes place giving rise to a sharp phase boundary, as seen in Figure 10.4a, between the core and shell. In contrast, lithiation results in a low-volume expansion (8.8%) and results in a smooth phase transition (single-phase reaction), as shown in Figure 10.4b. Figure 10.4c plots the concentration profile normalized by its maximum value (either at the fully sodiated or lithiation state) as a function of the radial distance that is normalized by the initial TiS_2 particle radius.

From Figure 10.4d,f it is concluded that for sodiation, σ_r and σ_θ have the same value that remains constant in the pristine TiS_2 core. Specifically, when 50% of the sodiated shell is at 0.25R, 0.5R, and 0.75R, the stress values are 2.13, 3.22, and 3.63 GPa, respectively. When 50% of the lithiated region is located at the same position, these values gradually decrease from the center to the surface of the particle. The stresses follow the same trends for both cases; in the shell, σ_r

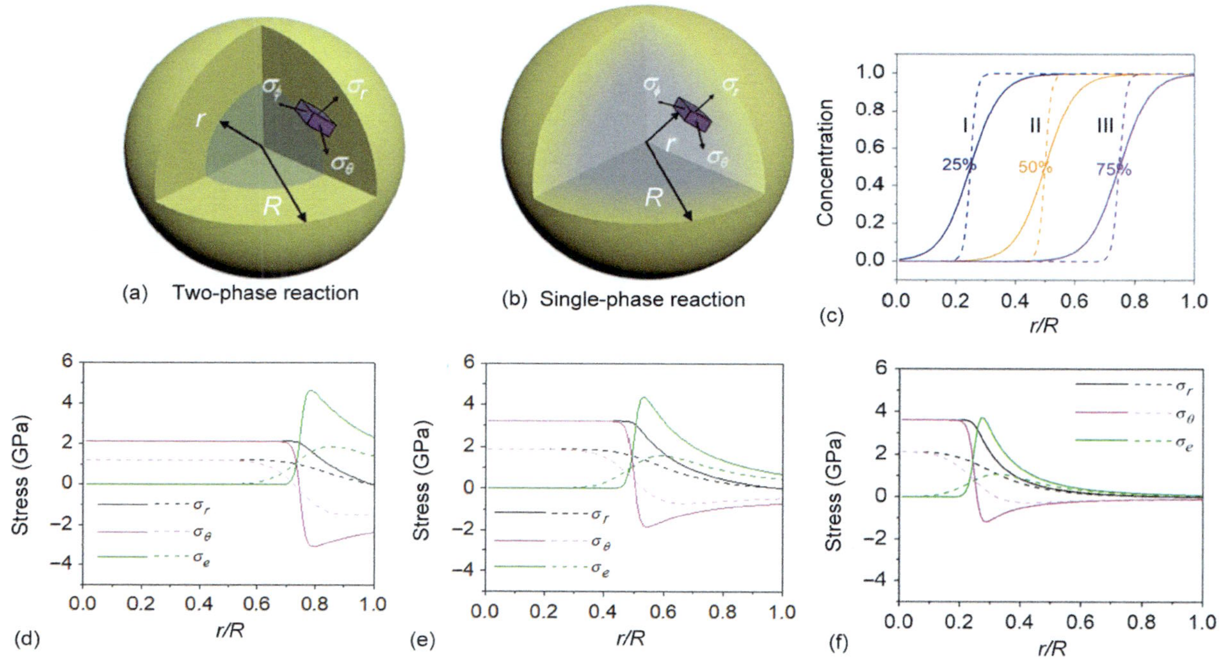

Figure 10.4 Schematic diagram of the stress components in a spherical particle undergoing (a) phase transformations (sodiation) and (b) a single-phase reaction (lithiation), where σ_θ, σ_φ are the tangential stress components, σ_r is the radial stress, R is the radius of the particle, and r is the distance from the representative material element to the center of the particle. (c) The normalized Li/Na concentration profile when 50% of the lithiated or sodiated region is located at 0.25R, 0.50R, and 0.75R from the center of the particle (from left to right), respectively. (d), (e), and (f) are the stress profiles of σ_r, and σ_θ and the von Mises effective stress σ_e, corresponding to the Li/Na concentration profiles of (I), (II), and (III) in (c). Source: Reproduced with permission from Hu et al. [34]/Elsevier.

gradually decreases to zero from the interface to the outer surface and σ_θ transitions from tension (positive value) to compression (negative value). Particularly, in lithium-poor and sodium-poor regions σ_θ is negative, but it becomes positive in lithium-rich and sodium-rich regions. Some brittle materials deform with visible shear bands. In this case, a shearing-type failure may occur along these bands/planes. It is assumed that the material fails when the shear stress acting on these planes is large enough to shear the material along them. This is the so-called Tresca yield criterion, which states that the material will fail at any location where the maximum shear stress reaches one-half the tensile yield stress of the material. In our case, due to spherical symmetry, the maximum shear stress is equal to the effective von Mises stress. The effective von Mises stress (σ_e) is defined as:

$$\sigma_e = |\sigma_r - \sigma_\theta| \tag{10.5}$$

The von Mises stress remains zero during sodiation, inside the pristine TiS_2 core, but in the shell, it rises to a maximum (near the interface) and then it slowly decreases. The maximum values of σ_e are 3.72, 4.38, and 4.65 GPa, which correspond to concentration profiles (I), (II), and (III) in Figure 10.4c, respectively. During lithiation, however, σ_e undergoes small changes over the whole volume. The maximum values of σ_e are 0.67, 1.07, and 1.37 GPa for the concentration profiles (I), (II), and (III) in Figure 10.4c, respectively; they are much lower than in the sodiation case.

It should be noted that the analytical solutions are only available for electrodes undergoing elastic deformations. When plastic deformation takes place, the analytical solution cannot be obtained. Also, ion insertion often leads to mechanical property changes in the electrodes, and analytical solutions cannot be obtained for a varying elastic modulus. For all such cases, more generalized models are needed, as will be discussed in Section 10.3.1.2.

10.3.1.2 Phase-Field Modeling of DISs

There are two different methods to model the sharp phase boundary that occurs during phase transformations: sharp interface models and diffusive interface models. For sharp interface models, concentration and stress discontinuities appear at the interface. The moving phase boundary needs to be tracked and specific boundary conditions need to be applied during insertion and de-insertion. Therefore, sharp interface models pose numerical difficulties for practical applications.

Diffusive models, such as phase-field models, are much more convenient as they use diffusive interfaces with a finite thickness (but very thin) to approximate sharp interfaces. By employing a diffusive interface model, the ion distributions can be modeled with a continuous concentration field. As a result, the reaction front does not need to be explicitly tracked. Such phase-field models have been successfully applied to study the stress generation in $LiFePO_4$ [37] and $LiMn_2O_4$ [38] particles.

Almost all articles, which use phase modeling (except for [39]), for studying DISs in electrode materials, choose to do so for cathode materials. The reason is that cathodes undergo small volume changes, i.e. $LiFePO_4$ contracts 7% upon complete Li de-insertion, and therefore small strain deformation can be used. As described

Table 10.2 Young's modulus of some cathode and anode materials in both lithiated and delithiated states.

Delithiated phase	Method	Young's modulus (GPa)	Lithiated phase	Young's modulus (GPa)
CoO_2	Computation [53]	59.8	$LiCoO_2$	264
C(graphite)	Computation [54]	32	LiC_6	109
Si(amorphous)	Computation [55]	96	$Li_{3.75}Si$	41
Si(crystalline)	Experiment [51]	90	$Li_{3.75}Si$	12
Sn(crystalline)	Computation [56]	51	$Li_{3.5}Sn$	24.7

in Chapter 6, most anode materials, particularly Si which is the most promising one, undergo large volume expansions up to ~400% ($Si_{4.4}Li$) upon complete Li-ion insertion. Although initially, plastic deformation was not considered in modeling, *in situ* measurements [40], first-principle studies [41], and molecular dynamics simulations [42] have shown that yielding and subsequent plastic deformation takes place in Si during Li-ion insertion. Models that account for plasticity (elastoplastic and viscoplastic) have been employed to capture DISs during lithiation in Si and Sn [36, 43, 44], and a common observation is that the hoop stress transitions from compressive to tensile during Li-ion insertion near the surface of the Si particles. This can explain the fracture that is observed experimentally upon the first lithiation for Sn nanoparticles [45], as well as Si nanospheres [35] and nanowires [46].

Despite the fact that there have been a lot of models on DISs of electrode materials, a comprehensive understanding has yet to be achieved. One aspect is the chemo-mechanical (CM) coupling effect. Initial works, particularly those using the "shrinking core" model, did not consider CM coupling. Since then, there have been some models that considered the coupling effect between chemical diffusion and mechanical deformation [32, 41, 47–52].

However, few works have taken into consideration the change that occurs in the elastic modulus of the electrode materials during Li^+ insertion. During insertion and de-insertion, elastic softening or hardening may occur. By softening we mean that the elastic modulus of the electrode is reduced upon Li insertion, while by hardening, that the elastic modulus increases upon lithiation. Both experiments [40, 51] and first-principle studies [52–54] have shown that there is a strong dependency of the material's elastic properties on concentration. Table 10.2 shows the elastic (Young's) modulus of the lithiated and delithiated phases of some electrode materials. A decrease (softening) or increase (hardening) in the elastic modulus has been observed experimentally [40, 51] and theoretically verified [52–54].

An initial study that considered a varying Young's modulus throughout Li-ion insertion employed linear and gradient elasticity [57]. When the elastic modulus variation was only 10% and the insertion reaction single phase, the results were in close agreement with the constant modulus case. For such cases, assuming a constant modulus is a valid approximation. However, very few studies [58], [32] have looked at the effect of a varying elastic modulus in the DISs of electrodes that undergo phase transformations. Herein, the results of a recent systematic study that accounted for both CM coupling and a concentration-dependent modulus, both for elasticity and plasticity, will be summarized. Since only the first lithiation will be considered, the elastic case is appropriate for cathodes, which experience low-volume expansions, while the plastic case is for anodes, such as Si.

Phase-Field Modeling of DISs Under Elastic Deformation Based on a variational formulation, a theoretical framework for the evolution of DIS can be established [59]. The effect of concentration-dependent material properties on DISs will be studied for electrodes undergoing both elastic and plastic deformations during ion insertion.

When the elastic modulus of the material is strongly dependent on concentration, a simple linear relation can be assumed as:

$$E(c^*) = E_0 + kc^* \tag{10.6}$$

where k is a softening or hardening parameter, E_0 is the Young's modulus of the Li-poor (delithiated) phase, and c^* is the relative concentration, ranging from 0 (no ions) to 1 (full insertion). The chemical potential μ is given by:

$$\begin{aligned}\mu = \mu_0 + RT\ln\left(\frac{c^*}{1-c^*}\right) + RT\chi(1-2c^*) - \Xi\sigma_h \\ - c_{\max}\,\nabla\cdot(\kappa\nabla c^*) - \frac{1}{2c_{\max}}\frac{dS_{ijkl}}{dc^*}\sigma_{ij}\sigma_{kl}\end{aligned} \tag{10.7}$$

where R is the gas constant (8.314 J K^{-1} mol^{-1}); T is the absolute temperature (300 K); μ_0 is a reference value of the chemical potential of the diffusing species, set as 0 since it will not affect the diffusion behavior; χ is the constant partial molar volume, which is dimensionless (to ensure phase separation, χ must be greater than 2); κ is the gradient energy coefficient, with units of energy per unit volume times a length squared; Ξ is the partial molar volume of Li ions; $\sigma_h = \frac{1}{3}\mathrm{tr}(\sigma)$ is the hydrostatic stress; and S_{ijkl} is the concentration-dependent compliance tensor. Therefore, the chemical potential relies on both the Li concentration and the local stress state, as well as the concentration gradient and the concentration-dependent material properties due to lithium insertion/de-insertion. For linear elastic and isotropic materials, the compliance tensor is given by:

$$S_{ijkl} = \frac{1-2\upsilon}{3E}\delta_{ij}\delta_{kl} + \frac{1+\upsilon}{2E}\left(\delta_{ik}\delta_{jl} + \delta_{il}\delta_{jk} - \frac{2}{3}\delta_{ij}\delta_{kl}\right) \tag{10.8}$$

As a result,

$$\frac{dS_{ijkl}}{dc^*}\sigma_{ij}\sigma_{kl} = \frac{1}{E^2}\frac{dE}{dc^*}(\upsilon\sigma_{ii}\sigma_{kk} - (1+\upsilon)\sigma_{jl}\sigma_{jl}) \tag{10.9}$$

The details regarding the complete model (governing equations) and numerical implementation (solving the weak form of the governing equations by finite-element method) are given in Ref. [59]. It should be noted that this work can be applied to other electrochemical systems that operate under the diffusion of ions, such as Na-ion batteries.

Table 10.3 Material properties of $LiFePO_4$.

Name	Symbol and unit	Value
Diffusion coefficient	D ($m^2 s^{-1}$)	10^{-14}
Young's modulus	E_0 (GPa)	124.5
Poisson's ratio	v	0.25
Partial molar volume	Ξ ($m^3 mol^{-1}$)	2.9×10^{-6}
Gradient coefficient	$\overline{\kappa}$ ($J m^{-1}$)	5.02×10^{-10}
Expansion coefficient	β	2.21×10^{-2}
Maximum concentration	c_{max} ($mol m^{-3}$)	2.29×10^4

To study the stress evolution of a spherical $LiFePO_4$ electrode, the parameters in Table 10.3 were employed. The other parameters used were: $\rho = 100$ nm, $E_0 = 124.5$ GPa, and $T = 300$ K, $\Delta t = 0.0005$, where Δt is the normalized time step. By using these parameters, finite-element computations were performed to obtain the evolution of the concentration, radial stress, hoop stress, and hydrostatic stress, as shown in Figure 10.5. In the plots (Figure 10.5), the solid lines stand for the case when the chemo-mechanical (CM) coupling effect is considered and dashed lines are for the case without CM coupling. Here we consider the evolution of the DISs during the Li-ion insertion process only and a constant Young's modulus ($k = 0$) is assumed.

It takes approximately 6000 Δt to reach fully lithiated state for the $LiFePO_4$ particle. Concentration distributions at different lithiation stages (500 Δt, 2000 Δt, and 4000 Δt, respectively) are as shown in Figure 10.5a. A sharp interface separates the Li-rich and Li-poor phases. At the early stage of lithiation (500 Δt), the concentration and stress (Figure 10.5b–d) profiles almost coincide with each other. As lithiation proceeds (2000 Δt and 4000 Δt), the Li-poor phase tends to have a slightly higher concentration and the ion diffusion tends to move faster when CM coupling is considered.

Both the radial σ_r (Figure 10.5b) and hoop σ_θ (Figure 10.5c) stresses are almost constant in the core (Li-poor) region. σ_r gradually decreases to zero from the center toward the outer surface and σ_θ transitions from tension in the core to compression in the shell. The hydrostatic stress σ_h (Figure 10.5d) shows a similar trend, transitioning from constant tension in the Li-poor region to compression in the Li-rich shell. In Eq. (10.7), the compressive/tensile hydrostatic stress leads to a higher/lower chemical potential. Under such a hydrostatic stress state, a chemical potential gradient is induced, which acts as a driving force for diffusion from the compressive to the tensile side. It is reflected on the concentration profiles, showing that the phase-boundary interface moves faster when CM coupling is considered. When radial, hoop, and hydrostatic stresses are compared for both the cases, with and without CM coupling, higher values are predicted for all these stresses in the coupling cases at the same time step. This is because a higher equilibrium

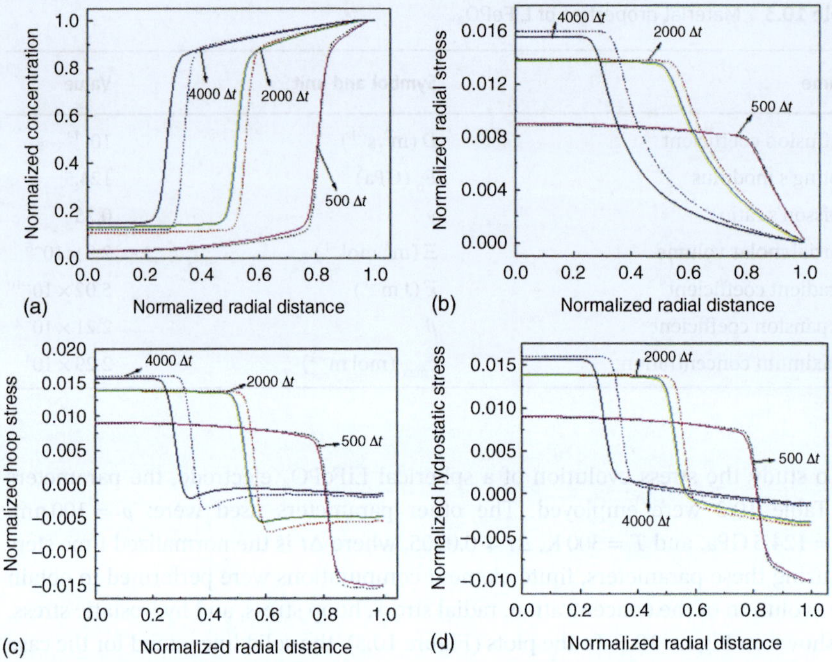

Figure 10.5 Radial distribution of (a) concentration profiles, (b) normalized radial stress, (c) normalized hoop stress, and (d) normalized hydrostatic stress at time steps 500 Δt, 2000 Δt, and 4000 Δt with/without chemo-mechanical-coupling effects. The solid and dashed lines are for with and without the chemo-mechanical-coupling effect, respectively. In this simulation, a constant modulus is assumed, i.e. $k = 0$. Source: Reproduced with permission from Wang et al. [59]/Elsevier.

concentration is obtained for the Li-poor phase in the coupling case, while equilibrium concentrations are almost the same for the Li-rich phase. The coupling model predicts a smaller miscibility gap (concentration differences between the Li-rich phase and Li-poor phase), thus, less volume changes and DISs are induced.

In order to study the influence of material property changes due to lithiation, the parameter k is set to be $k = -0.1E_0$ and $k = +0.1E_0$ (corresponding to a 10% reduction or increase in modulus with increasing ion concentration, respectively). All the other parameters are the same and CM coupling is considered. Since the stress and strain profiles have similar trends, only the profiles at the time step 2000 Δt are plotted and analyzed, as shown in Figure 10.6.

It should be noted that when a concentration-dependent modulus is considered, the concentration profile (Figure 10.6a) is almost the same. Similarly, the distributions of the stresses (Figure 10.6b–d) are the same but the magnitudes are different. This is because a higher stress is induced for materials with an increasing modulus as a function of Li-concentration when compared to the cases where the modulus is either not dependent on concentration or it decreases as a function of it.

To understand the stress evolution, shown in Figure 10.6, we need to correlate these images with the physical problem of lithiation. It has been mentioned that

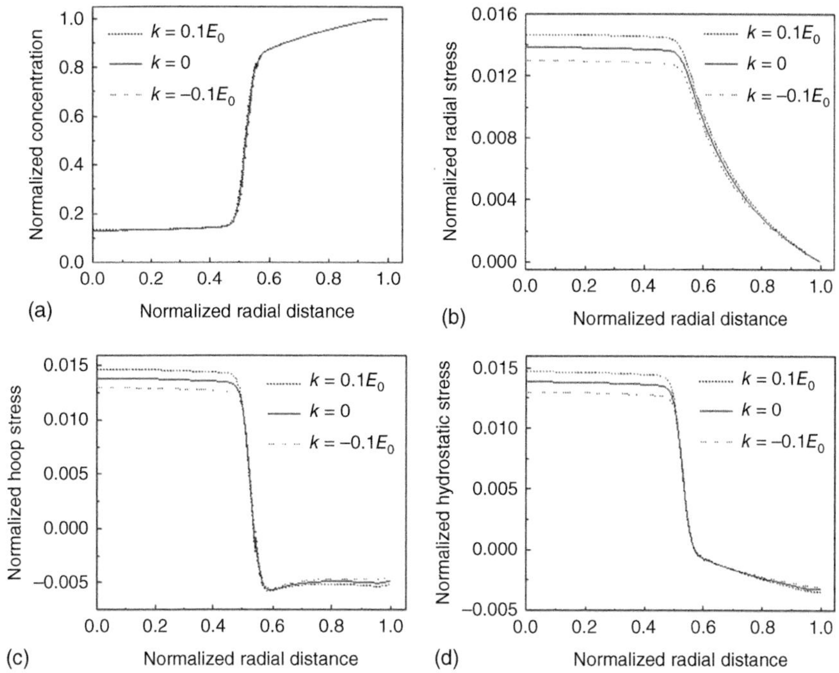

Figure 10.6 Radial distribution of (a) concentration profiles, (b) normalized radial stress, (c) normalized hoop stress, and (d) normalized hydrostatic stress at time step 2000 Δt with chemo-mechanical-coupling effects under elastic deformation. Constant modulus ($k = 0$), reduced modulus ($k = -0.1E_0$), and increased modulus ($k = 0.1E_0$) during ion insertion are considered. The solid, dotted, and dashed lines are for the three cases, respectively. All the stress components are normalized by E_0. Source: Reproduced with permission from Wang et al. [59]/Elsevier.

spherical particles attain a core–shell structure upon lithiation. The shell is Li-rich and therefore undergoes large volume expansions, while the core does not contain much Li and therefore expands very little, and even this small expansion is further constrained by the shell. This can explain why the hoop stress is compressive at the core–shell interface. However, tensile stresses also arise at this interface due to the volume mismatch between the core and shell. Since it is assumed that there is no matrix surrounding the particle, the radial stress tends to zero at the shell surface. We also assume spherical symmetry and a uniform Li concentration in the core, and therefore the radial and hoop stress are equal there. Since the Li concentration in the core increases with continuous Li insertion, both the radial and hoop stresses increase in the core for the elastic case. With continuous lithiation, the phase interface (at the shell) migrates toward the particle center and therefore the hoop stress decreases at the outer surface. As a result, the core minimizes, to a lesser extent, the expansion of the outer surface of the shell. The radial and hoop stresses in the center are in tension, while the hoop stress at the outer surface is always in compression. The radial stress at the outer surface is always zero (corresponding to the traction-free boundary conditions at the outer surface). As a result, cracks are not likely to occur and propagate during lithiation.

Phase-Field Modeling of DISs Under Plastic Deformation and Comparison with Elastic Case

In alloy-based electrode systems, such as $Li_{4.4}Si$, a large volumetric change may be induced (over 400% for $Li_{4.4}Si$), which can give rise to plastic deformation. When plasticity is taken into account, the constitutive relations for a spherical particle can be given by:

$$\left.\begin{aligned}\sigma_r &= \frac{E}{(1+v)(1-2v)}\left[(1-v)\left(\frac{du}{dr}-\varepsilon_r^p-\beta c^*\right)+2v\left(\frac{u}{r}-\varepsilon_\theta^p-\beta c^*\right)\right]\\ \sigma_\theta &= \frac{E}{(1+v)(1-2v)}\left[\left(\frac{u}{r}-\varepsilon_\theta^p-\beta c^*\right)+v\left(\frac{du}{dr}-\varepsilon_r^p-\beta c^*\right)\right]\end{aligned}\right\} \tag{10.10}$$

Here E is the Young's modulus, v is Poisson's ratio, $\varepsilon_r = \frac{du}{dr}$ is the radial strain, $\varepsilon_\theta = \frac{u}{r}$ is the tangential strain, and u is the radial displacement, ε_r^p and ε_θ^p are the radial and plastic strain components, respectively.

As for the elasticity case of the previous section, we consider here plasticity with and without CM coupling. The details regarding the complete model (governing equations) and numerical implementation (solving the weak form of the governing equations by finite-element method) are given in Ref [59].

For the sake of simplicity, concentration dependence of the modulus is not considered in this section, and therefore here the elastic modulus is kept the same for all cases; it is noted that the parameters for Si were used in implementing this model. Figure 10.7 depicts the evolution of concentration, stress, and plastic strain during the first Li-ion insertion.

At the initial stage, the differences in the concentration profiles (Figure 10.7a) are not distinct. Both the radial stress σ_r (Figure 10.7b) and the hoop stress σ_θ (Figure 10.7c) are almost constant in the core (Li-poor) region. σ_r gradually decreases to zero from the center toward the outer surface, σ_θ transitions from tension in the core to compression in the shell, and the hydrostatic stress, σ_h, (Figure 10.7d) shows similar trends as the hoop stress. The resulting von Mises stress, σ_v, (Figure 10.7e) gradually increases from the core towards the shell; when σ_v is greater than the yield stress, plastic deformation occurs. The von Mises stress, σ_v, (Figure 10.7e) is relatively small in the Li-poor core and reaches a local maximum near the phase (core–shell) interface, and then σ_v drops to a local minimum near the interface, but it increases rapidly towards the outer lithiated shell afterwards. The local stress changes can be also found in σ_θ (Figure 10.7c) and σ_h (Figure 10.7d). As lithiation proceeds, the core–shell interface (Figures 10.7a and 10.8a) moves from the shell exterior towards the center. When CM coupling is considered, a smaller miscibility gap is predicted. But contrary to what happens in the electrodes undergoing elastic deformation during lithiation, the diffusion speed for the electrodes undergoing plastic deformation during lithium insertion is slower when CM coupling is considered. This agrees with the in situ experiments for crystalline Si nanoparticles, where it is shown that a slower migration velocity occurs as lithiation progresses [60]. At later lithiation stages, the radial stress (Figure 10.8b), hoop stress (Figure 10.8c), and hydrostatic stress (Figure 10.8d) show similar trends for the three cases considered: strain hardening without CM

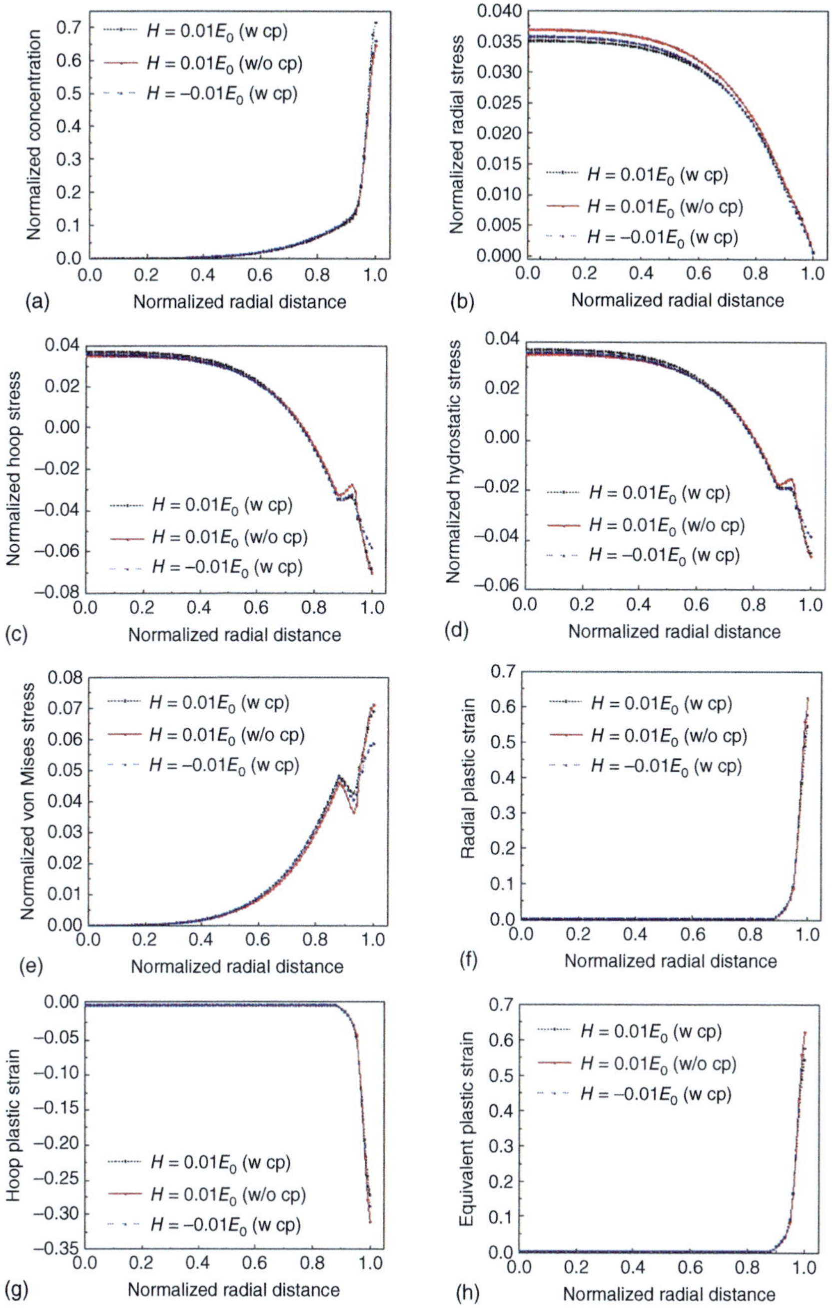

Figure 10.7 Radial distribution of (a) concentration profiles, (b) normalized radial stress, (c) normalized hoop stress, (d) normalized hydrostatic stress, (e) normalized von Mises stress, (f) radial plastic strain, (g) hoop plastic strain, and (h) equivalent plastic strain at time step 100 Δt with/without chemo-mechanical-coupling ("w cp"/"w/o cp") effects under elastoplastic deformation. Both strain-hardening and strain-softening cases are considered. All the stress components are normalized by E_0 (*the modulus for the pristine Si*). Source: Reproduced with permission from Wang et al. [59]/Elsevier.

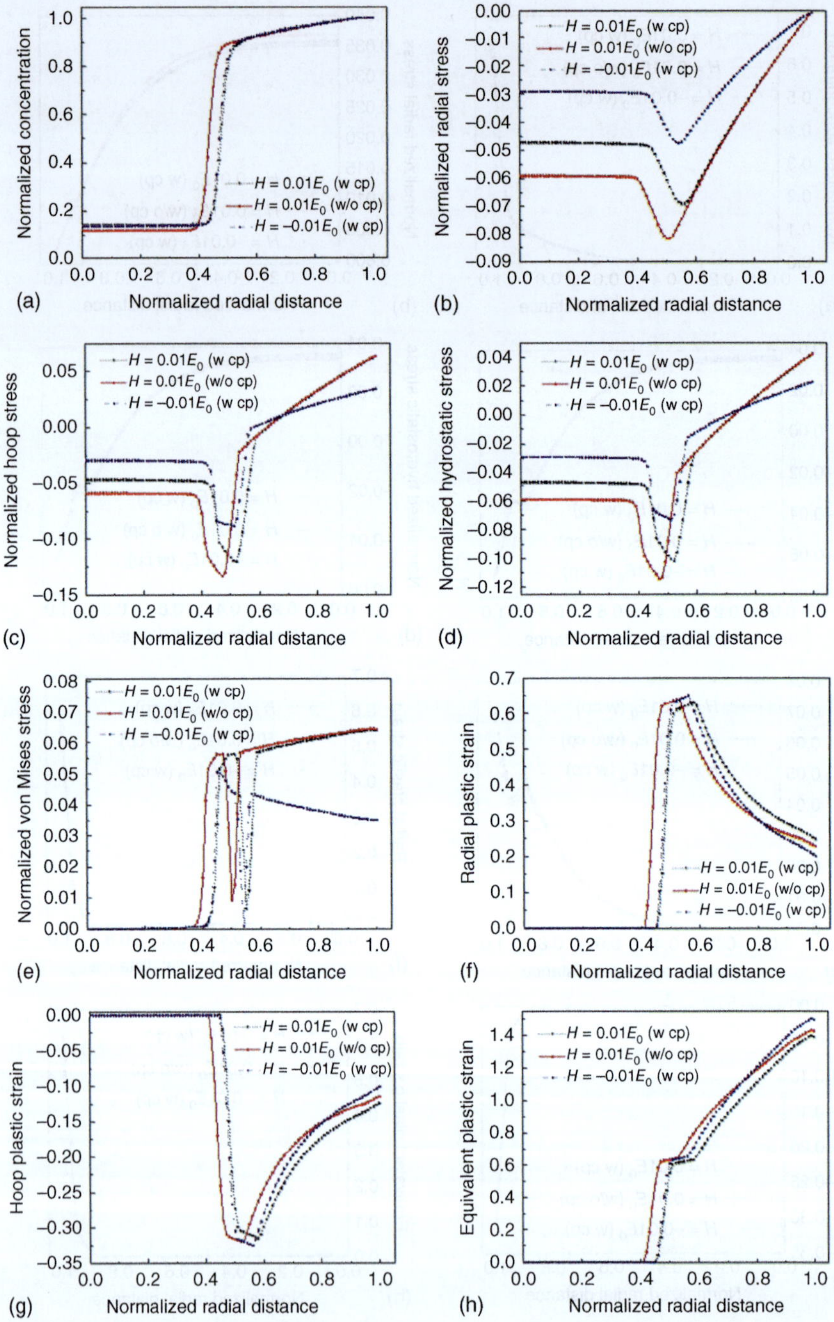

Figure 10.8 Radial distribution of (a) concentration profiles, (b) normalized radial stress, (c) normalized hoop stress, (d) normalized hydrostatic stress, (e) normalized von Mises stress, (f) radial plastic strain, (g) hoop plastic strain, and (h) equivalent plastic strain at time step 3000 Δt with/without chemo-mechanical-coupling ("w cp"/"w/o cp") effects under elastoplastic deformation. Both strain-hardening and strain-softening cases are considered. All the stress components are normalized by E_0. Source: Reproduced with permission from Wang et al. [59]/Elsevier.

coupling ("w/o cp"), strain hardening with CM coupling ("w cp"), and strain softening without CM coupling. The stress/strain profiles remain constant in the Li-poor core. The stress/strain/concentration values have the higher values for the strain-hardening case without CM coupling, while they have the lowest values for the strain-softening cases without CM coupling at the same position and same time step. The radial stress is compressive in the whole particle, while the hoop and hydrostatic stresses are compressive in the Li-poor core and transition from compression near the interface to tension in the shell exterior. This hydrostatic stress state hinders the diffusion of Li ions, which is the reason for the slower diffusion speed. Due to plastic deformation near the core–shell interface, the radial, hoop, and hydrostatic stress profiles show stress drops. The von Mises stresses (Figures 10.7e and 10.8e) are almost zero in the core regions. As a result, the core remains elastic and no plastic deformation (Figures 10.7f–h and 10.8f–h) occurs there. There are abrupt changes in the von Mises stress profiles near core/shell interfaces; first it increases, then decreases and increases again quickly and it continues to increase toward the shell for the strain-hardening case, while it decreases for the strain-softening case.

To have a better understanding of the differences in the stress evolution in an electrode particle undergoing elastic deformation versus one undergoing plastic deformation, the constitutive relation (Eq. (10.10)) may be revisited. The total stress is the sum of the diffusion-induced stress and the stress induced due to plastic deformation. If there is no plastic deformation, the hoop stress at the outer surface is always compressive. When plastic deformation takes place, the stress state depends on both lithiation-induced strain components and plastic strain components. When the electrode particle undergoes plastic deformation during the ion-insertion process, the core remains elastic all the time and the shell begins to yield. There is no plastic deformation in the core, as a result, both radial and hoop plastic strain components are zero, whereas the radial plastic strain, ε_r^p, remains positive and the hoop plastic strain, ε_θ^p, remains negative in the shell, and $\varepsilon_\theta^p = -\frac{1}{2}\varepsilon_r^p$. Positive radial plastic strain means radial expansion and negative hoop plastic strain means tangential shrinkage. As a result, the hoop stress can be rewritten as:

$$\sigma_\theta = \frac{E}{(1+\upsilon)(1-2\upsilon)}\left[\left(\frac{u}{r} - \beta c^*\right) + \upsilon\left(\frac{\mathrm{d}u}{\mathrm{d}r} - \beta c^*\right)\right] + \frac{E}{2(1+\upsilon)}\varepsilon_r^p \tag{10.11}$$

The first part of Eq. (10.11) is the constitutive equation for the hoop stress under elastic deformation only. Due to the concentration gradient, the Li-rich regions (shell) expand to a larger degree than the Li-poor regions (core). However, the expansion is constrained by the surrounding matrix materials. As a result, the hoop stress at the outer surface is compressive initially. At the initial stage, the von Mises stress is less than the yield stress. As lithiation proceeds, the effective stress (von Mises stress) becomes large enough, so that plastic deformation takes place near the phase interface.

The second part on the right-hand side of Eq. (10.11) is the component due to plastic deformation and it remains positive in the shell. During lithiation, the Li-rich shell undergoes plastic deformation and as Li is continuously inserted, plasticity accumulates. A transition from compression to tension in the hoop stress of the shell occurs when the plastic component of the stress exceeds the stress component induced only by diffusion. The development of tensile stresses can result in crack formation and fracture, which has been observed in Si, Sn, and SnO_2 particles during lithiation [35, 44, 60]. Specifically, the cracks grow toward the center of the particle, in the radial direction.

The DISs that develop in the electrode due to inhomogeneous volume changes can lead to fracture of the electrode particle. The formation and propagation of cracks, in turn, lead to degradation of the battery performance. To have a better understanding of the failure mechanism in the electrode, modeling fracture in the electrode particle will be discussed in Section 10.3.2.

10.3.2 Modeling Fracture in Single Particles

The ion transport that takes place during the operation of batteries is usually governed by diffusion, which can result in phase segregation, especially for active materials with a high theoretical capacity, e.g. Si electrodes. The large stresses that develop in the electrode due to inhomogeneous volume changes lead to mechanical failure, capacity loss, and ultimately the failure of the battery. Many theoretical and numerical studies have been carried out regarding the damage and failure of the electrodes.

Diffusion was initially incorporated into classical fracture mechanics models (similar to the early ones developed by Aifantis and applied to anodes) in order to try and predict the particle size that would inhibit fracture, for cathode materials, such as $LiCoO_2$ [61] and $LiFePO_4$ [62]. Anisotropic deformation and concentration-dependent diffusivity coefficients were taken into consideration. A main limitation of those models is that preexisting flaws are required to calculate the energy-release rate, and therefore they cannot capture the onset of fracture and crack propagation during the lithiation and delithiation process.

The J-integral, which is another common framework used in classical fracture mechanics, has also been used to couple mechano-diffusional driving forces to model fracture for Si anodes [63–65]. It was possible to couple the softening behavior of Li–Si alloys, mechanical deformation, and chemical diffusion together. An interesting conclusion of that work is that operation at higher concentrations is an effective means to minimize failure of thin-film Li–Si alloy electrodes. One of the drawbacks is that the proposed J-integral for the energy-release rate is not path independent, which may pose some challenges and difficulties when dealing with the coupled systems. Furthermore, such models could not capture the formation of multiple cracks that are observed in thin-film anodes, but rather examined the stress field in front of a hypothetical single crack. Figure 10.9 depicts an example of such a model.

10.3.2.1 Modeling of Damage by the Phase-Field Method

Currently, the phase-field method is gaining more and more popularity for capturing fracture. Unlike the classical sharp interface models, the phase-field method treats

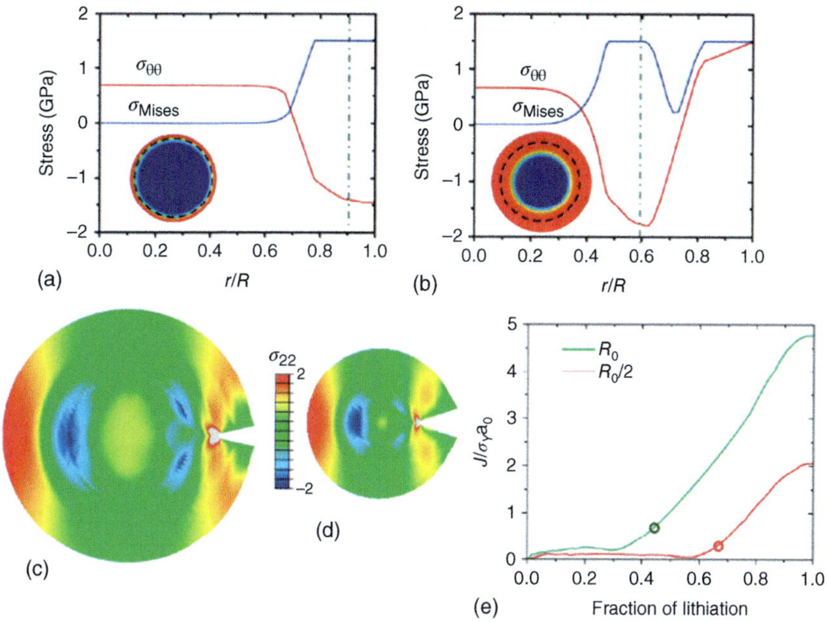

Figure 10.9 Stress generation in a Si nanoparticle. (a) and (b) Radial distribution of the hoop stress $\sigma_{\theta\theta}$ and von Mises stress at different lithiation stages ((a) is at an early stage and (b) later); (c) and (d) stress relaxation due to surface cracking with two different radii; (e) J-integral near the crack tip as a function of fraction of lithiation. Source: Reproduced with permission from Liu et al. [35]/American Chemical Society.

fracture boundaries as diffuse/smeared interfaces. The main difficulty, in the classical sharp-interface crack approaches, arises in the determination of the crack path, their kinking and splitting during their propagation. These events are commonly handled either by imposing unrealistic underlying physics into the constitutive models or computationally very cumbersome and demanding algorithms. In contrast, phase-field descriptions do not require explicitly tracking the discontinuities in the displacement field. This significantly reduces the implementation complexity.

A novel finite strain theory has been proposed [66] for chemo-elasticity (classical Cahn–Hilliard diffusion) coupled with phase-field modeling of fracture. This model could simulate crack nucleation and propagation under charging–discharging cycles; however, it is fully coupled and therefore consists of numerous governing equations that may pose difficulties for numerical implementation. Another deficiency is that there are too many parameters in the model and some of them cannot be obtained easily, and the effects of some key parameters are not known. Therefore, it was not applied to a specific electrode material, or compared with experiments. Another phase field coupled model [67] of Li diffusion, mechanical stress, and crack growth has been applied for modeling fracture in low volume expanding spherical particles. The model uses small strain formulation and has not been directly correlated with experiments.

Capturing Fracture in Si Film Anodes Phase-field modeling for batteries, was first experimentally validated by applying it [68] to successfully simulate the crack patterns of silicon film anodes. Dry bed-lake crack patterns [69, 70] for Si thin-film anodes have been observed for a long time and were the key motivation for the mechanics community to investigate electrode materials. Nevertheless, it wasn't until 2018 that it was possible to capture this fracture behavior. The model is summarized here as follows:

The diffusion equation is given by Fick's law as [68]:

$$\begin{cases} \dfrac{\partial c}{\partial t} = D\nabla^2 c \ \text{ in } \Omega \quad \text{where } D = D_0\left(\dfrac{1}{1-c} - 2\times 1.95\times c\right) \\ D\dfrac{\partial c}{\partial n} = q \quad \text{on } \partial\Omega_1 \\ c = 0 \qquad \text{on } \partial\Omega_2 \end{cases} \tag{10.12}$$

where c denotes the normalized Li-ion concentration, ranging between 0 and 1, corresponding to full de-insertion (pure Si) and full insertion ($Li_{3.75}Si$), respectively. q is the prescribed constant flux of Li ions over the upper electrode surface, $\boldsymbol{n}$ is the unit normal vector of $\partial\Omega_1$, and D is the diffusivity coefficient parameter. Here a nonlinear diffusivity is used to produce a sharp phase boundary, accounting for the saturation in Li ions during insertion when c reaches 1, and D_0 is the initial diffusivity when $c=0$. $\partial\Omega_1$ and $\partial\Omega_2$ are Neumann and Dirichlet boundaries, respectively. The above diffusion equation is for Li insertion only. For de-insertion, the same expression is adopted but replacing c with $1-c$ and using the initial condition $c=1$.

Small strain formulation is adopted. The total strain $\boldsymbol{\varepsilon}_t$ of the material is decomposed as the sum of the chemically induced strain $\boldsymbol{\varepsilon}_c$ and the elastic strain $\boldsymbol{\varepsilon}_e$:

$$\boldsymbol{\varepsilon}_t = \frac{1}{2}\{(\nabla\mathbf{u}) + (\nabla\mathbf{u})^T\} = \boldsymbol{\varepsilon}_e + \boldsymbol{\varepsilon}_c \tag{10.13}$$

where

$$\boldsymbol{\varepsilon}_c = \alpha c\boldsymbol{I} \tag{10.14}$$

with α being the dilatation coefficient that sets the strain induced for full insertion ($c=1$) and a negative α is adopted for de-insertion; $\boldsymbol{I}$ is the identity matrix and $\boldsymbol{u}$ is the displacement field.

The equations of equilibrium can be stated as:

$$\begin{cases} \nabla\cdot\boldsymbol{\sigma} + \mathbf{b} = \mathbf{0} \ \text{ in } \Omega \\ \mathbf{t} = \boldsymbol{\sigma}\cdot\mathbf{n} \ \text{ on } \Gamma_N \\ \mathbf{u} = \overline{\mathbf{u}} \ \text{ on } \Gamma_D \end{cases} \tag{10.15}$$

where the solid body occupies the domain Ω with Neumann boundary Γ_N and Dirichlet boundary Γ_D; $\boldsymbol{\sigma}$ is the stress tensor, $\mathbf{b}$ is the body force, and $\mathbf{t}$ and $\overline{\mathbf{u}}$ are prescribed values for the traction and displacement on the respective boundaries, Γ_N and Γ_D.

To ensure irreversibility of the crack evolution and handle loading and unloading histories, a history variable representing the maximum elastic strain energy density is introduced [71]:

$$\mathcal{H} = \mathcal{H}(\mathbf{u}, t) = \max_{\tau\in[0,t]} \psi(\mathbf{u}, \tau) \tag{10.16}$$

Here ψ is the stored elastic energy of the undamaged elastic solid and is a function of the strain tensor ε.

As for damage evolution, the governing equations are given by:

$$\begin{cases} G_c(-l^2\Delta d + d) = 2l(1-d)\mathcal{H} \ \text{ in } \Omega \\ \nabla d \cdot \mathbf{n} = \mathbf{0} \ \text{ on } \partial\Omega \\ d = 1 \ \text{ on } \Gamma \end{cases} \tag{10.17}$$

where G_c is the failure energy for unstable crack/damage extension, Δ is the Laplacian operator, ℓ is an internal length characteristic of the material, $\partial\Omega$ is the outer boundary of Ω, and Γ is a possible initial crack discontinuity surface within Ω. The phase-field variable d can only grow from 0 to 1, $d=0$ for the unbroken/intact state, $d=1$ for the fully broken/damaged state.

To account for contributions from tensile and compressive deformations, respectively, the elastic energy ψ is decomposed into positive and negative parts. The positive part is assumed to contribute to damage, and the negative part resists damage. With this split, only the tensile stress is degraded as follows [71]:

$$\begin{cases} \psi(\boldsymbol{\varepsilon}, d) = (1-d)^2\psi^+(\boldsymbol{\varepsilon}) + \psi^-(\boldsymbol{\varepsilon}) \\ \boldsymbol{\sigma}(\boldsymbol{\varepsilon}, d) = \dfrac{\partial\psi}{\partial\boldsymbol{\varepsilon}} = (1-d)^2\boldsymbol{\sigma}^+ + \boldsymbol{\sigma}^- \\ \mathcal{H} = \mathcal{H}(\mathbf{u}, t) = \max\limits_{\tau\in[0,t]} \psi^+(\mathbf{u}, \tau) \end{cases} \tag{10.18}$$

where the positive and negative parts of the energy are defined by using an orthogonal decomposition of the strain tensor. The strain energy density associated to the extensive part of the elastic deformation is defined as [68]:

$$\psi^+ = \frac{1}{2}k(\langle \mathrm{tr}(\boldsymbol{\varepsilon}_e)\rangle_+)^2 + \mu\left\{\mathrm{tr}\left((\boldsymbol{\varepsilon}_e)^2\right) - \frac{1}{3}(\mathrm{tr}(\boldsymbol{\varepsilon}_e))^2\right\} \tag{10.19}$$

where $\langle x\rangle_\pm = (x \pm |x|)/2$, $k = \lambda + \frac{2}{3}\mu$ is the bulk modulus and λ and μ are Lame constants.

Then the Cauchy stress tensor is given by [68]:

$$\boldsymbol{\sigma} = \{(1-d)^2\langle \mathrm{tr}(\boldsymbol{\varepsilon}_e)\rangle_+ + \langle \mathrm{tr}(\boldsymbol{\varepsilon}_e)\rangle_-\}k\boldsymbol{I} + 2\mu(1-d)^2\left\{\boldsymbol{\varepsilon}_e - \frac{1}{3}\mathrm{tr}(\boldsymbol{\varepsilon}_e)\boldsymbol{I}\right\} \tag{10.20}$$

Unlike the models in [67] and [66], which are fully coupled, the proposed formulation allows to recast the problem in this multiphysics model as two linear problems instead of solving a nonlinear two-field problem for the mechanical equilibrium and damage evolution. From a computational viewpoint, this is extremely convenient. It reduces the complexity.

To perform the simulation for a thin film, plane stress is assumed, and the simulation parameters are shown in Table 10.4. In Figure 10.10b, it is seen that the dry bed-lake crack pattern can be captured. Since Si thin-films fracture in a random pattern, to further validate and test the model, a patterned Si film was prepared by etching in it 1-μm holes/rounded edge squares. As shown in Figure 10.10c this resulted in an "ordered" fracture pattern of diagonal cracks after the first lithiation, which was maintained until the 10th cycle, after which multiple cracking

Table 10.4 Simulation parameters for Si anodes.

Parameters	Description	Value	Unit
E	Young's modulus	130	GPa
v	Poisson's ratio	0.2	—
G_c	Energy-release rate	2 000	$J\,m^{-2}$
l	Length scale	100	nm
α	Expansion coefficient	0.3	—
D_0	Initial diffusivity	25	$nm^2\,ms^{-1}$
q_0	Li-ion flux	1.4	$nm\,ms^{-1}$
L	Length of the film	10 000	nm
B	Width of the film	10 000	nm
h	Height of the film	500	nm
Δt	Size of time step	400	s

took place (Figure 10.10e). In applying the model, for the de-insertion process, the energy-release rate is taken as one-fifth of that of insertion, since de-insertion of Li-ions produces vacancies in the lattice structure, and it is assumed to reduce the fracture energy. Another reason is that the material after full insertion is completely different from pure Si. Therefore, a different G_c is adopted. As seen in Figure 10.10d,f, the model was able to accurately predict the fracture patterns after 1 and 10 cycles.

Capturing Fracture in Si Spherical Particles Capturing fracture of thin film anodes was a long-standing problem; however, thin films are not practical for commercial purposes, even if their fracture problem is solved (which can be for very thin films). The model previously used in the sub-section "Capturing fracture in Si film anodes" was, therefore, applied to predict fracture in different particle sizes in order to determine which particle size should be used in commercial porous type anodes.

The configuration modeled was that of five big Si particles with a diameter of 0.95 μm and four smaller particles with a diameter of 450 nm embedded in a square polymer binder with side length of 5550 nm. The initial conditions were $c = 0$ and $c = 1$, for Li insertion and de-insertion, respectively. The boundary conditions for this system were a constant flux of Li ions (1.4 $nm\,ms^{-1}$) [68] on the outer surface of the electrode and fixed displacement boundary conditions [72] for the exterior surfaces of the polymer. It took about 10 hours to finish the full Li insertion; as a result, the C-rate was 0.1C. The Young's modulus of the Si electrode and polymer were 130 and 2 GPa, respectively, and the Poisson's ratio was 0.2 and 0.34, respectively. The initial diffusion coefficient was 25 $nm^2\,s^{-1}$ [68] inside the Si particle and the value in the polymer was taken to be 1000 times as in the active material. By doing so, it ensured that the ion diffusion inside the polymer was immediate. The expansion coefficient α of the Si was taken to be 0.3 [68]. As for the binder, it was assumed to be zero since it could not host Li ions. The energy-release rate for the Si and internal

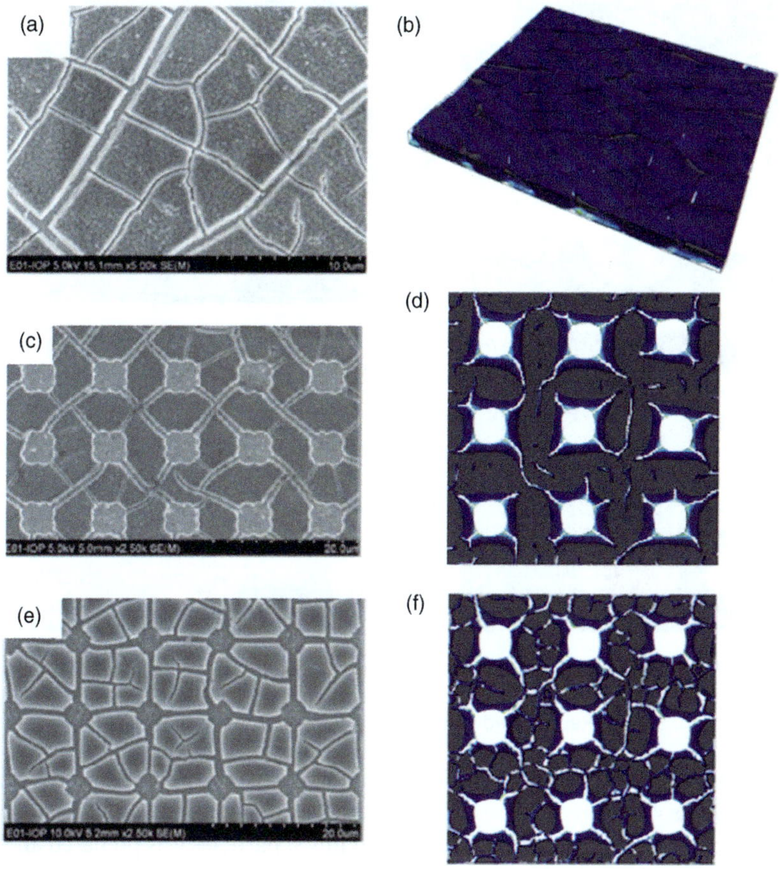

Figure 10.10 Comparison of scanning electron microscopy (SEM) experimental observations with simulation predictions. Crack patterns in a Si thin film after the first de-insertion: (a) SEM, and (b) simulation. Crack patterns in patterned Si thin films after the 1st de-insertion: (c) SEM, and (d) simulation, and after the 10th de-insertion (e) SEM, (f) simulation. Source: Reproduced with permission from Réthoré et al. [68]/Elsevier.

length was taken to be $10\,\mathrm{J\,m^{-2}}$ [73] and 10 nm, respectively. It was assumed that the polymer does not fracture, and for the model to capture this, the energy-release rate for the polymer had to be taken 1000 times higher than that of Si. The governing equations were solved in the open-source software FEniCS, by employing an unstructured mesh that had a size of approximately 1/100 of the side length of the polymer. Damage profiles after insertion and de-insertion are shown in Figure 10.11. Based on the simulation results, it is seen that cracks began to initiate at the ~1 μm particles after complete insertion and continued to propagate during the de-insertion process, cutting the particles in half.

To investigate further the relationships between particle size and its fracture pattern, more simulations were performed by changing only the Si particle size in the simulations. Fracture was always observed until a critical size of 95 nm, below which fracture did not occur. This is illustrated in the damage profile shown

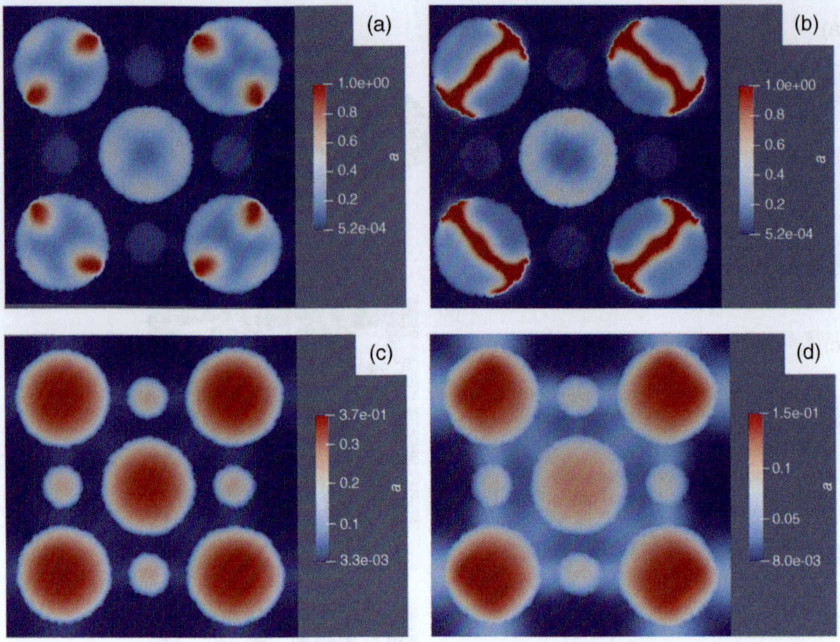

Figure 10.11 Damage profile of Si particles of diameter 0.95 μm after (a) full insertion and (b) de-insertion; damage profile of Si particle with diameter of 95 nm after (c) full insertion and (d) de-insertion. Source: Reproduced with permission from Ahuja et al. [74]/Elsevier.

in Figure 10.11c,d, where the diameter of the larger particles is 95 nm and of the smaller particles is 45 nm.

To verify the model, a porous electrode with ~1 μm Si particles was prepared and after the 1st and 2nd electrochemical cycle, it was opened and scanning electron microscopy was performed to observe fracture. Similarly, another set of electrodes using ~50–100 nm Si particles were prepared and cycled, but as the scale was smaller, transmission electron microscopy was employed to observe fracture. Figure 10.12 reveals that as predicted the micron-size Si particle began to fracture in half after one electrochemical cycle and the crack began to cleave the particle after two electrochemical cycles, while the ~100 nm particles did not fracture. Therefore, this model can capture size effects and the results are consistent with experimental observations. Further experimental results on the electrochemical performance of the Si-based electrodes, as shown in Figure 10.12, were included in Chapter 6.

10.3.2.2 Phase-Field Modeling for Capturing Stress Evolution and Fracture in Na-Ion Batteries

Similar phase-field models as those in Section 10.3.2.1 can also be applied to Na-ion batteries. For instance, phase field modeling [49] was employed to study the concentration and stress evolution in Na_xFePO_4 electrodes during sodiation, and it was found that there was a visible difference for the concentration and stress between the small deformation theory and the finite deformation theories. Phase-field models were also used to study the stress evolution in FeS_2 electrodes undergoing two-phase lithiation/sodiation [75]. Damage, however, was not modeled until the previously

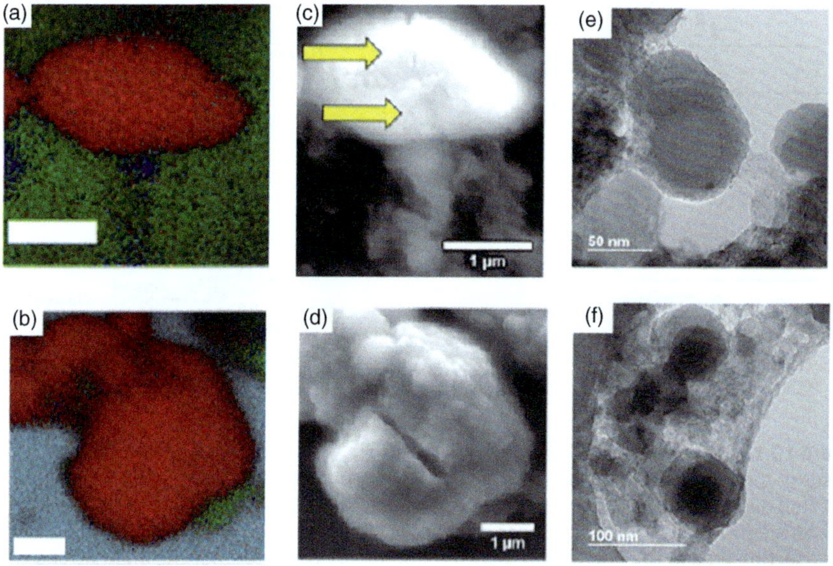

Figure 10.12 Microscale Si electrode after 1 cycle (a) SEM image, (b) EDS map (scale bar 1 μm); after 2 cycles (c) SEM image, (d) EDS map. In (b) and (d), the regions in red represent silicon, green for carbon, and blue for oxygen. TEM image of (e) nano Si electrode after 1 cycle, and (f) nano-Si electrode after 10 cycles. Source: Reproduced with permission from Ahuja et al. [74]/Elsevier.

mentioned formulation presented in the sub-section "Capturing fracture in Si film anodes" was applied for predicting fracture in Prussian-blue analogues (PBAs) during sodiation. PBAs are chemically expressed as $A_xM_yFe(CN)_6{\cdot}nH_2O$, with A being an alkali cation and M a divalent or trivalent transition metal.

As described in Chapter 9, PBAs are promising electrode materials due to their good electrochemical performance and ease of synthesis. $Na_xMn_{1-y}Fe(CN)_6{\cdot}nH_2O$ allows for a high capacity of 171 mAh g^{-1}, however, it cannot be retained. $Na_2NiFe(CN)_6$ allows for a much lower capacity of ~60 mAh g^{-1}, which, however, can be retained for ~1000 cycles. Microstructural analysis after electrochemical cycling has not been performed on such materials, however, $Na_xMn_{1-y}Fe(CN)_6{\cdot}nH_2O$ experiences a much higher volume expansion (~8.5%) than $Na_2NiFe(CN)_6$ (0.29%). To limit the volume expansion of $Na_xMn_{1-y}Fe(CN)_6{\cdot}nH_2O$, a gradient microstructure (g-NiMnHCF) was proposed [39] by which the non-expanding Ni-rich shell could constrain the expansion of the Mn-rich core. The previous model of sub-section "Capturing fracture in Si film anodes" was, therefore, applied to this system, which was introduced in Chapter 9.

PBA has a face-centered cubic structure, hence, a cubic unit cell was chosen and only one-eighth of the particle was considered due to symmetry. The model here was essentially the same as in the previous section. The details can be referred from the Ref. [39].

The particle size is chosen as 1.2 μm × 1.2 μm × 1.2 μm. The configurations and typical boundary conditions are shown in Figure 10.13.

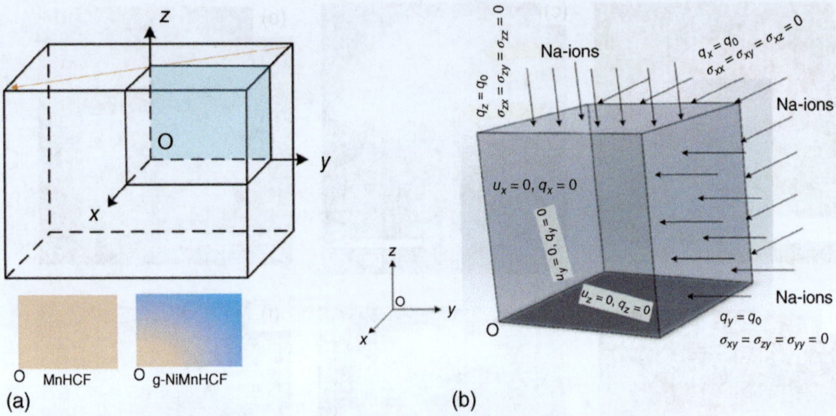

Figure 10.13 (a) Schematic diagram of a PBA particle with cubic structure, where the center of the particle is marked by the letter "O." Only one-eighth of the cubic particle is modeled due to symmetry. Two different compositions MnHCF and g-NiMnHCF are taken into consideration. The colored images represent the composition of the shaded cross-section area in the one-eighth model. (b) Boundary conditions for the analysis of a one-eighth PBA particle subjected to Na-ion insertion/de-insertion. The three sides subjected to Na-ion insertion are submitted to a constant Na-ion flux and are free of tractions. The other three sides are bound by symmetry boundary conditions and are free of ion flux. Source: Reproduced with permission from Hu et al. [39]/ACS.

For g-NiMnHCF, the composition changes gradually from the center to the exterior particle surface. Due to different ratios of Ni and Mn in the electrode, it is assumed that the material properties, such as Young's modulus E, Poisson's ratio ν, expansion coefficient α, and initial diffusion coefficient D_0, depend on the distance from the center. A simple linear assumption is adopted and the details can be referred from the appendix of Ref. [40]. Relevant parameters are given in Table 10.5. The diffusion coefficient ranges between 2.4×10^{-11}–3.8×10^{-11} cm^2 s^{-1} and 4.8×10^{-11}–5.9×10^{-11} cm^2 s^{-1} for MnHCF and in g-NiMnHCF, respectively, and therefore the values used in the model are 2.4×10^{-11} and 7.2×10^{-11} cm^2 s^{-1}.

The governing equations can be solved discretely by the finite-element method and the open-source platform FEniCS (https://fenicsproject.org/) is used for solving the equations. Two different compositions were considered: the homogeneous MnHCF and the gradient g-NiMnHCF. Contour plots (Figure 10.14a,b) of the normalized Na^+ concentration show a two-phase reaction with a steep reaction front. This is because both MnHCF and g-NiMnHCF undergo a monoclinic to cubic phase transition during Na-ion insertion/de-insertion. This steep concentration distribution also leads to an abrupt change in the stress field; where the core region is under compressive (negative) hydrostatic stress while the shell is under tensile (positive) hydrostatic stress during desodiation (as shown in Figure 10.14d,e). Therefore, there is a jump of the hydrostatic stress ($\underline{\sigma_h} = \mathrm{tr}(\boldsymbol{\sigma})/3$) at the phase boundary. For both MnHCF and g-NiMnHCF, the core region is under compression while the shell is under tension during desodiation. The stress states are reversed during sodiation. Figure 10.14c compares the diagonal distributions of concentration in MnHCF and g-NiMnHCF. g-NiMnHCF has a less steep reaction front than MnHCF due to the

Table 10.5 Simulation parameters for PBA cathodes.

Parameters	Description	Value	Unit
E_n	Young's modulus of $Na_2NiFe(CN)_6$	24	GPa
ν_n	Poisson's ratio of $Na_2NiFe(CN)_6$	0.34	—
E_m	Young's modulus of $Na_2MnFe(CN)_6$	35	GPa
ν_m	Poisson's ratio of $Na_2MnFe(CN)_6$	0.21	—
G_c	Failure energy	50	$J\,m^{-2}$
l	Length scale	50	nm
α_n	Expansion coefficient of $Na_2NiFe(CN)_6$	0.003	—
α_m	Expansion coefficient of $Na_2MnFe(CN)_6$	0.028	—
D_n	Initial diffusivity of $Na_2NiFe(CN)_6$	7.2×10^{-11}	$cm^2\,s^{-1}$
D_m	Initial diffusivity of $Na_2MnFe(CN)_6$	2.4×10^{-11}	$cm^2\,s^{-1}$
q_0	Na-ion flux	1.4	$nm\,ms^{-1}$
ρ	Half side length of the particle	600	nm
Δt	Size of time step	160	s

faster diffusion of Na^+ in g-NiMnHCF. It should also be noted that g-NiMnHCF has a much lower volume expansion (0.29%) than MnHCF (8.5%).

The maximum value for the von Mises stress σ_v is located near the core(pristine)–shell(sodiated) interface. Figure 10.14f compares the modulated von Mises stress σ_{vm} distributions along the diagonal direction on the surfaces of the particle ($\sigma_{vm} = \text{sign}(\sigma_h) \times \sigma_v$, where $\text{sign}(\sigma_h)$ represents the sign function of σ_h and its value is 1 when $\sigma_h > 0$ and -1 when $\sigma_h < 0$). The maximum σ_v in MnHCF (607 MPa) is about four times that in g-NiMnHCF (165 MPa), which shows that the stress level is significantly reduced by the gradient microstructure.

The damage evolution is also studied to better understand the influence of the different compositions on the mechanical performance. The damage profiles during initial cycling for both MnHCF and g-NiMnHCF are shown in Figure 10.14g,h. Damage in MnHCF was localized near the center of the external surfaces as well as near the center of the particles, while damage in the g-NiMnHCF particles was confined to the center of the cell. Figure 10.14i shows the damage distribution of MnHCF and g-NiMnHCF along the diagonal direction. Both MnHCF and g-NiMnHCF particles demonstrate about a threefold increase in maximum damage over the 1st and 10th cycle, even though for the g-NiMnHCF it is negligible. In Figure 10.14, it is seen that damage in the homogeneous particle accumulates at the inner corner of the unit cell, which corresponds to the center of the particle. Therefore, fracture would likely occur in those severely affected regions after many cycles, which is in accordance with experimental observations, which show that the cracks initiate at the center of the homogeneous particles, as shown in Figure 10.15a–g. For the gradient microstructure, no fracture was observed after 100 cycles, as seen in Figure 10.15h, which is in agreement with the simulation predictions.

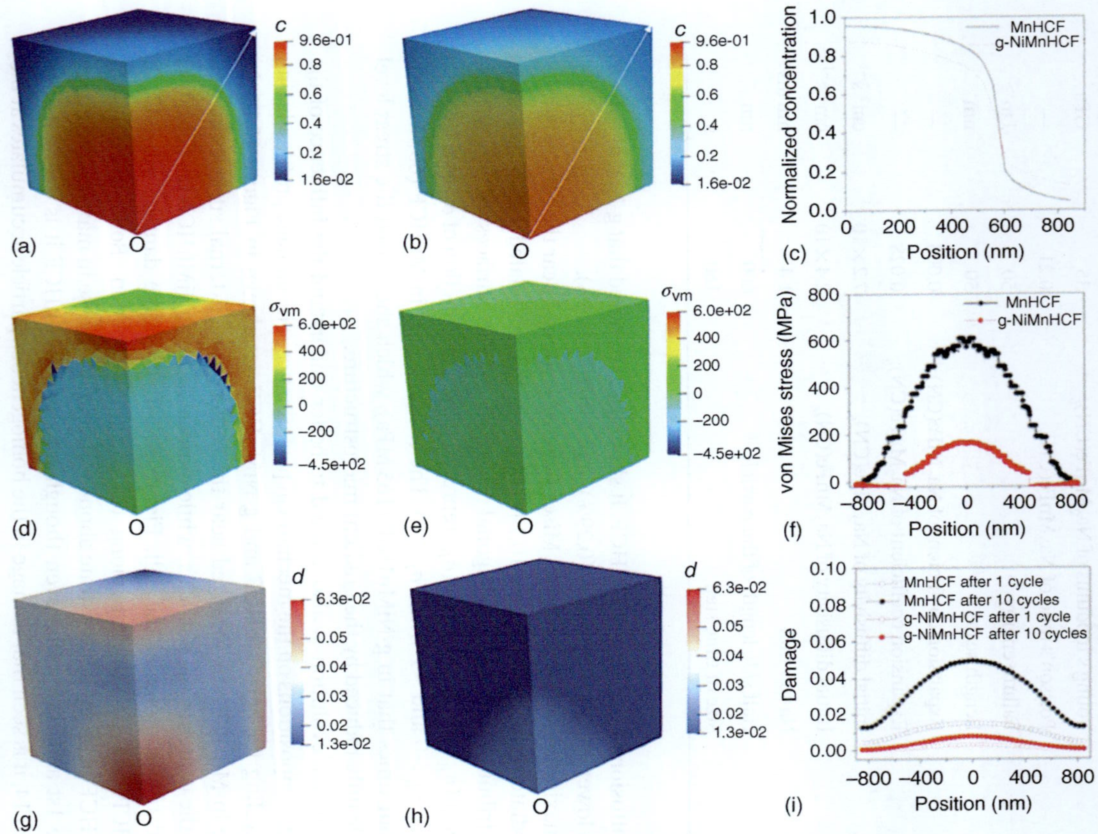

Figure 10.14 (a) Concentration distribution, (d) modulated von Mises stress (σ_{vm}) distribution, and (g) damage distribution in one-eighth unit cell at 50% state of charge (SOC) for MnHCF. (b) Concentration distribution, (e) modulated von Mises stress (σ_{vm}) distribution, and (h) damage distribution in one-eighth unit cell at 50% state of charge (SOC) for g-NiMnHCF. (c) Concentration profiles and (f) von Mises stress profiles from the center of the particle to the edge center of the top surface (along white arrows in a and b). (i) Comparison of damage profiles along the diagonal line on the top surface of MnHCF and g-NiMnHCF after 10 cycles. Source: Reproduced with permission from Hu et al. [39]/American Chemical Society.

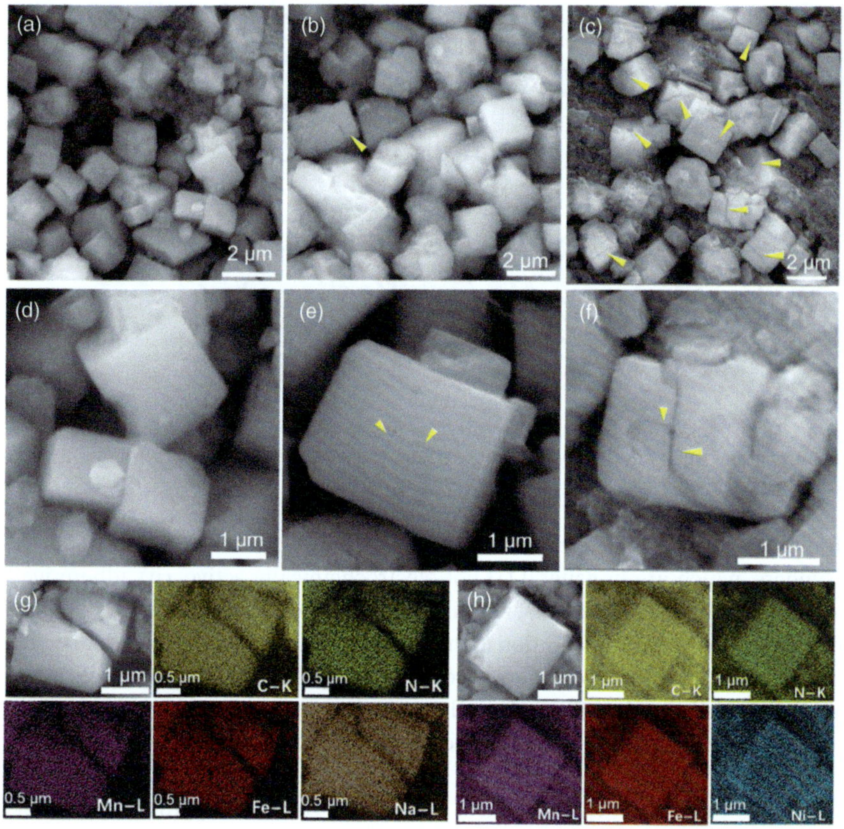

Figure 10.15 (a–c) Low-magnification SEM images of MnHCF materials after (a) 10, (b) 20, and (c) 50 cycles. (d–f) High magnification SEM images of MnHCF materials after (d) 10, (e) 20, and (f) 50 cycles. (g–h) SEM images and element mapping of (g) MnHCF and (h) g-($Ni_{0.1}Mn_{0.9}$)HCF after 100 cycles. Source: Reproduced with permission from Hu et al. [39]/American Chemical Society.

10.4 Full Electrode Modeling

Sections 10.3.1.1, 10.3.1.2, 10.3.2.1 were concerned with modeling the mechanical behavior of single active particles during Li-ion or Na-ion insertion. However, the porous electrode does not consist only of the active material but also of conducting material (e.g. carbon black), and binder (e.g. carbon-doped polyvinylidene fluoride [PVDF]). The electrolyte penetrates the pores of the solid matrix, leading to interconnected void space that constitutes a significant portion of the volume. Porous electrodes are considered advantageous because they greatly increase the electrolyte available active area, allowing the charge-transfer process to be substantially improved. The electrochemical behavior of porous electrodes is also enhanced due to good wetting of the electrolyte on the surface, and available space for storing dissolution products.

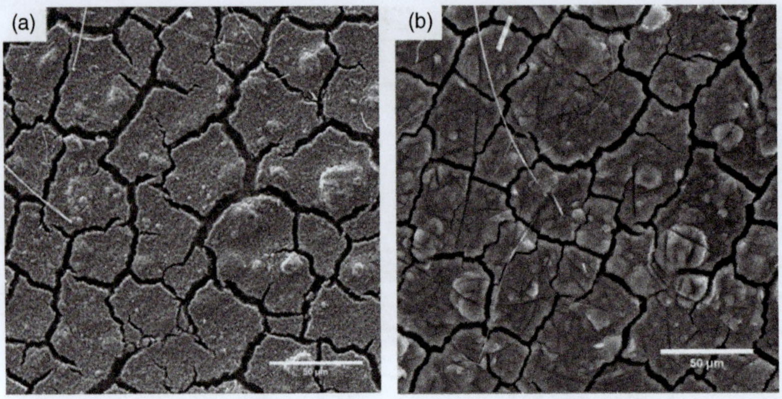

Figure 10.16 Fracture of porous electrode after electrochemical cycling. SEM image (a) Si after 1 cycle and (b) Si after 50 cycles. Source: Reproduced with permission from Ahuja et al. [74]/Elsevier.

In porous electrodes, active particles are mechanically connected and supported by binders, which are in general (i) not electrochemically active, (ii) adhere well to particles and current collectors, and (iii) are sufficiently compliant to compensate for particle deformation. In addition to the volume changes that the individual active particles undergo, many researchers have reported significant volume changes in the overall porous electrodes. The huge volume variations can lead to severe fracture of the electrode that disrupts the connectivity with the current collector as seen in Figure 10.16.

Modeling, therefore, the mechanical behavior of the whole porous electrode is necessary. The earliest papers dealing with porous electrode theory were published by Stender and Ksenzhek [76] and Euler and Nonnenmacher [77]. The governing equations were obtained from volume averaging. A few years later, Newman and Tobias extended the analysis to account for the effects of concentration variations on kinetics with concentration-independent electrolyte properties [78]. This paper also introduced the well-known equation for mass conservation inside a porous electrode undergoing reactions. Ion transport inside a porous electrode was also studied via volume-averaged conservation equations [79]. Specifically, a mean-field porous electrode theory was developed [48, 80–84], where the complex porous structure is simplified into a quasi-1D model and mean values of the relevant electrical, chemical, mechanical, and physical properties were considered in the macro 1D dimension. Other variables, such as the concentration of Li in the solid phase, were solved only at the micro/particle level. For phase separating materials, such as $LiFePO_4$, each particle was assumed to have a spherical phase boundary that moved as a "shrinking core," as one phase displaced the other. The model was general enough to include a wide range of polymeric separator materials, lithium salts, and composite insertion cathodes. Later on, 1D porous electrode theory was combined with the diffusion-induced stress model for spherical particles to develop a multiscale model [85]. The governing equations pertaining to the particle domain were solved with the boundary conditions formulated using the pore wall flux obtained from the porous electrode model. On the other hand, the Butler–Volmer kinetics were used to link the surface Li concentration of individual particles with that of the porous electrode model. This allowed the evaluation of the potential and stress response of the macroscopic system. A variety of macro homogeneous models with different configurations have been proposed [86–88].

A mathematical model to simulate the generation of mechanical stress during the discharge process in a dual porous insertion electrode cell sandwich comprised lithium cobalt oxide and carbon was presented in Ref. [87]. The simulation results indicated that the particles can be damaged even if they were cycled at intermediate or even low rates of discharge, which was due to the residual strain buildup caused by the phase transformation in the material [87]. In Ref. [88], an integrated 2D model of a Li-ion battery was developed to study the mechanical stress in storage particles as a function of material properties, and storage material properties were shown to influence the maximal stress experienced by particles.

In addition to the 1D homogenous type model, studies have also been conducted in higher dimensions that relied on the finite-element method. The stresses within a battery electrode were studied using a three-dimensional model [89], which accounted for active particles ($LiMn_2O_4$), binders, and electrolytes. It was found that the surfaces of the particles of active material were under compressive stress during discharge, while their interiors were in tension, and the stresses were intensified at the particle contacts. Limited by the computation power, 3D modeling was confined to electrodes of small size.

Models have also been developed for coupling the volume expansion of the active material and stresses during electrochemical cycling, which is of particular importance for Si-based anodes. A model to describe the volume change in porous electrodes was developed allowing for both porosity and changes in the dimensions of the electrode [90], by introducing a so-called swelling coefficient. This led to the development of a general model to simulate the stresses that build up in a battery during cycling [91], due to volume changes. It was found that the effects of volume changes on the electrode were twofold, i.e. change in electrode dimensions and change in porosity, both of which can significantly affect the resistance of the battery during cycling, which in turn alters the reaction distributions in the porous electrodes. By employing the model, the battery performance was predicted based on state of discharge and discharge rate [92]. This was achieved by using porous rock mechanics coupled with porous electrode theory. It was found that volume changes can affect the porosity and strain distributions across the electrode thickness. The porosity changes were accompanied with concentration gradients, which in turn led to the associated drop in performance based on the resistance to the electrode expansion and discharge rate.

Single-particle models have been used to examine the response of storage particles in detail, with stress generation and its effect on lithium diffusion coupled together. In these models, each electrode is represented by a single spherical particle, a core–shell particle, or a hollow sphere. These models use a simplified representation of the electrode and do not account for all the physical processes, for example, the electrolyte transport and associated macroscopic gradients in porous electrodes undergoing phase transformations are ignored. The porous electrode models have the advantage of providing a comprehensive account of the various physical processes occurring in a battery, as a result, can provide the stress distribution, concentration of chemical species, and charge/discharge current densities of the system being studied. The drawback is that solving the models is very time-consuming.

10.5 MD Simulations for Li-Ion Batteries

10.5.1 The Role of MD Simulations in LIBs

Another computational technique that has been used to capture Li-ion insertion is atomistic or molecular dynamics (MD) simulations. Unlike continuum chemo-mechanical modeling that requires a priori constitutive parameters in these simulations use empirical potentials to examine the fracture mechanisms of Li_xSi alloys [93–95], the stress effect on lithiation [96], [97], and the structural changes of Si anodes due to delithiation. The fundamental properties of crystalline and amorphous Li_xSi alloys, including the elastic modulus, yield strength [98], and Li-ion diffusion [92, 99], have been studied based on the second nearest-neighbor modified embedded atom method (2NN MEAM) interatomic potential [98].

MDs have also been applied to capture the isotropic and anisotropic deformation of Si nanoparticles (NPs) with a diameter of 10.0 nm [100], [101], but the generated stresses were discrete and different from previous theoretical predictions [102–104]. Simulations have also been used to determine the yield strength [37] of Li–Si alloys, Li-ion diffusion [38, 39], and fracture mechanisms of lithiated Si [94, 105]. Another MD study has shown that there exists a brittle-to-ductile transition as the Li concentration increases [94], which agrees with the high damage tolerance of electrochemically lithiated Si observed through *in situ* experiments [106]. Based on the reactive force field (ReaxFF), it was shown that the external stress significantly affects the lithiation behavior [97], which can interpret the bending-induced symmetry that takes place during lithiation of germanium nanowires [107]. The influence of the delithiation rate on the Si anodes was also studied by an "artificial" simulation approach, and it was shown that fast delithiation caused higher irreversible capacity loss compared to slow delithiation [108].

A more recent study was able to successfully simulate the initiation of plasticity and fracture in different types of Si nanopillars (amorphous, crystalline, and hollow) during Li insertion [42], and the results were in agreement with experimental data. In the remainder of this chapter, a summary of MD results on capturing plasticity and fracture will be given [42], since these processes are difficult to observe experimentally and MD studies are crucial in order to understand them.

10.5.2 MD Simulations of Lithiated Si Nanopillars

10.5.2.1 Simulation Setup and Empirical Potential

The MD setup of Shuang and Aifantis [42] was selected to match a set of experimentally studied Si microstructures depicted in Figure 10.17: crystalline Si nanopillars (c-SiNPs) with different axial orientations. Nanopillars were selected instead of

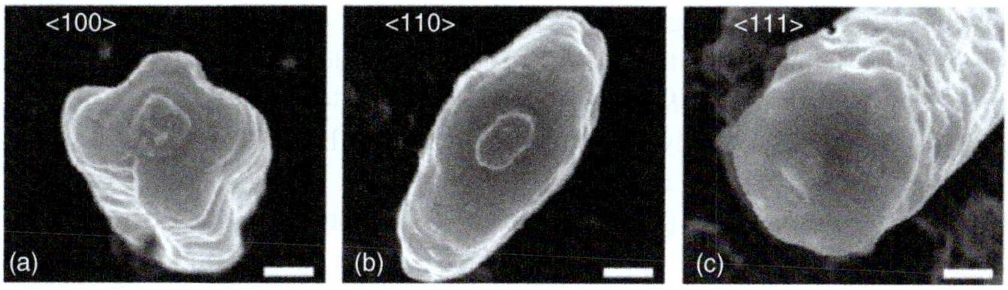

Figure 10.17 Experimental observations of c-SiNPs after lithiation, with different initial orientations (a) <100>, (b) <110> and (c) <111>. Scale bar is 200 nm. Source: Reproduced with permission from Lee et al. [109]/American Chemical Society.

nanospheres since to simulate a pillar, less atoms are required than for a sphere of similar diameter. The stress and deformation result from the reaction front at the curved surface, hence, the results can also be extended to spherical Si particles, which are the preferred experimental geometry. The hollow NPs were amorphous, while for the solid ones both c-SiNPs and amorphous Si nanopillars (a-SiNPs) were considered. Recent experimental evidence [109], depicted in Figure 10.17, has shown that the Si crystal orientation plays a significant role in the direction of the volume expansion, therefore three different axial directions were considered: <100>, <110>, and <111> when the diameter was 30 nm.

The size effect was studied by considering different diameters (D = 10.0, 20.0, and 30.0 nm) for a-SiNP. The largest a-SiNP contained 299 200 Si atoms. The initial radius of the a-SiNP before lithiation was denoted with R_0, and after lithiation, it had increased to r_0, during which material points at a radius R moved to a new location at a radius of r (Figure 10.18a), where A is the location of the lithiation front. Hollow pillars were created by satisfying the condition $R_{out}^2 - R_{in}^2 = R_0^2$ (R_0 = 30.0 nm), where R_{out} and R_{in} represent the initial inner radius and outer radius, such that hollow and solid a-SiNPs contained the same number of Si atoms (keeping the volume constant for different a-SiNPs). After deformation/lithiation,

Figure 10.18 (a) Solid and (b) hollow SiNPs before and after lithiation. Source: Reproduced with permission from Shuang and Aifantis [42]/American Chemical Society.

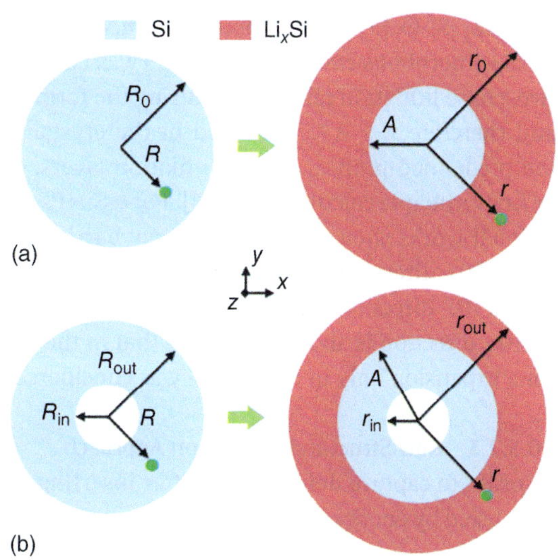

the inner and outer radii were denoted as, r_{in} and r_{out}, respectively (Figure 10.18b). For each outer diameter mentioned for the hollow NPs, four hollow different inner diameters were considered: R_{in} = 5.0, 10.0, 20.0, and 30.0 nm (R_{out} = 30.4, 31.6, 36.1, and 42.4 nm, respectively). It follows that the wall thickness ($R_{out} - R_{in}$) of the hollow SiNPs decreased with increasing R_{in}.

Free boundary conditions (BCs) were assumed in both the *x*- and *y*-directions, while periodic BCs were used for the *z*-direction, indicating an infinite length. The height (dimension in the *z*-direction) was around 2.2 nm for all SiNPs. All simulations were performed by the open-source package LAMMPS [110], and the second nearest-neighbor modified embedded atom method (2NN MEAM) presented in Ref. [98] was used to describe the interactions between Li and Si. The ReaxFF potential used in previous research was not considered here due to its high computational cost. The OVITO software [111] was used to visualize the simulation snapshots, and the evolution of deformation during Li-ion insertion was determined via the von Mises local shear strain [112].

10.5.2.2 Lithiation Process

Experimentally, Li-ion diffusion occurs under room temperature, at much greater timescales than what conventional MD simulations can capture [113]. An efficient method to simulate this in MD is to insert Li atoms into predefined regions of the SiNPs, layer by layer, as shown in Figure 10.19. The layers had a thickness of 1 nm, based on experiments which showed that during the formation of Li_xSi, the interface between Li_xSi and Si had a thickness of 1 nm [114]. Therefore, the number of steps needed to achieve complete lithiation equaled the diameters (the thicknesses for hollow ones) of SiNPs, for example, 30 steps were needed when R_0 = 30.0 nm.

For a-SiNPs, since experiments have captured that lithiation is isotropic [115], the Li_xSi-Si interfaces were taken to have a circular shape (Figure 10.19a). Experiments have also shown that for fracture to occur in a-SiNPs, $x = 2.5$, since $Li_{2.5}Si$ alloys form by two-phase lithiation that could cause a tensile hoop stress. Continuous lithiation from $Li_{2.5}Si$ to $Li_{3.75}Si$ has no effect since it occurs by one-phase lithiation [103]. In the amorphous case, hence, $x = 2.5$ was considered. On the contrary, for c-SiNPs two-phase lithiation takes place during the formation of $Li_{3.75}Si$, leading to fracture, and therefore $x = 3.75$ was used in the crystalline cases. Figure 10.19b–d shows the predefined profiles for the lithiation layers, observed either from experiments or first-principle calculations [116]. For c-SiNPs, previous experimental and computational works showed that the energy barrier for Si lithiation is the lowest along the <110> direction, such that Li atoms will diffuse into the Si crystal mostly along the <110> direction, but rarely along the other lateral directions (expansion in the <110> direction is about nine times that in the <111> direction) [109, 114]. Therefore, expansion only in the <110> was considered.

10.5.2.3 New Structural Relaxation Approach

In order to capture deformation, after inserting Li-ions, the whole system needs to be relaxed by energy minimization. For nanomaterials, the minimization algorithm

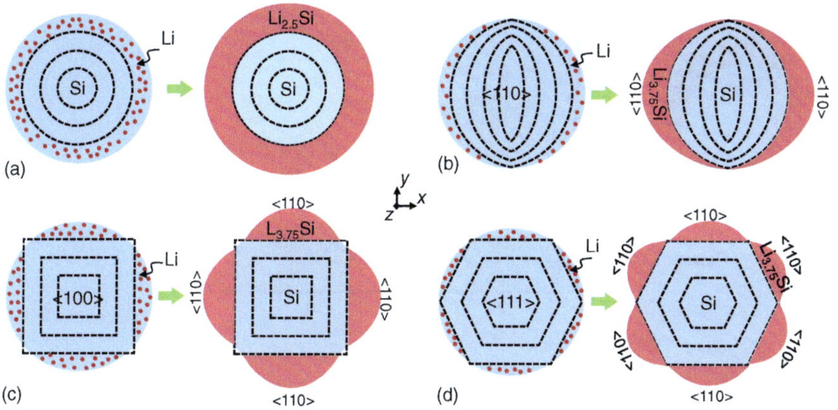

Figure 10.19 Lithiation protocol for different SiNPs. (a) a-SiNPs and (b–d) c-SiNPs with different axial orientations. Source: Reproduced with permission from Shuang and Aifantis [42]/American Chemical Society.

employed is of great importance for the efficiency and fidelity of simulation results [117]. A new structural relaxation approach was introduced in Ref. [42] and as it is more efficient than other methods, it is summarized herein, instead of just referring to the respective article. According to previous experimental observations, three different regions with different relaxation processes were defined in Figure 10.20a: region I denotes the Li_xSi alloy; region II denotes the total Li_xSi alloy that formed; and region III denotes both the lithiated and unlithiated phases, corresponding to the whole SiNP. To realize natural lithiation, four stages of relaxation were needed, as shown in Figure 10.20b. In the first stage, only atoms in region I were relaxed using 500 steps of the conjugate gradient (CG) method and then 500 steps of the fast inertial relaxation engine (FIRE) method with the Euler explicit integrator (the most efficient integrator in the FIRE method [118–120]). In continuing to stage II, only the atoms in region II were relaxed using 2500 FIRE steps. In stage III, the fully lithiated SiNP was relaxed with 1500 CG steps. Significantly high repulsive forces were present in unrelaxed Li_xSi, which could lead to nonphysical deformation; therefore, some regions had to be fixed in stages I and II. The CG method was used first in stage I since it is stable and can optimize very ill-conditioned nanostructures.

To test the necessity of further relaxation in stage IV and the performance of different relaxation methods, the microcanonical ensemble (NVE), CG, and FIRE were then used to relax the whole SiNP after stage III, and the results for a solid a-SiNP, when $R_0 = 10.0\,\text{nm}$, are shown in Figure 10.20c. It is seen that the most efficient method was NVE with temperature control (1 K) since the system's energy decreased to a very low level and reached a stable state after 15 000 steps. This method was performed by the LAMMPS commands "fix nve/limit" and "fix temp/rescale." We simulated the lithiation of various SiNPs with and without the fourth stage of relaxation and the results are presented in Figure 10.21, illustrating that the fourth stage of relaxation is necessary for capturing the experimental behavior. In the sequel hence, the full four-stage minimization strategy, shown in

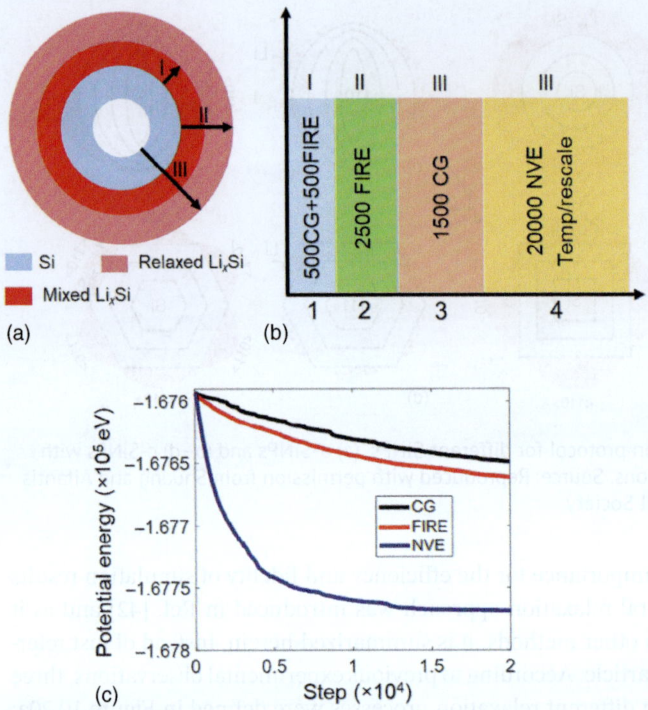

Figure 10.20 Energy minimization strategy used during lithiation of SiNPs. (a) Three regions defined in a lithiated SiNP; (b) schematic diagram for four-stage minimization; and (c) variation of potential energy using four different minimization methods during the fourth stage. Source: Reproduced with permission from Shuang and Aifantis [42]/American Chemical Society.

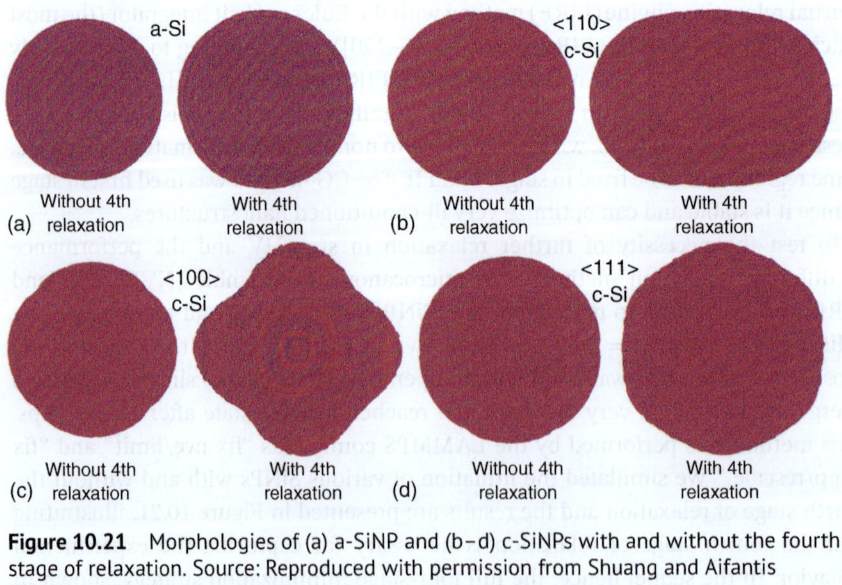

Figure 10.21 Morphologies of (a) a-SiNP and (b–d) c-SiNPs with and without the fourth stage of relaxation. Source: Reproduced with permission from Shuang and Aifantis [42]/American Chemical Society.

Figure 10.20b, is employed. NVE relaxation is applicable for nonzero temperatures and therefore the proposed method can be used for simulating lithiation at finite temperature.

It should be noted that recent studies indicated that the FIRE method with the Euler explicit integrator is more efficient than others when dealing with nanostructures with free surfaces and dislocation cores [119, 120]. In such cases, the simulation starts from a structure that is near equilibrium, and therefore analyzing the energy profile by the CG and FIRE methods can accelerate the minimization. For the present case, however, the initial energy is very high since Li insertion reduces the atom distances and the simulation begins at a state that is quite far from equilibrium. NVE simulation with a constant temperature can effectively skip intermediate states and decrease the system's energy. It can, hence, be said that NVE with temperature control is preferred for coarse energy minimization when there is significant energy change, while CG and FIRE should be used for fine energy minimization. It is interesting to emphasize here that the two MD studies [100, 101] whose results were not consistent with theoretical predictions [17] used the CG method.

10.5.2.4 Deformation and Stress Evolution During Lithiation

After simulating the lithiation process, the software LAMMPS can calculate the stress tensor (S_{ab}) through

$$S_{\mathrm{ab}} = -mv_a v_b - W_{\mathrm{ab}} \tag{10.21}$$

where a and b take on values x, y, and z to generate the six stress components. The first term in Eq. (10.21) is the kinetic energy contribution, and the second term is the virial contribution due to intra- and intermolecular interactions. The atomic stress can then be obtained by dividing with the atomic volume Ω

$$\sigma_{\mathrm{ab}} = \frac{S_{\mathrm{ab}}}{\Omega} \tag{10.22}$$

Ω is calculated by the Voronoi analysis in the OVITO software and it is ill-defined for the atoms at the outer surface, so the atomic stresses are not available for the outer surface, as seen in Figure 10.22. Applying coordinate transformation allows the hoop, radial, and axial stresses to be written, respectively, as:

$$\sigma_\theta = \sigma_x \sin^2\theta + \sigma_y \cos^2\theta - 2\tau_{xy} \sin\theta\cos\theta \tag{10.23}$$

$$\sigma_r = \sigma_x \cos^2\theta + \sigma_y \sin^2\theta + 2\tau_{xy} \sin\theta\cos\theta \tag{10.24}$$

$$\sigma_z = \sigma_z \tag{10.25}$$

Since all atomic quantities are discrete, it is necessary to smooth the data spatially by taking the average in 100×100 bins for all SiNPs. An example is given in Figure 10.22 for a-SiNP with $R_0 = 30.0$ nm. It is seen that the distribution of the hoop stress is well computed.

Solid a-SiNPs The lithiation of a-SiNPs with different radii is first studied. Figure 10.23a depicts the fully lithiated state of a 20.0 nm a-SiNP, illustrating that due to the lithium-induced volume expansion, the diameter increased from 20.0 to 36.4 nm. Figure 10.23b shows that the volume expansion increases with lithiation

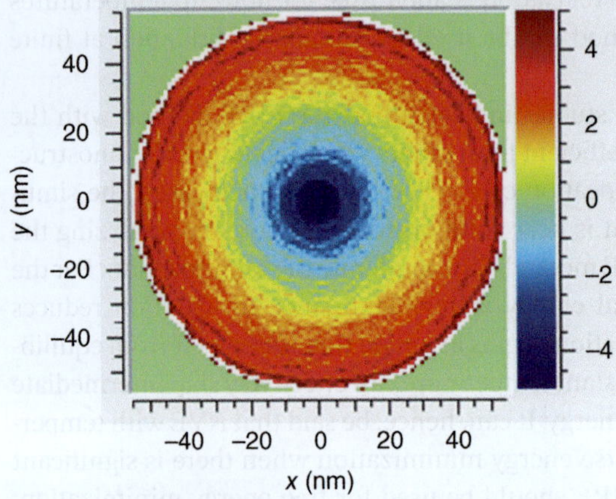

Figure 10.22 Distribution of hoop stress for a-SiNP with initial radius of 30.0 nm by using 100 × 100 bins. Source: Reproduced with permission from Shuang and Aifantis [42]/American Chemical Society.

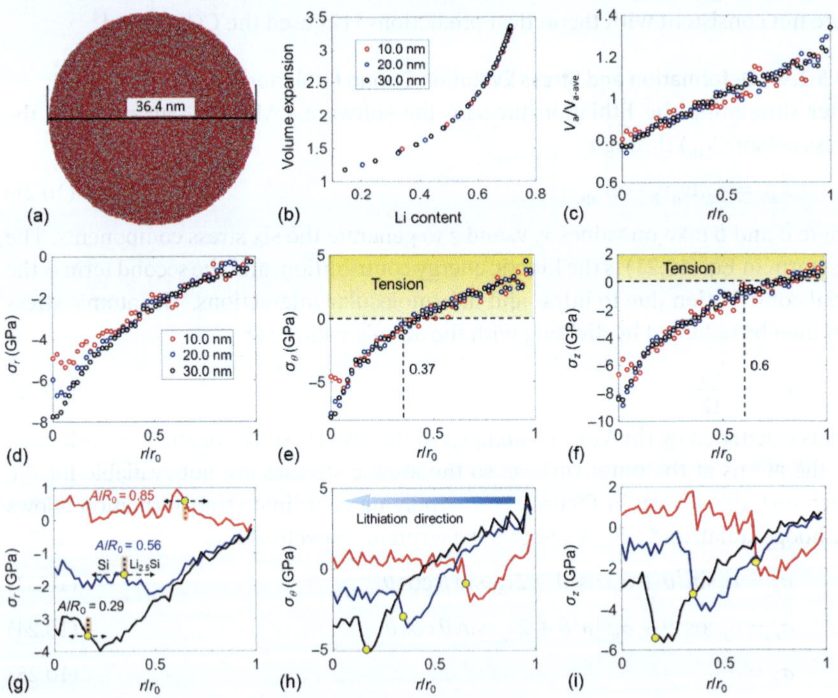

Figure 10.23 Simulation results of solid a-SiNPs with different diameters. (a) Final shape of lithiated a-SiNP with an initial radius of 10.0 nm; (b) variation of volume expansion of a-SiNP with respect to Li content; (c–f) distribution of dimensionless atomic volume, atomic radial stress (σ_r), hoop stress (σ_θ), and axial stress (σ_z) after full lithiation along the dimensionless radial distance; and (g–i) stress distribution at different lithiation stages. Source: Reproduced with permission from Shuang and Aifantis [42]/American Chemical Society.

in a similar manner for the different size a-SiNPS considered, with 336% being the expansion upon full lithiation for all cases, which is consistent with experiments [121] and diffusive MD simulations [122]. The inner and outer atoms of the NPs experienced different deformations and this is reflected in the gradient distribution of the atomic volume in the radial direction for maximum lithiation, as seen in Figure 10.23c.

The hoop (σ_θ), radial (σ_r), and axial (σ_z) stresses upon maximum lithiation are shown in Figures 10.23d–f, and it is seen that for all SiNP sizes both the hoop stress (σ_θ) and axial stress (σ_z) underwent a compressive to tensile transition at $r_1/r_0 = 0.37$ and $r_2/r_0 = 0.6$. As it was mentioned earlier, tensile stresses are required for crack initiation and therefore this behavior agrees with the experimental observations, which show that cracks initiate from the outer surface of particles upon lithiation [123]. Furthermore, this compressive to tensile transition is similar to that predicted by the previously described continuum mechanics models in Section 10.3.1.2. The radial stress (σ_r) was always below zero, implying a compressive state along the SiNPs axis. Despite the occurrence of tensile stresses, fracture cannot be observed for SiNPs with current MD since according to experimental observations cracks are inhibited when the radius is below 180.0 nm in c-SiNPs [123].

The significant volume expansion gives rise to plastic deformation, and the stresses are plotted for different lithiation stages ($A/R_0 = 0.85$, 0.56, and 0.29) in Figure 10.23g–i. Examining the stress evolution shows that in the pristine Si core the stresses are compressive (negative values) and constant and decrease with continuous Li-ion insertion. In the lithiated shell, the hoop stress (σ_θ) and axial stress (σ_z) transition from compression to tension due to the volume expansion, while the radial stress (σ_r) increases with continuous lithiation. These results agree well with the theoretical predictions in Section 10.3.1.2.

Hollow a-SiNPs The geometry of hollow a-SiNPs allows Li insertion to take place at both the inner and outer surfaces. The purpose of the work in Ref. [42] was to extend the pillar results to spherical particles, therefore, Li-insertion was allowed to occur only from the outer surface, since that would be the case for hollow spherical particles.

Figure 10.24 presents the changes in the radii, volume, and stresses that lithiation had in hollow a-SiNPs. For comparison purposes, the results for the solid a-SiNP with $R_0 = 30.0$ nm are also plotted. An interesting observation is that when $R_{in} = 5.0$ and 10.0 nm ($R_{out} = 30.4$ and 31.6 nm, respectively), the hollow structure of the SiNPs was lost after full lithiation since the inner radius size had collapsed to zero, as indicated by arrows. This illustrates that the volume expansion due to lithiation occurs in all directions and completely closes the inner hole. This may not seem consistent with experiments that showed no change in the inner radius of Si nanotubes during lithiation [124]; however, a native oxide had formed in the experiments preventing lithiation and expansion, whereas the simulations are for pure Si.

When the inner radius was small $R_{in} = 5.0$ and 10.0 nm, the behavior of the hollow SiNPs was similar to the solid ones, however, for $R_{in} = 20.0$ and 30.0 nm size effects were observed. Similar trends as those for the solid a-SiNPs were observed, since the

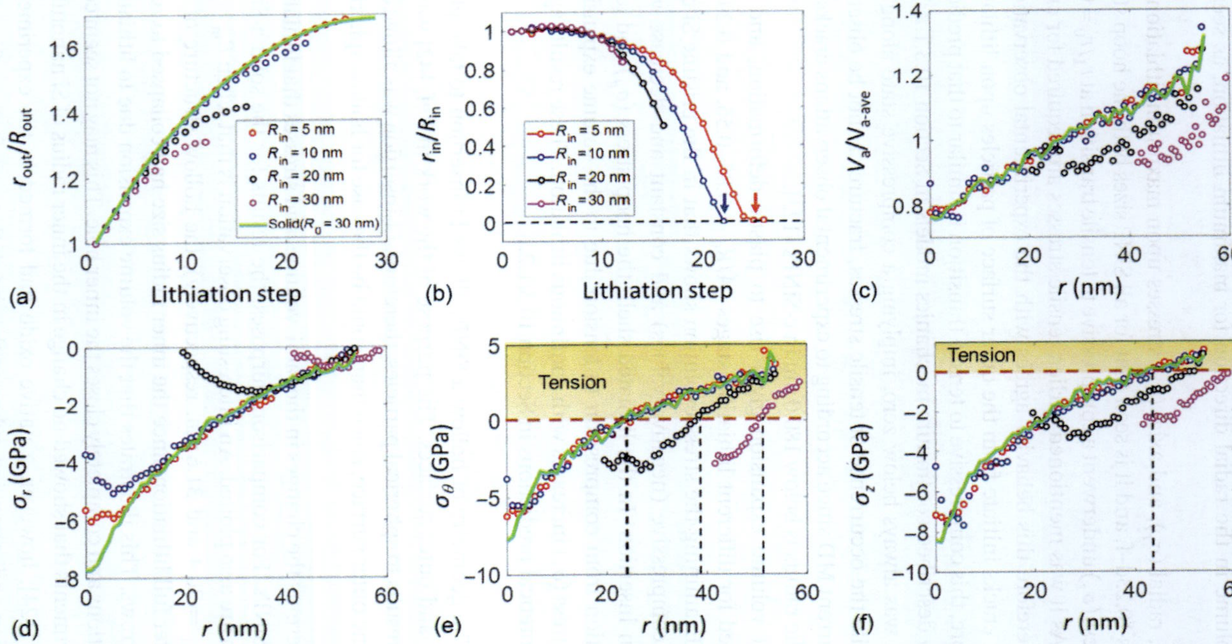

Figure 10.24 Simulation results of hollow a-SiNPs with different R_{in}. (a, b) Variation of R_{out} and R_{in} during the lithiation process. (c–f) Distribution of the atomic volume, atomic radial stress, hoop stress, and axial stress after full lithiation. Source: Reproduced with permission from Shuang and Aifantis [42]/American Chemical Society.

radial stress (σ_r) is always compressive, while both the hoop stress (σ_θ) and the axial stress (σ_z) undergo a compression to tensile transition, which is size dependent. The results show that the volume of $Li_{2.5}Si$ alloy that is under tension is reduced as R_{in} increases. A lower volume of $Li_{2.5}Si$ in tension suggests a lower chance for fracture, and this is consistent with experimental results that show good mechanical stability for hollow SiNPs [53].

Solid c-SiNPs In concluding the examination of different SiNPs, MD results are presented for the case where they were crystalline and non-hollow for a pillar diameter of 30.0 nm. Experiments have shown that the expansion of c-SiNPs upon lithiation is highly anisotropic, as seen in Figure 10.17. Hence MDs were performed for the SiNPs with orientations along the <110>, <100>, and <111>, the results of which are depicted in Figure 10.25. Direct comparison of the MDs with the scanning electron microscopy images shows that despite the difference in scale, there is absolute agreement between the two. The simulations allowed the calculation of the volume expansion upon maximum lithiation as 450% for all orientations. As for the a-SiNPs (hollow and solid), a transition from compression to tension occured for the hoop stress. The radial and axial stresses can be easily obtained but are not shown here because they are not important for fracture. An interesting observation is that in the MDs the highest value of the hoop stress was found to occur at the same location at which crack initiation may occur in the SEM images, as shown in Figure 10.17, which again validates the proposed relaxation method.

10.5.2.5 Plastic Flow of Lithiated SiNPs

To better observe the plastic flow during lithiation, the trajectories for specific Si atoms, whose initial position is indicated by blue, are shown in Figure 10.26a–d, for the solid a-SiNP and c-SiNPs; the discontinuous regions correspond to plastic flow between the Si atoms, and not crack propagation. The obtained trajectories illustrate that the a-SiNP deformed homogeneously along all directions, while the three c-SiNPs undergo significant anisotropic plastic flow due to the anisotropic lithiation rate. Analysis of these images shows (see arrows in Figure 10.26b–d) that deformation incompatibility is observed in two, four, and six locations for the <110>, <100>, and <111> orientations, respectively. The <111> c-SiNP is therefore anticipated to be more resistant to fracture since deformation can be distributed to six locations rather than accumulating in two or four.

A final illustration of plastic flow is shown in Figure 10.26e, where the atomic shear strain is shown at different lithiation states for specific atoms. The lower part of the selected area underwent plastic deformation, while the rest of the atoms formed shear transformation zones (STZs), indicative of the high damage tolerance of lithiated Si [106].

10.5.2.6 Fracture Analysis of Si Nanopillars Due to Lithiation

In concluding the summary of the work in Ref. [42], it is shown how fracture can be crudely modeled through MD. The stress intensity factor is defined as:

$$K_I = \beta\sigma\sqrt{\pi a_c} \tag{10.26}$$

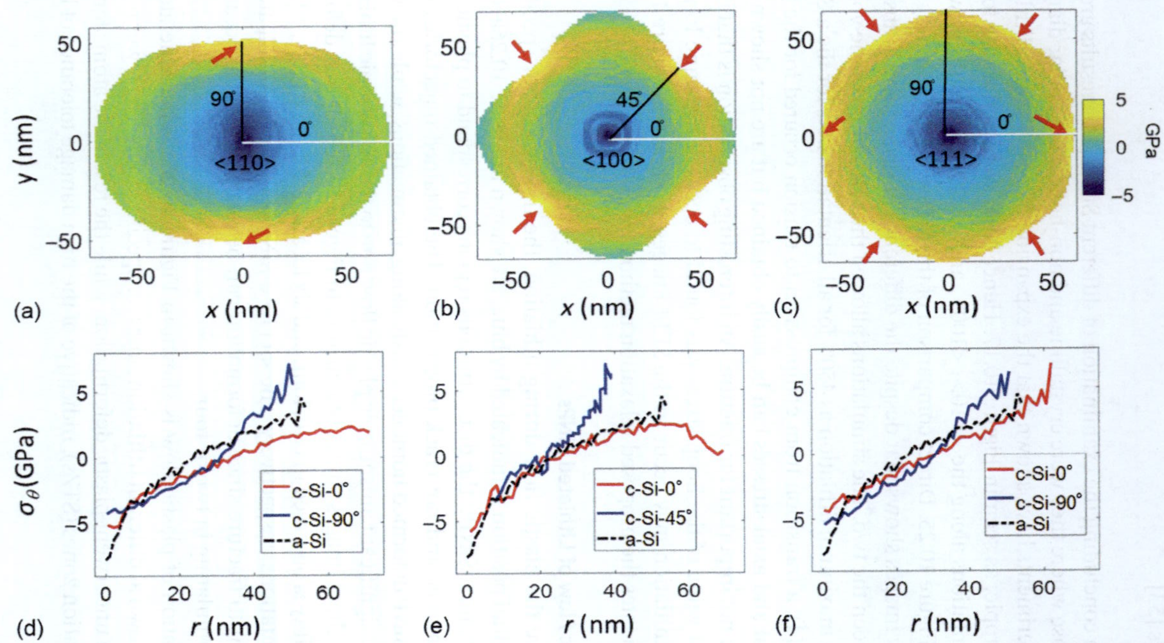

Figure 10.25 (a–c) Deformed shapes and hoop stress distribution in the whole region of c-SiNPs with axial orientations of <110>, <100>, and <111>. (d–f) Hoop stress distribution in particular directions. Source: Reproduced with permission from Shuang and Aifantis [42]/American Chemical Society.

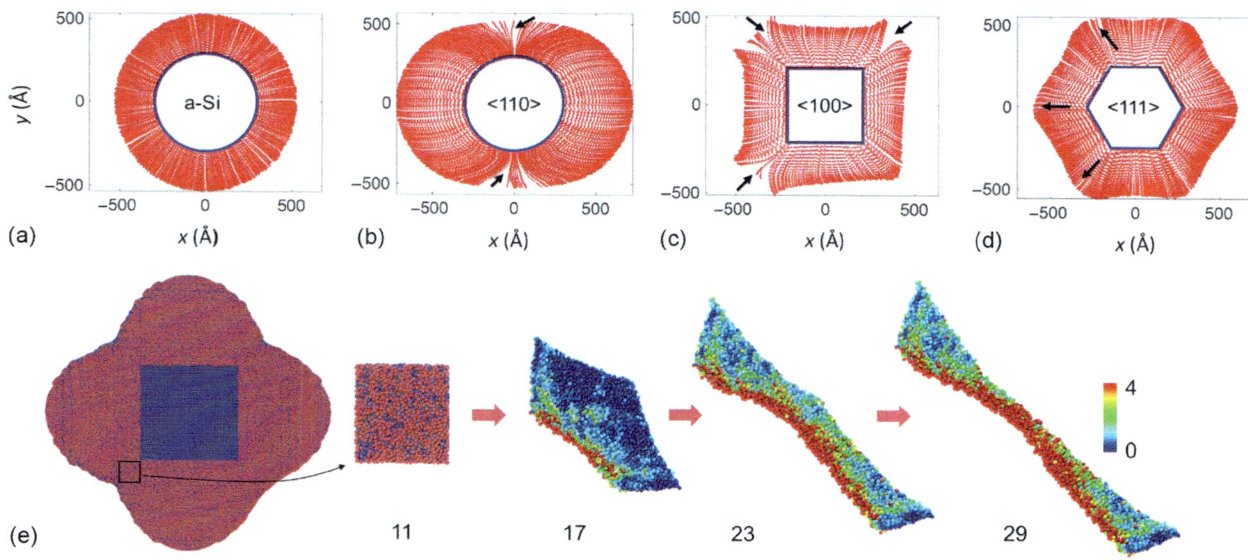

Figure 10.26 Plastic flow in c-SiNP. (a–d) Trajectories of selected atoms in c-SiNPs. (e) Atomic shear strain of a deformed $Li_{3.75}Si$ alloy block at different lithiation stages. 11, 17, 23, and 29 denote lithiation steps. Source: Reproduced with permission from Shuang and Aifantis [42]/American Chemical Society.

where β is the stress intensity modification factor and was taken to be 1 for all SiNPs, σ is the applied stress, and α_c is the critical crack length. It is well known that Eq. (10.26) was computed for an infinitely long plate, however, in Ref. [42] it was successfully used in employing MD to study fracture in SiNPs. Letting K_I equal to the fracture toughness (K_{Ic}) and $a_c = \alpha r_0$ gives that the critical radius at which a SiNP will fracture is (for details see Ref. [42]):

$$R_{0c} = \frac{1}{1.15\pi}\left(\frac{K_{\text{Ic}}}{\sigma}\right)^2 \tag{10.27}$$

For amorphous Si, $K_{\text{Ic}} = 4.62\ \text{MPa}\sqrt{\text{m}}$ for $\text{Li}_{2.5}\text{Si}$ [30] and according to Figure 10.23 σ_θ = 3.8 GPa; these values predict that the critical radius for fracture is R_{0c} = 409.1 nm, and is consistent with experimental data, which indicate that a-Si nanoparticles that were below 438 nm did not fracture [125]. For the case of c-Si $K_{\text{Ic}} = 7.12\ \text{MPa}\sqrt{\text{m}}$ [106] and according to Figure 10.25 $\sigma_\theta = 7.1$ GPa, which indicate that the critical radius below which fracture will not occur is 278.4 nm for c-SiNPs, which is higher than the value of 180 nm observed experimentally [123]. Such discrepancies are due to the simplified fracture model that was employed. Further, the factor of safety or fracture resistance (n) can be defined as [42]:

$$n = \frac{K_{\text{Ic}}}{K_I} = \frac{K_{\text{Ic}}}{\sigma\sqrt{1.15\pi R_0}} \tag{10.28}$$

Inserting the respective parameters for a-SiNPs and c-SiNPs, for the same R_0 = 30.0 nm, gave a higher n for a-SiNP, indicating that the amorphous structure is more resilient to fracture during lithiation. Moreover, Eq. (10.28) also indicates that hollow a-SiNPs have a higher n (better fracture resistance) than solid a-SiNPs because the former experience lower tensile stresses than the latter, in line with the existing experiments and chemomechanical modeling results that showed higher mechanical stability for hollow SiNPs [126, 127]. These results suggest that a hollow a-SiNP is the optimal anode design.

To study lithiation-induced fracture in a solid SiNP (R_0 = 30.0 nm), a 7.0 nm crack was inserted on its outer surface, and the crack length and crack tip opening displacement (CTOD) during lithiation are shown in Figure 10.27. The results suggest that the crack was blunt and that plastic flow was predominant, since the crack length decreased with lithiation, even though the CTOD increased. Analysis of the atomic shear strain showed that shear bands/STZs formed at the crack tip, which is similar to the STZ observations of Figure 10.26, even though a crack was not present in that case. Note that the crack length increased from points 1–3 in Figure 10.27b, indicative of the occurrence of crack propagation in these lithiation steps. Figure 10.27d–f shows the Li and Si atoms with bonds near the crack tip at points 1–3. One can see that the observed crack propagation resulted from the formation of a cavity due to Li—Si bonds breaking near a Si-rich region (Figure 10.27d,e). In previous work, it has been shown that Li—Li and Li—Si bond-switching processes play a crucial role in accommodating the lithiation-induced deformation in $\text{Li}_{2.5}\text{Si}$ alloys [106]. Such a bond-switching process is inhibited in the Si-rich region in Figure 10.27d–f, leading to crack propagation. Such crack growth can provide an interpretation for the different critical radii that the simulations predict over experiments for c-SiNPs. In experiments,

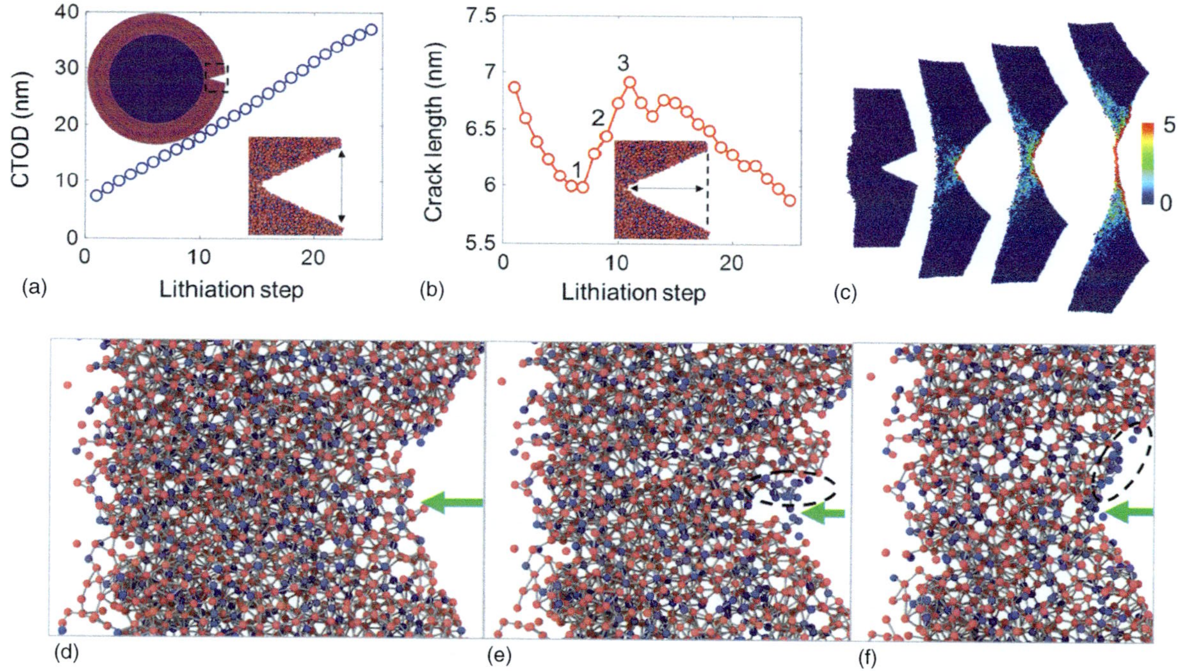

Figure 10.27 Simulation results of lithiation in a solid a-SiNP with $R_{in} = 30.0$ nm after inserting a crack. (a, b) Variation of CTOD and crack length during lithiation. (c) Atomic shear strain at the crack tip during lithiation. (d–f) Atomic bonds at the crack tip at points 1–3 in (b). Source: Reproduced with permission from Shuang and Aifantis [42]/American Chemical Society.

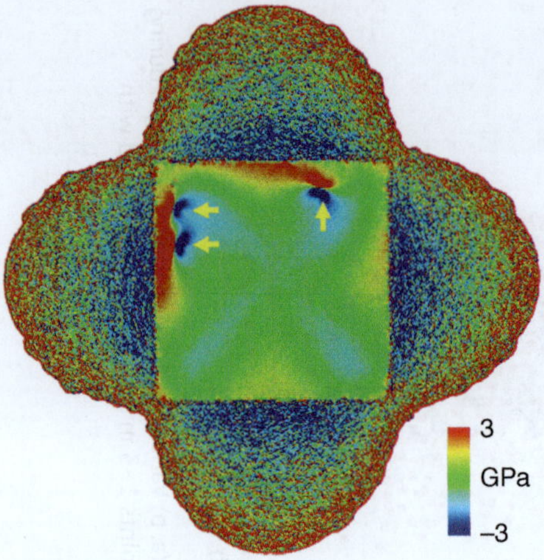

Figure 10.28 Von Mises stress near three dislocations in the c-Si core during the lithiation. Source: Reproduced with permission from Shuang and Aifantis [42]/American Chemical Society.

c-SiNPs may contain defects, such as dislocations, grain boundaries, or microcracks, which can induce a high local stress as shown in Figure 10.28. The nonuniform stress could cause different lithiation rates at different locations [96, 97, 107], further leading to the formation of Si-rich regions. As a result, c-SiNPs were observed to fracture with a lower critical radius in experiments.

10.6 Conclusions

Mechanics and atomistic simulations can reveal the deformation and failure mechanisms of ion insertion batteries. In this chapter, we reviewed existing key studies in this field and focused on recent works that successfully capture the stress evolution, plastic flow, and fracture behaviors of Si during lithiation for Li-ion batteries, and Prussian blue analogues during sodiation for Na-ion batteries. The simulation methods implemented here can be used to study other battery systems and can provide design criteria for the most promising material selections and configurations that prolong the battery lifetime.

Acknowledgment

The authors are grateful to the National Science Foundation for supporting this work through the CMMI grant [CMMI-1762602].

References

1 Li, J.-Y., Xu, Q., Li, G. et al. (2017). Research progress regarding Si-based anode materials towards practical application in high energy density Li-ion batteries. *Mater. Chem. Front.* 1 (9): 1691–1708. https://doi.org/10.1039/C6QM00302H.

2 Zhu, G., Yang, S., Wang, Y. et al. (2019). Dimethylacrylamide, a novel electrolyte additive, can improve the electrochemical performances of silicon anodes in lithium-ion batteries. *RSC Adv.* 9 (1): 435–443. https://doi.org/10.1039/C8RA07988A.

3 Mukhopadhyay, A. and Sheldon, B.W. (2014). Deformation and stress in electrode materials for Li-ion batteries. *Prog. Mater Sci.* 63 (February): 58–116. https://doi.org/10.1016/j.pmatsci.2014.02.001.

4 Wolfenstine, J. (1999). Critical grain size for microcracking during lithium insertion. *J. Power Sources* 79 (1): 111–113.

5 Huggins, R.A. and Nix, W.D. (2000). Decrepitation model for capacity loss during cycling of alloys in rechargeable electrochemical systems. *Ionics* 6 (1–2): 57–63.

6 Nix, W.D. (1989). Mechanical properties of thin films. *Metall. Trans. A* 20 (11): 2217.

7 Aifantis, K. and Hackney, S.A. (2003). An ideal elasticity problem for Li-batteries. *J. Mech. Behav. Mater.* 14 (6): 413–427.

8 Aifantis, K.E. and Dempsey, J.P. (2005). Stable crack growth in nanostructured Li-batteries. *J. Power Sources* 143 (1–2): 203–211. https://doi.org/10.1016/J.JPOWSOUR.2004.11.037.

9 Aifantis, K.E., Hackney, S.A., and Dempsey, J.P. (2007). Design criteria for nanostructured Li-ion batteries. *J. Power Sources* 165 (2): 874–879.

10 Aifantis, K.E. and Hackney, S.A. (2011). Mechanical stability for nanostructured Sn-and Si-based anodes. *J. Power Sources* 196 (4): 2122–2127.

11 Aifantis, E.C. (1984). On the microstructural origin of certain inelastic models. *J. Eng. Mater. Technol.* 106 (4): 326–330. https://doi.org/10.1115/1.3225725.

12 Fleck, N.A., Muller, G.M., Ashby, M.F., and Hutchinson, J.W. (1994). Strain gradient plasticity: theory and experiment. *Acta Metall. Mater.* 42 (2): 475–487.

13 Aifantis, K.E. and Willis, J.R. (2005). The role of interfaces in enhancing the yield strength of composites and polycrystals. *J. Mech. Phys. Solids* 53 (5): 1047–1070. https://doi.org/10.1016/J.JMPS.2004.12.003.

14 Ru, C.Q. and Aifantis, E.C. (1993). A simple approach to solve boundary-value problems in gradient elasticity. *Acta Mech.* 101 (1): 59–68.

15 Dimitrijevic, B.J., Aifantis, K.E., and Hackl, K. (2012). The influence of particle size and spacing on the fragmentation of nanocomposite anodes for Li batteries. *J. Power Sources* 206: 343–348.

16 Prussin, S. (1961). Generation and distribution of dislocations by solute diffusion. *J. Appl. Phys.* 32 (10): 1876–1881.

17 Li, J.C.-M. (1978). Physical chemistry of some microstructural phenomena. *Metall. Trans. A* 9 (10): 1353–1380.

18 Larché, F.C. and Cahn, J.L. (1982). The effect of self-stress on diffusion in solids. *Acta Metall.* 30 (10): 1835–1845.

19 Larché, F.C. and Cahn, J.W. (1984). The interactions of composition and stress in crystalline solids. *J. Res. Nat. Bur. Stand.* 89 (6): 467–500.

20 Lee, S., Wang, W.L., and Chen, J.R. (2000). Diffusion-induced stresses in a hollow cylinder. *Mater. Sci. Eng., A* 285 (1–2): 186–194.

21 Wang, W.L., Lee, S., and Chen, J.R. (2002). Effect of chemical stress on diffusion in a hollow cylinder. *J. Appl. Phys.* 91 (12): 9584–9590.

22 Ko, S.-C., Lee, S., and Chou, Y.T. (2005). Chemical stresses in a square sandwich composite. *Mater. Sci. Eng., A* 409 (1–2): 145–152.

23 Deshpande, R., Qi, Y., and Cheng, Y.-T. (2010). Effects of concentration-dependent elastic modulus on diffusion-induced stresses for battery applications. *J. Electrochem. Soc.* 157 (8): A967–A971.

24 Song, Y., Lu, B., Ji, X., and Zhang, J. (2012). Diffusion induced stresses in cylindrical lithium-ion batteries: analytical solutions and design insights. *J. Electrochem. Soc.* 159 (12): A2060–A2068.

25 Deshpande, R., Cheng, Y.-T., and Verbrugge, M.W. (2010). Modeling diffusion-induced stress in nanowire electrode structures. *J. Power Sources* 195 (15): 5081–5088.

26 Verbrugge, M.W. and Cheng, Y.-T. (2009). Stress and Strain-Energy Distributions within Diffusion-Controlled Insertion-Electrode Particles Subjected to Periodic Potential Excitations. *J. Electrochem. Soc.* 156 (11): A927–A937. https://doi.org/10.1149/1.3205485.

27 Cheng, Y.-T. and Verbrugge, M.W. (2010). Diffusion-induced stress, interfacial charge transfer, and criteria for avoiding crack initiation of electrode particles. *J. Electrochem. Soc.* 157 (4): A508–A516.

28 Cheng, Y.-T. and Verbrugge, M.W. (2008). The influence of surface mechanics on diffusion induced stresses within spherical nanoparticles. *J. Appl. Phys.* 104 (8): 83521.

29 Zhang, J., Lu, B., Song, Y., and Ji, X. (2012). Diffusion induced stress in layered Li-ion battery electrode plates. *J. Power Sources* 209: 220–227.

30 Bohn, E., Eckl, T., Kamlah, M., and McMeeking, R. (2013). A model for lithium diffusion and stress generation in an intercalation storage particle with phase change. *J. Electrochem. Soc.* 160 (10): A1638–A1652.

31 Anand, L. (2012). A Cahn–Hilliard-type theory for species diffusion coupled with large elastic–plastic deformations. *J. Mech. Phys. Solids* 60 (12): 1983–2002.

32 Chen, L.Q., Fan, F., Hong, L. et al. (2014). A phase-field model coupled with large elasto-plastic deformation: application to lithiated silicon electrodes. *J. Electrochem. Soc.* 161 (11): F3164–F3172. https://doi.org/10.1149/2.0171411jes.

33 di Leo, C.V., Rejovitzky, E., and Anand, L. (2015). Diffusion–deformation theory for amorphous silicon anodes: the role of plastic deformation on electrochemical performance. *Int. J. Solids Struct.* 67: 283–296.

34 Hu, P., Wang, B., Xiao, D., and Aifantis, K. (2019). Capturing the differences between lithiation and sodiation of nanostructured TiS_2 electrodes. *Nano Energy* 63: 103820.

35 Liu, X.H., Zhong, L., Huang, S. et al. (2012). Size-dependent fracture of silicon nanoparticles during lithiation. *ACS Nano* 6 (2): 1522–1531.

36 Huang, S., Fan, F., Li, J. et al. (2013). Stress generation during lithiation of high-capacity electrode particles in lithium ion batteries. *Acta Mater.* 61 (12): 4354–4364.

37 Cogswell, D.A. and Bazant, M.Z. (2012). Coherency strain and the kinetics of phase separation in LiFePO4 nanoparticles. *ACS Nano* 6 (3): 2215–2225.

38 Huttin, M. and Kamlah, M. (2012). Phase-field modeling of stress generation in electrode particles of lithium ion batteries. *Appl. Phys. Lett.* 101 (13): 133902.

39 Hu, P. et al. (2019). Concentration-gradient Prussian blue cathodes for Na-ion batteries. *ACS Energy Lett.* 5 (1): 100–108.

40 Sethuraman, V.A., Chon, M.J., Shimshak, M. et al. (2010). In situ measurement of biaxial modulus of Si anode for Li-ion batteries. *Electrochem. Commun.* 12 (11): 1614–1617.

41 Zhao, K., Pharr, M., Cai, S. et al. (2011). Large plastic deformation in high-capacity lithium-ion batteries caused by charge and discharge. *J. Am. Ceram. Soc.* 94: s226–s235.

42 Shuang, F. and Aifantis, K.E. (2021). A first molecular dynamics study for modeling the microstructure and mechanical behavior of Si nanopillars during lithiation. *ACS Appl. Mater. Interfaces* 13 (18): 21310–21319.

43 Ryu, I., Choi, J.W., Cui, Y., and Nix, W.D. (2011). Size-dependent fracture of Si nanowire battery anodes. *J. Mech. Phys. Solids* 59 (9): 1717–1730.

44 Chen, C.-H., Chason, E., and Guduru, P.R. (2017). Numerical solution of moving phase boundary and diffusion-induced stress of Sn anode in the lithium-ion battery. *J. Electrochem. Soc.* 164 (11): E3661–E3670.

45 Aifantis, K.E., Huang, T., Hackney, S.A. et al. (2012). Capacity fade in Sn–C nanopowder anodes due to fracture. *J. Power Sources* 197: 246–252.

46 Misra, S., Liu, N., Nelson, J. et al. (2012). In situ X-ray diffraction studies of (de) lithiation mechanism in silicon nanowire anodes. *ACS Nano* 6 (6): 5465–5473.

47 Tsagrakis, I. and Aifantis, E.C. (2018). Gradient elasticity effects on the two-phase lithiation of LIB anodes. In: *Generalized Models and Non-classical Approaches in Complex Materials*, vol. 2 (ed. H. Altenbach, J. Pouget, M. Rousseau, et al.), 221–235. Springer.

48 Christensen, J. and Newman, J. (2006). Stress generation and fracture in lithium insertion materials. *J. Solid State Electrochem.* 10 (5): 293–319.

49 Zhang, T. and Kamlah, M. (2018). Sodium ion batteries particles: phase-field modeling with coupling of Cahn–Hilliard equation and finite deformation elasticity. *J. Electrochem. Soc.* 165 (10): A1997–A2007.

50 Hao, F. and Fang, D. (2013). Diffusion-induced stresses of spherical core-shell electrodes in lithium-ion batteries: the effects of the shell and surface/interface stress. *J. Electrochem. Soc.* 160 (4): A595–A600.

51 Hertzberg, B., Benson, J., and Yushin, G. (2011). Ex-situ depth-sensing indentation measurements of electrochemically produced Si–Li alloy films. *Electrochem. Commun.* 13 (8): 818–821. https://doi.org/10.1016/J.ELECOM.2011.05.011.

52 Maxisch, T. and Ceder, G. (2006). Elastic properties of olivine Li_x $FePO_4$ from first principles. *Phys. Rev. B* 73 (17): 174112.

53 Qi, Y., Hector, L.G., James, C., and Kim, K.J. (2014). Lithium concentration dependent elastic properties of battery electrode materials from first principles calculations. *J. Electrochem. Soc.* 161 (11): F3010–F3018.

54 Qi, Y., Guo, H., Hector, L.G., and Timmons, A. (2010). Threefold increase in the Young's modulus of graphite negative electrode during lithium intercalation. *J. Electrochem. Soc.* 157 (5): A558–A566.

55 Shenoy, V.B., Johari, P., and Qi, Y. (2010). Elastic softening of amorphous and crystalline Li–Si phases with increasing Li concentration: a first-principles study. *J. Power Sources* 195 (19): 6825–6830.

56 Stournara, M.E., Guduru, P.R., and Shenoy, V.B. (2012). Elastic behavior of crystalline Li–Sn phases with increasing Li concentration. *J. Power Sources* 208: 165–169.

57 Natarajan, S. and Aifantis, K. (2020). Size effects in nanostructured Li-ion battery cathode particles. *J. Mech. Behav. Mater.* 29 (1): 36–43.

58 Lu, Y., Soh, A.K., Ni, Y., and He, L. (2019). Understanding size-dependent migration of a two-phase lithiation front coupled to stress. *Acta Mech.* 230 (1): 303–317.

59 Wang, B., Réthoré, J., and Aifantis, K.E. (2021). Capturing the stress evolution in electrode materials that undergo phase transformations during electrochemical cycling. *Int. J. Solids Struct.* 224: 111032.

60 McDowell, M.T., Ryu, I., Lee, S.W. et al. (2012). Studying the kinetics of crystalline silicon nanoparticle lithiation with in situ transmission electron microscopy. *Adv. Mater.* 24 (45): 6034–6041.

61 Zhao, K., Pharr, M., Vlassak, J.J., and Suo, Z. (2010). Fracture of electrodes in lithium-ion batteries caused by fast charging. *J. Appl. Phys.* 108 (7): 073517. https://doi.org/10.1063/1.3492617.

62 Hu, Y., Zhao, X., and Suo, Z. (2010). Averting cracks caused by insertion reaction in lithium–ion batteries. *J. Mater. Res.* 25 (6): 1007–1010.

63 Gao, Y.F. and Zhou, M. (2013). Coupled mechano-diffusional driving forces for fracture in electrode materials. *J. Power Sources* 230: 176–193.

64 Liu, Y. et al. (2011). Lithiation-induced embrittlement of multiwalled carbon nanotubes. *ACS Nano* 5 (9): 7245–7253.

65 Chew, H.B., Hou, B., Wang, X., and Xia, S. (2014). Cracking mechanisms in lithiated silicon thin film electrodes. *Int. J. Solids Struct.* 51 (23–24): 4176–4187.

66 Miehe, C., Dal, H., Schänzel, L.-M., and Raina, A. (2016). A phase-field model for chemo-mechanical induced fracture in lithium-ion battery electrode particles. *Int. J. Numer. Methods Eng.* 106 (9): 683–711.

67 Klinsmann, M., Rosato, D., Kamlah, M., and McMeeking, R.M. (2016). Modeling crack growth during Li extraction in storage particles using a fracture phase field approach. *J. Electrochem. Soc.* 163 (2): A102–A118.

68 Réthoré, J., Zheng, H., Li, H. et al. (2018). A multiphysics model that can capture crack patterns in Si thin films based on their microstructure. *J. Power Sources* 400: 383–391.

69 Beaulieu, L.Y., Eberman, K.W., Turner, R.L. et al. (2001). Colossal reversible volume changes in lithium alloys. *Electrochem. Solid-State Lett.* 4 (9): A137–A140.

70 Wang, Y.H., He, Y., Xiao, R.J. et al. (2012). Investigation of crack patterns and cyclic performance of Ti–Si nanocomposite thin film anodes for lithium ion batteries. *J. Power Sources* 202: 236–245.

71 Miehe, C., Hofacker, M., and Welschinger, F. (2010). A phase field model for rate-independent crack propagation: Robust algorithmic implementation based on operator splits. *Comput. Meth. Appl. Mech. Eng.* 199 (45–48): 2765–2778.

72 Rahani, E.K. and Shenoy, V.B. (2013). Role of plastic deformation of binder on stress evolution during charging and discharging in lithium-ion battery negative electrodes. *J. Electrochem. Soc.* 160 (8): A1153.

73 Choi, Y.S., Pharr, M., Oh, K.H., and Vlassak, J.J. (2015). A simple technique for measuring the fracture energy of lithiated thin-film silicon electrodes at various lithium concentrations. *J. Power Sources* 294: 159–166.

74 Ahuja, U., Wang, B., Hu, P. et al. (2021). Polydopamine coated Si nanoparticles allow for improved mechanical and electrochemical stability. *Electrochim. Acta* 392: 138993.

75 Boebinger, M.G. et al. (2018). Avoiding fracture in a conversion battery material through reaction with larger ions. *Joule* 2 (9): 1783–1799.

76 Stender, V.V. and Ksenzhek, O.S. (1959). Grafitirovannye anody pri elekrolize vodnykh rastvorov khloristykh solei. *Zh. Prikl. Khim.* 32: 110.

77 Euler, J. and Nonnenmacher, W. (1960). Stromverteilung in porösen elektroden. *Electrochim. Acta* 2 (4): 268–286.

78 Newman, J.S. and Tobias, C.W. (1962). Theoretical analysis of current distribution in porous electrodes. *J. Electrochem. Soc.* 109 (12): 1183.

79 de Vidts, P. and White, R.E. (1997). Governing equations for transport in porous electrodes. *J. Electrochem. Soc.* 144 (4): 1343.

80 Doyle, M. and Newman, J. (1995). The use of mathematical modeling in the design of lithium/polymer battery systems. *Electrochim. Acta* 40 (13–14): 2191–2196.

81 Doyle, M., Fuller, T.F., and Newman, J. (1993). Modeling of galvanostatic charge and discharge of the lithium/polymer/insertion cell. *J. Electrochem. Soc.* 140 (6): 1526.

82 Doyle, M. and Newman, J. (1997). Analysis of capacity–rate data for lithium batteries using simplified models of the discharge process. *J. Appl. Electrochem.* 27 (7): 846–856.

83 Devan, S., Subramanian, V.R., and White, R.E. (2005). Transient analysis of a porous electrode. *J. Electrochem. Soc.* 152 (5): A947.

84 Christensen, J. and Newman, J. (2006). A mathematical model of stress generation and fracture in lithium manganese oxide. *J. Electrochem. Soc.* 153 (6): A1019–A1030.

85 Christensen, J. (2010). Modeling diffusion-induced stress in Li-ion cells with porous electrodes. *J. Electrochem. Soc.* 157 (3): A366.

86 Golmon, S., Maute, K., and Dunn, M.L. (2009). Numerical modeling of electrochemical–mechanical interactions in lithium polymer batteries. *Comput. Struct.* 87 (23–24): 1567–1579.

87 Renganathan, S., Sikha, G., Santhanagopalan, S., and White, R.E. (2009). Theoretical analysis of stresses in a lithium ion cell. *J. Electrochem. Soc.* 157 (2): A155.

88 Purkayastha, R.T. and McMeeking, R.M. (2012). An integrated 2-D model of a lithium ion battery: the effect of material parameters and morphology on storage particle stress. *Comput. Mech.* 50 (2): 209–227.

89 Garca, R.E., Chiang, Y.-M., Carter, W.C. et al. (2004). Microstructural modeling and design of rechargeable lithium-ion batteries. *J. Electrochem. Soc.* 152 (1): A255.

90 Garrick, T.R., Kanneganti, K., Huang, X., and Weidner, J.W. (2014). Modeling volume change due to intercalation into porous electrodes. *J. Electrochem. Soc.* 161 (8): E3297.

91 Gomadam, P.M. and Weidner, J.W. (2005). Modeling volume changes in porous electrodes. *J. Electrochem. Soc.* 153 (1): A179.

92 Wang, H., Ji, X., Chen, C. et al. (2013). Lithium diffusion in silicon and induced structure disorder: A molecular dynamics study. *AIP Adv.* 3 (11): https://doi.org/10.1063/1.4829440.

93 Fan, F. Huang, S., Yang, H. et al. (2013). Mechanical properties of amorphous Li_xSi alloys: A reactive force field study. *Modell. Simul. Mater. Sci. Eng.* 21 (7): 074002.

94 Ding, B., Li, X., Zhang, X. et al. (2015). Brittle versus ductile fracture mechanism transition in amorphous lithiated silicon: From intrinsic nanoscale cavitation to shear banding. *Nano Energy* 18: 89–96. https://doi.org/10.1016/j.nanoen.2015.10.002.

95 Khosrownejad, S.M. and Curtin, W.A. (2017). Crack growth and fracture toughness of amorphous Li-Si anodes: Mechanisms and role of charging/discharging studied by atomistic simulations. *J. Mech. Phys. Solids* 107: 542–559. https://doi.org/10.1016/j.jmps.2017.06.010.

96 Pan, J., Zhang, Q., Li, J. et al. (2015). Effects of stress on lithium transport in amorphous silicon electrodes for lithium-ion batteries. *Nano Energy* 13: 192–199. https://doi.org/10.1016/j.nanoen.2015.02.020.

97 Ding, B., Wu, H., Xu, Z. et al. (2017). Stress effects on lithiation in silicon. *Nano Energy* 38 (May): 486–493. https://doi.org/10.1016/j.nanoen.2017.06.021.

98 Cui, Z., Gao, F., Cui, Z., and Qu, J. (2012). A second nearest-neighbor embedded atom method interatomic potential for Li-Si alloys. *J. Power Sources* 207: 150–159. https://doi.org/10.1016/j.jpowsour.2012.01.145.

99 Yan, X., Gouissem, A., and Sharma, P. (2015). Atomistic insights into Li-ion diffusion in amorphous silicon. *Mech. Mater.* 91: 306–312. https://doi.org/10.1016/j.mechmat.2015.04.001.

100 Fan, F., Yang, H., and Zeng, Z. (2018). An atomistic perspective on lithiation-induced stress in silicon nanopillars. *Scr. Mater.* 152: 74–78. https://doi.org/10.1016/j.scriptamat.2018.04.014.

101 Cao, L., Yang, H., and Fan, F. (2019). Stress generation during anisotropic lithiation in silicon nanopillar electrodes: a reactive force field study. *Phys. Lett., Sect. A Gen. At. Solid State Phys.* 383 (33): 125955. https://doi.org/10.1016/j.physleta.2019.125955.

102 Jia, Z. and Li, T. (2015). Stress-modulated driving force for lithiation reaction in hollow nano-anodes. *J. Power Sources* 275: 866–876. https://doi.org/10.1016/j.jpowsour.2014.11.081.

103 Jia, Z. and Li, T. (2016). Intrinsic stress mitigation via elastic softening during two-step electrochemical lithiation of amorphous silicon. *J. Mech. Phys. Solids* 91: 278–290. https://doi.org/10.1016/j.jmps.2016.03.014.

104 Zhao, K., Pharr, M., Wan, Q. et al. (2012). Concurrent reaction and plasticity during initial lithiation of crystalline silicon in lithium-ion batteries. *J. Electrochem. Soc.* 159 (3): A238–A243. https://doi.org/10.1149/2.020203jes.

105 Wang, H. and Chew, H.B. (2016). Molecular dynamics simulations of plasticity and cracking in lithiated silicon electrodes. *Extreme Mech. Lett.* 9: 503–513. https://doi.org/10.1016/j.eml.2016.02.020.

106 Wang, X., Fan, F., Wang, J. et al. (2015). High damage tolerance of electrochemically lithiated silicon. *Nat. Commun.* 6 (1): 1–7. https://doi.org/10.1038/ncomms9417.

107 Gu, M., Yang, H., Perea, D.E. et al. (2014). Bending-induced symmetry breaking of lithiation in germanium nanowires. *Nano Lett.* 14 (8): 4622–4627. https://doi.org/10.1021/nl501680w.

108 Kim, K.J., Wortman, J., Kim, S.Y., and Qi, Y. (2017). Atomistic simulation derived insight on the irreversible structural changes of Si electrode during fast and slow delithiation. *Nano Lett.* 17 (7): 4330–4338. https://doi.org/10.1021/acs.nanolett.7b01389.

109 Lee, S.W., McDowell, M.T., Choi, J.W., and Cui, Y. (2011). Anomalous shape changes of silicon nanopillars by electrochemical lithiation. *Nano Lett.* 11 (7): 3034–3039. https://doi.org/10.1021/nl201787r.

110 Plimpton, S. (1995). Fast parallel algorithms for short-range molecular dynamics. *J. Comput. Phys.* 117 (1): 1–19. https://doi.org/10.1006/jcph.1995.1039.

111 Stukowski, A. (2010). Visualization and analysis of atomistic simulation data with OVITO-the open visualization tool. *Modell. Simul. Mater. Sci. Eng.* 18 (1): 015012. https://doi.org/10.1088/0965-0393/18/1/015012.

112 Shimizu, F., Ogata, S., and Li, J. (2007). Theory of shear banding in metallic glasses and molecular dynamics calculations. *Mater. Trans.* 48 (11): 2923–2927. https://doi.org/10.2320/matertrans.MJ200769.

113 Lee, H.S. and Lee, B.J. (2014). Structural changes during lithiation and delithiation of Si anodes in Li-ion batteries: a large scale molecular dynamics study. *Met. Mater. Int.* 20 (6): 1003–1009. https://doi.org/10.1007/s12540-014-6002-x.

114 Liu, X.H., Wang, J.W., Huang, S. et al. (2012). *In situ* atomic-scale imaging of electrochemical lithiation in silicon. *Nat. Nanotechnol.* 7 (11): 749–756. https://doi.org/10.1038/nnano.2012.170.

115 Wang, J.W., He, Y., Fan, F. et al. (2013). Two-phase electrochemical lithiation in amorphous silicon. *Nano Lett.* 13 (2): 709–715. https://doi.org/10.1021/nl304379k.

116 Cubuk, E.D., Wang, W.L., Zhao, K. et al. (2013). Morphological evolution of Si nanowires upon lithiation: a first-principles multiscale model. *Nano Lett.* 13 (5): 2011–2015. https://doi.org/10.1021/nl400132q.

117 Shuang, F., Xiao, P., Ke, F. et al. (2017). Efficiency and fidelity of molecular simulations relevant to dislocation evolutions. *Comput. Mater. Sci.* 139: 266–272.

accessed May 05 2019. [Online]. Available: https://www.sciencedirect.com/science/article/pii/S0927025617304068.

118 Bitzek, E., Koskinen, P., Gähler, F. et al. (2006). Structural relaxation made simple. *Phys. Rev. Lett.* 97 (17): 170201. https://doi.org/10.1103/PhysRevLett.97.170201.

119 Guénolé, J., Nöhring, W.G., Vaid, A. et al. (2020). Assessment and optimization of the fast inertial relaxation engine (fire) for energy minimization in atomistic simulations and its implementation in LAMMPS. *Comput. Mater. Sci.* 175: 109584. https://doi.org/10.1016/j.commatsci.2020.109584.

120 Shuang, F., Xiao, P., Shi, R. et al. (2019). Influence of integration formulations on the performance of the fast inertial relaxation engine (FIRE) method. *Comput. Mater. Sci.* 156: 135–141. https://doi.org/10.1016/j.commatsci.2018.09.049.

121 Schmidt, H., Jerliu, B., Hüger, E., and Stahn, J. (2020). Volume expansion of amorphous silicon electrodes during potentiostatic lithiation of Li-ion batteries. *Electrochem. Commun.* 115: 106738. https://doi.org/10.1016/j.elecom.2020.106738.

122 Mendez, J.P., Ponga, M., and Ortiz, M. (2018). Diffusive molecular dynamics simulations of lithiation of silicon nanopillars. *J. Mech. Phys. Solids* 115: 123–141. https://doi.org/10.1016/j.jmps.2018.03.008.

123 Lee, S.W., Shin, J.Y., Lee, A. et al. (2012). Counting single photoactivatable fluorescent molecules by photoactivated localization microscopy (PALM). *Proc. Natl. Acad. Sci. U.S.A.* 109 (43): 17436–17441. https://doi.org/10.1073/pnas.1201088109.

124 Wang, J., Luo, H., Liu, Y. et al. (2016). Tuning the outward to inward swelling in lithiated silicon nanotubes via surface oxide coating. *Nano Lett.* 16 (9): 5815–5822. https://doi.org/10.1021/acs.nanolett.6b02581.

125 McDowell, M.T., Lee, S.W., Harris, J.T. et al. (2013). *In situ* TEM of two-phase lithiation of amorphous silicon nanospheres. *Nano Lett.* 13 (2): 758–764. https://doi.org/10.1021/nl3044508.

126 Yao, Y., McDowell, M.T., Ryu, I. et al. (2011). Interconnected silicon hollow nanospheres for lithium-ion battery anodes with long cycle life. *Nano Lett.* 11 (7): 2949–2954. https://doi.org/10.1021/nl201470j.

127 Wang, C., Wen, J., Luo, F. et al. (2019). Anisotropic expansion and size-dependent fracture of silicon nanotubes during lithiation. *J. Mater. Chem. A* 7 (25): 15113–15122.

11

Future Ion-Battery Technologies

Pu Hu[1] *and Katerina E. Aifantis*[2]

[1] *Wuhan Institute of Technology, School of Materials Science and Engineering, LiuFang Campus, No. 206, Guanggu 1st road, Donghu New & High Technology Development Zone, Wuhan 430205, China*
[2] *University of Florida, Department of Mechanical and Aerospace Engineering, 1064 Center Drive, Gainesville, FL 32603, USA*

In addition to Li- and Na-ion batteries, which have been summarized in the previous chapters, there exist other elements, such as Mg, Zn, Ca, and Al, that can also be used as anodes in forming respective secondary batteries.

Figure 11.1 presents the mass specific capacity, volume specific capacity, and reduction potential of various such alternative elements. Zinc-ion batteries use the Zn^{2+}/Zn two-electron reaction to store energy, with a low-oxidation–reduction potential (−0.763 V vs. standard hydrogen electrode [SHE]) and high theoretical capacity (820 mAh g^{-1} and 5855 mAh cm^{-3}). An aqueous solution can be used as the electrolyte (unlike Li-ion and Na-ion that mostly use non-aqueous), which has the advantages of low cost and good safety. Therefore, Mg-based and Zn-based rechargeable batteries have drawn increasing attention in energy storage, and will be briefly summarized in this chapter.

11.1 Magnesium-Based Batteries

The magnesium secondary battery (or magnesium battery) is a new type of battery, based on divalent magnesium ions, that has been developed recently and has great potential. Compared with lithium-ion batteries, magnesium batteries have many advantages:

(1) Similarly to Na, there exist abundant reserves of Mg in the earth's crust, and the price is much lower than that for Li.
(2) Mg metal as an anode allows for a two-electron reaction, since it has two valence electrons, and it has a relatively high negative electrode potential (−2.37 V vs. SHE). As a result, Mg anodes have a high mass energy density (2205 mAh g^{-1}), and a volume energy density (3833 mAh cm^{-3}) that is even higher than that of Li anodes (2046 mAh cm^{-3}).

Rechargeable Ion Batteries: Materials, Design, and Applications of Li-Ion Cells and Beyond, First Edition. Edited by Katerina E. Aifantis, R. Vasant Kumar, and Pu Hu.

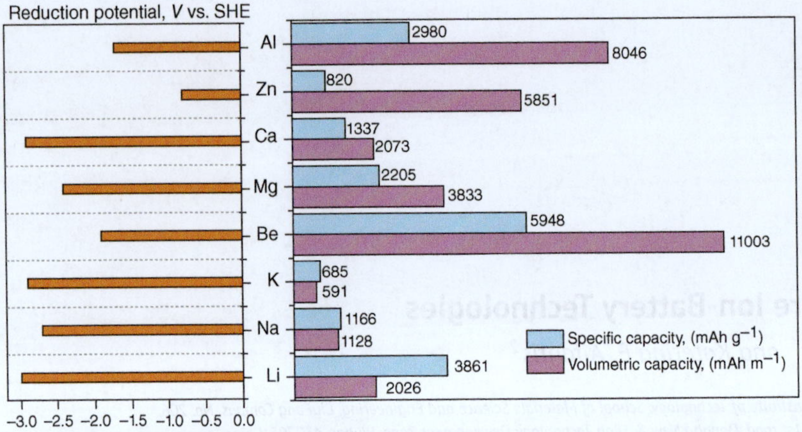

Figure 11.1 Reduction potential, specific capacity, and volumetric capacity of various ions used in rechargeable batteries.

(3) The deposition of magnesium during the charge/discharge process is dendritic-free, and the melting point of Mg metal (648 °C) is much higher than that of Li metal (180 °C). Also, although Mg has a certain reactivity with air at room temperature, it is still relatively stable. Therefore, the Mg-metallic anodes are safer than Li-metal anodes.

The most important difference between Mg batteries and Li-ion batteries is the difference in the chemical properties of Li^+ and Mg^{2+}. Although the high energy density and capacity of Mg cells are attributed to the two-electron reaction that Mg allows, the divalent change has severe drawbacks in the kinetics. Therefore, even though Li^+ (0.76 Å) and Mg^{2+} (0.72 Å) have similar ionic radii, the higher charge density of Mg^{2+} cations leads to a sluggish diffusion rate in electrode materials during electrochemical cycling due to a stronger Coulomb effect. Also, the divalent charge of Mg^{2+} makes it difficult to retain the local neutral balance during the diffusion process. This poses severe challenges in developing stable electrolytes and cathodes for Mg cells.

11.1.1 Electrolytes of Mg Batteries

The electrolyte, which was initially comprised of polar aprotic solvents and salt anions (e.g. ClO_4^-, BF_4^-, and $SO_3CF_3^-$–), formed a solid electrolyte interface (SEI) that was not a conductor of Mg^{2+}, and lead to the irreversible dissolution/deposition reaction of magnesium, without forming dendrites. Such passivation films can form either by the reduction of the salt in the electrolyte or by the reduction of the electrolyte solvent. A suitable electrolyte, therefore, will allow for the reversible deposition and dissolution of magnesium ions without the formation of a passivation film. Current research on Mg batteries, therefore, focuses on the development of new electrolytes, since not only they directly determine the materials that can be used as cathodes, but the main issues of Mg-ion batteries are directly related to the electrolyte, such as a low voltage window (less than 3.0 V), severe corrosion of non-precious metal current collectors, and poor reversibility of magnesium deposition and dissolution.

Simple inorganic salts cannot be used in Mg electrolytes since they have been shown to form a passivation film on the surface of magnesium metal. The Grignard reagent RMgX (R=alkyl, X=Cl, Br), as a typical reducing reagent, does not react with magnesium metal and was initially considered. Specifically, the Grignard reagent is Mg-ion conductivite, can reversibly deposit and dissolve magnesium ions, and has a high coulombic efficiency [1, 2]. However, the deposition of Mg is not simply a two-electron migration of Mg-ions, but also involves the adsorption of components formed by the reversible decomposition of the Grignard reagent salt solution. Mg does not react with groups such as MgR_2, $RMgX^{2-}$, or RX in the Grignard reagent, so it is highly stable. Therefore, the first step of the polarization reaction of the magnesium electrode is that the MgX_2, RMg, R_2Mg and other groups are first adsorbed on the electrode surface to accept electrons. In fact, the Mg deposit will always be covered by the adsorption layer composed of these groups. The dissolution of Mg and the anode polarization of the electrode will cause the desorption of the adsorption layer. Therefore, the deposition and dissolution of Mg are carried out through a certain surface film. Although the ether solution of the Grignard reagent allows for a high coulombic efficiency, the oxidation stability and conductivity of Grignard reagents are poor, and cannot match with those of most cathode materials, especially high-voltage cathode materials. Therefore, its application in magnesium ion battery electrolytes are limited.

In 2000, Doron Aurbach's research group reported the first rechargeable magnesium battery model based on an organic halogenated aluminate solution as the electrolyte, in which the cathode was the Chevrel phase Mo_6S_8, and Mg metal was used as the anode [3]. The electrolyte was prepared by the Lewis acid, RnAlCl $> n$ (R = alkyl, aryl), and Lewis base, $RmMgCl_2$, through an acid–base reaction in tetrahydrofuran (THF). The general formula is $Mg(AlCl_2RR')_2$ Rnl, Rn_2 = alkyl or aryl. By optimizing the combination of different Lewis acids and Lewis bases and the reaction concentration, the best electrochemical performance can be obtained. The electrolyte was prepared by using Bu_2Mg and EtAlCh (Et = ethtl) at a molar ratio of 1 : 2, and the resulting solution was named "the first-generation electrolyte" (0.25 M $Mg(AlCHBuEt)_2$, butyl–ethyl complex (BEC) electrolyte). The ionic conductivity of this electrolyte is as high as $1.4\,mS\,cm^{-1}$, the dissolution/deposition efficiency of magnesium can reach 100%, and the electrochemical window is also increased to 2.4 V vs. Mg^{2+}/Mg (Figure 11.2).

The effects of the organic substituents, the aluminum–magnesium element ratio, and solvent in this type of electrolyte on the electrochemical window and ionic conductivity of the reagent derivative electrolyte were carefully studied. It was found that the Lewis acid–base ratio, chlorine content, and substituent size have important effects on the overpotential of dissolved magnesium in the electrolyte deposition and the electrochemical window. The higher the Lewis acid–base ratio and the greater the chlorine content, the greater the overpotential of the deposited dissolved magnesium; while the smaller the alkyl substituent group, the wider the electrochemical window.

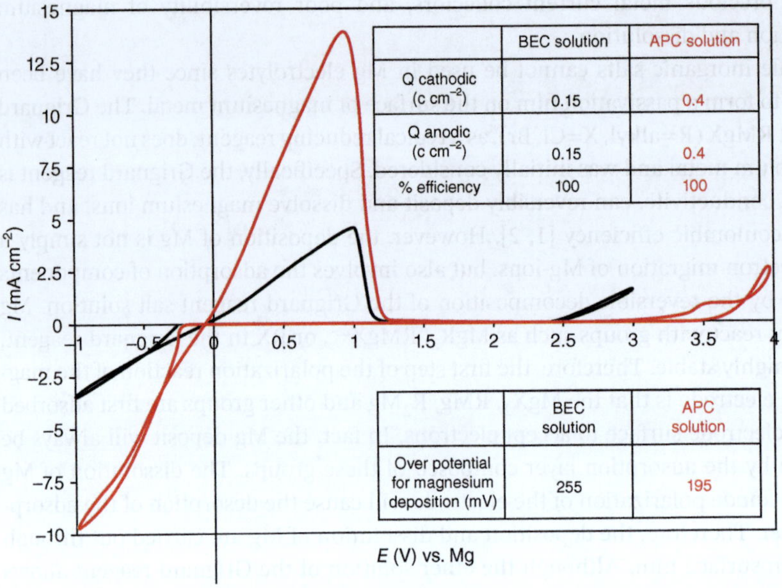

Figure 11.2 Steady-state cyclic voltammograms of 0.25 M BEC and 0.4 M APC solutions, measured with Pt electrodes at 25 mV s^{-1}. The charge involved, Mg cycling efficiency, and overpotential for Mg deposition are given as well. Source: Reproduced with permission from Mizrahi et al. [4]/IOP Publishing.

This work lead to the development of a per-phenyl organochloroaluminate magnesium salt electrolyte by substituting a phenyl group without β-H for an alkyl group. As shown in Figure 11.2, the electrochemical window of the APC electrolyte ($PhMgCl_2$–$AlCl_3$) is as high as 3.3 V vs. Mg^{2+}/Mg , the ion conductivity is about 2×10^{-3} S cm^{-1}, and the Coulombic efficiency of deposited dissolved magnesium is about 100% [4].

11.1.2 Cathode Material

The Chevrel phase Mo_6S_8 was the first cathode material reported in the literature that can reversibly intercalate and release Mg^{2+}, and in combination with an organic halogenated aluminate solution, it gave an average working voltage of about 1.1 V vs. Mg^{2+}/Mg, a specific capacity of 75 mAh g^{-1}, and a specific energy density of ~82 Wh kg^{-1}, as seen in Figure 11.3 [3]. This behavior is very poor compared to that of Li cells, and therefore even though since 2000 research on Mg batteries gradually increased, it did not attract the attention of Li-based battery systems as they were already used in almost all portable cell phones and laptops by early 2000.

The general formula of the Chevrel phase is $M_xMo_6T_8$, where M is a metal ion, such as Li, Mg, Zn, and T is S, Se, or Te. As shown in Figure 11.4, the Chevrel phase crystal structure is composed of Mo_6S_8 cubes; six molybdenum atoms form Mo_6 octahedrons and occupy the center of the cube, while eight S anions occupy eight

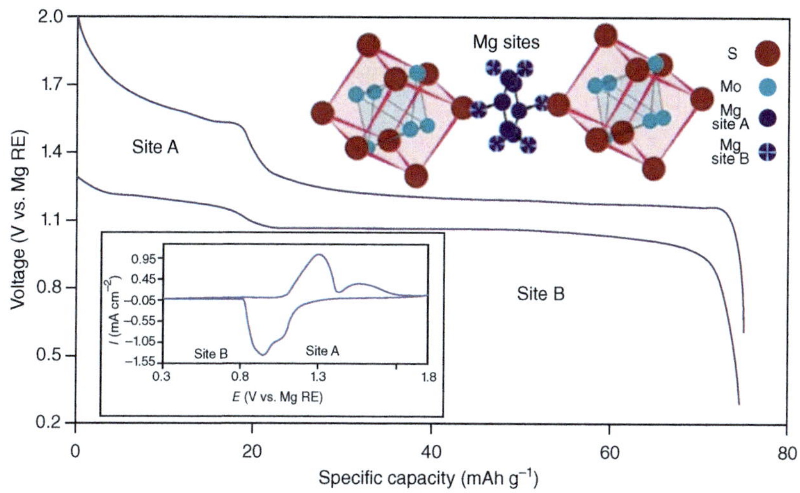

Figure 11.3 Typical electrochemical behavior and the basic structure of the $Mg_xMo_6S_8$ cathodes, $0 < x < 1$, corresponding to a maximal charge capacity of 122 mAh g^{-1}. Source: Reproduced with permission from Aurbach et al. [3]/IOP Publishing.

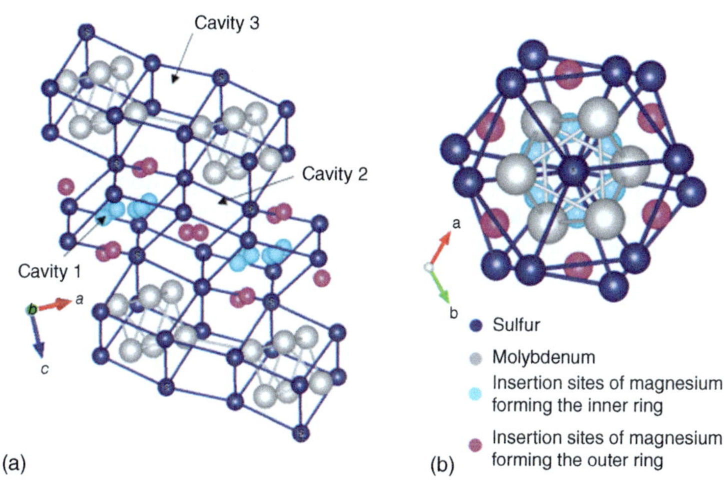

Figure 11.4 Structure of the Chevrel phase $Mg_2Mo_6S_8$ after Mg insertion: (a) the three types of cavities (1, 2, and 3) with the magnesium insertion sites and (b) projection on the perpendicular plan of the three-coordinated sulfurs axis. Source: Reproduced with permission from Richard et al. [5]/American Chemical Society.

fixed-point positions. The supercell formed by the cubes has a 3D channel formed by site A and site B, which can be used for Mg^{2+} diffusion. In the Mo_6 cluster, the electron bonding is non-directional and the orbital is highly delocalized. Therefore, the special structure of Mo_6S_8 makes it easy to receive electrons, which are not all scattered on the Mo_6 cluster, but on the S_8 cluster. The positive charge of Mg^{2+} is basically neutralized by the S anion, so the positive charge of Mg^{2+} can be effectively masked and it can be quickly transmitted in this structure.

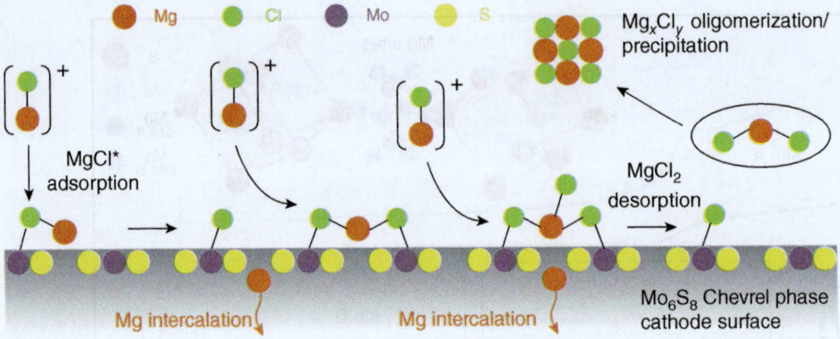

Figure 11.5 The desolvation and intercalation mechanism of APC electrolyte at the surface of Mo_6S_8. Source: Reproduced with permission from Wan et al. [9]/American Chemical Society.

The theoretical specific capacity of 2 Mg^{2+} ions (corresponding to four-electron reactions) embedded in Mo_6S_8 is 128.6 mAh g^{-1}. However, in reality, Mo_6S_8 cannot undergo the reversible insertion/extraction process of 2 Mg^{2+} ions (corresponding to the 4-electron reaction) mainly due to the fact that Mg^{2+} is not easy to escape from the vacancies in the inner ring sites of the host material [6–8]. For $Mg_xMo_6S_8$, when $1<x<2$, the diffusion speed of magnesium ions in $Mg_xMo_6S_8$ is fast, while when $0<x<1$, it is difficult, and this process limits the capacity of the cathode material.

In addition to faster diffusion kinetics, the special structure of Mo_6S_8 also facilitates the charge-transfer process at the cathode/electrolyte interface (Figure 11.5). In electrolytes containing a lot of chlorine, the active ions are Mg_2Cl_3 or $MgCl^-$ ions. The first step in the Mg^{2+}-intercalation reaction is to dissociate Mg^{2+} from the active ions. Mo atoms on the surface of Mo_6S_8 play a catalytic role in the process of breaking the Mg—Cl chemical bond. By chemically interacting with Cl^-, the activation energy can be reduced from 3000 to 200 meV. When Mg^{2+} is embedded, Cl^- is left to form a bond with Mo. The remaining Cl^- will continue to interact with the $MgCl^+$ at the interface to form a neutral $MgCl_2$, so that the Mo site is re-released for the next step of Cl^- adsorption.

Due to the rapid diffusion kinetics of Mg^{2+} and the interface charge-transfer process, Mo_6S_8 is by far the best cathode material at room temperature. However, due to the low energy density, complex preparation process, and high energy consumption, it is necessary to develop other high energy density cathode materials.

Researchers have tried to replace the S anions with Se, which has a larger ionic radius and lower charge density to generate Mo_6Se_8. Although Mo_6Se_8 can reversibly cause two Mg^{2+} insertion/extraction reactions, the addition of Se atoms with a larger mass fraction reduces the theoretical specific capacity of the material [10].

11.2 Zinc-Based Batteries

Zn-based anodes are the oldest active materials that have been continuously used in energy storage since the commercialization of batteries. As seen in Chapter 2, Zn/MnO_2 dry batteries have a history of more than 100 years and are currently the most widely used primary batteries in the market, with a market share of approximately US$ 100 million per year. Zn, however, has many advantages for secondary battery systems as well. The obvious one is that Zn is a non-toxic and low-cost metal, with reserves in the earth's crust about 10 times that of lithium resources and an annual output of 12 million tons of products [11]. Another unique feature is that Zn can be used in combination with a water electrolyte, and its standard electrode potential is −0.78 V vs. SHE, which is lower than the commonly used water-based anode materials, such as $LiTi_2(PO_4)_3$ and $NaTi_2(PO_4)_3$ and can, therefore, provide a higher working voltage potential [12, 13]. Similarly, like Mg-ion batteries Zn also has two valence electrons, allowing for a two-electron transfer reaction and providing a high theoretical capacity of 820 mAh g^{-1}. In the alkaline system of a primary Zn/MnO_2 dry battery, γ-MnO_2 or δ-MnO_2 is used as the cathode. The first electron undergoes the reaction:

Cathode:

$$MnO_2 + H_2O + e^- \rightarrow MnOOH + OH^- \tag{11.1}$$

Anode:

$$Zn + 4OH^- \rightarrow [Zn(OH)_4]^{2-} + 2e^- \tag{11.2}$$

As the battery continues to discharge, the cathode and anode will undergo irreversible transformations due to the second electron reaction

Cathode:

$$MnOOH + H_2O + e^- \rightarrow Mn(OH)_2 + OH^-$$

Anode:

$$Zn + 2OH^- \rightarrow Zn(OH)_2 + 2e^- \text{ or } Zn(OH)_2 \rightarrow ZnO + H_2O$$

Therefore, Zn/MnO_2 dry batteries are used as disposable batteries, and the strong alkaline electrolyte will seriously corrode devices or equipment when the alkaline electrolyte leaks, due to corrosion [14, 15].

The first Zn secondary battery using a neutral water system, with $Zn(NO_3)_2$, as the electrolyte was developed in 2012 [16]; pure Zn was the anode and a-MnO_2 was the cathode. As shown in Figure 11.6a,b, Zn^{2+} shuttles back and forth between the a-MnO_2 cathode and the anode, with the following reaction mechanism:

Cathode:

$$2\alpha - MnO_2 + Zn^{2+} + 2e^- \leftrightarrow ZnMn_2O_4$$

Anode:

$$Zn \leftrightarrow Zn^{2-} + 2e^-$$

This reaction principle is similar to the rocking-chair Li-ion, Na-ion, and Mg-ion batteries. It was shown [16] that at 0.5 C the charge–discharge capacity was 210 mAh g^{-1}, however, the number of cycles was not given. At 6 C, it was possible to retain a capacity of 100 mAh g^{-1} for 100 cycles at nearly 100%. Increasing the C-rate to 126 C gave a reversible capacity of 68 mAh g^{-1}, illustrating excellent rate capability. So far, there are few reports on cathodes suitable for Zn^{2+} de-intercalation in aqueous solutions; for example, $\alpha - MnO_2$ can be used in a neutral water system to intercalate Zn^{2+}.

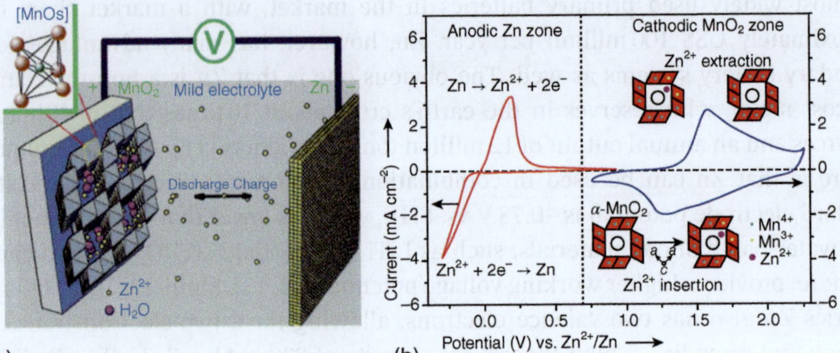

(a) (b)

Figure 11.6 Schematics of the chemistry of the zinc-ion battery. (b) Cyclic voltammogram of the Zn anode (left) and the α-MnO_2 cathode (right) at 2 mV s^{-1} in 0.1 mol l^{-1} $Zn(NO_3)_2$ aqueous electrolyte. Source: Reproduced with permission from Xu et al. [16]/John Wiley & Sons.

11.2.1 Dendrite Formation

When a zinc-ion battery, using a pure Zn anode, is charged in an alkaline electrolyte, the formed $Zn(OH)_4{}^{2-}$ should theoretically be completely reduced to Zn. However, the formation of a soluble intermediate product of $Zn(OH)_4{}^{2-}$, leads to the formation of dendrites. That is, the concentration gradient of $Zn(OH)_4{}^{2-}$ controls the concentration gradient of the Zn deposition. When the metal deposition area extends beyond the diffusion boundary of $Zn(OH)_4{}^{2-}$, Zn dendrites will be produced. In other words, due to the influence of $Zn(OH)_4{}^{2-}$ and Zn^{2+}, ion reduction will preferentially occur along a specific crystal orientation to form dendrites. It follows that the dendrite formation also occurs since Zn is not re-deposited in the same regions of the anode from which it dissolved into the electrolyte, leading to a significant shape change in the anode in addition to dendrite formation. This results in a non-uniform current distribution in the reaction zone, which leads to a capacity loss after continuous cycling.

When the zinc-ion battery is charged in a neutral or weakly acidic electrolyte, there is still a soluble intermediate product of Zn^{2+}. In general, the formation of Zn dendrites is mainly affected not only by the concentration of $Zn(OH)_4{}^{2-}$ and Zn^{2+}, but also by the transport process in the electrolyte, the kinetics/diffusion velocity, the current density distribution during charge/discharge, and the distribution of basal nucleation sites. One way to prevent dendrite formation is by controlling the current density. For example, when the current density is greater than 10 mA cm^{-2}, the Zn anode is likely to produce Zn dendrites during cycling. It is worth mentioning that Zn dendrites will also form at lower deposition overpotentials during a longer initial time. Strategies to suppress the formation of dendrites mainly include the research of new separators and electrode/electrolyte additives. It was reported [17] that low-cost glucose was used as an additive to modulate the typical $ZnSO_4$ electrolyte system for improving the reversible plating/stripping on the Zn anode. Glucose molecules are capable of replacing one H_2O molecule in the primary solvation shell of Zn^{2+} and thus effectively hinder the severe by-reactions derived from too much active water around the surface of the Zn anode. Besides, the surface of Zn metal prefers to absorb glucose over H_2O, therefore, promoting the Zn^{2+} transfer and changing the electric field distribution around the deposited Zn layer. Concomitantly, glucose molecules are inclined to adsorb on the surface of Zn anode, suppressing the random growth of Zn dendrites [17].

In zinc batteries, due to the higher ion conductivity of alkaline electrolytes, 6–7 M KOH electrolyte is usually used. However, the solubility of discharge products increases with the increase of electrolyte concentration, leading to a high amount of Zn to be dissolved and redeposited during charge/discharge, which also leads to serious changes in the shape of the zinc electrode and poor cycle performance.

11.2.2 Hydrogen Evolution

Under alkaline conditions, the standard reduction potential of Zn/ZnO in a solution of pH = 14 is −1.26 V (relative to the standard potential of hydrogen). This voltage is lower than that at which the hydrogen evolution reaction occurs (−0.83 V, relative to the standard potential of hydrogen); therefore when ZnO reacts with H_2O in the water-based electrolyte, OH^- forms. Additionally, a large amount of OH^- is produced when the hydrogen evolution reaction occurs.

$$ZnO + 2e^- + H_2O \leftrightarrow Zn + 2OH^-, \quad -1.26\,V \tag{11.3}$$

$$2H_2O + 2e^- \leftrightarrow H_2 \uparrow + 2OH^-, \quad -0.83\,V \tag{11.4}$$

The standard reduction potential of Zn/Zn^{2+} in a neutral electrolyte environment is −0.76 V (relative to the standard potential of hydrogen), which is higher than that required for the hydrogen evolution reaction. Therefore, the oxidation–reduction reaction of zinc occurs preferentially to the hydrogen evolution reaction.

$$Zn^{2+} + 2e^- \leftrightarrow Zn, \quad -0.76\,V \tag{11.5}$$

$$2H_2O + 2e^- \leftrightarrow H_2 \uparrow + 2OH^-, \quad -0.83\,V \tag{11.6}$$

The standard reduction potential of Zn/Zn^{2+} in a weakly acidic electrolyte environment is also −0.76 V (relative to the standard potential of hydrogen), which is lower than that required for the hydrogen evolution reaction (H^+/H_2 requires 0 V in a weakly acidic electrolyte), so the hydrogen evolution reaction has priority over the oxidation–reduction reaction of zinc. The hydrogen evolution process consumes hydrogen ions and makes the solution gradually neutral. Therefore, the hydrogen evolution reaction under weak-acid conditions is not as violent as the hydrogen evolution reaction under alkaline conditions. The hydrogen evolution reactions in a weakly acidic electrolyte are:

$$Zn^{2+} + 2e^- \leftrightarrow Zn, \quad -0.76\,V \tag{11.7}$$

$$2H^+ + 2e^- \leftrightarrow H_2 \uparrow, \quad 0\,V \tag{11.8}$$

Hydrogen evolution is inevitable in both alkaline and acidic electrolytes, so the zinc electrode cannot be charged with 100% Coulombic efficiency, because the hydrogen evolution reaction consumes part of the electrons provided to the zinc anode during the charging process. Due to the hydrogen evolution reaction, the water content in the electrolyte decreases and the OH^- concentration near the electrode increases, which will aggravate the corrosion of the zinc anode.

The bubbles generated by the hydrogen gas will also cause insufficient contact between the electrode and the electrolyte, increasing the internal resistance, gas evolution, and causing the battery to expand, leading hence, possibly, to explosions. In acidic electrolytes, ZnOH does not form, and hence dendrites are not a concern.

11.3 Dual-Ion Hybrid Batteries

Multivalent metal-ion batteries may share the successful reversible operating mechanism with lithium-based batteries while using much more widely available elements as ionic-charge carriers. Compared to Li-ion batteries, metal anodes for such systems provide competitive energy densities and are thus suitable for large-scale energy storage and even for some propulsion applications. In developing a multivalent metal-ion battery, Zn^{2+} or Mg^{2} can be used as the anode, however, cathode materials suited for reversibly intercalating/de-intercalating the Zn^{2+} or Mg^{2+} remain a major challenge. Herein some preliminary, unpublished, data for such systems are given.

Constructing a dual-salt hybrid battery (such as Na^{+}/Zn^{2+} hybrid battery) assembled with a metallic Zn anode and a Na-intercalation cathode (such as Na^{+} superionic conductor (NASCION)-type cathode $Na_3V_2(PO_4)_3/C$) in a Zn^{2+}/Na^{+} acetate mixture electrolyte (1 M NaAC and ZnAC) may be an effective method to simultaneously combine the superiority of low cost, high safety, and stability.

The Zn and $Na_3V_2(PO_4)_3/C$ (NVP) electrodes were tested separately by using a saturation calomel electrode (SCE) and Pt as the reference and counter electrodes in the mixed electrolyte containing 1 M NaAc and $Zn(Ac)_2$ (Znacetate). In the cyclic voltammetry curves of Figure 11.7a, it is shown that the redox peaks occurred around −0.9 V (vs. SCE) for the Zn electrode and can be attributed to the reversible dissolution and deposition of Zn with the redox reaction of Zn^{2+}/Zn in the hybrid

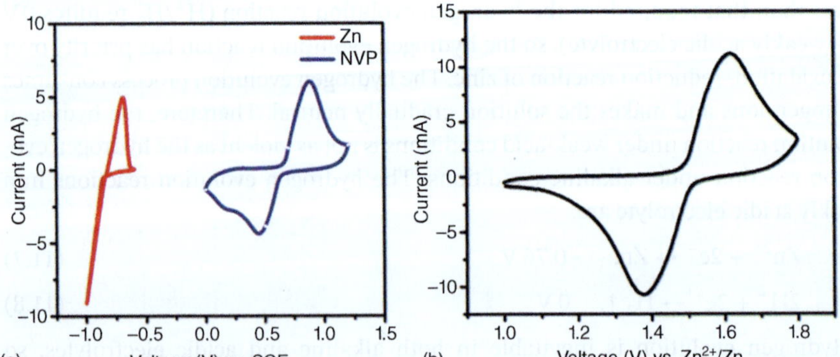

Figure 11.7 (a) Cyclic voltammetry (CV) of the Zn and NVP in 1 M NaAC and $Zn(Ac)_2$ at the scan rate of $5\,mV\,s^{-1}$ using Pt and SCE as the counter and reference electrodes, respectively. (b) Cyclic voltammetry scan results using $Na_3V_2(PO_4)_3/C$ as the working electrode and Zn as the counter electrode and reference electrode.

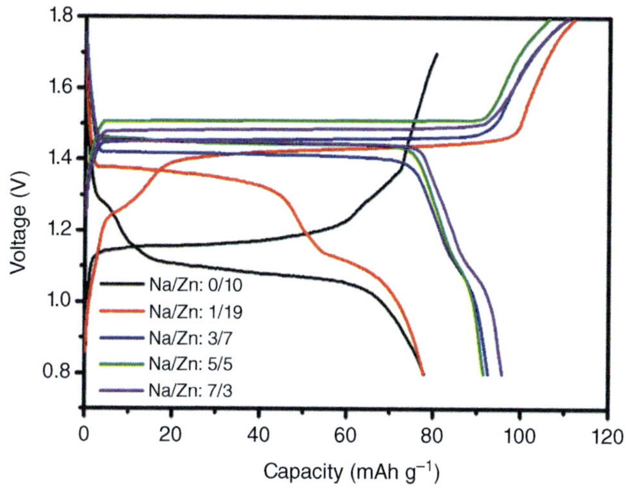

Figure 11.8 The charge–discharge curve of battery in electrolyte with different Na^+/Zn^{2+}.

electrolyte. The redox peaks of the NVP electrode appeared at the potential of 0.5–0.8 V, from the redox reaction of V^{3+}/V^{4+}. Both electrodes exhibited little separation between cathodic and anodic peaks, suggesting the weak polarization of the electrochemical reaction. The large potential difference between the Zn and NVP electrode suggests that the battery assembled with Zn and NVP as the anode and cathode, respectively, could deliver a high-operating voltage. Figure 11.7b shows the cyclic voltammetry scan results using $Na_3V_2(PO_4)_3$/C as the working electrode and Zn as the counter electrode and reference electrode. It can be seen that a pair of obvious redox peaks are located at 1.6 and 1.4 V, and the peak potential difference of 0.2 V indicates that the cathode and anode reactions of the battery have a high reversibility.

Figure 11.8 shows the battery charge and discharge curves in electrolytes with different Na^+/Zn^{2+} ratios at 0.2 C. When pure $Zn(Ac)_2$ is used as the electrolyte, Na^+ is de-intercalated from NVP and dissolved in the electrolyte during the first charge, so the solution is not pure $Zn(Ac)_2$ at the end of charge. Therefore, new electrolyte needs to be used in order to ensure that the subsequent intercalation happens only from Zn^{2+} during discharge. In the pure $Zn(Ac)_2$ electrolyte, it can be found that the charge and discharge plateau of the battery occurs at about 1.1 V. When the Na concentration increases, the low-potential charge–discharge plateau at ~1.1 V disappears, and there is only one voltage plateau at ~1.4 V. The value of the discharge plateau increases as the Na concentration increases.

Figure 11.9a shows the cycling performance of the Na^+/Zn^{2+} dual-salt hybrid battery (1 M NaAC and $Zn(Ac)_2$). The initial discharge capacity was 90 mAh g^{-1}, and after 200 cycles it had decreased slightly to 80 mAh g^{-1}, showing a very good capacity retention of 89%. Figure 11.9b depicts the charge–discharge curve at different C rates, and as expected, the capacity gradually decreases as the rate increases from 1 to 25 C. When the rate increased to 20 and 25 C, the capacity was 86 and 85 mAh g^{-1}, respectively, which is equivalent to 85% of the 1 C capacity.

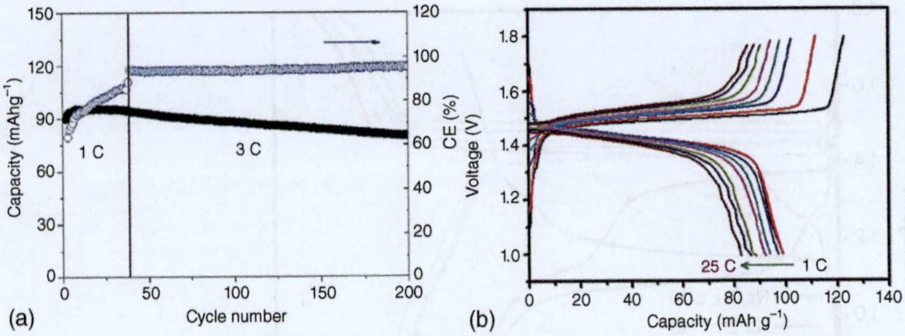

Figure 11.9 (a) Cycling stability of battery with Na^+/Zn^{2+} hybrid electrolyte. (b) Capacity and charge–discharge curve of battery at different rates in Na^+/Zn^{2+} hybrid electrolyte.

This is considered an excellent rate performance, which is superior to the rate characteristics of $Na_3V_2(PO_4)_3$ in organic electrolytes used in Na-ion batteries due to the higher ion conductivity in aqueous solutions.

11.4 Conclusions

A brief overview was given in this concluding chapter of some other secondary battery systems that are still at very early developmental stages. Mg batteries were chosen to be discussed as the fact that dendrites do not form on pure Mg anodes is very promising. Zn batteries exhibit similar dendritic issues as when Li anodes are used; therefore, a breakthrough in solving the dendrite issue in Li solid-state and Li–S systems will result in the advancement of Zn systems. Perhaps the most exciting future electrochemical cells are multivalent metal-ion batteries, even though they are at an early stage.

References

1 Gaddum, L.W. and French, H.E. (1927). *J. Am. Chem. Soc.* 49 (5): 1295–1299.
2 Genders, J.D. and Pletcher, D. (1986). *J. Electroanal. Chem. Interfacial Electrochem.* 199 (1): 93–100.
3 Aurbach, D., Lu, Z., Schechter, A. et al. (2000). *Nature* 407 (6805): 724–727.
4 Mizrahi, O., Amir, N., Pollak, E. et al. (2008). *J. Electrochem. Soc.* 155 (2): A103.
5 Richard, J., Benayad, A., Colin, J.F., and Martinet, S. (2017). *J. Phys. Chem. C* 121 (32): 17096–17103.
6 Levi, E., Levi, M.D., Chasid, O., and Aurbach, D. (2009). *J. Electroceram.* 22 (1): 13–19.
7 Levi, E., Gershinsky, G., Aurbach, D. et al. (2009). *Chem. Mater.* 21 (7): 1390–1399.
8 Kaewmaraya, T., Ramzan, M., Osorio-Guillén, J.M., and Ahuja, R. (2014). *Solid State Ionics* 261: 17–20.

9 Wan, L.F., Perdue, B.R., Apblett, C.A., and Prendergast, D. (2015). *Chem. Mater.* 27 (17): 5932–5940.

10 Levi, E., Lancry, E., Mitelman, A. et al. (2006). *Chem. Mater.* 18 (23): 5492–5503.

11 Wang, X., Li, M., Wang, Y. et al. (2015). *J. Mater. Chem. A* 3 (16): 8280–8283.

12 Li, Z., Young, D., Xiang, K. et al. (2013). *Adv. Energy Mater.* 3 (3): 290–294.

13 Park, S.I., Gocheva, I., Okada, S., and Yamaki, J. (2011). *J. Electrochem. Soc.* 158 (10): A1067–A1070.

14 Alfaruqi, M.H., Mathew, V., Gim, J. et al. (2015). *Chem. Mater.* 27 (10): 3609–3620.

15 Lee, J.W., Ahn, T., Soundararajan, D. et al. (2011). *Chem. Commun.* 47 (22): 6305–6307.

16 Xu, C.J., Li, B.H., Du, H.D., and Kang, F.Y. (2012). *Angew. Chem. Int. Ed.* 51 (4): 933–935.

17 Sun, P., Ma, L., Zhou, W.H. et al. (2021). *Angew. Chem. Int. Ed.* 60 (33): 18247–18255.

Index

Rechargeable Ion Batteries: Materials, Design, and Applications of Li-Ion Cells and Beyond, First Edition.
Edited by Katerina E. Aifantis, R. Vasant Kumar, and Pu Hu.

t